Lecture Notes in Computer Science   8494

Commenced Publication in 1973
Founding and Former Series Editors:
Gerhard Goos, Juris Hartmanis, and Jan van Leeuwen

Editorial Board

David Hutchison
  Lancaster University, UK

Takeo Kanade
  Carnegie Mellon University, Pittsburgh, PA, USA

Josef Kittler
  University of Surrey, Guildford, UK

Jon M. Kleinberg
  Cornell University, Ithaca, NY, USA

Alfred Kobsa
  University of California, Irvine, CA, USA

Friedemann Mattern
  ETH Zurich, Switzerland

John C. Mitchell
  Stanford University, CA, USA

Moni Naor
  Weizmann Institute of Science, Rehovot, Israel

Oscar Nierstrasz
  University of Bern, Switzerland

C. Pandu Rangan
  Indian Institute of Technology, Madras, India

Bernhard Steffen
  TU Dortmund University, Germany

Demetri Terzopoulos
  University of California, Los Angeles, CA, USA

Doug Tygar
  University of California, Berkeley, CA, USA

Gerhard Weikum
  Max Planck Institute for Informatics, Saarbruecken, Germany

Jon Lee   Jens Vygen (Eds.)

Integer Programming and Combinatorial Optimization

17th International Conference, IPCO 2014
Bonn, Germany, June 23-25, 2014
Proceedings

 Springer

Volume Editors

Jon Lee
University of Michigan
Department of Industrial and Operations Engineering
1205 Beal Avenue, Ann Arbor, MI 48109-2117, USA
E-mail: jonxlee@umich.edu

Jens Vygen
University of Bonn
Research Institute for Discrete Mathematics
Lennéstr. 2, 53113 Bonn, Germany
E-mail: vygen@or.uni-bonn.de

ISSN 0302-9743   e-ISSN 1611-3349
ISBN 978-3-319-07556-3   e-ISBN 978-3-319-07557-0
DOI 10.1007/978-3-319-07557-0
Springer Cham Heidelberg New York Dordrecht London

Library of Congress Control Number: 2014939557

LNCS Sublibrary: SL 1 – Theoretical Computer Science and General Issues

Typesetting: Camera-ready by author, data conversion by Scientific Publishing Services, Chennai, India

Printed on acid-free paper

Springer is part of Springer Science+Business Media (www.springer.com)

Preface

This volume contains the 34 extended abstracts presented at IPCO 2014, the 17th Conference on Integer Programming and Combinatorial Optimization, held June 23-25, 2014, in Bonn, Germany.

The IPCO conference is under the auspices of the Mathematical Optimization Society. It is held every year, except for those in which the International Symposium on Mathematical Programming takes place. The conference is a forum for researchers and practitioners working on various aspects of integer programming and combinatorial optimization. The aim is to present recent developments in theory, computation, and applications in these areas. Traditionally, IPCO consists of three days of non-parallel sessions, with no invited talks. More information on IPCO and its history can be found at www.mathopt.org/?nav=ipco.

This year, there were 143 submissions, two of which were withdrawn before the review process started. Each reviewed submission was reviewed by at least three Program Committee members, often with the help of external reviewers. The Program Committee met in Aussois in January 2014 and, after thorough discussions, selected 34 papers to be presented at IPCO 2014 and included in this volume. The record number of submissions, their high quality, and the more or less constant number of papers that can be accepted made this IPCO even more competitive than previous editions, with an acceptance rate of less than 25%.

We would like to thank:

- All authors who submitted extended abstracts to IPCO; it is a pleasure to see how active all areas of integer programming and combinatorial optimization are
- The members of the Program Committee, who graciously gave their time and energy
- The external reviewers, whose expertise was instrumental in guiding our decisions
- The EasyChair developers for their excellent platform making many things so much easier
- Springer for their efficient cooperation in producing this volume
- The members of the Organizing Committee and all people in Bonn who helped to make this conference possible
- The speakers of the summer school preceding IPCO: Gérard Cornuéjols, András Frank, Thomas Rothvoß, and David Shmoys

- The Mathematical Optimization Society and in particular the members of its IPCO Steering Committee: Andreas Schulz, Andrea Lodi, and David Williamson, for their help and advice.

March 2014

Jon Lee
Jens Vygen

Organization

Program Committee

Flavia Bonomo	Universidad de Buenos Aires, Argentina
Sam Burer	University of Iowa, USA
Gérard Cornuéjols	Carnegie Mellon University, USA
Satoru Fujishige	Kyoto University, Japan
Michael Jünger	Universität zu Köln
Matthias Köppe	University of California, Davis, USA
Jon Lee (chair)	University of Michigan, USA
Jeff Linderoth	University of Wisconsin, USA
Jean-Philippe Richard	University of Florida, USA
András Sebő	CNRS, Laboratoire G-SCOP, Grenoble, France
Maxim Sviridenko	University of Warwick, UK
Chaitanya Swamy	University of Waterloo, Canada
Jens Vygen	Universität Bonn, Germany
David P. Williamson	Cornell University, USA
Laurence Wolsey	Université catholique de Louvain, Belgium

Organizing Committee

Stephan Held (co-chair)	Bernhard Korte
Stefan Hougardy	Jens Vygen (chair)

Additional Reviewers

Aardal, Karen	Bornstein, Claudson
Aharoni, Ron	Boyar, Joan
Ahmadian, Sara	Brenner, Ulrich
Albers, Susanne	Brešar, Boštjan
An, Hyung-Chan	Buchbinder, Niv
Ando, Kazutoshi	Buchheim, Christoph
Argiroffo, Gabriela	Butenko, Sergiy
Badanidiyuru, Ashwinkumar	Byrka, Jarosław
Balasundaram, Baski	Chrobak, Marek
Baldacci, Roberto	Cornaz, Denis
Bansal, Nikhil	de Klerk, Etienne
Bhaskar, Umang	Dey, Santanu
Bökler, Fritz	Di Summa, Marco

Dourado, Mitre
Duarte Pinto, Paulo Eustáquio
Dvorak, Zdenek
Ehrgott, Matthias
Ene, Alina
Faenza, Yuri
Fanelli, Angelo
Feuerstein, Esteban
Fiorini, Samuel
Frank, András
Friggstad, Zachary
Gaspers, Serge
Gester, Michael
Gleixner, Ambros
Goemans, Michel
Grigoriev, Alex
Groß, Martin
Guenin, Bertrand
Günlük, Oktay
Gupta, Anupam
Gyárfás, András
Hähnle, Nicolai
Hajiaghayi, Mohammadtaghi
Harks, Tobias
Havet, Frédéric
Held, Stephan
Hirai, Hiroshi
Hoefer, Martin
Hungerländer, Philipp
Im, Sungjin
Imai, Hiroshi
Jansen, Klaus
Jeronimo, Gabriela
Jordán, Tibor
Kakimura, Naonori
Kamiński, Marcin
Kamiyama, Naoyuki
Karakostas, George
Karpinski, Marek
Kawahara, Jun
Kelner, Jonathan
Király, Tamás
Kitahara, Tomonari
Klewinghaus, Niko
Kobayashi, Yusuke

Kortsarz, Guy
Krumke, Sven
Kumar, Amit
Letchford, Adam
Li, Jian
Li, Shi
Liers, Frauke
Lin, Min Chih
Luedtke, James
Mądry, Aleksander
Maffray, Frédéric
Mallach, Sven
Manlove, David
Margot, François
Martin, Alexander
Marx, Dániel
Matuschke, Jannik
McCormick, S. Thomas
Megow, Nicole
Mestre, Julian
Mirrokni, Vahab
Miyazaki, Shuichi
Moldenhauer, Carsten
Moscardelli, Luca
Moseley, Ben
Müller, Dirk
Murota, Kazuo
Nagano, Kiyohito
Nagarajan, Viswanath
Newman, Alantha
Niedermeier, Rolf
Ochsendorf, Philipp
Olver, Neil
Oriolo, Gianpaolo
Ostrowski, Jim
Pferschy, Ulrich
Pilipczuk, Marcin
Pokutta, Sebastian
Queyranne, Maurice
Rautenbach, Dieter
Ravi, R.
Rinaldi, Giovanni
Röglin, Heiko
Rothvoß, Thomas
Rotter, Daniel

Sanità, Laura
Sassano, Antonio
Sau, Ignasi
Savelsbergh, Martin
Schäfer, Till
Schalekamp, Frans
Scheifele, Rudolf
Schieber, Baruch
Schmidt, Daniel
Schneider, Jan
Schorr, Ulrike
Shepherd, Bruce
Shigeno, Maiko
Shioura, Akiyoshi
Silvanus, Jannik
Silveira, Rodrigo
Sitters, René
Skopalik, Alexander
Skutella, Martin
Smriglio, Stefano
Soto, José
Spirkl, Sophie
Spisla, Christiane
van Stee, Rob
Svensson, Ola

Szigeti, Zoltán
Takazawa, Kenjiro
Tanigawa, Shin-ichi
Trotignon, Nicolas
Tunçel, Levent
Uetz, Marc
Van Vyve, Mathieu
Végh, László
Ventura, Paolo
Verschae, José
Vielma, Juan Pablo
Vishnoi, Nisheeth
Vondrák, Jan
Ward, Justin
Wiese, Andreas
Woeginger, Gerhard J.
Wollan, Paul
Wong, Prudence W.H.
Woods, Kevin
Wotzlaw, Andreas
Young, Neal
Zenklusen, Rico
van Zuylen, Anke
van der Zwaan, Ruben
Zwick, Uri

Table of Contents

The Cycling Property for the Clutter
of Odd *st*-Walks

Ahmad Abdi and Bertrand Guenin

Department of Combinatorics and Optimization, University of Waterloo
{a3abdi,bguenin}@uwaterloo.ca

Abstract. A binary clutter is cycling if its packing and covering linear
program have integral optimal solutions for all eulerian edge capacities.
We prove that the clutter of odd *st*-walks of a signed graph is cycling
if and only if it does not contain as a minor the clutter of odd circuits
of K_5 nor the clutter of lines of the Fano matroid. Corollaries of this
result include, of many, the characterization for weakly bipartite signed
graphs [5], packing two-commodity paths [7,10], packing T-joins with
small $|T|$, a new result on covering odd circuits of a signed graph, as
well as a new result on covering odd circuits and odd T-joins of a signed
graft.

1   Introduction

A *clutter* $\mathcal{C}$ is a finite collection of sets, over some finite ground set $E(\mathcal{C})$, with
the property that no set in $\mathcal{C}$ is contained in, or is equal to, another set of $\mathcal{C}$.
This terminology was first coined by Edmonds and Fulkerson [2]. A *cover* B is
a subset of $E(\mathcal{C})$ such that $B \cap C \neq \emptyset$, for all $C \in \mathcal{C}$. The *blocker* $b(\mathcal{C})$ is
the clutter of the minimal covers. It is well known that $b(b(\mathcal{C})) = \mathcal{C}$ ([8,2]). A clutter
is *binary* if, for any $C_1, C_2, C_3 \in \mathcal{C}$, their symmetric difference $C_1 \triangle C_2 \triangle C_3$
contains, or is equal to, a set of $\mathcal{C}$. Equivalently, a clutter is binary if, for every
$C \in \mathcal{C}$ and $B \in b(\mathcal{C})$, $|C \cap B|$ is odd ([8]). It is therefore immediate that a clutter
is binary if and only if its blocker is.

Let $\mathcal{C}$ be a clutter and $e \in E(\mathcal{C})$. The *contraction* $\mathcal{C}/e$ and *deletion* $\mathcal{C} \setminus e$ are
clutters on the ground set $E(\mathcal{C}) - \{e\}$ where $\mathcal{C}/e$ is the collection of minimal sets
in $\{C - \{e\} : C \in \mathcal{C}\}$ and $\mathcal{C} \setminus e := \{C : e \notin C \in \mathcal{C}\}$. Observe that $b(\mathcal{C}/e) = b(\mathcal{C}) \setminus e$
and $b(\mathcal{C} \setminus e) = b(\mathcal{C})/e$. Contractions and deletions can be performed sequentially
and the result does not depend on the order. A clutter obtained from $\mathcal{C}$ by a
sequence of deletions E_d and a sequence of contractions E_c ($E_d \cap E_c = \emptyset$) is
called a *minor* of $\mathcal{C}$ and is denoted $\mathcal{C} \setminus E_d/E_c$.

Given edge-capacities $w \in \mathbb{Z}_+^{E(\mathcal{C})}$ consider the linear program

$$(P) \quad \begin{cases} \min & \sum(w_e x_e : e \in E(\mathcal{C})) \\ \text{s.t.} & x(C) \geq 1, \quad C \in \mathcal{C} \\ & x_e \geq 0, \quad e \in E(\mathcal{C}), \end{cases}$$

J. Lee and J. Vygen (Eds.): IPCO 2014, LNCS 8494, pp. 1–12, 2014.

and its dual

$$(D) \quad \begin{cases} \max & \sum(y_C : C \in \mathcal{C}) \\ \text{s.t.} & \sum(y_C : e \in C \in \mathcal{C}) \leq w_e, \quad e \in E(\mathcal{C}) \\ & y_C \geq 0, \quad C \in \mathcal{C}. \end{cases}$$

A clutter is said to be *ideal* if, for every edge-capacities $w \in \mathbb{Z}_+^{E(\mathcal{C})}$, (P) has an optimal solution that is integral. A beautiful result of Lehman [9] states that a clutter is ideal if and only if its blocker is. Edge-capacities $w \in \mathbb{Z}_+^{E(\mathcal{C})}$ are said to be *eulerian* if, for every B and B' in $b(\mathcal{C})$, $w(B)$ and $w(B')$ have the same parity. Seymour [13] calls a binary clutter *cycling* if, for every eulerian edge-capacities $w \in \mathbb{Z}_+^{E(\mathcal{C})}$, (P) and (D) both have optimal solutions that are integral. It can be readily checked that if a clutter is cycling (or ideal) then so are all its minors ([13,14]). Therefore, one can characterize the class of cycling clutters by excluding minor-minimal clutters that are not in this class. In this paper, we will only focus on binary clutters.

$\mathcal{O}_5$ is the clutter of the odd circuits of K_5. Let $\mathcal{L}_7$ be the clutter of the lines of the Fano matroid, i.e. $E(\mathcal{L}_7) = \{1, 2, 3, 4, 5, 6, 7\}$ and

$$\mathcal{L}_7 := \{\{1, 2, 7\}, \{3, 4, 7\}, \{5, 6, 7\}, \{1, 3, 5\}, \{1, 4, 6\}, \{2, 3, 6\}, \{2, 4, 5\}\}.$$

Let $\mathcal{P}_{10}$ be the collection of the postman sets of the Petersen graph, i.e. sets of edges which induce a subgraph whose odd degree vertices are the (odd degree) vertices of the Petersen graph. Observe that the four clutters $\mathcal{O}_5, b(\mathcal{O}_5), \mathcal{L}_7, \mathcal{P}_{10}$ are binary, and moreover, it can be readily checked that none of these clutters is cycling. Hence, if a binary clutter is cycling then it cannot have any of these clutters as a minor. The following excluded minor characterization is predicted.

Conjecture 1 (Cycling Conjecture). *A binary clutter is cycling if, and only if, it has none of the following minors: $\mathcal{O}_5, b(\mathcal{O}_5), \mathcal{L}_7, \mathcal{P}_{10}$.*

The Cycling Conjecture, as stated, can be found in Schrijver [12]. However, this conjecture was first proposed by Seymour [13] and then modified by A.M.H. Gerards and B. Guenin. It is worth mentioning that this conjecture contains the *four color theorem* [15]. None of our results in this paper have any apparent bearings on this theorem.

Consider a finite graph G, where parallel edges and loops are allowed. A *cycle* of G is the edge set of a subgraph of G where every vertex has even degree. A *circuit* of G is a minimal cycle, and a *path* is a circuit minus an edge. We define an *st-path* as follows: if $s \neq t$ then it is a path where s and t are the degree one vertices of the path; otherwise, when $s = t$ then it is just the singleton vertex s. Let Σ be a subset of its edges. The pair (G, Σ) is called a *signed graph*. We say a subset S of the edges is *odd* (resp. *even*) in (G, Σ) if $|S \cap \Sigma|$ is odd (resp. even). Let s, t be vertices of G. We call a subset of the edges of (G, Σ) an *odd st-walk* if it is either an odd st-path, or it is the union of an even st-path P and an odd circuit C where P and C share at most one vertex. Observe that when $s = t$ then an odd st-walk is simply an odd circuit. It is easy to see that clutters of odd st-walks are closed under taking minors. As is shown in [6] the clutter of odd st-walks is binary, and it does not have a minor isomorphic to $b(\mathcal{O}_5)$ or $\mathcal{P}_{10}$. In this paper, we verify the Cycling Conjecture for this class of binary clutters:

Theorem 2. *A clutter of odd st-walks is cycling if, and only if, it has no $\mathcal{O}_5$ and no $\mathcal{L}_7$ minor.*

2 Restating Theorem 2

One can view Theorem 2 as a packing and covering result. We need the following definition: two edges of a signed graph are *parallel* if they have the same end-vertices as well as the same sign. Now let $(G = (V, E), \Sigma)$ be a signed graph without any parallel edges, and choose $s, t \in V$. Let $\mathcal{C}$ be the clutter of the odd st-walks, over the ground set E, and choose edge-capacities $w \in \mathbb{Z}_+^E$. An *odd st-walk cover of* (G, Σ) is simply a cover for $\mathcal{C}$. When there is no ambiguity, we refer to an odd st-walk cover as just a cover.

Proposition 3 (Guenin [6]). *If a subset of the edges is a minimal cover then it is either an st-bond (a minimal st-cut) or it is of the form $\Sigma \triangle C$, where C is a cut with s and t on the same shore.*

The minimal covers of the latter form above are called *signatures*. Notice that if Σ' is a signature, then (G, Σ) and (G, Σ') have the same clutter of odd st-walks.

Reset (G, Σ) as follows: replace each edge e of (G, Σ) with w_e parallel edges. The *packing number* $\nu(G, \Sigma)$ of (G, Σ) is the maximum number of pairwise (edge-)disjoint odd st-walks. A dual parameter to the packing number is the *covering number* $\tau(G, \Sigma)$, which records the minimum size of a cover of (G, Σ). Consider a packing of $\nu(G, \Sigma)$ pairwise disjoint odd st-walk and a cover of size $\tau(G, \Sigma)$. As the cover intersects every odd st-walk in the packing, it follows that $\tau(G, \Sigma) \geq \nu(G, \Sigma)$. A natural question arises: when does equality hold? Theorem 2 gives sufficient conditions for a signed graph to satisfy $\tau(G, \Sigma) = \nu(G, \Sigma)$. To elaborate, observe that $\tau(G, \Sigma)$ is the value of (P) and $\nu(G, \Sigma)$ is the value of (D). For w to be eulerian is to say that every two minimal covers of (G, Σ) have the same parity. Therefore, Proposition 3 implies the following.

Remark 4. *Edge-capacities $w = 1$ are eulerian if, and only if,*

(i) $s = t$ and the degree of every vertex is even, or
(ii) $s \neq t$, $\deg(s) - |\Sigma|$ and the degree of every vertex in $V - \{s, t\}$ are even.

We call such signed graphs *st-eulerian*.

Just like how we defined minor operations for clutters, we now define minor operations for signed graphs. Let $e \in E$. Then the minor operations for $\mathcal{C}$ correspond to the following minor operations for (G, Σ): (1) *delete* e: replace (G, Σ) by $(G \setminus e, \Sigma - \{e\})$, (2) *contract* e: replace (G, Σ) by $(G/e, \Sigma')$, where Σ' is a signature of (G, Σ) that does not use the edge e. Observe that vertices s and t move to wherever the edge contractions take them, and if s and t are ever identified then we say $s = t$. A signed graph (H, Γ) *is a minor of* (G, Σ) if it is isomorphic to a signed graph obtained from (G, Σ) by a sequence of edge deletions, edge contractions, and possibly deletion of isolated vertices and switching s and t. Note that if (H, Γ) is a minor of (G, Σ), then the clutter of odd st-walks of (H, Γ) is a minor of the clutter of odd st-walks of (G, Σ).

The two special clutters $\mathcal{O}_5$ and $\mathcal{L}_7$ that appear in Theorem 2 have the following representations: $\mathcal{O}_5$ is the clutter of odd st-walks of $\widetilde{K_5} := (K_5, E(K_5))$ where $s = t$ is one of the five vertices, and $\mathcal{L}_7$ is the clutter of odd st-walks of the signed graph F_7 with $s \neq t$, as shown in Figure 1. Observe that $\tau(\widetilde{K_5}) = 4 > 2 = \nu(\widetilde{K_5})$ and $\tau(F_7) = 3 > 1 = \nu(F_7)$. We can now restate Theorem 2 as follows, and in fact, we will prove this restatement instead of the original one:

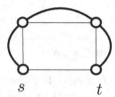

Fig. 1. Signed graph F_7: a representation of $\mathcal{L}_7$. Bold edges are odd.

Theorem 5. *Let* (G, Σ) *be a signed graph with* $s, t \in V(G)$. *If* (G, Σ) *is an st-eulerian signed graph that does not contain* $\widetilde{K_5}$ *or* F_7 *as a minor then* $\tau(G, \Sigma) = \nu(G, \Sigma)$.

3   Extensions of Theorem 2

Let $(G = (V, E), \Sigma)$ be a signed graph with $s, t \in V$. Suppose (G, Σ) is an st-eulerian signed graph that does not contain $\widetilde{K_5}$ or F_7 as a minor. If $s \neq t$ let τ_{st} be the size of a minimum st-bond, otherwise let $\tau_{st} := \tau(G, \Sigma)$. Observe that $\tau_{st} \geq \tau(G, \Sigma)$ as every st-bond is also a cover. Add $\tau_{st} - \tau(G, \Sigma)$ odd loops to (G, Σ) to obtain another st-eulerian signed graph (G', Σ'). Since neither $\widetilde{K_5}$ nor F_7 contain an odd loop, it follows that (G', Σ') also does not contain $\widetilde{K_5}$ or F_7 as a minor. Observe that $\tau(G', \Sigma') = \tau(G, \Sigma) + (\tau_{st} - \tau(G, \Sigma)) = \tau_{st}$ and so by Theorem 2, one can find a packing of τ_{st} pairwise disjoint odd st-walks in (G', Σ'). In (G, Σ) this packing corresponds to a collection of τ_{st} pairwise disjoint elements, $\tau(G, \Sigma)$ of which are odd st-walks and the remaining elements are even st-paths. Therefore, we get the following equivalent, and sharper, formulation of Theorem 5.

Theorem 6. *Let* (G, Σ) *be a signed graph with* $s, t \in V(G)$. *Suppose that* (G, Σ) *is an st-eulerian signed graph that does not contain* $\widetilde{K_5}$ *or* F_7 *as a minor. Then there exists a collection of* $\tau_{st}(G, \Sigma)$ *pairwise (edge-)disjoint elements,* $\tau(G, \Sigma)$ *of which are odd st-walks and the remaining elements are even st-paths.*

We can obtain a counterpart to Theorem 6 as follows: let τ_Σ be the size of a minimum signature. Observe that $\tau_\Sigma \geq \tau(G, \Sigma)$ and that $\tau(G, \Sigma) = \min\{\tau_{st}, \tau_\Sigma\}$. In contrast to above, this time we add $\tau_\Sigma - \tau(G, \Sigma)$ even edges between s and t to (G, Σ) to obtain another st-eulerian signed graph (G', Σ'). Notice, however, that we can no longer guarantee that (G', Σ') contains no $\widetilde{K}_5$ or F_7 minor. Observe that this is true if, and only if, (G, Σ) does not contain $\widetilde{K}_5, \widetilde{K}_5^{\,0}, \widetilde{K}_5^{\,1}, \widetilde{K}_5^{\,2}, \widetilde{K}_5^{\,3}$ or F_7^- as a minor, where

(i) for $i \in \{0, 1, 2, 3\}$, $\widetilde{K}_5^{\,i}$ is the signed graph obtained from splitting a vertex, and its incident edges, of $\widetilde{K}_5$ into two vertices s, t, where s has degree i and t has degree $4 - i$, and

(ii) F_7^- is the signed graph obtained from F_7 by deleting the edge between s and t.

Note that if we add an even edge to any of these signed graphs, then a $\widetilde{K}_5$ or an F_7 appears as a minor. It can be readily checked that if (G, Σ) does not contain any of these five signed graphs as a minor, then (G', Σ') contains no $\widetilde{K}_5$ or F_7 minor. Observe now that $\tau(G', \Sigma') = \tau(G, \Sigma) + (\tau_\Sigma - \tau(G, \Sigma)) = \tau_\Sigma$ and so by Theorem 2, one can find a packing of τ_Σ pairwise disjoint odd st-walks in (G', Σ'). In (G, Σ) this packing corresponds to a collection of τ_Σ pairwise disjoint elements, $\tau(G, \Sigma)$ of which are odd st-walks and the remaining elements are odd circuits. Thus, the following counterpart to Theorem 6 is obtained.

Theorem 7. *Let (G, Σ) be a signed graph with $s, t \in V(G)$. Suppose that (G, Σ) is an st-eulerian signed graph that does not contain $\widetilde{K}_5, \widetilde{K}_5^{\,0}, \widetilde{K}_5^{\,1}, \widetilde{K}_5^{\,2}, \widetilde{K}_5^{\,3}$ or F_7^- as a minor. Then in (G, Σ) there exists a collection of $\tau_\Sigma(G, \Sigma)$ pairwise (edge-)disjoint elements, $\tau(G, \Sigma)$ of which are odd st-walks and the remaining elements are odd circuits.*

4 Applications of Theorem 2

In this section, we discuss some applications of Theorem 2. Observe that a cycling clutter is also ideal. As a corollary, we get the following theorem:

Corollary 8 (Guenin [6]). *A clutter of odd st-walks is ideal if, and only if, it has no $\mathcal{O}_5$ and no $\mathcal{L}_7$ minor.*

When $s = t$ an odd st-walk is just an odd circuit. A signed graph is said to be *weakly bipartite* if the clutter of its odd circuits is ideal. The clutter of odd circuits does not contain an $\mathcal{L}_7$ minor [6]. Hence, we get the following two results as corollaries of Theorem 2:

Corollary 9 (Guenin [5]). *A signed graph is weakly bipartite if, and only if, it has no $\widetilde{K}_5$ minor.*

Corollary 10 (Geelen and Guenin [3]). *A clutter of odd circuits is cycling if, and only if, it has no $\mathcal{O}_5$ minor.*

Observe that $2w$ is eulerian for any $w \in \mathbb{Z}_+^{E(G)}$. As a result, the following result follows as a corollary of Theorem 2:

Theorem 11. *Suppose that $\mathcal{C}$ is a clutter of odd st-walks without an $\mathcal{O}_5$ or an $\mathcal{L}_7$ minor. Then, for any edge-capacities $w \in \mathbb{Z}_+^{E(G)}$, the linear program (P) has an optimal solution that is integral and its dual (D) has an optimal solution that is half-integral.*

To obtain more applications of Theorem 2, we will turn to its restatement Theorem 5, and naturally try to find nice classes of signed graphs without a $\widetilde{K}_5$ or an F_7 minor.

4.1 Signed Graphs without $\widetilde{K}_5$ and F_7 Minor

Let (G, Σ) be a signed graph with $s, t \in V$. Observe that if $s = t$ then (G, Σ) has no F_7 minor, and there are many classes of such signed graphs without a $\widetilde{K}_5$ minor. For instance, whenever G is planar or $|\Sigma| = 2$, (G, Σ) does not contain a $\widetilde{K}_5$ minor. Other classes of such signed graphs can be found in [4,3]. In this section, we focus only on signed graphs (G, Σ) with distinct $s, t \in V$.

A *blocking vertex* is a vertex v whose deletion removes all the odd cycles, and a *blocking pair* is a pair of vertices $\{u, v\}$ whose deletion removes all the odd cycles.

Remark 12. *The following classes of signed graphs with $s \neq t$ do not contain $\widetilde{K}_5$ or F_7 as a minor:*

(1) *signed graphs with a blocking vertex,*
(2) *signed graphs where $\{s, t\}$ is a blocking pair,*
(3) *plane signed graphs with at most two odd faces,*
(4) *signed graphs that have an even face embedding on the projective plane, and s and t are connected with an odd edge,*
(5) *signed graphs where every odd st-walk is connected, and*
(6) *plane signed graphs with a blocking pair $\{u, v\}$ where s, u, t, v appear on a facial cycle in this cyclic order.*

Observe that class (5) contains (2) and (4). We will apply Theorem 5 to the first three classes, and in the first two cases, we obtain two well-known results. However, the third class will yield a new and interesting result on packing odd circuit covers. Notice that one can even apply Theorem 6 to these classes.

Observe further that the signed graphs in (1) and (2) do not contain $\widetilde{K}_5^{\,0}, \widetilde{K}_5^{\,1}$, $\widetilde{K}_5^{\,2}, \widetilde{K}_5^{\,3}$ or F_7^- as a minor either, so one may even consider applying Theorem 7 to these classes. We leave it to the reader to find out what Theorems 6 and 7 applied to these classes imply.

4.2   Class (1): Packing T-joins with $|T| = 4$

Let H be a graph with vertex set W, and choose an even vertex subset T. A T-join of H is an edge subset whose odd degree vertices are (all) the vertices in T. A T-cut of H is an edge subset of the form $\delta(U)$ where $U \subseteq W$ and $|U \cap T|$ is odd. Observe that the blocker of the clutter of minimal T-joins is the clutter of minimal T-cuts.

We are now ready to prove the following result as a corollary of Theorem 2. However, it should be noted that this result (for T of size at most 8, in fact) is relatively easy to prove from first principles, as is shown in [1].

Corollary 13 (Cohen and Lucchesi [1]). *Let H be a graph and choose a vertex subset T of size 4. Suppose that every vertex of H not in T has even degree and that all the vertices in T have degrees of the same parity. Then the maximum number of pairwise (edge-) disjoint T-joins is equal to the minimum size of a T-cut.*

Proof. Suppose that $T = \{s, t, s', t'\}$. Identify s' and t' to obtain G, and let $\Sigma = \delta_H(s')$. Then the signed graph (G, Σ) contains a blocking vertex $s't'$, and so it belongs to class (i). By Remark 4, (G, Σ) is st-eulerian. Theorem 2 then implies that $\tau(G, \Sigma) = \nu(G, \Sigma)$. However, observe that an odd st-walk of (G, Σ) is a T-join of H, and a T-join in H contains an odd st-walk of (G, Σ). Hence, $\tau(G, \Sigma) = \nu(G, \Sigma)$ implies that the maximum number of pairwise disjoint T-joins is equal to the minimum size of a T-cut.   □

4.3   Class (2): Packing Two-commodity Paths

Corollary 14 (Hu [7], Rothschild and Whinston [10]). *Let H be a graph and choose two pairs (s_1, t_1) and (s_2, t_2) of vertices, where $s_1 \neq t_1$, $s_2 \neq t_2$, all of s_1, t_1, s_2, t_2 have the same parity, and all the other vertices have even degree. Then the maximum number of pairwise (edge-)disjoint paths, that are between s_i and t_i for some $i = 1, 2$, is equal to the minimum size of an edge subset whose deletion removes all $s_1 t_1$- and $s_2 t_2$-paths.*

Proof. Identify s_1 and s_2, as well as t_1 and t_2 to obtain G, and let $\Sigma = \delta_H(s_1) \triangle \delta_H(t_2)$. Let $s := s_1 s_2 \in V(G)$ and $t := t_1 t_2 \in V(G)$. Then the signed graph (G, Σ) has $\{s, t\}$ as a blocking pair, and so it belongs to class (2). Again by Remark 4 (G, Σ) is st-eulerian. Therefore, by Theorem 2 we get that $\tau(G, \Sigma) = \nu(G, \Sigma)$. However, observe that an odd st-walk of (G, Σ) is an $s_i t_i$-path of H, for some $i = 1, 2$, and such a path in H contains an odd st-walk of (G, Σ). Thus, $\tau(G, \Sigma) = \nu(G, \Sigma)$ proves the corollary.   □

4.4   Class (3): Packing Odd Circuit Covers

Theorem 15. *Let (H, Σ) be a plane signed graph with exactly two odd faces and choose distinct $g, h \in V(H)$. Let (G, Σ) be the signed graph obtained from identifying g and h in H, and suppose that every two odd circuits of (G, Σ) have the same size parity. Then in (G, Σ) the maximum number of pairwise disjoint odd circuit covers is equal to the size of a minimum odd circuit.*

(Here an odd circuit cover is simply a cover for the clutter of odd circuits.) As the reader may be wondering, what is the rationale behind the rather strange construction of (G, Σ) above? Interestingly, the clutter of minimal odd circuit covers is binary, and so the Cycling Conjecture predicts an excluded minor characterization for when this clutter is cycling. As we did with the clutter of odd st-walks, one can restate the Cycling Conjecture for the clutter of odd circuit covers as follows:

> *(?) for signed graphs (G, Σ) without a $\widetilde{K_5}$ minor such that every two odd circuits have the same parity, the maximum number of pairwise disjoint odd circuit covers is equal to the minimum size of an odd circuit. (?)*

The construction in the statement of Theorem 15 yields a signed graph (G, Σ) that has no $\widetilde{K_5}$ minor, and Theorem 15 verifies the restatement above for these classes of signed graphs.

Proof. Let H^* be the plane dual of H, and let P be an odd gh-path in (H, Σ). Let s and t be the two odd faces of (H, Σ). Consider the plane signed graph (H^*, P); note that this signed graph has precisely two odd faces, namely g and h, and so it belongs to (3). In particular, (H^*, P) contains no $\widetilde{K_5}$ and F_7 minor. Since every two odd circuits of (G, Σ) have the same parity, it follows from Remark 4 that (H^*, P) is st-eulerian. So Theorem 2 applies and we have $\tau(H^*, P) = \nu(H^*, P)$.

We claim that an odd cycle of (G, Σ) is an odd st-walk cover of (H^*, P), and vice-versa. Let L be an odd cycle of (G, Σ). If L is an odd cycle of (H, Σ) then L separates the two odd faces s and t, and so it is an st-cut in (H^*, P). Otherwise, L is an odd gh-path and so $L \triangle P$ is an even cycle of (H, Σ). However, an even cycle in (H, Σ) is a cut in (H^*, P) having s and t on the same shore. Hence, L is of the form $P \triangle \delta(U)$ where $s, t \in U \subseteq V(H^*)$. Therefore, in either cases, L is an odd st-walk cover of (H^*, P). Similarly, one can show that an odd st-walk cover of (H^*, P) is an odd cycle of (G, Σ). Therefore, since $b(b(\mathcal{C})) = \mathcal{C}$ for any clutter $\mathcal{C}$, it follows that an odd circuit cover of (G, Σ) is an odd st-walk of (H^*, P), and vice-versa.

Hence, $\tau(H^*, P)$ is the minimum size of an odd circuit of (G, Σ), and $\nu(H^*, P)$ is the maximum number of pairwise disjoint odd circuit covers of (G, Σ). Since $\tau(H^*, P) = \nu(H^*, P)$, the result follows.   □

4.5   Clutter of Odd Circuits and Odd T-joins

Here, we provide yet another application of Theorem 2. This result generalizes Theorem 15. Let $(G = (V, E), \Sigma)$ be a signed graph, and let $T \subseteq V$ be a subset of even size. We call the triple (G, Σ, T) a *signed graft*. Let $\mathcal{C}$ be the clutter over the ground set E that consists of odd circuits and minimal odd T-joins of (G, Σ, T). This minor-closed class of clutters is fairly large. For instance, if $T = \emptyset$ then $\mathcal{C}$ is the clutter of odd circuits, and if Σ is a T-cut then $\mathcal{C}$ is the clutter of T-joins.

Remark 16. $\mathcal{C}$ *is a binary clutter.*

Proof. Take any three elements C_1, C_2, C_3 of $\mathcal{C}$. If an even number of C_1, C_2, C_3 are odd circuits, then $C_1 \triangle C_2 \triangle C_3$ is an odd T-join and so it contains an element of $\mathcal{C}$. Otherwise, an odd number of C_1, C_2, C_3 are odd circuits, and so $C_1 \triangle C_2 \triangle C_3$ is an odd cycle and so it contains an element of $\mathcal{C}$. Since this is true for all C_1, C_2, C_3 in $\mathcal{C}$, it follows from definition that $\mathcal{C}$ is binary. □

Remark 17. *Minimal covers of* $\mathcal{C}$ *are of the form* $\Sigma \triangle \delta(U)$, *where* $U \subseteq V$ *and* $|U \cap T|$ *is even.*

Proof. Let B be a minimal cover of $\mathcal{C}$. Then B intersects every odd circuit of (G, Σ), and so $B \triangle \Sigma = \delta(U)$ for some $U \subseteq V$. The preceding remark showed $\mathcal{C}$ is binary, and so B intersects every odd T-join in an odd number of edges, so $|U \cap T|$ must be even. □

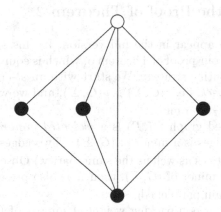

Fig. 2. Signed graft $\widetilde{F}_7$, where all edges are odd and filled-in vertices are in T. For this signed graft, the clutter of odd circuits and minimal odd T-joins isomorphic to $\mathcal{L}_7$.

Theorem 18. *Let* (G, Σ, T) *be a plane signed graft with exactly two odd faces that has no minor isomorphic to* $\widetilde{F}_7$, *depicted in Figure 2. Let* $\mathcal{C}$ *be the clutter of odd circuits and minimal odd* T-*joins, and suppose that every two elements of* $\mathcal{C}$ *have the same size parity. Then the maximum of pairwise disjoint minimal covers of* $\mathcal{C}$ *is equal to the minimum size of an element of* $\mathcal{C}$.

Proof. The proof is similar to the proof of Theorem 15. Let G^* be the plane dual of G, and let P be an odd T-join in (G, Σ, T). Let s and t be the two odd faces of (G, Σ, T). Since (G, Σ, T) has no minor isomorphic to $\widetilde{F}_7$, it follows that the signed graph (G^*, P) contains no F_7 minor, and since it is planar, it has no $\widetilde{K}_5$ minor either. Since every two elements of $\mathcal{C}$ have the same parity, it follows that (G^*, P) is st-eulerian. Hence, by Theorem 5, $\tau(G^*, P) = \nu(G^*, P)$.

We claim that $\mathcal{C}$ is the clutter of odd st-walk covers of (G^*, P), and vice-versa. Let $C \in \mathcal{C}$. If C is an odd circuit of (G, Σ, T), then C is an st-cut of G^*. Otherwise, C is an odd T-join and so $C \triangle P$ is an even cycle of (G, Σ). Thus, $C = P \triangle \delta(U)$ for some $U \subseteq V(G^*) - \{s, t\}$, i.e. C is a signature of (G^*, P).

Hence, $\tau(G^*, P)$ is the minimum size of an element of $\mathcal{C}$, and $\nu(G^*, P)$ is the maximum number of pairwise disjoint covers of $\mathcal{C}$. Since $\tau(G^*, P) = \nu(G^*, P)$, the result follows.   □

Let us explain how this result implies Theorem 15. In the context of Theorem 15, let $T = \{g, h\}$. Observe that (H, Σ, T) is a plane signed graft with exactly two odd faces, and it has no minor isomorphic to $\widetilde{F}_7$ (for $|T| = 2$). However, the clutter of odd circuits and minimal odd T-joins of (H, Σ, T) is isomorphic to the clutter of odd circuits of (G, Σ). It is now easily seen that Theorem 18 implies Theorem 15.

5   Overview of the Proof of Theorem 2

A complete proof will appear in the full version. In this section, however, we provide an overview of our proof of Theorem 5, which is equivalent to Theorem 2. The proof follows a routine strategy. We start with an st-eulerian signed graph (G, Σ) that does not *pack*, i.e. $\tau(G, \Sigma) > \nu(G, \Sigma)$, and we will look for either of the *obstructions* $\widetilde{K}_5$, F_7 as a minor.

We say that a signed graph (H, Γ) is a *weighted minor* of (G, Σ) if (H, Γ) minus some parallel edges is a minor of (G, Σ). (Two edges are parallel if they have the same end vertices as well as the same parity.) Observe that if $\widetilde{K}_5$ or F_7 appears as a weighted minor of (G, Σ), then it is also present as a minor since neither of $\widetilde{K}_5$, F_7 contain parallel edges.

Among all st-eulerian non-packing weighted minors of (G, Σ), we pick one (G', Σ') with smallest $\tau(G', \Sigma')$, smallest $|V(G')|$ and largest $|E(G')|$, in this order of priority. Such a non-packing weighted minor exists. Indeed, if an edge has sufficiently many parallel edges, then it may be contracted while keeping (G', Σ') non-packing and $\tau(G', \Sigma')$ unchanged. Reset $(G, \Sigma) := (G', \Sigma')$ and let $\tau := \tau(G, \Sigma)$, $\nu := \nu(G, \Sigma)$. By identifying a vertex of each (connected) component with s, if necessary, we may assume that G is connected. (Notice that neither of the obstructions $\widetilde{K}_5$, F_7 has a cut-vertex.)

Remark 19. *There do not exist $\tau - 1$ pairwise disjoint odd st-walks in (G, Σ).*

Proof. Suppose otherwise. Remove some $\tau - 1$ pairwise disjoint odd st-walks in (G, Σ). Observe that what is left is an odd $\{s, t\}$-join because $|\Sigma|$, $\deg(s)$, $\deg(t)$ and τ all have the same parity and all vertices other than s, t have even degree. Hence, since every odd $\{s, t\}$-join contains an odd st-walk, one can actually find τ pairwise disjoint odd st-walks in (G, Σ), contradicting the fact that (G, Σ) is non-packing.   □

Let B be a cover of (G, Σ) of size τ. Choose an edge Ω as follows. If $s = t$ then let $\Omega \in E - B$, and since label s is irrelevant to our problem in this case, we may as well assume $\Omega \in \delta(s)$. Otherwise, when $s \neq t$, let $\Omega \in (\delta(s) \cup \delta(t)) - B$. Indeed, if such an edge does not exist, then $\delta(s) \cup \delta(t)$ is contained in the minimum cover B, implying that $\delta(s) \cup \delta(t) = \delta(s) = \delta(t)$, but this cannot be the case as G is connected and non-packing. Again, we may assume that Ω is incident to s. Let s' be the other end-vertex of Ω. Add two parallel edges Ω_1, Ω_2 to Ω to obtain (K, Γ); this st-eulerian signed graph must pack since $\tau(K, \Gamma) = \tau$ as B is also a minimum cover for (K, Γ), $V(K) = V(G)$ but $|E(K)| > |E(G)|$. Hence, (K, Γ) contains a collection $\{L_1, L_2, \ldots, L_\tau\}$ of pairwise disjoint odd st-walks. Observe that all of Ω, Ω_1 and Ω_2 must be used by the odd st-walks in $\{L_1, L_2, \ldots, L_\tau\}$, say by L_1, L_2, L_3, since otherwise one finds at least $\tau - 1$ disjoint odd st-walks in (G, Σ), which is not the case by the preceding remark. As a result, the sequence $(L_1, L_2, L_3, \ldots, L_\tau)$ corresponds to an Ω-*packing of odd st-walks* in (G, Σ), described as follows:

   (i) $L_1, \ldots, L_\tau$ are odd st-walks in (G, Σ),
   (ii) $\Omega \in L_1 \cap L_2 \cap L_3$ and $\Omega \notin L_4 \cup \cdots \cup L_\tau$, and
   (iii) $(L_j - \{\Omega\} : 1 \leq j \leq \tau)$ are pairwise disjoint subsets of edges.

We fix an Ω-packing $(L_1, L_2, L_3, \ldots, L_\tau)$ having a minimum number of edges in their union.

We call T a *transversal* of a collection of sets if T picks exactly one element from each of the sets. For an odd st-walk L, we say that a minimal cover B is a *mate* of L if $|B - L| = \tau - 3$.

Lemma 20. *Let L be an odd st-walk such that $(G, \Sigma) \setminus L$ contains at least $\tau - 3$ pairwise disjoint odd st-walks collected in $\mathcal{L}$. Then L has a mate B, and $B - L$ is a transversal of $\mathcal{L}$.*

Proof. The signed graph $(G, \Sigma) \setminus L$ packs as it is st-eulerian and $\tau((G, \Sigma) \setminus L) < \tau$. Let B' be one of its minimum covers. By our assumption, $\tau((G, \Sigma) \setminus L) \geq \tau - 3$. Since both (G, Σ) and $(G, \Sigma) \setminus L$ are st-eulerian, it follows that $\tau((G, \Sigma) \setminus L)$ and τ have different parities, and so $\tau((G, \Sigma) \setminus L)$ is either $\tau - 3$ or $\tau - 1$. However, observe that the latter is not possible due to Remark 19 and the fact that $(G, \Sigma) \setminus L$ packs. As a result $|B'| = \tau((G, \Sigma) \setminus L) = \tau - 3$. It is now clear that $B' \cup L$ contains a mate for L, and that B' is a transversal of $\mathcal{L}$. $\square$

Observe that if $L \subseteq L_1 \cup L_2 \cup L_3$ or $L \in \{L_4, \ldots, L_\tau\}$, then $(G, \Sigma) \setminus L$ does contain at least $\tau - 3$ pairwise disjoint odd st-walks. Thus, the preceding lemma guarantees the existence of a mate for any such odd st-walk. Vaguely speaking, mates are used as means to build connectivity, with appropriate signing, between the odd st-walks.

Let us call an odd st-walk L *simple* if it is an odd st-path P; otherwise when L is the union of an odd circuit C and an even st-path P, we call L a *non-simple* odd st-walk. By our definition then, when $s = t$ all the odd st-walks are non-simple. For each $1 \leq i \leq \tau$, either L_i is a simple odd st-walk P_i, or it is a non-simple odd st-walk $C_i \cup P_i$, where C_i is an odd circuit and P_i is an even st-path.

Lemma 21. *One of the following holds:*

(i) L_1, L_2 and L_3 are simple,
(ii) at least one of L_1, L_2, L_3 is non-simple, and whenever L_k is non-simple for some $1 \le k \le 3$, then $\Omega \in C_k$,
(iii) at least two of L_1, L_2, L_3 are non-simple, and $\Omega \in P_1 \cap P_2 \cap P_3$.

We analyze each of the three cases separately, and the techniques used to tackle each case are different. A major difference between our proof and the ones for Corollaries 8, 9 and 10 (see [6,5,11,3]) is in where an obsruction is looked for. In any of the aforementioned proofs, only the first three sets of the Ω-packing assisted in finding an obstruction. For our proof, however, this is no longer the case; some of the odd st-walks in $L_4, \ldots, L_\tau$, as well as their mates, help us in finding either of the obstructions. This concludes our overview of the proof of Theorem 5.

References

1. Cohen, J., Lucchesi, C.: Minimax relations for T-join packing problems. In: Proceedings of the Fifth Israeli Symposium on Theory of Computing and Systems (ISTCS 1997), pp. 38–44 (1997)
2. Edmonds, J., Fulkerson, D.R.: Bottleneck Extrema. J. Combin. Theory Ser. B 8, 299–306 (1970)
3. Geelen, J.F., Guenin, B.: Packing odd circuits in eulerian graphs. J. Combin. Theory Ser. B 86, 280–295 (2002)
4. Gerards, A.M.H.: Multicommodity flows and polyhedra. CWI Quart. 6, 281–296 (1993)
5. Guenin, B.: A characterization of weakly bipartite graphs. J. Combin. Theory Ser. B 83, 112–168 (2001)
6. Guenin, B.: Integral polyhedra related to even-cycle and even-cut matroids. Math. Oper. Res. 27(4), 693–710 (2002)
7. Hu, T.C.: Multicommodity network flows. Oper. Res. 11, 344–360 (1963)
8. Lehman, A.: A solution of the Shannon switching game. Society for Industrial Appl. Math. 12(4), 687–725 (1964)
9. Lehman, A.: On the width-length inequality. Math. Program. 17(1), 403–417 (1979)
10. Rothschild, B., Whinston, A.: Feasibility of two-commodity network flows. Oper. Res. 14, 1121–1129 (1966)
11. Schrijver, A.: A short proof of Guenin's characterization of weakly bipartite graphs. J. Combin. Theory Ser. B 85, 255–260 (2002)
12. Schrijver, A.: Combinatorial optimization. Polyhedra and efficiency, pp. 1408–1409. Springer (2003)
13. Seymour, P.D.: Matroids and multicommodity flows. Europ. J. Combinatorics 2, 257–290 (1981)
14. Seymour, P.D.: The forbidden minors of binary matrices. J. London Math. Society 2(12), 356–360 (1976)
15. Seymour, P.D.: The matroids with the max-flow min-cut property. J. Combin. Theory Ser. B 23, 189–222 (1977)

On Simplex Pivoting Rules
and Complexity Theory

Ilan Adler, Christos Papadimitriou*, and Aviad Rubinstein*

University of California, Berkeley CA 94720, USA

Abstract. We show that there are simplex pivoting rules for which it is PSPACE-complete to tell if a particular basis will appear on the algorithm's path. Such rules cannot be the basis of a strongly polynomial algorithm, unless P = PSPACE. We conjecture that the same can be shown for most known variants of the simplex method. However, we also point out that Dantzig's shadow vertex algorithm has a polynomial path problem. Finally, we discuss in the same context randomized pivoting rules.

Keywords: linear programming, the simplex method, computational complexity.

1   Introduction

Linear programming was famously solved in the late 1940s by Dantzig's simplex method [8]; however, many variants of the simplex method were eventually proved to have exponential worst-case performance [21], while, around the same time, Karp's 1972 paper on NP-completeness [18] mentions linear programming as a rare problem in NP which resists classification as either NP-complete or polynomial-time solvable. Khachiyan's ellipsoid algorithm [20] resolved positively this open question in 1979, but was broadly perceived as a poor competitor to the simplex method. Not long after that, Karmarkar's interior point algorithm [19] provided a practically viable polynomial alternative to the simplex method. However, there was still a sense of dissatisfaction in the community: The number of iterations of both the ellipsoid algorithm and the interior point method depend not just on the dimensions of the problem (the number of variables d and the number of inequalities n) but also on the number of bits needed to represent the numbers in the input; such algorithms are sometimes called "weakly polynomial".

A *strongly polynomial algorithm* for linear programming (or any problem whose input is an array of integers) is one that is a polynomial-time algorithm in the ordinary sense (always stops within a number of steps that is polynomial in the total number of bits in the input), but it also takes a number of elementary arithmetic operations that is polynomial in the dimension of the input array.

---

* The research of Christos Papadimitriou and Aviad Rubinstein is supported by NSF Grant CCF-0964033.

J. Lee and J. Vygen (Eds.): IPCO 2014, LNCS 8494, pp. 13–24, 2014.

Strongly polynomial algorithms exist for many network-related special cases of linear programming, as was first shown in [11]. This was extended by Tardos [30] who established the existence of such an algorithm for "combinatorial" linear programs, that is, linear programs whose constraint matrix is 0-1 (or, more generally, contains integers that are at most exponentially large in the dimensions). However, no strongly polynomial algorithm is known for general linear programming.

The following summarizes one of the most important open problems in optimization and the theory of algorithms and complexity:

Conjecture 1. There is a strongly polynomial algorithm for linear programming.

One particularly attractive direction for a positive answer for this conjecture is the search for polynomial variants of the simplex method. It would be wonderful to discover a pivoting rule for the simplex method which (unlike all known such methods) always finds the optimum after a number of iterations that is polynomial in d and n. Hence the following is an interesting speculation:

Conjecture 2. There is a pivoting rule for the simplex method that terminates after a number of iterations that is, in expectation, polynomial in d and n.

In relation to Conjecture 2, clever randomized pivoting rules of a particular recursive sort were discovered rather recently, with worst-case number of iterations that has a subexponential dependence on d [16,23]. Other recent results related to Conjecture 1 can be found in [6,34].

In the next section we formalize the concept of a *pivoting rule:* A method for jumping from one basic solution to an adjacent one that (1) is strongly polynomial per iteration; (2) is guaranteed to increase a potential function at each step; and (3) is guaranteed to always terminate at the optimum (or certify infeasibility or unboundedness). We also give several examples of such rules. It is important to note that in our definition we allow pivoting rules to jump to *infeasible bases* in order to include pivoting rules other than of the primal type. Also, our original definition in Section 2 restricts pivoting rules to be deterministic; we discuss the important subject of randomized rules in Section 5.

Recently there has been a glimmer of hope that some stronger forms of the two conjectures could be disproved, after the disproof of the Hirsch Conjecture [27]. The Hirsch conjecture [9] posited that the diameter of a d-dimensional polytope with n facets is at most $n - d$, the largest known lower bound. The best known upper bound for this diameter is the quasi-polynomial bounds of [17]. But even a super-polynomial lower bound would only falsify the conjectures for *primal* pivoting rules (ones going through only feasible bases, i.e., vertices of the polytope), but *not* for the many other kinds of pivoting rules (see the next section). Furthermore, it is now clear that the techniques involved in the disproof of the Hirsch conjecture are incapable of establishing a nonlinear lower bound on the diameter of polytopes, and it is widely believed that there is a polynomial upper bound on the diameter of polytopes.

In this paper we contemplate whether the concepts and methods of complexity theory can be applied productively to illuminate the problem of strongly polynomial algorithms for linear programming and Conjecture 1. We show a result suggesting that PSPACE-completeness may be relevant.

In particular, we propose to classify deterministic pivoting rules by the complexity of the following problem, which we call THE PATH PROBLEM of a pivoting rule: Given a linear program and a basic solution, will this latter one appear on the pivot rule's path? Recall that PSPACE is the class of problems solvable in polynomial *memory*. This class contains NP, and it is strongly believed to contain it strictly. The PATH PROBLEM of a pivoting rule is clearly in PSPACE, because it can be solved by following the (possibly exponentially long) path of the rule, reusing space; if it is PSPACE-complete, then the pivoting rule cannot be polynomial (unless, of course, P = PSPACE).

But it is not a priori clear that there are pivoting rules for which the path problem is PSPACE-complete. We show (Theorem 1) that they do exist; unfortunately, we prove this not for one of the many classical pivoting rules, but for a new, explicitly constructed — and fairly unnatural — one. We conjecture that the same result holds for essentially all known deterministic pivoting rules; such a proof looks quite challenging; obviously, in such a proof much more will need to be encoded in the linear program (which, in the present proof, is of logarithmic complexity and isomorphic to $\{0,1\}^n$). However, we do exhibit (Theorem 2) a pivoting rule whose path problem is in P: It is Dantzig's well-known *self-dual simplex* [9] (also known as *shadow vertex algorithm*), which is known to be exponential in the worst case [24], but has been used in several sophisticated algorithmic upper bounds for linear programming, such as average-case analysis and smoothness [5,28,2,1,33,29]. We briefly discuss the apparent connection between the average-case performance of a pivoting rule and the complexity of its path problem.

The motivation for our approach came from recent results establishing that it is PSPACE-complete to compute the final result of certain well known algorithms for finding fix points and equilibria [13]. However, the proof techniques used here are completely different from those in [13].

2   Definitions

Consider an algorithm whose input is an array of n integers. The algorithm is called *strongly polynomial* if

- it is polynomial-time as a Turing machine, and
- if one assumes that all elementary arithmetic operations have cost one, the worst-case complexity of the algorithm is bounded by a polynomial in n, and is therefore independent of the size of the input integers.

In linear programming one seeks to maximize $c^T x$ subject to $Ax = b, x \geq 0$, where A is $m \times n$. An $m \times m$ nonsingular submatrix B of A is a *basis*. A *feasible basis* B is one for which the system $Bx_B = b$ (where by x_B we denote vector x

restricted to the coordinates that correspond to B) has a nonnegative solution; in this case, x_B is called a *basic feasible solution*. Basic feasible solutions are important because they render linear programming a combinatorial problem, in that the optimum, if it exists, occurs at one of them. We say that two bases are *adjacent* if they differ in only one column.

There are many versions of linear programming (with inequality constraints, minimization, unrestricted in sign variables, etc.) but they are all known to be easily interreducible. We shall feel free to express linear programs in the most convenient of these.

We shall assume that the linear programs under consideration are non-degenerate (no two bases result in the same basic solution). Detecting this condition is nontrivial (NP-hard, as it turns out). However, there are several reasons why this very convenient assumption is inconsequential. First, a random perturbation of a linear program (obtained, say, by adding a random small vector to b) is non-degenerate with probability one. And second, simplex-like algorithms can typically be modified to essentially perform (deterministic versions of) this perturbation on-line, thus dealing with degeneracy.

We next define a class of algorithms for linear programming that are variants of the simplex method, what we call *pivoting rules*. To start, we recall from linear programming theory three important kinds of bases B, called *terminal bases*:

- optimality: $B^{-1}b \geq 0, c^T - c_B^T B^{-1}A \leq 0$. B is the optimal feasible basis of the linear program.
- unboundedness: $B^{-1}A_j \leq 0, c_j - c_B^T B^{-1}A_j > 0$ for some column A_j of A. This implies that the linear program is unbounded if feasible.
- infeasibility: $(B^{-1})_i A \geq 0, (B^{-1})_i b < 0$ for some row $(B^{-1})_i$ of B^{-1}. This means the linear program is infeasible.

Notice that, given a basis, it can be decided in strongly polynomial time whether it is terminal (and of which kind).

Definition 1. *A pivoting rule R is a strongly polynomial algorithm which, given a linear program (A, b, c):*

- *produces an initial basis B_0;*
- *given in addition a basis B that is not terminal, it produces an adjacent basis $n_R(B)$ such that $\phi_R(n_R(B)) > \phi_R(B)$, where ϕ_R is a potential function.*

The path of pivoting rule R for the linear program (A, b, c) is the sequence of bases $(B_0, n_R(B_0), n_R^2(B_0), \ldots, , n_R^k(B_0))$, ending at a terminal basis, produced by R.

Obviously, any pivoting rule constitutes a correct algorithm for linear programming, since it will terminate (by monotonicity and finiteness), and can only terminate at a terminal basis. Notice that pivoting rules may pass through infeasible basic solutions (for example, they can start with one). Incidentally, the inclusion of infeasible bases implies that such rules operate not on the linear program's polytope, but on its *linear arrangement*. Since the latter has diameter

$O(mn)$, even the existence of polytopes with super-polynomial diameter will not rule out strongly polynomial pivoting rules.

There are many known deterministic pivoting rules (ties are broken lexicographically, say):

1. **Dantzig's rule (steepest descent).** In this rule (as well as in all other primal rules that follow), given a feasible basis B we first calculate, for each index j not in the basis the objective increase gradient $c_j^B = c_j - c_B^T B^{-1} A_j$. Define $J(B) = \{j : c_j^B > 0\}$. Dantzig's rule selects the $j \in J(B)$ with largest c_j^B and brings it in the basis. By non-degeneracy (if not a terminal basis), this completely determines the next basis. As with all primal pivoting rules, the potential function ϕ_R is the objective.

2. **Steepest edge rule.** Instead of the maximum c_j^B, select the largest $\frac{c_j^B}{\|B^{-1}A_j\|}$.

3. **Greatest improvement rule.** We bring in the index that results in the largest increment of the objective.

4. **Bland's rule.** Select the smallest $j \in J(B)$.
   For all these rules, however, we have not specified the original basis B_0. This is obviously a problem, since all these rules are primal and need feasible bases, and a feasible basis may not be a priori available. Primal pivoting rules such as these are best applied not on the original $m \times n$ linear program (A, b, c), but to a simple $m \times 2n$ variant called "the big M version," defined as $(A | - A), b, (c | - M, \ldots, -M)$, where M is a large number (M can either be handled symbolically, or be given an appropriate value computed in strongly polynomial time). It is trivial now to find an initial feasible basis. In fact, the pivoting rule running on the new linear program can be thought of as a slightly modified pivoting rule acting on the original linear program (when $j \in J(B)$, A_j is negated, and c_j is replaced by $-M$).

5. **Shadow vertex rule.** Here B_0 is any basis. Given B_0, we construct two vectors c_0 and b_0 such that B_0 is a feasible basis, and also a dual feasible basis, of the relaxed linear program $\max c_0^T x$ subject to $Ax = b_0, x \geq 0$. Now consider the line segment between these two linear programs, with right-hand side and objective $\lambda b + (1 - \lambda)b_0$ and $\lambda c + (1 - \lambda)c_0$, respectively. Moving on this line segment from $\lambda = 0$, we have both primal-feasible and dual-feasible (and hence optimal) solutions. At some point, one of the two will become infeasible (and only one, by non-degeneracy). We find a new basic solution by exchanging variables as dictated by the violation, and continue. The potential function is the current λ. When $\lambda = 1$ we are at the optimum.

6. **Criss-cross rules.** A class of pivoting rules outside our framework, whose first variant appeared in [35], goes from one (possibly infeasible) basis to the other and convergence to a terminal basis is proved through a combinatorial argument that does not involve an explicit potential function. However, certain such rules (such as the criss-cross pivoting rule suggested in [32]) have been shown ([12]) to possess a monotone potential function, and so they can be expressed within our framework.

7. **Dual pivoting rules.** Naturally, any of the primal pivoting rules can work on the dual.

8. **Primal-dual pivoting rule.** This classical algorithm [10] is an important tool for developing simplex-inspired combinatorial algorithms for a broad set of network problems, acting as a reduction from weighted to unweighted combinatorial problems. It does not conform to our framework, because it involves an inner loop solving a full-fledged linear program.

9. **Pivoting rules with state.** Finally, also outside our framework are pivoting rules relying on data other than A, b, c, and B, for example a pivoting rule relying on statistics of the history of pivoting such as selecting to include the index which has in the past been selected least frequently.

10. **Randomized pivoting rules.** There are several proposed randomized pivoting rules. The ambition here is that the rule's expected path length is polynomial. The simplest one [9] is to pick a random index in $J(B)$. Another important class of randomized rules are the *random facet* rules used in the proofs of subexponential diameter bounds [16,17,23]. We discuss randomized pivoting rules in Section 5.

A pivoting rule is *strongly polynomial* if for any linear program the length of the path is bounded above by a polynomial in m and n. All pivoting rules within our framework mentioned above are known *not* to be strongly polynomial, in that for each one of them there is an explicit family of linear programs with non-polynomial path length, see [3] for a unifying survey.

Explicit constructions are one way of ruling out pivoting rules. *But is there a complexity-theoretic way?* Our interest was sparked by the story of a well-known pivoting rule for a problem other than linear programming: The Lemke-Howson algorithm for two-player Nash equilibrium, discovered in the 1960s [22]. The first explicit construction was obtained decades later [31] and was extremely complicated. More recently, it was established that the problem of finding the Nash equilibrium discovered by the Lemke-Howson algorithm is PSPACE-complete [13] (and thus the algorithm cannot be polynomial, as long as P $\neq$ PSPACE). Remarkably, the PSPACE-completeness proof was much simpler than the explicit construction. We are led to the main definition of this paper:

Definition 2. *The* path problem *associated with a pivoting rule R is the following: Given a linear program and a basis B, does B appear on the path of R for this linear program?*

A pivoting rule is called intractable *if its path problem is PSPACE-complete. A pivoting rule is* tractable *if its path problem can be solved in strongly polynomial time.*

The reason why this concept may be useful in understanding the complexity of linear programming is the following straightforward result:

Proposition 1. *If an intractable pivoting rule is strongly polynomial, then PSPACE = P.*

But are there examples of these two categories? This is the subject of the next two sections.

3   An Intractable Pivoting Rule

This section is devoted to the proof of the following theorem.

Theorem 1. *There is an intractable pivoting rule R.*

The PSPACE-completeness reduction is based on the Klee-Minty construction, the original explicit exponential example for a variant of the simplex method [21], which we recall next.

The *d-dimensional Klee-Minty cube* is the following linear program:

$$\max x_1$$
$$0 \le x_d \le 1$$
$$\epsilon x_{i+1} \le x_i \le 1 - \epsilon x_{i+1}, i = 1, \ldots, d - 1$$
$$x_i \ge 0, i = 1, \ldots, d$$

The feasible region of this linear program is a distorted d-hypercube (it obviously describes precisely the d-hypercube when $\epsilon = 0$): A polytope whose vertices are within a radius of ϵ from those of a hypercube, and are therefore in one-to-one correspondence with the elements of $\{0,1\}^d$. Thus the feasible bases will also be represented as bit strings in $\{0,1\}^d$. The objective function has a minimum at 0^d (a string of d 0's) and a maximum at 10^{d-1}.

Let us now recall a well-known order on $\{0,1\}^d$ called *Gray code* and denoted G_d. G_1 is simply the order $(0,1)$. Inductively, the Gray code G_{i+1} is $(0G_i, 1G_i^R)$, by which we mean, the sequence G_i with each bit string preceded by a 0, followed by the *reverse* of the order G_i, this time with each bit string preceded by 1. If $0 \le k < 2^d$, we denote by $G_d[k]$ the k-th bit string in G_d.

G_d is a bijection between $\{0, 1, \ldots, 2^{d-1}\}$ and $\{0,1\}^d$, and therefore we can define the *successor function* $S_d : \{0,1\}^d \mapsto \{0,1\}^d$ as follows: $S_d(x) = G_d[G_d^{-1}(x) + 1]$. The following is straightforward:

Lemma 1. *S_d can be computed in polynomial time.*

Consider a vertex of the Klee-Minty cube of dimension d — equivalently, a bit string $(b_1, \ldots, b_d) \in \{0,1\}^d$. This vertex has d adjacent vertices, each obtained by flipping one of the b_i's. We call the i-th coordinate *increasing* at this vertex if the objective increases by flipping b_i. The following are known important properties of the Klee-Minty cube:

Lemma 2. *(a) The i-th coordinate is increasing if and only if $\sum_{j=1}^{i} b_j$ is even.*
(b) Therefore the sequence of the vertices sorted in increasing objective is precisely G_d.

We next describe the starting PSPACE-complete problem (see e.g., [25] for definitions regarding PSPACE and Boolean circuits).Suppose that we are given a Boolean circuit C with n input bits and n output bits, such that for all inputs x,

x and $C(x)$ always differ in one bit. The *path* of C is the sequence $(x_i, i = 0, \ldots)$, where $x_0 = 0^n$ and $x_{i+1} = C(x_i)$. Consider now this problem: C-PATH: Given C and $x_C \in \{0,1\}^n$, is x_C on the path of C? It is obviously in PSPACE (one need only try the first 2^n bit strings in the path of C, reusing space; if x_C is not reached by that time, we are in a loop and x_C will never be reached). The following is straightforward:

Lemma 3. *There is a family of circuits C of size polynomial in the number of inputs and of polynomial complexity such that C-PATH is PSPACE-complete.*

The reduction proceeds as follows: Given an input $x_C \in \{0,1\}^n$, we shall construct a linear program and a basis $\hat{B}$ such that $\hat{B}$ lies on the path of rule R (yet to be described) if and only if x_C lies on the path of C. The linear program is the Klee-Minty cube of dimension $2n$. The last (least significant) n coordinates of the cube will serve to encode the current bit string on the path of C, while the first n coordinates will maintain a counter in Gray code. We denote the last string of the Gray code, 10^{n-1}, by x_G. The sought basis $\hat{B}$ is taken to be $\hat{B} = x_G x_C$.

Next we describe the pivoting rule R. In fact, it suffices to define R only on Klee-Minty cubes of even dimension — on any other linear program, R can be any pivoting rule, say steepest descent. First, the initial basis of R is $B_0 = 0^{2n}$. Second, here is the description of how R modifies the current basis B (which, since the linear program is the Klee-Minty cube of dimension $2n$ is represented by a bit string of length $2n$):

Pivoting Rule R on basis B:

1. If $B = 10^{2n-1}$, this is a terminal basis and we are done. Otherwise, let $B = (B_1, B_2)$, each a string of length n.
2. If $B_2 = x_C$ then $R(B) = (S_n(B_1), B_2)$.
3. Otherwise, if $B_1 = x_G$ then $R(B) = (B_1, S_n(B_2))$.
4. Otherwise, construct the circuit with n inputs in the family C.
5. Compute $C(B_2)$; suppose that B_2 and $C(B_2)$ differ in the i-th place (by assumption, they only differ in one).
6. If the $n + i$-th coordinate of B is increasing, then $R(B) = (B_1, C(B_2))$.
7. Otherwise, $R(B) = (S_n(B_1), B_2)$.

To explain the workings of R, the first n bits are a counter, and the last n bits encode the current bit string on the path of C from 0^n. If either the first n bits are x_G or the last n bits are x_C, then R just counts up in the other counter (Steps 2 and 3). Otherwise, (Steps 4 and 5), $C(B_2)$ is computed. The intention now is to update the last n bits to be $C(B_2)$. If the flipped coordinate happens to be increasing in B, then this is done immediately (Step 6). But if it is not, then we do the following maneuver: We increment the B_1 counter by flipping the bit in B_1 that leads to the next string in the Gray code (Step 7). This way, in the next invocation of the pivot rule the flipped bit *will* be increasing (by Lemma 2(a)).

To show that R is a pivoting rule, it remains to show that it is strongly polynomial, and that there is a potential function ϕ_R such that the pivot step of R is always monotonically increasing. The former is immediate. For the latter, $\phi_R(B)$ is the value of the objective x_1 in the basic feasible solution represented by B. It is easy to see by inspection of Steps 2, 3, 6, and 7 that in each of these four cases $\phi_R(R(B)) > \phi_R(B)$.

Finally, we must show that $\hat{B}$ is on the path of R if and only if x_C is on the path of C. If x_C is on the path of C then eventually B_2 will be x_C, after at most $2^n - 1$ steps, and from then on Step 2 will be executed to increment the counter B_1. This counter must go through $\hat{B}$ just before arriving at the terminal basis. If x_C is not on the path of C then the path of C will cycle until eventually Step 7 will be executed for a 2^n-th time (it can be easily checked that the cycling of the path of C does not avoid Step 7), at which point $B_1 = x_G$. From then on $\hat{B}$ cannot be reached. This completes the proof of Theorem 1.   □

4 A Tractable Pivoting Rule

The pivoting rule we proved intractable is not a natural one. We conjecture that essentially all the pivoting rules described in the last section are intractable (even though proving such a result seems to us challenging). However, here we point out that there is a natural, classical pivoting rule that is tractable:

Theorem 2. *The shadow vertex pivoting rule is tractable.*

Proof. Given a linear program (A, b, c), let B_0 be the initial basis, and let b_0 and c_0 be the corresponding initial values of the primal and dual right-hand-side vectors. Given a basis B, we claim that the following is a necessary and sufficient condition that B lies on the path of shadow vertex:

There is a real number $\lambda \in [0, 1]$ such that $(1 - \lambda)B^{-1}b_0 + \lambda B^{-1}b \geq 0$ and $(1 - \lambda)(c_0^T - (c_0)_B^T B^{-1}A) + \lambda(c^T - c_B^T B^{-1}A) \leq 0$.

In proof, any basis on the path has a non-empty interval of λ's for which these inequalities hold. And if for a given basis B this condition is satisfied, then the inequalities are satisfied for a subinterval of $[0, 1]$. If we assume, for contradiction, that B is not on the path of shadow vertex, then we can run the shadow vertex pivoting rule forward and backward from B, and eventually arrive from a different path to the beginning and end, contradicting non-degeneracy. As the condition is a system of $2n$ linear inequalities with one unknown, this completes the proof.   □

There is an interesting story here, connecting tractability of pivoting rules and the saga of the average-case analysis of the simplex method. During the early 1980s, and in the wake of the ellipsoid algorithm, average analysis of the simplex method (under some reasonable distribution of linear programs) was an important and timely open question, and indeed there was a flurry of work on that problem [5,28,2,1,33]. It was noticed early by researchers working on this problem that one obstacle in analyzing the average complexity of various versions of the simplex method was a complete inability to predict the path of pivoting rules — that is, the apparent intractability of the path problem we are studying here. And this makes sense: If one cannot characterize well the circumstances under which a vertex will appear on the path, it is difficult to deduce the average performance of the algorithm by adding expectations over all vertices. Once Borgwardt [5] and Smale [28] had the idea of using the shadow vertex pivoting rule in this context, further progress ensued [2,1,33].

5   Randomized Pivoting Rules

Many pivoting rules are explicitly randomized, aiming at good expected performance. Our definition can easily be extended to include randomization: In the definition of a pivoting rule, $R(B)$ is not a single adjacent basis, but a *distribution* on the set of adjacent bases (naturally, this set is polynomially small). Any basis B' in the support of $R(B)$ must satisfy $\phi_R(B') > \phi_R(B)$. Obviously, deterministic pivoting rules are a special case, and therefore Theorems 1 and 2 trivially apply here too.

What is slightly nontrivial is to define what "intractable" means in this case. That is, what is the "path problem" for a randomized pivoting rule R? We believe that the right answer is the following "promise" problem:

Definition 3. *Fix a polynomial p and a function $f : Z^2 \mapsto [0, 1 - \frac{1}{p(m,n)}]$. The (f, p)-path problem associated with a randomized pivoting rule R is the following: Given an $m \times n$ linear program and a feasible basis B, distinguish between these two cases: B appears in the execution of R with probability (a) at most $f(m,n)$; and (b) at least $f(m,n) + \frac{1}{p(m,n)}$.*

The analog of Proposition 1 is now:

Proposition 2. *If a randomized pivoting rule R is strongly polynomial in expectation, then the (f, p)-path problem of rule R is in BPP, for all f and p.*

Recall that BPP is the class of all problems that can be solved by randomized algorithms, possibly with a small probability or error, see Chapter 10 in [25].

6   Discussion

Pivoting rules constitute a rich and interesting class of algorithmic objects, and here we focused on one important attribute: whether or not the path problem of a pivoting rule is tractable. We have pointed out that there is an intractable pivoting rule, whereas a well-known classical pivoting rule is tractable. The most important problem we are leaving open is to exhibit a natural intractable pivoting rule. For example, establishing the following would be an important advance:

Conjecture 3. Steepest descent is intractable.

This looks quite challenging. Obviously, in such a proof much more will need to be encoded in the linear program (which, in the present proof, was of logarithmic complexity). The ultimate goal is a generic intractability proof that works for a large class of pivoting rules, thus delimiting the possibilities for a strongly polynomial algorithm. For example: Are all primal pivoting rules (the ones using only feasible bases) intractable?

There are pivoting rules beyond linear programming, usually associated with the linear complementarity problem (LCP, see [7]). They generally do not have a potential function, and termination is proved (when it is proved) by combinatorial arguments. Lemke's algorithm is a well-known general pivoting rule, known to terminate with a solution (or with a certification that no solution exists) in several special cases. It is known to be intractable in general [13], but it can be shown to be tractable when the matrix is positive definite. We conjecture that it is intractable when the matrix is positive semidefinite.

References

1. Adler, I., Karp, R.M., Shamir, R.: A Family of Simplex Variants Solving an m x d Linear Program in Expected Number of Pivot Steps Depending on d Only. Mathematics of Operations Research 11(4), 570–590 (1986)
2. Adler, I., Megiddo, N.: A Simplex Algorithm whose Average Number of Steps is Bounded between Two Quadratic Functions of the Smaller Dimension. Journal of the ACM 32(4), 471–495 (1985)
3. Amenta, N., Ziegler, G.: Deformed products and maximal shadows of polytopes. In: Advances in Discrete and Computational Geometry, pp. 57–90 (1996)
4. Avis, D., Chvatal, C.: Notes on Bland's Pivoting Rule. Math. Programming Study 8, 24–34 (1978)
5. Borgwardt, K.H.: The Average Number of Steps Required by the Simplex Method is Polynomial. Zeitschrift fur Operations Research 26(1), 157–177 (1982)
6. Chubanov, S.: A strongly polynomial algorithm for linear systems having a binary solution. Math. Programming 134(2), 533–570 (2012)
7. Cottle, R., Pang, J.-S., Stone, R.E.: The linear complementarity problem. Academic Press (1992)
8. Dantzig, G.B.: Maximization of a Linear Function of Variables subject to Linear Inequalities (1947); Published in Koopmans, T.C. (ed.): Activity Analysis of Production and Allocation, pp. 339–347. Wiley & Chapman-Hall (1951)
9. Dantzig, G.B.: Linear Programming and Extensions. Princeton University Press and the RAND Corporation (1963)
10. Danzig, G.B., Ford, L.R., Fulkerson, D.R.: A Primal-Dual Algorithm for Linear Programming. In: Kuhn, H.W., Tucker, A.W. (eds.) Linear Inequalities and Related Systems, pp. 171–181. Princeton University Press (1954)
11. Edmonds, J., Karp, R.M.: Theoretical improvements in algorithmic efficiency for network flow problems. Journal of the ACM 19(2), 248–264 (1972)
12. Fukuda, K., Matsui, T.: On the Finiteness of the Criss-cross Method. European Journal of Operations Research 52, 119–124 (1991)

13. Goldberg, P.W., Papadimitriou, C.H., Savani, R.: The Complexity of the Homotopy Method, Equilibrium Selection, and Lemke-Howson Solutions. ACM Transactions on Economics and Computation 1(2), Article 9 (2013)
14. Goldfarb, D., Reid, J.K.: A Practical Steepest-Edge Simplex Algorithm. Mathematical Programming 12, 361–371 (1977)
15. Jeroslow, R.G.: The Simplex Algorithm with the Pivot Rule of Maximizing Criterion Improvement. Discrete Mathematics 4, 367–377 (1973)
16. Kalai, G.: A Subexponential Randomized Simplex Algorithm. In: ACM 24th Symposium on Theory of Computing, pp. 475–482 (1992)
17. Kalai, G., Kleitman, D.: Quasi-polynomial Bounds for the Diameter of Graphs and Polyhedra. Bull. Amer. Math. Soc. 26, 315–316 (1992)
18. Karp, R.M.: Reducibility Among Combinatorial Problems. In: Miller, R.E., Thatcher, J.W. (eds.) Complexity of Computer Computations, pp. 85–103. Plenum Press, New York (1972)
19. Karmarkar, N.K.: A New Polynomial-time Algorithm for Linear Programming. Combinatorica 4, 373–395 (1984)
20. Khachian, R.M.: A Polynomial Algorithm in Linear Programming. Doklady Akad. Nauk SSSR 244(5), 1093–1096; Translated in Soviet Math. Doklady, 20, 191–194 (1979)
21. Klee, V., Minty, G.J.: How Good is the Simplex Algorithm? In: Shisha, O. (ed.) Inequalities III, pp. 159–175. Academic Press, New York (1972)
22. Lemke, C.E., Howson, J.T.: Equilibrium points of bimatrix games. SIAM Journal on Applied Mathematics 12(2), 413–423 (1996)
23. Matousek, J., Sharir, M., Welzl, E.: A Subexponential Bound for Linear programming. Algorithmica 16(4-5), 498–516 (1996)
24. Murty, K.G.: Computational Complexity of Parametric Linear Programming. Mathematical Programming 19, 213–219 (1980)
25. Papadimitriou, C.: Computational Complexity. Addison Wesley (1994)
26. Ross, C.: An Exponential Example for Terlaky's Pivoting Rule for the Criss-cross Simplex Method. Mathematical Programming 46, 78–94 (1990)
27. Santos, F.: A Counterexample to the Hirsch Conjecture. Annals of Mathematics 176(1), 383–412 (2013)
28. Smale, S.: On the Average Number of Steps of the Simplex Method of Linear Programming. Mathematical Programming 27, 241–267 (1983)
29. Spielman, D.A., Teng, S.-H.: Smoothed Analysis of Algorithms: Why the Simplex Algorithm Usually Takes Polynomial Time. Journal of the ACM 51(3), 385–463 (2004)
30. Tardos, E.: A Strongly Polynomial Algorithm to Solve Combinatorial Linear Programs. Operations Research 34, 250–256 (1986)
31. Savani, R., von Stengel, B.: Hard to Solve Bimatrix Games. Econometrica 74(2), 397–429 (2006)
32. Terlaky, T.: A Convergent Criss-cross Method. Optimization 16, 683–690 (1990)
33. Todd, M.J.: Polynomial Expected Behvior of Pivoting Algorithm for Linear Complementarity and Linear Programming Problems. Mathematical Programming 35, 173–192 (1986)
34. Ye, Y.: The Simplex and Policy-Iteration Methods Are Strongly Polynomial for the Markov Decision Problem with a Fixed Discount Rate. Mathematics of Operations Research 36(4), 593–603 (2011)
35. Zionts, S.: The Criss-cross Method for Solving Linear Programming Problems. Management Science 15, 426–445 (1979)

A Strongly Polynomial Time Algorithm
for Multicriteria Global Minimum Cuts

Hassene Aissi[1], A. Ridha Mahjoub[1], S. Thomas McCormick[2],
and Maurice Queyranne[2]

[1] PSL, Université Paris-Dauphine, LAMSADE, France
{aissi,mahjoub}@lamsade.dauphine.fr
[2] Sauder School of Business, University of British Columbia, Canada
{tom.mccormick,maurice.queyranne}@sauder.ubc.ca

Abstract. We investigate the bicriteria global minimum cut problem where each edge is evaluated by two nonnegative cost functions. The *parametric complexity* of such a problem is the number of linear segments in the parametric curve when we take all convex combinations of the criteria. We prove that the parametric complexity of the global minimum cut problem is $O(|V|^3)$. As a consequence, we show that the number of non-dominated points is $O(|V|^7)$ and give the first strongly polynomial time algorithm to compute these points. These results improve on significantly the super-polynomial bound on the parametric complexity given by Mulmuley [11], and the pseudo-polynomial time algorithm of Armon and Zwick [1] to solve this bicriteria problem. We extend some of these results to arbitrary cost functions and more than two criteria, and to global minimum cuts in hypergraphs.

1 Introduction

We consider the multicriteria version of the global minimum cut problem in undirected graphs. This problem is extensively studied in combinatorial optimization since many practical problems in, e.g., communications and electrical networks, contain it as a subproblem [1]. Let $G = (V, E)$ be an undirected graph, and $c^1, \ldots, c^k : E \to \mathbb{R}_+$ be k nonnegative cost functions, or *criteria*, defined on its edges. A *cut* C of G is a subset $C \subseteq V$ such that $\emptyset \neq C \neq V$, and it contains the set of edges $\delta(C)$ with exactly one end in C. The *cost* of cut C w.r.t. criterion j is $c^j(C) \equiv c^j(\delta(C))$. We would like a cut that simultaneously minimizes all criteria, but such a solution usually does not exist.

Therefore, we focus on *Pareto optimal solutions*, i.e., solutions that cannot be improved upon in any criterion without degrading another criterion. Each cut C is associated with its criteria vector (or point) $y(C) = (c^1(C), \ldots, c^k(C))$ in the criteria space $\mathbb{R}^k$. Let $Y = \{y(C) : \emptyset \neq C \subset V\}$ be the set of all criteria points associated with cuts (note that different cuts might give rise to the same criteria point). Given points $y(C), y(C') \in Y$, $y(C')$ *dominates* $y(C)$ if $y(C')_i \leqslant y(C)_i$, for all $i = 1, \ldots, k$, and $y(C')_j < y(C)_j$ for at least one j. If there exists no

J. Lee and J. Vygen (Eds.): IPCO 2014, LNCS 8494, pp. 25–36, 2014.
© Springer International Publishing Switzerland 2014

$y(C') \in Y$ that dominates $y(C)$, then $y(C)$ is *non-dominated*. Let Y_{ND} be the set of non-dominated points in Y.

A vector of multipliers $\mu \in \mathbb{R}^k$ forms a convex combination if $\mu \geq 0$ and $\sum_{i=1}^{k} \mu_i = 1$; the set of all such multipliers is the *simplex* S^k. Given $y(C') \in Y_{ND}$, if there exists $\mu \in S^k$ such that $C' \in \arg\min_C\{\sum_{i=1}^{k} \mu_i c^i(C) : \emptyset \neq C \subset V\}$ then $y(C')$ is called a *Supported Non-Dominated* (SND) point. The non-dominated points that are not SND points are called *Unsupported Non-Dominated* (UND) points. By "solving" a multicriteria discrete optimization problem we mean generating all SND and UND points.

The computation of SND points is related to the field of *parametric optimization*. The function $f : S^k \mapsto \mathbb{R}$ defined by $f(\mu) = \min_C\{\sum_{i=1}^{k} \mu_i c^i(C) : \emptyset \neq C \subset V\}$ is piecewise linear and concave; the facets of its graph correspond to SND points [3]. The *parametric complexity* of a multicriteria problem is the maximum number of facets. Our main interest here is to study the parametric complexity of global minimum cut, mainly for the case where $k = 2$.

A natural subproblem of parametric minimum cut is solving single-criterion (ordinary) minimum cut, e.g., for some fixed value of μ. The fastest deterministic algorithms for this problem run in $O(|E| \cdot |V| + |V|^2 \log |V|)$ time (Nagamochi and Ibaraki [13] and Stoer and Wagner [19]). The fastest randomized algorithm runs in $O(|E| \log^3 |V|)$ time (Karger [7]). These algorithms are faster than minimum s–t-cut algorithms that are based on network flows. See [14] for a detailed treatment of graph connectivity problems.

The multicriteria versions of several combinatorial optimization problems has been extensively studied (see Ehrgott [3]). These problems are often *intractable* in the sense that the cardinality of the set of (supported) non-dominated points grows exponentially in the input size. Furthermore, it is often hard even to verify if a given point is non-dominated. Multicriteria global minimum cut is an exception to the above intractability results. Indeed, Armon and Zwick [1] show that the decision version of the global multicriteria problem can be solved in polynomial time. The proof relies on the fact that the single-criterion global minimum cut problem has at most a strongly polynomial number of near-optimal solutions. More precisely, given $\alpha \geq 1$, a cut is called α-*approximate* if its cost is less than α times the minimum. Karger and Stein [8] show that there are $O(|V|^{2\alpha})$ α-approximate cuts, and they give a randomized algorithm for finding them in $\tilde{O}(|V|^{2\alpha})$ time. Nagamochi et al. [16] give a deterministic algorithm to find them in $O(|E||V|^{2\alpha})$ time, and they prove that there are $O(|V|^2)$ $\frac{4}{3}$-approximate cuts. Henzinger and Williamson [4] improve on this result by proving that there are $O(|V|^2)$ $\frac{3}{2}$-approximate cuts; they also show that $\frac{3}{2}$ is the largest possible approximation factor α for which there exist $O(|V|^2)$ α-approximate cuts.

For multicriteria global minimum cut, Armon and Zwick [1] used the result of [8] to give a pseudo-polynomial time algorithm to compute all the non-dominated points. Carstensen [2] shows that the parametric complexity of the s–t minimum cut problem is exponential for one parameter. Mulmuley [11] gives a simpler proof of this result, and studies the parametric complexity of the global minimum cut problem for $k = 2$. He considers the case where i) cost functions c^1 and

c^2 may be negative, ii) the parametric edge costs are positive and iii) the bit size of the values of c^1 and c^2 are at most a polynomial in $|V|$. And he shows in Theorem 3.10 that the parametric complexity is polynomial in this case. However, if iii) is relaxed, the proof of his theorem implies that the parametric complexity is $O(|V|^{19} \log |V| \log C_{max})$, where C_{max} is the maximum cost over all edges. This is surprising since the parametric function f is the minimum of the parametric functions of $O(|V|)$ minimum s–t cut problems (by fixing s and letting t vary over the other nodes), each of which could have an exponential number of breakpoints.

In this paper we give a much smaller, strongly polynomial upper bound on the parametric complexity of minimum cut, which leads to a strongly polynomial time algorithm for parametric global minimum cut, and hence a strongly polynomial time algorithm for the multicriteria version. In Section 2 we study in detail the bicriteria case, $k = 2$. In Section 3 we consider extensions, including arbitrary cost functions and more than two criteria, and global cuts in hypergraphs.

2   Complexity and Algorithms for $k = 2$

2.1   Parametric Complexity of the Global Min Cut Problem

We are given a graph $G = (V, E)$, and two nonnegative cost functions c^1, c^2 : $E \to \mathbb{R}_+$. For $\mu \in [0, 1]$, define the parametric cost function $c_\mu = \mu c^1 + (1 - \mu)c^2$. Let $\mathcal{S}(G)$ denote the set of cuts which are optimal solutions for some $\mu \in [0, 1]$. Our main result is a $O(|V|^3)$ upper bound on $\mathcal{S}(G)$. For every cut $X \in \mathcal{S}(G)$, let $I(G, X)$ denote the largest sub-interval of $[0, 1]$ such that X is optimal for all $\mu \in I(G, X)$.

Theorem 1 *Assume that $\mu c^1(X) + (1 - \mu)c^2(X) > 0$ for every $X \in \mathcal{S}(G)$ and $\mu \in I(G, X)$. Then the parametric complexity of the global min cut problem is $O(|V|^3)$.*

The proof of Theorem 1 is non-constructive and relies on the following definitions. Let $H = (W, F)$ denote a graph (which may be a subgraph of G); c^1, c^2 : $F \to \mathbb{R}_+$ two nonnegative edge cost functions; and X a cut in H. If lines $\mu c^1(X) + (1 - \mu)c^2(X)$ and $\frac{\mu c^1(F)}{|W|} + \frac{(1-\mu)c^2(F)}{|W|}$ intersect in $[0, 1]$, let $INT(H, X) \in [0, 1]$ denote their intersection point. For optimal $X \in \mathcal{S}(H)$, let $I'(H, X) \subseteq I(H, X)$ be a maximal subinterval satisfying

$$\frac{\mu c^1(F)}{|W|} + \frac{(1 - \mu)c^2(F)}{|W|} \leqslant \mu c^1(X) + (1 - \mu)c^2(X), \text{ for every } \mu \in I'(H, X). \quad (1)$$

Note that $I'(H, X)$ might be empty. Let $\mathcal{S}^{\geqslant}(H)$ denote the set of optimal solutions satisfying (1); $\mathcal{S}^{\geqslant}(H)$ might also be empty.

The set of solutions $\mathcal{S}(H) \setminus \mathcal{S}^{\geqslant}(H)$ can be partitioned into three subsets:

1. $\mathcal{S}_1^<(H) = \{X \in \mathcal{S}(H) \setminus \mathcal{S}^{\geqslant}(H) \colon I(H, X) \subseteq [0, INT(H, X)]\}$,
2. $\mathcal{S}_2^<(H) = \{X \in \mathcal{S}(H) \setminus \mathcal{S}^{\geqslant}(H) \colon I(H, X) \subseteq [INT(H, X), 1]\}$,
3. $\mathcal{S}_3^<(H) = \{X \in \mathcal{S}(H) \setminus \mathcal{S}^{\geqslant}(H) \colon$ function $\mu c^1(X) + (1 - \mu)c^2(X)$ is below $\frac{\mu c^1(F)}{|W|} + \frac{(1-\mu)c^2(F)}{|W|}$ in $[0, 1]\}$.

Figure 1 depicts an example of function f having six facets associated to optimal solutions $\mathcal{S}(H) = \{X_1, \ldots, X_6\}$. Parametric function $\frac{\mu c^1(F)}{|W|} + \frac{(1-\mu)c^2(F)}{|W|}$ intersects the facets of f corresponding to X_2 and X_4. For X_4, for instance, we have $I(H, X_4) = [\mu_1, \mu_3]$, $INT(H, X_4) = \mu_2$, and $I'(H, X_4) = [\mu_1, \mu_2]$. However, for X_3 we have $I(H, X_3) = I'(H, X_3)$. Here we have, $\mathcal{S}^{\geqslant}(H) = \{X_2, X_3, X_4\}$, $\mathcal{S}_1^<(H) = \{X_1\}$, $\mathcal{S}_2^<(H) = \{X_5, X_6\}$ and $\mathcal{S}_3^<(H) = \emptyset$.

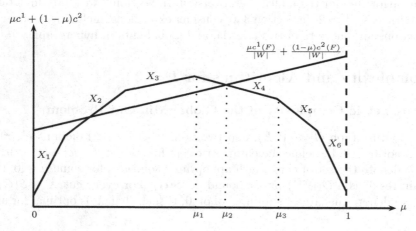

Fig. 1. Functions $f(\mu)$ and $\frac{\mu c^1(F)}{|W|} + \frac{(1-\mu)c^2(F)}{|W|}$

The proof of Theorem 1 uses the following lemma, whose proof is omitted.

Lemma 1 $|\mathcal{S}^{\geqslant}(H)| = O(|W|^2)$.

Proof. (of Theorem 1)

It suffices by Lemma 1 to give an upper bound on the cardinality of $\mathcal{S}_1^<(G) \cup \mathcal{S}_2^<(G) \cup \mathcal{S}_3^<(G)$. We focus on $\mathcal{S}_0 = \mathcal{S}_2^<(G) \cup \mathcal{S}_3^<(G)$ and show that its cardinality is $O(|V^3|)$. Set $\mathcal{S}_1^<(G)$ can be handled similarly. Assume that we have an oracle **O** that computes $\mathcal{S}(H)$ for any graph H.

In what follows we proceed in two steps in order to show that $|\mathcal{S}_0| \leqslant O(|V|^3)$. We will show that there exist two subsets $\mathcal{S}$ and $\mathcal{S}'$ such that $\mathcal{S}_0 \subseteq \mathcal{S} \cup \mathcal{S}'$, and $O(|\mathcal{S}|) \leqslant |V|^3$ and $O(|\mathcal{S}'|) \leqslant |V|$. This is a consequence of Algorithms 1 and 2 given below.

Figure 2 depicts the behavior of Algorithm 1. In each iteration r of this algorithm, an edge with large c^1-cost is contracted. Let $G_r = (V_r, E_r)$ denote the resulting graph obtained at iteration r. Note that the loops that arise by the contractions are kept in G_r. The idea is to have $|E_r| = |E_{r-1}| - 1$, and to ensure that $c^1(E_r)$ is not too small in comparison with $c^1(E_{r-1})$.

Consider the first iteration of the repeat loop. We will show next that once an edge is contracted in Step 4, $\mathcal{S}_0$ is partitioned into $\mathcal{S}^{\geqslant}(G_1)$, $\mathcal{S}_1^{<}(G_1)$ and $\mathcal{S}_2^{<}(G_1) \cup \mathcal{S}_3^{<}(G_1)$. By Lemma 1, we know that $|\mathcal{S}^{\geqslant}(G_1)| \leqslant O(|V_1|^2)$. We will show that the cardinality of $\mathcal{S}_1^{<}(G_1) \cap \mathcal{S}_0$ is also $O(|V_1|^2)$.

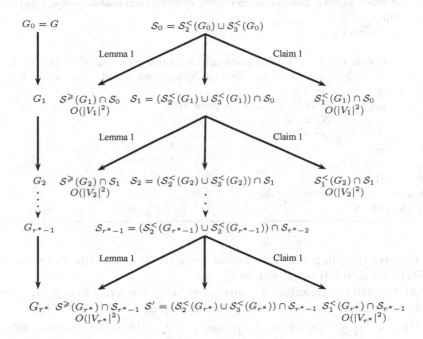

Fig. 2. The behavior of Algorithm 1

An upper bound on the cardinality of the remaining set $\mathcal{S}_1 = (\mathcal{S}_2^{<}(G_1) \cup \mathcal{S}_3^{<}(G_1)) \cap \mathcal{S}_0$ will be computed recursively. The algorithm keeps contracting edges until either the residual graph only contains two nodes or the cost $c^1(e)$ of every non-loop edge e is not in $\left[\frac{c^1(E_r)}{|V_r|}, \frac{c^1(E_r)}{2}\right]$. Let r^* denote the number of iterations of Algorithm 1. The algorithm returns the cardinality of the set $\mathcal{S} = (\mathcal{S}_0 \cap (\mathcal{S}^{\geqslant}(G_1) \cup \mathcal{S}_1^{<}(G_1))) \cup \cdots \cup (\mathcal{S}_{r^*-1} \cap (\mathcal{S}^{\geqslant}(G_{r^*}) \cup \mathcal{S}_1^{<}(G_{r^*})))$. If the last graph G_{r^*} contains more than two nodes, then additional work needs to be done in order to give an upper bound on the cardinality of the set $\mathcal{S}' = (\mathcal{S}_2^{<}(G_{r^*}) \cup \mathcal{S}_3^{<}(G_{r^*})) \cap \mathcal{S}_{r^*-1}$.

The following claim establishes a relation between $\mathcal{S}_0$ and intermediate sets $\mathcal{S}(G_r), \mathcal{S}_1^{<}(G_r), \mathcal{S}_2^{<}(G_r)$ and $\mathcal{S}_3^{<}(G_r)$ generated by the algorithm (proof omitted).

Claim 1. For iteration r, we have

$$\mathcal{S}_0 \subseteq (\mathcal{S}_{r-1} \cap (\mathcal{S}_2^{<}(G_r) \cup \mathcal{S}_3^{<}(G_r))) \cup (\cup_{l=1}^{r}(\mathcal{S}_{l-1} \cap (\mathcal{S}^{\geqslant}(G_l) \cup \mathcal{S}_1^{<}(G_l)))), \quad (2)$$

and

$$|\mathcal{S}_0| \leqslant |\mathcal{S}_{r-1} \cap (\mathcal{S}_2^{<}(G_r) \cup \mathcal{S}_3^{<}(G_r))| + O(r|V|^2). \quad (3)$$

Algorithm 1.

1: let $E_0 \leftarrow E$, $V_0 \leftarrow V$, $G_0 \leftarrow G$, $r \leftarrow 0$, $\mathcal{S} \leftarrow \emptyset$, $\mathcal{S}_0 \leftarrow \mathcal{S}_2^<(G_0) \cup \mathcal{S}_3^<(G_0)$, $test \leftarrow$ true
2: **repeat**
3:   **if** there exists a non-loop edge $e = (u, v) \in E_r$ such that $\frac{c^1(E_r)}{|V_r|} \leqslant c^1(e) \leqslant \frac{c^1(E_r)}{2}$ **then**
4:     contract e
5:     $r \leftarrow r + 1$
6:     denote by $G_r = (V_r, E_r)$ the resulting graph such that $V_r = (V_{r-1} \setminus \{u, v\}) \cup \{w\}$ where w is the node obtained by merging u and v, and $E_r = E_{r-1} \setminus \{e\}$ (the loops are kept)
7:     apply oracle **O** for G_r and compute $\mathcal{S}(G_r)$
8:     denote by $\mathcal{S}_r = (\mathcal{S}_2^<(G_r) \cup \mathcal{S}_3^<(G_r)) \cap \mathcal{S}_{r-1}$
9:     $\mathcal{S} \leftarrow \mathcal{S} \bigcup ((\mathcal{S}^{\geqslant}(G_r) \cup \mathcal{S}_1^<(G_r)) \cap \mathcal{S}_{r-1})$
10: **else**
11:    $test \leftarrow$ false
12: **end if**
13: **until** $|V_r| = 2$ or $test =$ false
14: Output: $|\mathcal{S}|$

Observe that, in general, the bound given in (3) is not tight since sets $\mathcal{S}_{r-1} \cap (\mathcal{S}^{\geqslant}(G_r) \cup \mathcal{S}_1^<(G_r))$ might not be disjoint.

At the end of Algorithm 1, three cases may happen. First, if G_{r^*} contains only two nodes, then by Claim 1, we have $|\mathcal{S}_0| \leqslant |\mathcal{S}| + 1 \leqslant O(|V|^3)$. Next, if $c^1(e) < \frac{c^1(E_{r^*})}{|V_{r^*}|}$ for all non-loop edge $e \in E_{r^*}$, then the problem reduces to computing an upper bound for $|\mathcal{S}'|$. Here we can show that $|\mathcal{S}'| \leqslant \frac{|V_{r^*}|}{2}$. Finally, if both previous cases do not hold, then there exist a non-loop edge $\bar{e} \in E_{r^*}$ such that $c^1(\bar{e}) > \frac{c^1(E_{r^*})}{2}$ and $c^1(e) < \frac{c^1(E_{r^*})}{|V_{r^*}|}$ for all non-loop edges $e \in E_{r^*} \setminus \{\bar{e}\}$. This case can be handled in a similar way as the previous one. Therefore, we only focus in the rest of the proof on the second case.

Algorithm 2. (G_{r^*})

1: $E_0' \leftarrow E_{r^*}$, $V_0' \leftarrow V_{r^*}$, $G_0' \leftarrow G_{r^*}$, $r \leftarrow 0$,
2: **while** $|V_r'| > 2$ **do**
3:   choose an edge $e = (u, v) \in E_r'$ with probability $\frac{c^1(e)}{c^1(E_r')}$
4:   **if** e is not a loop **then**
5:     contract e and remove it from the residual graph
6:     $r \leftarrow r + 1$
7:     denote by $G_r' = (V_r', E_r')$ the graph such that $V_r' = (V_{r-1}' \setminus \{u, v\}) \cup \{w\}$ where w is a node obtained from merging u and v, and $E_r' = E_{r-1}' \setminus \{e\}$ (the loops are kept)
8:   **end if**
9: **end while**
10: return the unique cut

For this purpose, it is sufficient to show that $|\mathcal{S}_2^<(G_{r^*}) \cup \mathcal{S}_3^<(G_{r^*})| \leqslant \frac{|V_{r^*}|}{2}$. This is done using Algorithm 2 based on probabilistic arguments similar to Karger's algorithm [6].

Algorithm 2 has as input graph G_{r^*} provided by Algorithm 1. In each iteration r, the current graph is denoted by $G'_r = (V'_r, E'_r)$ and the algorithm randomly chooses an edge e with probability $\frac{c^1(e)}{c^1(E'_r)}$. If e is a non-loop edge, then it is contracted. This process continues until the last graph only contains two nodes. Then the algorithm returns the unique cut in this graph. By hypothesis, $\mu c^1(X) + (1 - \mu)c^2(X) > 0$ for every $X \in \mathcal{S}(G)$ and $\mu \in I(G, X)$. Thus $c^1(E'_r) > 0$ for $r \leqslant V_{r^*} - 2$ and the probability of edge selection is always defined. As in Algorithm 1, loops resulting from contractions are not removed from residual graphs. However, by contrast to [6,7], Algorithm 2 has a pseudo-polynomial expected running time.

Claim 2. Algorithm 2 has a pseudo-polynomial expected running time.

Proof. Given an integer $r \in \{1, \ldots, |V_{r^*}| - 2\}$, let N_r denote a random variable defining the number of iterations of the while loop separating two consecutive contraction operations, say the rth and $r + 1^{\text{st}}$, and let N be a random variable defining the total number of iterations of the algorithm. We have $E(N) = |V_{r^*}| - 2 + \sum_{l=1}^{|V_{r^*}|-2} E(N_l)$. Let $\bar{E}_r$ denote the set of loops in the current graph G'_r. N_r is a geometric random variable with parameter $p_r = \frac{\sum_{e \in E'_r \setminus \bar{E}_r} c^1(e)}{c^1(E'_r)}$. Thus $E(N_r) = \frac{1}{p_r} \leqslant c^1(E'_r) \leqslant c^1(E)$ and $E(N) \leqslant (|V_{r^*}| - 2)(1 + c^1(E))$. ∎

Claim 3. Algorithm 2 returns any solution in $\mathcal{S}_2^<(G_{r^*}) \cup \mathcal{S}_3^<(G_{r^*})$ with probability at least $\frac{2}{|V_{r^*}|}$.

Proof. Consider any solution $X \in \mathcal{S}_2^<(G_{r^*}) \cup \mathcal{S}_3^<(G_{r^*})$. Algorithm 2 returns X only if none of its edges has been contracted. Therefore, no error occurs if a loop is selected. In the rest of the proof, we only focus on iterations of the while loop where a non-loop edge is chosen. Suppose that r edges not in X have been contracted through the algorithm. Since loops are not removed, $|E'_r| = |E'_{r-1}| - 1$. The probability that an edge from X is selected in the $r + 1^{th}$ contraction operation is $\frac{c^1(X)}{c^1(E'_r)} \leqslant \frac{c^1(E'_0)}{|V'_0|c^1(E'_r)}$. Since $c^1(e) < \frac{c^1(E'_0)}{|V'_0|}$ for every non-loop edge $e \in E'_0$, it follows that $c^1(E'_r) \geqslant c^1(E'_0)(1 - \frac{r}{|V'_0|})$. Therefore, the probability for error is at most $\frac{1}{|V'_0|-r}$, and the probability that no edge of X is chosen after the $r + 1^{th}$ contraction operation is at least $\frac{|V'_0|-r-1}{|V'_0|-r}$. Hence, the probability that no edge from X is never chosen (after $|V'_0| - 2$ contraction operations) is at least $\frac{|V'_0|-1}{|V'_0|} \cdot \frac{|V'_0|-2}{|V'_0|-1} \cdot \frac{|V'_0|-3}{|V'_0|-2} \cdots \frac{2}{3} = \frac{2}{|V'_0|} = \frac{2}{|V_{r^*}|}$. ∎

Since the probability that a cut in $\mathcal{S}_2^<(G_{r^*}) \cup \mathcal{S}_3^<(G_{r^*})$ survives all the contraction operations is at least $\frac{2}{|V_{r^*}|}$, and that no two cuts can survive simultaneously, it follows that $|\mathcal{S}_2^<(G_{r^*}) \cup \mathcal{S}_3^<(G_{r^*})| \leqslant \frac{|V_{r^*}|}{2}$. Therefore, $|\mathcal{S}_0| \leqslant O(|V|^3)$.

Using a similar argument, we can also show that $|\mathcal{S}_1^<(G)|$ is $O(|V|^3)$, and the proof is complete. □

As a consequence of Theorem 1, we will show that the number of non-dominated points is also strongly polynomial.

Corollary 1 *The number of SND and UND points of the global minimum cut problem are $O(|V|^5)$ and $O(|V|^7)$, respectively.*

Proof. As it is proved in [6,8], the number of optimal solutions of the global minimum cut problem is $O(|V|^2)$. Thus, the bound on the number of SND points follows from combining this result and Theorem 1.

Now consider two SND points X_1, X_2. Suppose that they are optimal for some $\mu = \mu_1$. Let X_3 be a UND point dominated by a convex combination of X_1 and X_2. W.l.o.g., suppose that $c^1(X_1) < c^1(X_2)$ and $c^2(X_2) < c^2(X_1)$. We then have $c^1(X_3) < c^1(X_2)$ and $c^2(X_3) < c^2(X_1)$. Therefore,

$$\mu_1 c^1(X_3) + (1 - \mu_1)c^2(X_3) < \mu_1 c^1(X_2) + (1 - \mu_1)c^2(X_1)$$
$$\leqslant \mu_1 c^1(X_2) + (1 - \mu_1)c^2(X_2) + \mu_1 c^1(X_1) +$$
$$(1 - \mu_1)c^2(X_1).$$

Thus, X_3 is a 2-approximate solution for $\mu = \mu_1$. The bound on the number of UND points follows from [6] and Theorem 1. □

2.2 Efficient Algorithms for f and the Non-dominated Points Set

We are in the single-parameter case ($k = 2$; the full paper shows how to extend these algorithms for $k > 2$), and we want to compute f for all $\mu \in [0, 1]$. Our strongly polynomial algorithm is based on the Discrete Newton Algorithm [10,18]. For a fixed $\mu_1 \in [0, 1]$, using a cactus representation of minimum cuts [15] we compute optimal cuts $X_+^*(\mu_1)$ and $X_-^*(\mu_1)$ at μ_1 such that $X_+^*(\mu_1)$ ($X_-^*(\mu_1)$) is the optimal cut X whose line $\mu_1 c^1(X) + (1 - \mu_1)c^2(X)$ has the largest (smallest) slope among all optimal cuts at μ_1. Let $Steepest^+(\mu_1)$ and $Steepest^-(\mu_1)$ be the lines associated with $X_+^*(\mu_1)$ and $X_-^*(\mu_1)$. Consider Algorithm 3.

Algorithm 3. Discrete Newton method

1: $L \leftarrow \{[0, 1]\}, \mathcal{B} \leftarrow \emptyset$
2: **while** $L \neq \emptyset$ **do**
3:     choose an interval $[\mu_1, \mu_2] \in L$ and compute $Steepest^-(\mu_1)$ and $Steepest^+(\mu_2)$
4:     compute $\mu_3 \in [\mu_1, \mu_2]$ corresponding to the intersection of $Steepest^-(\mu_1)$ and $Steepest^+(\mu_2)$
5:     **if** $\min_C\{\mu_3 c^1(C) + (1 - \mu_3)c^2(C) : \emptyset \neq C \neq V\} = \mu_3 c^1(X_-^*(\mu_1)) + (1 - \mu_3)c^2(X_-^*(\mu_1)) = \mu_3 c^1(X_+^*(\mu_2)) + (1 - \mu_3)c^2(X_+^*(\mu_2))$ **then**
6:         $L \leftarrow L \setminus \{[\mu_1, \mu_2]\}$ and $\mathcal{B} \leftarrow \mathcal{B} \cup \{\mu_3\}$
7:     **else**
8:         $L \leftarrow L \setminus \{[\mu_1, \mu_2]\} \cup \{[\mu_1, \mu_3], [\mu_3, \mu_2]\}$
9:     **end if**
10: **end while**
11: Return $\mathcal{B}$

Algorithm 3 manages a list L of unexplored intervals containing at least one breakpoint and a list $\mathcal{B}$ of breakpoints. In each iteration the algorithm chooses an interval $[\mu_1, \mu_2] \in L$ and computes μ_3 as the intersection of the lines $Steepest^-(\mu_1)$ and $Steepest^+(\mu_2)$. It is clear that these lines are part of the function f. If the condition in Step 5 holds, then μ_3 is the unique breakpoint in $[\mu_1, \mu_2]$ and so we can fathom $[\mu_1, \mu_2]$ and add μ_3 to $\mathcal{B}$. Otherwise, $[\mu_1, \mu_3]$ and $[\mu_3, \mu_2]$, each contains at least one breakpoint, and $[\mu_1, \mu_2]$ is replaced by them. Using this, we obtain the following.

Proposition 1 *Algorithm 3 has $O(|E||V|^4 + |V|^5 \log |V|)$ running time.*

Proof. For any facet of function f, Algorithm 3 computes a global minimum cut and finds a cactus representation for at most three values: the two extremities and an intermediate value. Therefore, the total number of iterations is at most twice the number of facets. A cactus representation can be obtained in time $O(|E||V| + |V|^2 \log |V|)$ [15] and a global minimum cut can be computed in the same time complexity [13]. Therefore, the time complexity follows by Theorem 1.   □

Since the time required to compute UND points dominates that of SND points, we have:

Proposition 2 *The time required to compute all the non-dominated points of the global minimum cut problem is $O(|E||V|^7)$.*

Proof. Computing all the 2-approximate solutions can be performed in time $O(|E||V|^4)$ [16]. The result follows by combining this and Theorem 1.   □

3 Extensions

3.1 Parametric Complexity with More Than Two Criteria and Arbitrary Cost Functions

First, suppose we are given a graph $G = (V, E)$, and two cost functions $c^1, c^2 : E \rightarrow \mathbb{R}$. Here we allow some cost components to be negative. For $\mu \in \mathbb{R}$ define the parametric edge costs $f_\mu(e) = \mu c_e^1 + c_e^2$. We suppose that there exists an interval $[\alpha, \beta]$ where all $f_\mu(e)$ are positive. Here Lemma 1 and Step 3 of Algorithm 2 are no longer applicable and thus, Theorem 1 does hold in this case. Using ideas in [11] and avoiding unnecessary subdivisions of the interval $[\alpha, \beta]$, we obtain a strongly polynomial bound. The next result is also non-constructive (proof omitted).

Theorem 2 *Assume that the parametric edge costs $f_\mu(e)$ are positive for all $\mu \in [\alpha, \beta]$. The parametric complexity of the global min cut problem is $O(|E|^2|V|^2 \log |V|)$.*

It is natural to wonder if Theorem 3 could be extended to a constant number of parameters at least equal to two. More precisely, we are given $k \geqslant 3$ cost functions $c^1, \ldots, c^k : E \rightarrow \mathbb{R}$. For $\mu = (\mu_1, \mu_2) \in \mathbb{R}^{k-1}$, define the parametric edge costs $f_\mu(e) = \sum_{i=1}^{k-1} \mu_i c_e^i + c_e^k$ for $e \in E$. Assume that there exists an hyper-rectangle $I = [\alpha_1, \beta_1] \times \cdots \times [\alpha_{k-1}, \beta_{k-1}]$ such that $f_\mu(e)$ are positive for all $\mu \in I$ and $e \in E$. The next result shows that the parametric complexity is again strongly polynomial in this case (proof omitted).

Theorem 3 *Assume that the parametric edge costs $f_\mu(e)$ are positive for all $\mu \in I$. Then the parametric complexity of the global min cut problem is $O(|E|^k |V|^2 \log^{k-1} |V|)$.*

3.2 Hypergraphs

We consider finite hypergraphs $H = (V, E)$, where V is a finite set of *nodes* and each *edge* $e \in E$ is a subset of V. Hypergraph H is *rank-ρ* if every edge in H has cardinality at most ρ (e.g., a graph is a rank-ρ hypergraph for every $\rho \geq 2$).

A *cut* C in $H = (V, E)$ is any nontrivial node subset, i.e., satisfying $\emptyset \neq C \subset V$. Let $\Delta(C) = \{e \in E : e \cap C \neq \emptyset \neq e \setminus C\}$ denote the set of edges crossed by the cut C. Given nonnegative edge costs $c(e) \geq 0$ ($e \in E$), let $c(F) = \sum_{e \in F} c(e)$ be the total cost of all edges in subset $F \subseteq E$, and let $c(C) \equiv c(\Delta(C))$ denote the total cost of all the edges crossed by the cut $C \subset V$. Further define $c(H) = \min_C \{c(C) : \emptyset \neq C \subset V\}$ as the minimum cost of a cut in H. There exist polynomial time algorithms for finding a minimum cost cut in a hypergraph, see [9,12,17].

The technique used to derive an upper bound on the parametric complexity can be generalized to hypergraphs. In order to extend Lemma 1 to hypergraphs, we now bound the number of approximate cuts in hypergraphs.

Theorem 4 *For any fixed integer $\rho \geq 2$ and scalar $\alpha \geq 1$, and rank-ρ hypergraph $H = (V, E)$ with nonnegative edge costs c and positive minimum cut cost, the number of cuts C with cost $c(C) \leq \alpha c(H)$ is $O\left(|V|^{B(\rho, \alpha)}\right)$ where*

$$
B(\rho, \alpha) = \begin{cases} 2 & \text{if } \rho \leq 3 \text{ and } \alpha < \frac{3}{2}; \\ 2\alpha & \text{if } \rho \leq 3 \text{ and } \alpha \geq \frac{3}{2}; \\ \frac{\rho}{2} + \frac{2}{3} & \text{if } \rho \geq 4 \text{ and } \alpha < \frac{3}{2}; \\ \left(\frac{\rho}{2} + \frac{2}{3}\right)\alpha & \text{otherwise, i.e., if } \rho \geq 4 \text{ and } \alpha \geq \frac{3}{2}. \end{cases}
$$

Proof. W.l.o.g., assume that every edge $e \in E$ has cardinality $|e| \geq 2$ (since edges e with $|e| \leq 1$ are not crossed by any cut). We prove the result by approximating the minimum cut problem in hypergraph $H = (V, E)$ with edge costs c by the minimum cut problem in the complete graph $K(V) = (V, E_{K(V)})$ with edge costs c', whereby each edge $e \in E$ is replaced by a clique on the nodes in e, where each edge in the clique has cost $c_e/(|e| - 1)$, i.e., by letting

$$
c'_{i,j} = \sum_{e \in E : \{i,j\} \subseteq e} \frac{c_e}{|e| - 1} \quad \text{for all } \{i, j\} \in E_{K(V)}.
$$

In particular, every cardinality-2 edge $e = \{i, j\} \in E$ contributes its full cost c_e to $c'_{i,j}$, and thus to the cost $c'(C)$ of every cut C in $K(V)$ that crosses it. Note also that every cardinality-3 edge $e \in E$ that is crossed by cut C has two of its nodes on one side of the cut and the other node on the other side, and thus also contributes its exact cost $2(c_e/2) = c_e$ to $c'(C)$. Therefore (as it is well known, e.g., Ihler et al. [5]), when $\rho \leq 3$ this transformation is exact, i.e., $c'(C) = c(C)$ for every cut C. The first two cases in the definition of $B(\rho, \alpha)$ then follow from Henzinger and Williamson [4] and Karger and Stein [8], respectively.

Now assume that $\rho \geq 4$. A cut C that crosses an edge $e \in E$ with cardinality $|e| \geq 4$ crosses at least $|e| - 1$ edges in the clique $K(e)$ (when exactly one node of e is on one side of the cut), and at most $|e|^2/4$ such edges (when half the nodes of e are on either side). Thus every edge $e \in E$ crossed by C contributes at least c_e and at most $(|e|^2/4)\frac{c_e}{|e|-1} \leq \frac{\rho^2/4}{\rho-1} c_e$ to the cost $c'(C)$. Therefore,

$$c(C) \leq c'(C) \leq \frac{\rho^2/4}{\rho - 1} c(C) = \left(\frac{\rho}{4} + \frac{1}{4} \left(1 + \frac{1}{\rho - 1} \right) \right) c(C) \leq \beta(\rho) c(C),$$

where $\beta(\rho) = \frac{\rho}{4} + \frac{1}{3}$ and the last inequality follows from $\rho \geq 4$. Let C' denote a minimum cut for $(K(V), c')$. If C is an α-optimal cut for (H, c), i.e., $c(C) \leq \alpha c(H)$, we have

$$c'(C) \leq \beta(\rho) c(C) \leq \beta(\rho) \alpha c(H) \leq \beta(\rho) \alpha c(C') \leq \beta(\rho) \alpha c'(C'),$$

implying that C is a $(\beta(\rho) \alpha)$-optimal cut for $(K(V), c')$. Then the last two cases in the definition of $B(\rho, \alpha)$ again follow from Henzinger and Williamson [4] and Karger and Stein [8], respectively.   □

Theorem 5 *For any fixed scalar $\rho \geq 2$, let $H = (V, E)$ be a rank-ρ hypergraph with nonnegative edge costs c and $c^1, c^2 : E \to \mathbb{R}_+$ two nonnegative cost functions defined on its edges. Assume that the edge costs $f_\mu(e) = \mu c_e^1 + (1 - \mu)c_e^2$, for all $e \in E$, are functions of a parameter $0 \leq \mu \leq 1$, and $\mu c^1(X) + (1-\mu)c^2(X) > 0$ for any $X \in \mathcal{S}(G)$ and $\mu \in I(G, X)$. The parametric complexity of global minimum cut is $O(|V|^{B(\rho, \frac{3}{2})+1})$.*

Proof. The proof is an adaptation of that of Theorem 1 to hypergraphs. By Theorem 4 and Lemma 1, we have $\mathcal{S}^\geq(H) \leq O(|V|^{B(\rho, \frac{3}{2})})$. Now one can extend Algorithm 1 to hypergraphs, by contracting hyperedges instead of edges, and obtain that (2) holds in this case and (3) extends to

$$|\mathcal{S}_0| \leq |\mathcal{S}_{r-1} \cap (\mathcal{S}_2^<(G_r) \cup \mathcal{S}_3^<(G_r))| + O(r|V|^{B(\rho, \frac{3}{2})}). \tag{4}$$

Similarly, still three cases have to be considered. If G_{r^*} contains only two nodes, then by (4) we have $|\mathcal{S}_0| \leq O(|V|^{B(\rho, \frac{3}{2})+1})$. If $c^1(e) < \frac{c^1(E_{r^*})}{|V_{r^*}|}$ for all $e \in E_{r^*}$, then the problem reduces to computing an upper bound for $|\mathcal{S}'|$. Algorithm 2 and Claims 2-3 apply in this case and yield $|\mathcal{S}'| \leq \frac{|V_{r^*}|}{2}$. Therefore by (4), we have $|\mathcal{S}_0| \leq O(|V|^{B(\rho, \frac{3}{2})+1})$. Finally, the case where there exists a non-loop hyperedge $\bar{e} \in E_{r^*}$ such that $c^1(\bar{e}) > \frac{c^1(E_{r^*})}{2}$ (and $c^1(e) < \frac{c^1(E_{r^*})}{|V_{r^*}|}$ for all non-loop hyperedge $e \in E_{r^*} \setminus \{\bar{e}\}$) can be handled in a similar way as the previous one. Therefore, the result follows.   □

References

1. Armon, A., Zwick, U.: Multicriteria global minimum cuts. Algorithmica 46(1), 15–26 (2006)
2. Carstensen, P.: Complexity of some parametric integer and network programming problems. Mathematical Programming 26, 64–75 (1983)
3. Ehrgott, M.: Multicriteria Optimization. Springer (2005)
4. Henzinger, M., Williamson, D.P.: On the number of small cuts in a graph. Information Processing Letters 59(1), 41–44 (1996)
5. Ihler, E., Wagner, D., Wagner, F.: Modeling hypergraphs by graphs with the same mincut properties. Information Processing Letters 45(4), 171–175 (1993)
6. Karger, D.R.: Global min-cuts in RNC, and other ramifications of a simple min-cut algorithm. In: Proceedings of the Fourth Annual ACM-SIAM Symposium on Discrete Algorithms, pp. 21–30 (1993)
7. Karger, D.R.: Minimum cuts in near-linear time. Journal of the ACM 47(1), 46–76 (2000)
8. Karger, D.R., Stein, C.: A new approach to the minimum cut problem. Journal of the ACM 43(4), 601–640 (1996)
9. Klimmek, R., Wagner, F.: A simple hypergraph min cut algorithm. Freie Univ., Fachbereich Mathematik (1996)
10. McCormick, S.T., Ervolina, T.R.: Computing maximum mean cuts. Discrete Applied Math 52, 53–70 (1994)
11. Mulmuley, K.: Lower bounds in a parallel model without bit operations. SIAM Journal on Computing 28(4), 1460–1509 (1999)
12. Mak, W.K., Wong, D.F.: A fast hypergraph min-cut algorithm for circuit partitioning. Integration, the VLSI Journal 30(1), 1–11 (2000)
13. Nagamochi, H., Ibaraki, T.: Computing edge-connectivity in multigraphs and capacitated graphs. SIAM Journal on Discrete Mathematics 5(1), 54–66 (1992)
14. Nagamochi, H., Ibaraki, T.: Algorithmic aspects of graph connectivity. Cambridge University Press (2008)
15. Nagamochi, H., Nakamura, S., Ishii, T.: Constructing a cactus for minimum cuts of a graph in $O(mn + n^2 \log n)$ time and $O(m)$ space. Inst. Electron. Inform. Comm. Eng. Trans. Inform. Systems, 179–185 (2003)
16. Nagamochi, H., Nishimura, K., Ibaraki, T.: Computing all small cuts in undirected networks. SIAM Journal of Discrete Mathematics 10, 469–481 (1997)
17. Queyranne, M.: Minimizing symmetric submodular functions. Mathematical Programming 82(1-2), 3–12 (1998)
18. Radzik, T.: Newton's Method for Fractional Combinatorial Optimization. In: Proceedings of IEEE Annual Symp. of Foundations of Computer Science, pp. 659–669 (1992)
19. Stoer, M., Wagner, F.: A simple min-cut algorithm. Journal of the ACM 44(4), 585–591 (1997)

Integer Programs with Prescribed Number of Solutions and a Weighted Version of Doignon-Bell-Scarf's Theorem

Iskander Aliev[1], Jesús A. De Loera[2], and Quentin Louveaux[3]

[1] Cardiff University, UK
AlievI@cardiff.ac.uk
[2] University of California, Davis
deloera@math.ucdavis.edu
[3] Université de Liège, Belgium
q.louveaux@ulg.ac.be

Abstract. In this paper we study a generalization of the classical feasibility problem in integer linear programming, where an ILP needs to have a prescribed number of solutions to be considered solved.

We first provide a generalization of the famous Doignon-Bell-Scarf theorem: Given an integer k, we prove that there exists a constant $c(k, n)$, depending only on the dimension n and k, such that if a polyhedron $\{x : Ax \le b\}$ contains exactly k integer solutions, then there exists a subset of the rows of cardinality no more than $c(k, n)$, defining a polyhedron that contains exactly the same k integer solutions.

The second contribution of the article presents a structure theory that characterizes precisely the set $\mathrm{Sg}_{\ge k}(A)$ of all vectors b such that the problem $Ax = b, x \ge 0, x \in \mathbb{Z}^n$, has *at least* k-solutions. We demonstrate that this set is finitely generated, a union of translated copies of a semigroup which can be computed explicitly via Hilbert bases computation. Similar results can be derived for those right-hand side vectors that have *exactly* k solutions or *fewer than* k solutions.

Finally we show that, when n, k are fixed natural numbers, one can compute in polynomial time an encoding of $\mathrm{Sg}_{\ge k}(A)$ as a generating function, using a short sum of rational functions. As a consequence, one can identify all right-hand side vectors that have exactly k solutions (similarly for at least k or less than k solutions). Under the same assumptions we prove that the k-Frobenius number can be computed in polynomial time.

1 Introduction

Given a matrix $A \in \mathbb{Z}^{d \times n}$ and a vector $b \in \mathbb{Z}^d$, the classical integer linear feasibility problem asks whether the system $IP_A(=, b)$

$$Ax = b, \quad x \ge 0, \quad x \in \mathbb{Z}^n, \tag{1}$$

has a solution or not. There is of course a slightly more general form $IP_A(\le, b)$ of the problem above

$$Ax \le b, \quad x \in \mathbb{Z}^n. \tag{2}$$

J. Lee and J. Vygen (Eds.): IPCO 2014, LNCS 8494, pp. 37–51, 2014.
© Springer International Publishing Switzerland 2014

We refer to these two problems as $IP_A(b)$, unless specifying which of (1) or (2) is necessary.

For a given integer k there are three natural variations of the feasibility problem that in some intuitive sense measure the strength of $IP_A(b)$ "being feasible":

- Are there *at least* k distinct solutions for $IP_A(b)$? If yes, we say that the problem is $\geq k$-*feasible*.
- Are there *exactly* k distinct solutions for $IP_A(b)$? If yes, we say that the problem is $= k$-*feasible*.
- Are there *less than* k distinct solutions for $IP_A(b)$? If yes, we say that the problem is $< k$-*feasible*.

We call these three problems, *the fundamental problems of k-feasibility in integer linear programs*. In this paper we investigate the question of, given a matrix A, determining for which right-hand-side vectors b are the problems $IP_A(b) = k$-feasible, $\geq k$-feasible, or $< k$-feasible. In what follows we say that b is $= k$-feasible (respectively, $\geq k$-feasible, $< k$-feasible) if the corresponding integer program is.

Clearly the classical feasibility problem is just the problem of deciding whether $IP_A(b)$ is ≥ 1-feasible. This indicates directly that all these problems are NP-hard in complexity. Recently Eisenbrand and Hähnle [18] showed that the related problem of finding the right-hand-side vector b that maximizes the number of lattice points solutions, when b is restricted to take values in a polyhedron, is NP-hard. The theory of k-feasibility is actually quite useful in applications where for some reason a given number of solutions k needs to be achieved to consider the problem solved or situations where one cannot allow too many solutions. Naturally this "weighted version" of the k-feasibility problem has some interesting applications in combinatorics, statistics, and number theory: Consider first the widely popular recreational puzzle *sudoku*, each instance can be thought of as an integer linear program where the hints provided in some of the entries are the given right-hand-sides of the problem. Of course in that case newspapers wish to give readers a puzzle where the solution is unique ($k = 1$). It is not difficult to see that this is a special case of a *3-dimensional transportation problem* that is, the question to decide whether the set of *integer* feasible solutions of the $r \times s \times t$-transportation problem

$$\left\{ x \in \mathbb{Z}^{rst} : \sum_{i=1}^{r} x_{ijk} = u_{jk}, \sum_{j=1}^{s} x_{ijk} = v_{ik}, \sum_{k=1}^{t} x_{ijk} = w_{ij}, x_{ijk} \geq 0 \right\}$$

has a unique solution given right-hand sides u, v, w. Another application of k-feasibility appears in statistics, concretely in application in the data security problem of *multi-way contingency tables*, because when the number of solutions is small, e.g. unique, the margins of the statistical table may disclose personal information which is illegal [16]. Consider next the k-*Frobenius problem*. Let a be a positive integral n-dimensional primitive vector, i.e., $a = (a_1, \ldots, a_n)^T \in \mathbb{Z}_{>0}^n$ with $\gcd(a_1, \ldots, a_n) = 1$. For a positive integer k the k-*Frobenius number* $F_k(a)$

is the largest number which cannot be represented in at least k different ways as a non-negative integral combination of the a_i's. Thus, putting $A = a^T$,

$$F_k(a) = \max\{b \in \mathbb{Z} : IP_A(b) \text{ is } < k \text{ feasible}\}.$$

When $k = 1$ this has been studied by a large number of authors and both the structure and algorithmic properties are well-understood. Computing $F_1(a)$ when n is not fixed is an NP-hard problem (Ramirez Alfonsin [26]). On the other hand, for any fixed n the classical Frobenius number can be found in polynomial time by sophisticated deep algorithms due to Kannan [22] and Barvinok and Woods [6]. The general problem of finding $F_1(a)$ has been traditionally referred to as the *Frobenius problem*. There is a rich literature on the various aspects of this question. For a comprehensive and extensive survey we refer the reader to the book of Ramirez Alfonsin [27]. More recently a k-feasibility generalization of the Frobenius number was introduced and studied by Beck and Robins [8]. They give formulas for $n = 2$ of the k-Frobenius number, but for general n and k only bounds on the k-Frobenius number $F_k(a)$ are available (see [3],[4] and [19]).

Finally, other areas in which polyhedra with fixed number of (interior) lattice points play a role are algebraic and discrete geometry. Indeed, there has been a lot of work, going back to classical results of Minkowski and van der Corput, to show that the volume of a lattice polytope P with $k = \text{card}(\mathbb{Z}^n \cap \text{int} P) \geq 1$ is bounded above by a constant that only depends in n and k (see e.g., [23,24]). Similarly, the supremum of the possible number of points of $\mathbb{Z}^n$ in a lattice polytope in $\mathbb{R}^n$ containing precisely n points of $\mathbf{Z}^d$ in its interior, can be bounded by a constant that only depends in n and k. Such results play an important role in the theory of toric varieties and the structure of lattice polyhedra (see e.g., [20] and the references therein).

Our Results

This paper has three main contributions to the study of k-feasibility:

1. One of the most famous results in the theory of integer programming is the theorem of Doignon [17] (later reproved by Bell and Scarf [7,29]). This theorem has played an interesting role in many papers, including Clarkson's probabilistic algorithm for integer linear programming [11]:

   Theorem [*Doignon 1973*] Let A be a $d \times n$ matrix and b a vector of $\mathbb{R}^d$. If the problem $IP_A(\leq, b)$ is infeasible, then there is a subset S of the rows of A of cardinality no more than 2^n, with the property that the smaller integer program $IP_S(\leq, b)$ is also infeasible.

   Our first contribution is to prove a $= k$-feasibility version of Doignon's theorem:

   Theorem 1. *Given n, k two non-negative integers there exists a universal constant $c(k, n)$ depending only on k and n such that for any $d \times n$ integral matrix A, and d-vector b if $IP_A(\leq, b)$ has exactly k integral solutions, then there is a subset S of the rows of A of cardinality no more than $c(k, n)$, with* the property that the smaller integer program $IP_S(\leq, b)$ has exactly the same k solutions as $IP_A(\leq, b)$.

We will use this theorem later on in some applications. Our technique to prove this theorem is quite close to the proof of Doignon in [28] with some twists. In addition our initial estimation of the constant $c(k, n)$ appears to be loose, thus in the extended journal version of this paper we will include better estimations in low dimension. It should be remarked that the $\geq k$ version of the problem is not interesting.

2. Second, we prove a structural result that implies that the set of b's that provide a $\geq k$-feasible $IP_A(b)$ is finitely generated.

Let $\mathrm{Sg}_{\geq k}(A)$ (respectively $\mathrm{Sg}_{=k}(A)$ and $\mathrm{Sg}_{<k}(A)$) be the set of right-hand side vectors $b \in \mathrm{cone}(A) \cap \mathbb{Z}^d$, where $\mathrm{cone}(A)$ is the cone generated by the columns of A, that make $IP_A(b) \geq k$-feasible (respectively $= k$-feasible, $< k$-feasible). Note that $\mathrm{Sg}(A) := \mathrm{Sg}_{\geq 1}(A)$ is the semigroup generated by the column vectors of the matrix A.

The first structural result of this paper gives an algebraic description of the sets $\mathrm{Sg}_{\geq k}(A)$ and $\mathrm{Sg}_{<k}(A)$. Let $e_1, \ldots, e_n$ be the standard basis vectors in $\mathbb{Z}_{\geq 0}^n$. We define the *coordinate subspace of* $\mathbb{Z}_{\geq 0}^n$ of dimension $r \geq 1$ determined by $e_{i_1}, \ldots, e_{i_r}$ with $i_1 < \cdots < i_r$ as the set $\{e_{i_1} z_1 + \cdots + e_{i_r} z_r : z_j \in \mathbb{Z}_{\geq 0}$ for $1 \leq j \leq r\}$. By the 0-dimensional coordinate subspace of $\mathbb{Z}_{\geq 0}^n$ we understand the origin $0 \in \mathbb{Z}_{\geq 0}^n$.

Theorem 2. *(i) There exists a monomial ideal $I(A) \subset \mathbb{Q}[x_1, \ldots, x_n]$ such that*

$$\mathrm{Sg}_{\geq k}(A) = \{A\lambda : \lambda \in E(A)\}, \tag{3}$$

where $E(A)$ is the set of exponents of monomials of $I(A)$.
(ii) The set $\mathrm{Sg}_{<k}(A)$ can be written as a finite union of translates of the sets $\{A\lambda : \lambda \in S\}$, where S is a coordinate subspace of $\mathbb{Z}_{\geq 0}^n$.

By the Gordan-Dickson lemma, the ideal $I(A)$ is finitely generated, so that $\mathrm{Sg}_{\geq k}(A)$ is a finite union of translated copies of a semigroup. The proof of Theorem 2 relies on some basic facts on lattice points when we think of them as generators of monomial ideals. The basic tool is a characterization of the complement of a monomial ideal (see [12]). Some of the arguments are of interest for the study of affine semigroups and toric varieties [9,31].

Our results extend the decomposition theorem of Hemmecke, Takemura and Yoshida [21] for $k = 1$. They investigated the semigroup $\mathrm{Sg}(A)$ and the vectors that are not in the semigroup but still lie within $\mathrm{cone}(A)$. Note even when there exists a real nonnegative solution for $Ax = b$, there may not exist an integral nonnegative solution. Those authors studied $Q_{\mathrm{sat}} = \mathrm{cone}(A) \cap \mathrm{lattice}(A)$, where $\mathrm{lattice}(A)$ is the lattice generated by the columns of A. They called $H = Q_{\mathrm{sat}} \setminus \mathrm{Sg}(A)$ the set of *holes* of $\mathrm{Sg}(A)$ (in the context of numeric semigroups and the Frobenius number, holes have also been called *gaps*, see [25]) The set of holes H may be finite or infinite, but their main result is to give a finite description of the holes as a finitely-generated set.

Our Theorem 2 was inspired by theirs. For us the holes of [21] are just a special case for $k = 1$. We can generalize this notion to consider k-holes, namely those right hand-sides b for which $Ax = b$ has *less than* k non-negative integer solutions.

In the last part of the article we show how to make "effective" the decomposition theorem above via Hilbert bases computations.

3. Third, for n and k fixed integer numbers, our first algorithmic result establishes a way to compute all the $\geq k$-feasible vectors b's, not explicitly one by one, but rather the $\geq k$-feasible b's are encoded as a *generating function*, $\sum_{\geq k-\text{feasible}} t^b$.

Theorem 3. *Let $A \in \mathbb{Z}^{d \times n}$. Assuming that n and k are fixed, there is a polynomial time algorithm to compute a short sum of rational function $G(t)$ which efficiently represents the formal sum $\sum_{\geq k-\text{feasible}} t^b$. Moreover, from the algebraic formula, one can perform the following tasks in polynomial time:*

(a) Count the number of $\geq k$-feasible vectors (if finite).
(b) Extract the lexicographic-smallest b, $\geq k$-feasible vector.
(c) Find the $\geq k$-feasible vector b that maximizes the dot product $c^T b$.
(d) Similar generating function descriptions, with same computational properties, hold for the sets of b which are $= k$-feasible or $< k$-feasible.
(e) Identical results hold for problems in the inequality form $IP_A(\leq, b)$.

Let us explain a bit the philosophy of such theorem for those not familiar with this point of view: In 1993 A. Barvinok [5] gave an algorithm for counting the lattice points inside a polyhedron P in polynomial time when the dimension of P is a constant. The input of the algorithm is the inequality description of P, the output is a polynomial-size formula for the multivariate generating function of all lattice points in P, namely $f(P) = \sum_{a \in P \cap \mathbb{Z}^n} x^a$ where x^a is an abbreviation of $x_1^{a_1} x_2^{a_2} \ldots x_n^{a_n}$. Hence, a long polynomial with exponentially many monomials is encoded as a much shorter sum of rational functions of the form

$$f(P) \quad = \quad \sum_{i \in I} \pm \frac{x^{u_i}}{(1 - x^{c_{1,i}})(1 - x^{c_{2,i}}) \ldots (1 - x^{c_{n-d,i}})}. \tag{4}$$

Later on Barvinok and Woods [6] developed a set of powerful manipulation rules for using these short rational functions in Boolean constructions on various sets of lattice points, as well as a way to recover the lattice points inside the linear projection of a convex polytope. It is very interesting that to prove the last item of the theorem we will use Theorem 1. In this paper we apply Barvinok's theory to prove Theorem 3. From the results of Barvinok [5] for fixed n, but not necessarily fixed k, one can decide whether a particular b is k-feasible in polynomial time, but more strongly, as a corollary of Theorem 3, one can find more for knapsack problems.

Corollary 1. *Consider the knapsack problem $a^T x = b$ associated with $a = (a_1, \ldots, a_n)^T \in \mathbb{Z}_{>0}^n$ with $\gcd(a_1, \ldots, a_n) = 1$. For a fixed positive integer k and fixed n the k-Frobenius number can be computed in polynomial time.*

The paper is organized as follows. The first three sections propose proofs for the three main theorems, namely Section 2 gives a proof of Theorem 1, Section 3 gives a proof of Theorem 2, and Section 4 gives a proof of Theorem 3 (which in particular uses our version of Doignon-Bell-Scarf). In Section 5, we propose a more practical way to compute k-holes than what follows from Theorem 3, but without the computational complexity guarantees of Theorem 3.

2   A Generalization of Doignon-Bell-Scarf's theorem

In this section we will prove Theorem 1. The constant $c(n, k)$ we provide is $2^k 2^n$, but we will present improvements of this constant in the journal version of this paper.

[*Proof of Theorem* 1] The proof proceeds by contradiction. Consider a system of m linear inequalities,

$$a_1 x \leq \beta_1, \ldots, a_m x \leq \beta_m, x \in \mathbb{R}^n . \tag{5}$$

Suppose (5) has exactly k integral solutions and $m \geq 2^k 2^n + 1$. Suppose this system is a counterexample to Theorem 1 with $c(k, n) = 2^k 2^n + 1$. That is, if we delete any of the constraints in (5), the remaining system has at least $k + 1$ integral solutions.

Thus there exist integral vectors $x_1, \ldots, x_m$ such that x_j violates $a_j x \leq \beta_j$ but satisfies all other inequalities in (5). Consider the set of lattice points

$$H = \operatorname{conv}\{x_1, \ldots, x_m\} \cap \mathbb{Z}^n . \tag{6}$$

Consider the set $\Gamma \subset \mathbb{R}^m$ of the vectors $(\gamma_1, \ldots, \gamma_m)$ such that

$$\gamma_j \geq \min\{a_j z \mid z \in H, a_j z > \beta_j\} \tag{7}$$

and

the system $a_1 x < \gamma_1, \ldots, a_m x < \gamma_m$ has exactly k integral solutions in H. (8)

The set Γ is nonempty as we can take the equality in (7). Next, Condition (8), together with the lower bounds on the γ_i, implies that any integral solution of the system (5) remains feasible for the system $a_1 x < \gamma_1, \ldots, a_m x < \gamma_m$ for $\gamma \in \Gamma$. Thus, for all $\gamma \in \Gamma$, $a_1 x < \gamma_1, \ldots, a_m x < \gamma_m$ share exactly the same k integral solutions as (5).

Observe also that the set Γ is bounded, because if not γ_j for some j grows arbitrarily large, but then there exist z in H that satisfies $a_1 z < \gamma_1, \ldots, a_m z < \gamma_m$ which would be an additional integral feasible point and contradict Condition (8).

Claim 1. There is a point $(\nu_1, \ldots, \nu_m) \in \Gamma$ such that

for each $j = 1, \ldots, m$ there exists $y_j \in H$ so that $a_j y_j = \nu_j$ and $a_i y_j < \nu_i \, (i \neq j)$.
$$(9)$$
Proof of Claim: To see this, take any point $(\nu_1, \ldots, \nu_m) \in \Gamma$ and suppose that for some j this property does not hold. Consider

$$\nu'_j = \sup\{\nu : (\nu_1, \ldots, \nu_{j-1}, \nu, \nu_{j+1}, \ldots, \nu_m) \in \Gamma\}. \qquad (10)$$

The supremum in (10) is finite as the set Γ is bounded. Observe that there should exist $y_j \in H$ with $a_j y_j = \nu'_j$ and $a_i y_j < \nu_i (i \neq j)$. Otherwise $(\nu_1, \ldots, \nu_{j-1}, \nu'_j + \epsilon, \nu_{j+1}, \ldots, \nu_m) \in \Gamma$ for sufficiently small $\epsilon > 0$ as H is a finite set. Next, if $(\nu_1, \ldots, \nu_{j-1}, \nu'_j, \nu_{j+1}, \ldots, \nu_m) \notin \Gamma$ then, by (8) and (10), for any $\delta > 0$ there should exist a point $z \in H$ such that $\nu'_j - \delta \leq a_j z < \nu'_j = a_j y_j$. This is impossible as H is finite. Consequently, $(\nu_1, \ldots, \nu_{j-1}, \nu'_j, \nu_{j+1}, \ldots, \nu_m) \in \Gamma$ and we can replace ν_j by ν'_j. After at most m such replacements we will construct a point satisfying (9).

The property of the set $\{y_1, \ldots, y_m\}$ expressed by (9) is very important and as we will use it several times later, we formally name it.

Definition 1. *Let X be a finite subset of $\mathbb{Z}^n$. We say that X satisfies the support hyperplane property if for every $y \in X$, there exists a hyperplane $f^T x \leq g$ such that $f^T y = g$ and $f^T z < g$ for every $z \in X, z \neq y$. Furthermore, we say that the inequality $f^T x \leq g$ fulfills the support hyperplane property for y.*

Observe that the support hyperplane property is equivalent to saying that all members of X are vertices of $\mathrm{conv}(X)$. We will need the following two intermediate results.

Claim 2. Consider a set $X \subseteq \mathbb{Z}^n$ with $|X| \geq 2^n + 1$ that satisfies the support hyperplane property, i.e. such that for every member $y_i \in X$, there exists a hyperplane $f_i^T x \leq g_i$ such that $f_i^T y_i = g_i$ and $f_i^T y_j < g_i$ for $j \neq i$. Then there exists an integral point $z \in \mathbb{Z}^n$ that satisfies $f_i^T z < g_i$ for all $i = 1, \ldots, |X|$.

Proof of Claim: Since $|X| \geq 2^n + 1$, by the pigeonhole principle there exist $y_{i_1}, y_{i_2} \in X$ with $y_{i_1} \neq y_{i_2}$ and $y_{i_1} \equiv y_{i_2} (\mod 2)$ (that is all entries of $y_{i_1} - y_{i_2}$ are even). Therefore $z = \frac{1}{2}(y_{i_1} + y_{i_2}) \in \mathbb{Z}^n$. Obviously $f_i^T z < g_i$ for all $i = 1, \ldots, |X|$.

Claim 3. Consider a finite set $X \subseteq \mathbb{Z}^n$ that satisfies the support hyperplane property. Consider $z \in \mathrm{conv}(X) \cap \mathbb{Z}^n$. There exists a subset $\bar{X} \subseteq X$ with $|\bar{X}| \geq \lceil \frac{|X|}{2} \rceil$ such that $\bar{X} \cup \{z\}$ satisfies the support hyperplane property.

Proof of Claim: There exists a hyperplane $\bar{f}^T x = \bar{g}$ such that $\bar{f}^T z = \bar{g}$ and the equality does not hold for any other member of X. We can split the other members of X into two sets $X_< = X \cap \{x \in \mathbb{R}^n \mid \bar{f}^T x < \bar{g}\}$ and $X_> = X \cap \{x \in \mathbb{R}^n \mid \bar{f}^T x > \bar{g}\}$. Since the two sets are disjoint, one of them has cardinality at least $\lceil \frac{|X|}{2} \rceil$. The result follows since for every $x \in X$ that lies in $X_<$ (resp.

$X_>$), the inequality fulfilling the support hyperplane property still fulfills the hyperplane property in $X_<$ (resp. $X_>$). The inequality $\bar{f}^T x \leq \bar{g}$ (resp. $\bar{f}^T x \geq \bar{g}$) fulfills the support hyperplane property for z.

We will now construct $k + 1$ sets S_i, $i = 0, \ldots, k$ by induction. Throughout, the sets that are constructed have the following property.

Inductive Property. S_i has $2^{k-i} 2^n + 1$ integral points and satisfies the support hyperplane property.

We start with $S_0 = \{y_1, \ldots, y_m\}$. Observe that the inductive property is true for S_0. If the property is true for $i - 1$, and $i \leq k$, then the assumptions of the second claim are satisfied, and there exists an integral point z_{i-1} from which we can apply the third claim and obtain a subset $\bar{S}_{i-1} \subseteq S_{i-1}$ such that $S_i = \bar{S}_{i-1} \cup \{z_{i-1}\}$ satisfies the support hyperplane property which implies that the inductive property is satisfied.

Following the construction, z_i satisfies $a_j^T z_j < \nu_j$ for all j as it is obtained as a convex combination of points $y_1, \ldots, y_m$ with at least two points having a positive multiplier in the combination. Furthermore, we must have $z_i \neq z_j$ for $i < j$. Indeed, if $z_i \in S_j$, by construction there exists a hyperplane that separates them and they are clearly different if $z_i \notin S_j$.

This is now a contradiction since we have constructed $k + 1$ different integral points $z_0, \ldots, z_k$ satisfying (8).

3   Proof of Theorem 2

For $f \in \operatorname{cone}(A) \cap \mathbb{Z}^d$ define

$$L_{A,f}^k = \{\lambda \in \mathbb{Z}_{\geq 0}^n : IP_A(f + A\lambda) \text{ is } \geq k \text{ feasible}\},$$

so that $\operatorname{Sg}_{\geq k}(A) = \{A\lambda : \lambda \in L_{A,0}^k\}$. Consider the monomial ideal

$$I(A) = \langle x^\lambda : \lambda \in L_{A,0}^k \rangle.$$

To see that (3) is satisfied it is enough to check that for any $\lambda_0 \in L_{A,0}^k$ the inclusion $\lambda_0 + \mathbb{Z}_{\geq 0}^n \subset L_{A,0}^k$ holds. We will prove the following more general statement. For any $f \in \operatorname{cone}(A) \cap \mathbb{Z}^d$ and $\lambda_0 \in L_{A,f}^k$ we have the inclusion

$$\lambda_0 + \mathbb{Z}_{\geq 0}^n \subset L_{A,f}^k. \tag{11}$$

Let $\lambda_0 \in L_{A,f}^k$, so that there exist k distinct vectors $\lambda_1, \ldots, \lambda_k \in \mathbb{Z}_{\geq 0}^n$ with

$$f + A\lambda_0 = A\lambda_1 = \cdots = A\lambda_k.$$

Take any vector $\mu \in \mathbb{Z}_{\geq 0}^n$ and set $\nu = \lambda_0 + \mu$. Then, clearly, we have

$$f + A\nu = A(\lambda_1 + \mu) = \cdots = A(\lambda_{k-1} + \mu),$$

where all vectors $\lambda_1 + \mu, \ldots, \lambda_k + \mu \in \mathbb{Z}_{\geq 0}^n$ are distinct. Consequently, $IP_A(f + A\nu)$ is $\geq k$ feasible and, thus, $\nu \in L_{A,f}^k$. Hence (11) holds and we have proved the first claim of Theorem 2.

Let us now prove the second claim. Recall that the elements of the set $\mathrm{Sg}_{<k}(A)$ are also called k-*holes*. A k-hole f is called *fundamental* if there is no other k-hole $h \in \mathrm{Sg}_{<k}(A)$ such that $f - h \in \mathrm{Sg}_{\geq 1}(A)$.

Lemma 1. *The set of fundamental k-holes is a subset of the zonotope*

$$P = \{A\lambda : \lambda \in [0,1)^n\}.$$

Proof. Let $f \in \mathrm{Sg}_{<k}(A)$ be a fundamental hole. We can write

$$f = A\lambda, \ \lambda \in \mathbb{Q}_{\geq 0}^n.$$

Suppose $f \notin P$. Then for some j we must have $\lambda_j \geq 1$. Thus, denoting by A_j the jth column vector of A, the element $f' = f - A_j$ is a k-hole as any k distinct solutions for $IP_A(f')$ would correspond to k distinct solutions for $IP_A(f)$. Thus we get a contradiction with our choice of f as a fundamental k-hole. This implies $\lambda_j < 1$ for all j and, consequently, $f \in P$. The lemma is proved.

Lemma 1 shows, in particular, that the number of fundamental k-holes is finite. Furthermore, any k-hole can be represented as $f + A\lambda$ for some fundamental hole f and $\lambda \in \mathbb{Z}_{\geq 0}^n$. Let us fix a fundamental k-hole f and consider the monomial ideal $I_{A,f}^k \subset \mathbb{Q}[x_1, \ldots, x_n]$ defined as

$$I_{A,f}^k = \langle x^\lambda : \lambda \in L_{A,f}^k \rangle.$$

Then, in view of (11), $f + A\lambda$ is not a k-hole if and only if $x^\lambda \in I_{A,f}^k$.

Thus we need to write down the set $C(I_{A,f}^k)$ of exponents of *standard monomials* for the ideal $I_{A,f}^k$. Any such exponent $\lambda \in C(I_{A,f}^k)$ corresponds to the k-hole $f + A\lambda$.

By Theorem 3 in Chapter 9 of [12], the set $C(I_{A,f}^k)$ can be written as a finite union of translates of coordinate subspaces of $\mathbb{Z}_{\geq 0}^n$. Since the number of fundamental k-holes is finite, the second claim of Theorem 2 is proved.

4   Proof of Theorem 3

We use the technics of rational generating functions developed by Barvinok and Woods in [5,6]. We wish to prove a representation theorem of a set of lattice points as a sum $\sum_{\geq k-\text{feasible}} t^b$. Recall that A is an integral $d \times n$ matrix and k is a constant. For a subset of indices $I \subset \{1, 2, \ldots, n\}$ we can define the polyhedron (note X_i denotes an n-dimensional vector):

$$Q_I(A,k) = \{(X_1, X_2, \ldots, X_k) : \ AX_1 = AX_2 = \cdots = AX_k, \ X_i = X_j \text{ for } i,j \in I \text{ and } X_i \geq 0\}.$$

Clearly if $I = \emptyset$, then

$$Q_\emptyset(A,k) = \{(X_1, X_2, \ldots, X_k) : \ AX_1 = AX_2 = AX_3 = \cdots = AX_k \text{ and } X_i \geq 0\}.$$

In other words $Q_\emptyset(A, k)$ contains precisely k-tuples of n-vectors (possibly repeated) that give the same right-hand-side vector. More generally $Q_I(A, k)$ contains as lattice points the vectors b such that $b = AX_j$ for X_j $j = 1 \ldots k$ integer non-negative vectors, but with exactly $|I|$ of the vectors X_j being identical.

Using Barvinok's algorithm in [5], we can compute in polynomial time the generating function of the lattice points in the polyhedron $Q_I(A, k)$ which lives in fixed dimension kn. The resulting expression is the sum over all lattice points in a rational polytope $Q_I(A, k)$.

$$f(Q_I(A, k)) = \sum \left\{ z_1^{a_1} z_2^{a_2} \ldots z_k^{a_k} : (a_1, a_2, \ldots, a_k) \in Q_I(A, k) \cap \mathbb{Z}^{nk} \right\}$$

Next we will apply Boolean operations on generating functions $f(Q_I(A, k))$ in such a way that we are only left with the k-tuples of *distinct* non-negative vectors which satisfy $Aa_i = b$. We can do this by the following result:

Lemma 2 (Corollary 3.7 in [6]). *Let us fix l (the number of sets $S_i \subset \mathbb{Z}^d$) and r (the number of binomials in each fraction of the generating function $f(S_i)$). Then there exists an $s = s(l, r)$ and a polynomial time algorithm, which, for any l (finite) sets of lattice points $S_1, \ldots, S_l \subset \mathbb{Z}^d$ given by their generating functions $f(S_i)$ and a set $S \subset \mathbb{Z}^n$ defined as a Boolean combination of $S_1, \ldots, S_m$, computes $f(S)$ in the form*

$$f(S) = \sum_{i \in I} \gamma_i \frac{x^{u_i}}{(1 - x^{v_{i1}}) \cdots (1 - x^{v_{is}})},$$

where $\gamma_i \in \mathbb{Q}$, $u_i, v_{ij} \in \mathbb{Z}^n$ and $v_{ij} \neq 0$ for all i, j.

Now we can compute in polynomial time (because k is fixed) the following Boolean expression with 2^k summands

$$D(A, k) = Q_\emptyset(A, k) - \cup_{|I|=2} Q_I(A, k) + \cup_{|I|=3} Q_I(A, k) - \cdots - (-1)^k \cup_{|I|=k} Q_I(A, k).$$

Note that this is essentially the inclusion-exclusion principle applied to sets of lattice points, where each set is represented by a generating function (in rational function form). The new generating function $f(D(A, k))$ when expanded into monomials $z_1^{a_1} z_2^{a_2} \ldots z_k^{a_k}$ has only those where $a_i \neq a_j$. Namely, this is precisely the set of all k-tuples of distinct vectors in $\mathbb{Z}_{\geq 0}^n$ that give the same value $Aa_1 = Aa_2 = \ldots Aa_k$.

Finally another key subroutine introduced by Barvinok and Woods is the following *Projection Theorem*. In both Lemmas 2 and 3, the dimension n is assumed to be fixed.

Lemma 3 (Theorem 1.7 in [6]). *Assume the dimension n is a fixed constant. Consider a rational polytope $P \subset \mathbb{R}^n$ and a linear map $T : \mathbb{Z}^n \to \mathbb{Z}^k$. There is a polynomial time algorithm which computes a short representation of the generating function $f(T(P \cap \mathbb{Z}^n), x)$.*

In this case we apply a very simple linear map $(X_1, X_2, \ldots, X_k) \to AX_1$, by multiplication with A. This yields of course for each k-tuple (which has $X_i \neq X_j$) the corresponding right-hand side vector $b = AX_1$ that has at least k-distinct solutions. The final expression will look like $f = \sum_{b \in Q}$ with at least k-representations t^b. Which is the desired short rational function which efficiently represents the sum $\sum_{\geq k-\text{feasible}} t^b$. This proves the main result in the body of the paper for $\geq k$-feasible. Because if one knows a description for $\text{Sg}_{\geq k}(A)$ and $\text{Sg}_{\geq k+1}(A)$ one knows $\text{Sg}_{=k}(A) = \text{Sg}_{\geq k}(A) \backslash \text{Sg}_{\geq k+1}(A)$ and $\text{Sg}_{<k}(A) = \text{Sg}(A) \backslash \text{Sg}_{\geq k}(A)$, the Boolean properties of generating functions in Lemma 2 give the theorem in all three cases.

Now we move to prove Parts (a) to (d) of the theorem.

Part (a) If we have a generating function representation of

$$\sum_{\geq k-\text{feasible}} t^b,$$

it has the form

$$f(t) = \sum_{i \in I} \alpha_i \frac{t^{p_i}}{(1 - t^{a_{i1}}) \cdots (1 - t^{a_{ik}})}.$$

Note that by specializing at $t = (1, \ldots, 1)$, we can count how many b's are $\geq k$-feasible (when finite). Remark the substitution is not immediate since $t = (1, \ldots, 1)$ is a pole of each fraction in the representation of f. This problem is solvable because it has been shown by Barvinok and Woods that this computation can be handled efficiently (see Theorem 2.6 in [6] for details) and will prove Part (a).

Part (b) This item is a direct corollary of the following extraction lemma.

Lemma 4 (Lemma 8 in [14] or Theorem 7.5.2 in [15]). *Assume the dimension n is fixed. Let $S \subset \mathbb{Z}_+^n$ be nonempty and finite set of lattice points. Suppose the polynomial $f(S; z) = \sum_{\beta \in S} z^\beta$ is represented as a short rational function and let c be a cost vector. We can extract the (unique) lexicographic largest leading monomial from the set $\{x^\alpha : \alpha \cdot c = M, \alpha \in S\}$, where $M := \max\{\alpha \cdot c : \alpha \in S\}$, in polynomial time.*

Part (c) Barvinok and Woods developed a way to do monomial substitutions (not just $t_i = 1$ as we used in Part (a)), where the variable t_i in the current series, is replaced by a new monomial $z_1^{a_1} z_2^{a_2} \cdots z_r^{a_r}$. Note that the rational generating function $f = \sum_{b \in Q \cap \mathbb{Z}^d} b^b$ can give the evaluations of the b's for a given objective function $c \in \mathbb{Z}^d$. If we make the substitution $t_i = z^{c_i}$, the above equation yields a *univariate* rational function in z:

$$f(z) = \sum_{i \in I} E_i \frac{z^{c \cdot u_i}}{\prod_{j=1}^d (1 - z^{c \cdot v_{ij}})}. \tag{12}$$

Moreover $f(z) = \sum_{b \in Q \cap \mathbb{Z}^d} z^{c \cdot b}$. Thus we just need to find the (lexicographically) largest monomial in the sum in polynomial time. But this follows from Part (b).

Part (d) The reason the same generating function descriptions exist also for the sets those b which are $= k$-feasible, $\geq k$-feasible, or $< k$-feasible is because the sets can be obtained from the set we computed above as Boolean operations (intersection, unions, complements). Indeed using Barvinok Woods theory about such Boolean expressions, and the fact that $\mathrm{Sg}_{\geq k+1}(A) \setminus \mathrm{Sg}_{\geq k}(A) = \mathrm{Sg}_{=k}(A)$ and that $\mathrm{Sg}_{<k}(A) = \mathrm{Sg}_{\geq k}(A) \setminus \mathrm{Sg}_{=k}(A)$ the results follow.

Part (e) To prove this result we will use our generalization of Doignon-Bell-Scarf's theorem. Any problem of the form $Ax \leq b$ can be transferred to a problem of the form $Ax + Is = b$ by adding slack variables s. Then such a system is in the shape of the main part of Theorem 3 except we need a fixed number of rows. To see this is possible, by Theorem 1, if $Ax \leq b$ has k-solutions then, the same solutions appear in a subsystem $A_S x \leq b$ with no more than a constant $c(n, k)$ rows. Thus when we add slacks we will only add a constant number of slacks, only $n + c(k, n)$ many of them. Of course we do not know which rows form the system but there are only $\binom{d}{c(k,n)}$ possibilities for subsystems $A_S x + Is = b$ (each subsystem has a fixed number of columns now, thus it can be solved in polynomial time). Therefore, we can also decide for which b's the polyhedron has k points $Ax \leq b$ in polynomial time (again encoded in a rational function format).

To conclude we see how to compute the k-Frobenius number efficiently. We may see now that Corollary 1 follows directly from what we achieved in Theorem 3 and the Boolean operation Lemma of Barvinok and Woods. Indeed, from Theorem 3 we have a rational function representation of the k-feasible b for the Knapsack problem $f(t) = \sum_{i \in I} E_i \frac{t^{c \cdot u_i}}{\prod_{j=1}^{d}(1-t^{c \cdot v_{ij}})} = \sum_{b \in Q \cap \mathbb{Z}^d, \, k-\text{feasible}} t^{c \cdot b}$. Clearly the k-Frobenius number is simply the largest (lexicographic) b, such that t^b is **not** in $f(t)$, it is in its complement. Then, for the complement $\overline{S} = \mathbb{Z}_+ \setminus S$, we compute the generating function $f(\overline{S}; x) = (1-t)^{-1} - f(t)$ and then we compute the largest such t^b in the complement using Lemma 4.

5   Computing k-Holes via Hilbert Bases

In contrast to the *implicit* representation via rational generating functions that we saw in Section 4, we now present an algorithm to compute an *explicit* representation of $\mathrm{Sg}_{\geq k}(A)$, even for an infinite case. Such an explicit representation need not be of polynomial size in the input size of A, but will allow us to present some concrete computations and results for Knapsack problems in the extended version of this paper.

In this section we combine the results of Hemmecke et al. [21] with our techniques to computing the elements of $\mathrm{Sg}_{<k}(A)$. In view of the proof of Theorem 2

(ii), it is enough to compute all fundamental k-holes and then for each fundamental k-hole f compute the standard monomials of the ideal $I^k_{A,f}$. In view of Lemma 1, all fundamental k-holes are located in a zonotope $P = \{A\lambda : \lambda \in [0,1)^n\}$. Thus, with a straightforward generalization of the approach proposed in Hemmecke et al. [21], the fundamental k-holes the can be computed by using a Hilbert basis of the cone $\mathrm{cone}(A)$. In the special case $k = 1$ Hemmecke et al. [21] obtained the following result.

Theorem 4. *There exists an algorithm that computes for an integral matrix A a finite explicit representation for the set H of holes of the semigroup Q generated by the columns of A. The algorithm computes (finitely many) vectors $h_i \in \mathbb{Z}^d$ and monoids M_i, each given by a finite set of generators in $\mathbb{Z}^d$, $i \in I$, such that*

$$H = \bigcup_{i \in I} \left(\{h_i\} + M_i\right).$$

Here M_i could be trivial, that is, $M_i = \{0\}$.

Let f be a fundamental k-hole. Recall that the monomial ideal $I^k_{A,f} \subset \mathbb{Q}[x_1,\ldots,x_n]$ is defined as

$$I^k_{A,f} = \langle x^\lambda : \lambda \in L^k_{A,f} \rangle$$

and $f + A\lambda$ is not a k-hole if and only if $x^\lambda \in I^k_{A,f}$.

Thus we need to compute the exponents of standard monomials for the ideal $I^k_{A,f}$. Any such exponent $\lambda \in \mathbb{Z}^n_{\geq 0}$ corresponds to the k-hole $f + A\lambda$.

The exponents of standard monomials can be computed explicitly from a set of generators of the ideal. Hence, it is enough to find the generators of $I^k_{A,f}$. Let us fix an ordering $\prec$ in $\mathbb{Z}^n_{\geq 0}$. The minimal generators for the ideal $I^k_{A,f}$ correspond to the $\prec$-minimal elements of the set

$$L^k_{A,f} = \{\lambda \in \mathbb{Z}^n_{\geq 0} : \exists \text{ distinct } \mu_1,\ldots,\mu_k \in \mathbb{Z}^n_{\geq 0} \text{ such that}$$
$$f + A\lambda = A\mu_1 = \cdots = A\mu_k\}.$$

For computational purposes it is enough to compute a set of vectors of $L^k_{A,f}$ that contains all the $\prec$-minimal elements. We will proceed as follows. Let K be a complete graph with the vertex set $V = \{1, 2, \ldots, k\}$. By a weighted orientation H of K we will understand a weighted directed graph $H = (V, E)$ such that any two vertices of H are connected by a directed edge $e \in E$ with a weight $w(e) \in \{1,\ldots,n\}$. Let $\mathcal{S}$ be set of all weighted orientations of K.

For each $H \in \mathcal{S}$ we construct the following two auxiliary sets: the set

$$L_H = \{\lambda \in \mathbb{Z}^n_{\geq 0} : \exists \mu_1,\ldots,\mu_k \in \mathbb{Z}^n_{\geq 0} \text{ such that } f + A\lambda = A\mu_1 = \cdots = A\mu_k$$
$$\text{and } (\mu_i)_{w(e)} \leq (\mu_j)_{w(e)} - 1 \text{ for each } e = (i,j) \in E\}$$

and the set

$$M_H = \{(\lambda, \mu_1, \ldots, \mu_k) \in \mathbb{Z}^{(k+1)n}_{\geq 0} : f + A\lambda = A\mu_1 = \cdots = A\mu_k$$
$$\text{and } (\mu_i)_{w(e)} \leq (\mu_j)_{w(e)} - 1 \text{ for each } e = (i,j) \in E\}.$$

Then, in particular, $L_{A,f}^k = \bigcup_{H \in \mathcal{S}} L_H$, where the union is taken over all orientations in $H \in \mathcal{S}$.

We will need the following result.

Lemma 5. *Let λ_0 be a $\prec$-minimal element of L_H. Then there exists a $\prec$-minimal element of M_H of the form $(\lambda_0, \hat{\mu}_1, \ldots, \hat{\mu}_k)$.*

Let λ_0 be a $\prec$-minimal element of L_H. Suppose on contrary, for every $(\mu_1, \ldots, \mu_k) \in \mathbb{Z}_{\geq 0}^{kn}$ the vector $(\lambda_0, \mu_1, \ldots, \mu_k)$ is not a $\prec$-minimal element of M_H. Let $(\hat{\mu}_1, \ldots, \hat{\mu}_k)$ be a $\prec$-minimal element of the set

$$M_H|_{\lambda = \lambda_0} = \{(\mu_1, \ldots, \mu_k) \in \mathbb{Z}_{\geq 0}^{kn} : f + A\lambda_0 = A\mu_1 = \cdots = A\mu_k$$
$$\text{and } (\mu_i)_{w(e)} \leq (\mu_j)_{w(e)} - 1 \text{ for each } e = (i,j) \in E\}.$$

By the assumption, there exists a vector $(\lambda', \mu_1', \ldots, \mu_k') \in M_H$ such that $(\lambda', \mu_1', \ldots, \mu_k') \prec (\lambda_0, \hat{\mu}_1, \ldots, \hat{\mu}_k)$ and $(\lambda', \mu_1', \ldots, \mu_k') \neq (\lambda_0, \hat{\mu}_1, \ldots, \hat{\mu}_k)$. If $\lambda' \neq \lambda_0$ we get a contradiction to the $\prec$-minimality of λ_0 in L_H. On the other hand, if $\lambda' = \lambda_0$ we get a contradiction to the $\prec$-minimality of $(\hat{\mu}_1, \ldots, \hat{\mu}_k)$ in $M_H|_{\lambda = \lambda_0}$.

In view of Lemma 5, to compute a generating set for $L_{A,f}^k$ it is now enough to compute the set of all minimal elements for $M_H, H \in \mathcal{S}$ and remove the last kn components from each of them.

References

1. 4ti2 team. 4ti2–Software package for algebraic, geometric and combinatorial problems on linear spaces, http://www.4ti2.de/
2. Aardal, K., Lenstra, A.K.: Hard equality constrained integer knapsacks. In: Cook, W.J., Schulz, A.S. (eds.) IPCO 2002. LNCS, vol. 2337, pp. 350–366. Springer, Heidelberg (2002)
3. Aliev, I., Fukshansky, L., Henk, M.: Generalized Frobenius Numbers: Bounds and Average Behavior. Acta Arith. 155, 53–62 (2012)
4. Aliev, I., Henk, M., Linke, E.: Integer Points in Knapsack Polytopes and s-covering Radius. Electron. J. Combin. 20(2), Paper 42, 17 (2013)
5. Barvinok, A.I.: Polynomial time algorithm for counting integral points in polyhedra when the dimension is fixed. Math. of Operations Research 19, 769–779 (1994)
6. Barvinok, A.I., Woods, K.: Short rational generating functions for lattice point problems. J. Amer. Math. Soc. 16, 957–979 (2003)
7. Bell, D.E.: A theorem concerning the integer lattice. Studies in Applied Mathematics 56(1), 187–188 (1977)
8. Beck, M., Robins, S.: A formula related to the Frobenius problem in two dimensions. In: Number Theory (New York, 2003), pp. 17–23. Springer, New York (2004)
9. Bruns, W., Gubeladze, J., Trung, N.V.: Problems and algorithms for affine semigroups. Semigroup Forum 64, 180–212 (2002)
10. Bruns, W., Koch, R.: NORMALIZ, computing normalizations of affine semigroups, ftp://ftp.mathematik.uni-osnabrueck.de/pub/osm/kommalg/software/

11. Clarkson, K.L.: Las Vegas algorithms for linear and integer programming when the dimension is small. Journal of the ACM 42(2), 488–499 (1995)
12. Cox, D., Little, J., O'Shea, D.: Ideals, Varieties and Algorithms, Undergraduate Texts in Mathematics. Springer, New York (1992)
13. De Loera, J.A., Hemmecke, R., Tauzer, J., Yoshida, R.: Effective lattice point counting in rational convex polytopes. Journal of Symbolic Computation 38(4), 1273–1302 (2004)
14. De Loera, J.A., Haws, D.C., Hemmecke, R., Huggins, P., Sturmfels, B., Yoshida, R.: Short rational functions for toric algebra and applications. Journal of Symbolic Computation 38(2), 959–973 (2004)
15. De Loera, J.A., Hemmecke, R., Köppe, M.: Algebraic and geometric ideas in the theory of discrete optimization. MOS-SIAM Series on Optimization, vol. 14, p. xx+322. Society for Industrial and Applied Mathematics (SIAM), Philadelphia; Mathematical Optimization Society, Philadelphia (2013)
16. Dobra, A., Karr, A.F., Sanil, P.A.: Preserving confidentiality of high-dimensional tabulated data: statistical and computational issues. Stat. Comput. 13, 363–370 (2003)
17. Doignon, J.-P.: Convexity in cristallographical lattices. Journal of Geometry 3(1), 71–85 (1973)
18. Eisenbrand, F., Hähnle, N.: Minimizing the number of lattice points in a translated polygon. In: Proceedings of SODA, pp. 1123–1130 (2013)
19. Fukshansky, L., Schürmann, A.: Bounds on generalized Frobenius numbers. European J. Combin. 3, 361–368 (2011)
20. Haase, C., Nill, B., Payne, S.: Cayley decompositions of lattice polytopes and upper bounds for h^*-polynomials. J. Reine Angew. Math. 637, 207–216 (2009)
21. Hemmecke, R., Takemura, A., Yoshida, R.: Computing holes in semi-groups and its application to transportation problems. Contributions to Discrete Mathematics 4, 81–91 (2009)
22. Kannan, R.: Lattice translates of a polytope and the Frobenius problem. Combinatorica 12(2), 161–177 (1992)
23. Lagarias, J.C., Ziegler, G.M.: Bounds for lattice polytopes containing a fixed number of interior points in a sublattice. Canad. J. Math. 43(5), 1022–1035 (1991)
24. Pikhurko, O.: Lattice points in lattice polytopes. Mathematika 48(1-2), 15–24 (2001)
25. Ramírez Alfonsín, J.L.: Gaps in semigroups. Discrete Mathematics 308(18), 4177–4184 (2008)
26. Ramírez Alfonsín, J.L.: Complexity of the Frobenius problem. Combinatorica 16(1), 143–147 (1996)
27. Ramírez Alfonsín, J.L.: The Diophantine Frobenius Problem. Oxford Lecture Series in Mathematics and Its Applications. Oxford University Press, New York (2006)
28. Schrijver, A.: Theory of linear and integer programming. Wiley (1998)
29. Scarf, H.E.: An observation on the structure of production sets with indivisibilities. Proceedings of the National Academy of Sciences 74(9), 3637–3641 (1977)
30. Stanley, R.P.: Combinatorics and Commutative Algebra, 2nd edn. Progress in Mathematics, vol. 41. Birkhäuser, Basel (1996)
31. Sturmfels, B.: Gröbner Bases and Convex Polytopes. University Lecture Series, vol. 8. AMS, Providence (1995)
32. Takemura, A., Yoshida, R.: A generalization of the integer linear infeasibility problem. Discrete Optimization 5, 36–52 (2008)

Centrality of Trees for Capacitated k-Center

Hyung-Chan An[1], Aditya Bhaskara[2,*], Chandra Chekuri[3,**], Shalmoli Gupta[3], Vivek Madan[3], and Ola Svensson[1,***]

[1] École Polytechnique Fédérale de Lausanne, 1015 Lausanne, Switzerland
{hyung-chan.an,ola.svensson}@epfl.ch
[2] Google Research, NY 10011, USA
bhaskara@cs.princeton.edu
[3] University of Illinois at Urbana-Champaign, IL 61801, USA
{chekuri,sgupta49,vmadan2}@illinois.edu

Abstract. We consider the capacitated k-center problem. In this problem we are given a finite set of locations in a metric space and each location has an associated non-negative integer capacity. The goal is to choose (open) k locations (called centers) and assign each location to an open center to minimize the maximum, over all locations, of the distance of the location to its assigned center. The number of locations assigned to a center cannot exceed the center's capacity. The uncapacitated k-center problem has a simple tight 2-approximation from the 80's. In contrast, the first constant factor approximation for the capacitated problem was obtained only recently by Cygan, Hajiaghayi and Khuller who gave an intricate LP-rounding algorithm that achieves an approximation guarantee in the hundreds. In this paper we give a simple algorithm with a clean analysis and prove an approximation guarantee of 9. It uses the standard LP relaxation and comes close to settling the integrality gap (after necessary preprocessing), which is narrowed down to either 7, 8 or 9. The algorithm proceeds by first reducing to special *tree instances*, and then uses our best-possible algorithm to solve such instances. Our concept of tree instances is versatile and applies to natural variants of the capacitated k-center problem for which we also obtain improved algorithms. Finally, we give evidence to show that more powerful preprocessing could lead to better algorithms, by giving an approximation algorithm that beats the integrality gap for instances where all non-zero capacities are the same.

Keywords: approximation algorithms, capacitated network location problems, capacitated k-center problem, LP-rounding algorithms.

1 Introduction

Network location problems form a large and important class of problems in discrete and combinatorial optimization. Many of these problems can be phrased

---

* Work done while the author was at EPFL, Switzerland.
** Supported in part by NSF grants CCF-1016684 and CCF-1319376.
*** Supported in part by ERC Starting Grant 335288-OptApprox.
† The main result of this paper was obtained independently by An, Bhaskara and Svensson, and by Chekuri, Gupta and Madan. This paper is based on the manuscript of the first group.

in terms of choosing centers or facilities to best serve a given set of clients, typically under the assumption that the locations for the facilities and clients lie in a metric space. One can imagine several objective functions to measure the quality of service. Perhaps the most natural and well-studied ones are "social welfare", where we wish to *minimize the average* distance from a client to its assigned center, and "fairness", in which we wish to *minimize the maximum* distance from a client to its assigned center. Note that, once we have selected the centers, both of these objectives are minimized by assigning each client to its closest center. An inherent drawback of this strategy, however, is that it is unable to deal with centers of (different) capacities that limit the number of clients they can serve, which is a constraint present in typical applications. Capacity constraints in location problems pose difficult algorithmic challenges from both a theoretical and empirical point of view and our understanding continues to evolve despite a long history of work.

For uncapacitated network location problems, several beautiful algorithmic techniques, such as LP-rounding [5], primal-dual framework [14] and local search [16,4] have been used to obtain a fine-grained understanding of the approximability of the well-known variants: k-center, k-median, and facility location[1]. Already in the 80's, Gonzales [9] and Hochbaum & Shmoys [12] developed tight 2-approximation algorithms for the k-center problem. For facility location, the current best approximation algorithm is due to Li [18]. He combined an algorithm by Byrka [3] and an algorithm by Jain, Mahdian, and Saberi [13] to achieve an approximation guarantee of 1.488. This is nearly tight, as it is hard to approximate the problem within a factor of 1.463 [10]. The gap is slightly larger for k-median: a recent LP rounding [19] achieves an approximation guarantee of $1 + \sqrt{3} \approx 2.732$ improving upon a local search algorithm by Arya et al. [1]; and it is NP-hard to do better than $1 + 2/e \approx 1.736$ [13]. Although the different problems have algorithms with different approximation guarantees, they share many techniques, and improvements have often come hand in hand. In particular, most of the above progress relies on standard linear programming (LP) relaxations.

In contrast to the uncapacitated versions, the standard LP relaxations for the capacitated problems have unbounded integrality gaps and this is one reason for the coarser understanding we have. Apart from special cases, such as uniform capacities [15], soft capacities (a center can be opened several times) [21,15,14], and other variants [17,7], the only known constant factor approximation algorithm until recently, was for facility location. In a sequence of works, including Korupolu, Plaxton & Rajaraman [16], Pál, Tardos & Wexler [20], Chudak & Williamson [6], and Zhang, Chen & Ye[23] increasingly enhanced local search algorithms culminated in an approximation guarantee of 5 due to Bansal, Garg, and Gupta [2]. These methods are elegant but specialized to facility location and

---

[1] Recall that in k-center and k-median, we wish to select k centers so as to minimize the fairness and social welfare, respectively; facility location is similar to k-median but instead of having a constraint k on the number of centers to open, each center has an opening cost.

are not LP-based. In fact, finding a relaxation-based algorithm for capacitated facility location with a constant approximation guarantee remains a major open problem (see e.g. "Problem 5" of the ten open problems from the recent book by Williamson and Shmoys [22]). One of the motivations for finding algorithms based on relaxations is that the methods are often flexible and the developed techniques transfer to different settings, as has indeed been the case in the study of uncapacitated location problems.

In the quest to obtain a better understanding and more general (relaxation based) techniques for capacitated network location problems, it is natural to start with the capacitated k-center problem. Indeed, even though we have a good understanding of uncapacitated location problems in general, the uncapacitated k-center problem stands out, with an extremely simple greedy algorithm that gives a tight analysis of the LP relaxation. Our failure to understand the capacitated k-center problem is therefore solely due to the lack of techniques for analyzing capacity constraints. An important recent development in this line of research is due to Cygan, Hajiaghayi, and Khuller [8], who obtain the first constant factor approximation for the capacitated k-center problem. Their algorithm works by preprocessing the instance to overcome the unbounded integrality gap of the natural LP relaxation, followed by an intricate rounding procedure. The approximation factor is not computed explicitly, but is estimated to be roughly in the hundreds. This however, is still quite far off from the integrality gap lower bound of 7 (after preprocessing) [8] and the inapproximability results which rule out a factor better than 3 (see e.g. [8] for a simple proof).

In this paper, we develop novel techniques to further close the gap in our understanding of capacitated location problems. In particular, we present a simple algorithm for the capacitated k-center problem with a clean analysis that allows us to prove an approximation guarantee of 9. Our result is based on the standard LP relaxation and it almost settles its integrality gap (after the preprocessing of Cygan et al. [8]): it is either 7, 8 or 9 (both the integrality gap and approximation ratio can only take integral values; this is because the worst instances can easily be seen to be ones defined by the shortest-path metric on an unweighted graph). Due to the simplicity of our analyses, we hope that some of the ideas could be applied to other location problems, such as capacitated k-median, for which no constant factor approximation algorithms are known.

Main result and outline of algorithmic approach. Our main result is the following.

Theorem 1. *There exists a 9-approximation algorithm for the capacitated k-center problem.*

The algorithm guesses the optimal solution value τ and considers an unweighted graph $G_{\leq \tau}$ on the given set of vertices where two vertices are adjacent if and only if their distance is at most τ: the edges in this graph represent the assignments that are "admissible" with respect to τ. This graph can be assumed to be connected (see [8]). The algorithm then solves a natural and standard LP on $G_{\leq \tau}$. This determines if it is possible to (fractionally) open k vertices while assigning every vertex to a center that is adjacent in $G_{\leq \tau}$. If this LP is infeasible, we know that the optimum is larger than τ; otherwise, our algorithm will open k centers and find an assignment of every vertex to an open center that is within a distance of 9 in $G_{\leq \tau}$, and moreover the assignment respects the capacities of open centers. This leads to a 9-approximation algorithm.

The LP solution specifies a set of *opening variables* that indicate the fraction to which each vertex is to be opened. Our algorithm rounds these opening variables by "transferring" openings between vertices to make them integral. Since we do not create any new opening, our rounding will naturally open at most k centers; however, the challenge is to ensure that there exists a small-distance assignment of the vertices to open centers. If, for example, the opening of a vertex v is transferred to another vertex that is far away, the clients that were originally assigned to v may be unable to find an available center nearby. For another example, if the opening of a high-capacity vertex gets transferred to a low-capacity one, the low-capacity vertex may fail to provide sufficient capacity to cover the vertices in the neighborhood. Thus, we need to ensure that our rounding algorithm transfers openings only in small vicinity, and that "locally available capacity" of the graph does not decrease. (Definition 3 formalizes this concept as a *distance-r transfer*.)

We reduce the rounding problem to the special case of *tree instances*, and present a best-possible algorithm that rounds such instances. A tree instance is given by a set of opening variables defined on a rooted tree, where every non-leaf node has an opening variable of 1. Tree instances are generalizations of caterpillars used by Cygan et al. [8], which can be considered as tree instances whose non-leaf nodes form a path and have certain degree bounds. Suppose we have a tree instance where the capacities are uniform and there are exactly two leaves u and v each of which is opened by $1/2$, whereas every other vertex is opened by 1. If u and v are distant, this may appear problematic at a glance as we cannot transfer the opening of one to the other. However, there exists a (unique) path $u, w_1, \ldots, w_m, v$ in the tree, and we can transfer the opening of $1/2$ in a "chain" along this path: from u to w_1, from w_1 to w_2, ..., from w_m to v. This idea can in fact be carried through to give an algorithm for capacitated k-center when all capacities are equal.

Unfortunately, this chain of transfers causes a problem when the capacities are given arbitrarily: suppose in the previous example that u and v have very high capacities compared to the others. Then we will not be able to transfer the opening of u to w_1, since the open centers around u may not be able to provide sufficient capacity to cover the vertices that were originally assigned to u. However, from another angle, w_1 (or any other non-leaf vertex) is "wasting" the budget, since it opens a center while contributing relatively small capacity to the graph. This provides us some "slack" in the budget that we can utilize: in this particular example, by transferring an opening of $1/2$ *from w_1 to u*, and the other $1/2$ from w_1 to v in a chain, we can successfully round the given instance thanks to the decision of closing w_1 which had originally had its opening variable equal to one. This strategy of *closing a fully open center* is quite powerful, yet we need to ensure that its capacity can be accomodated by nearby centers if we want to close it. Thus, the viability of such a strategy tends to depend on several factors, including how its capacity compares to vertices in the neighborhood, which of these vertices are to be opened, and so on – all decisions which could depend on more and more distant vertices.

In contrast, our algorithm departs from previous works by using a simple *local* strategy that does not depend on distant vertices and applies to *every* non-leaf node. The reason our strategy works locally is that the decision of closing fully open centers is determined using solutions to subinstances, which are solved recursively. This key idea significantly eases the analysis and leads to our algorithm for tree instances that is the best possible. The simplicity of our analysis also helps us more carefully analyze the approximation ratio and extend our techniques to related problems. Section 4 formally presents our algorithm to round a tree instance; the full version of this extended abstract presents its extensions to two related problems: the *capacitated k-supplier problem* and the *budgeted opening problem with uniform capacity*.

Finally, we reduce the given problem to a tree instance by constructing a tree on a subset of vertices that are chosen as "candidates" to be opened. Non-leaf nodes will be carefully chosen, in order to yield a 9-approximation algorithm. Two adjacent vertices in the constructed tree instance will not necessarily be adjacent in the original graph, but will be in close proximity; hence, if the tree instance can be rounded using short transfers of openings, the original instance can also be rounded using only slightly longer transfers.

Additional results and future directions. We also explore future directions towards a better understanding of the problem. Recall that our algorithm proceeds in three steps: firstly, we preprocess the given instance as done by Cygan et al. [8]; secondly, we reduce the problem to a tree instance; lastly, we solve this tree instance. Given that our tree rounding algorithm is best-possible, it is natural to seek to improve the first two steps. The preprocessing step of Cygan et al. allows us to bring down the integrality gap from unbounded to 9; however, the integrality gap after the basic preprocessing is known to be at least 7 [8], which is larger than the best known inapproximability result that rules out a better factor than 3. The instance showing the integrality gap of 7 (and also that of the inapproximability result) has a special structure that every capacity is either 0 or L for some constant L. In order to understand the potential of stronger preprocessing methods, we investigate this $\{0, L\}$-*case* and show that additional preprocessing and a sophisticated rounding gives a 6-approximation algorithm (see the full paper).The interesting fact is that we obtain an approximation ratio which *surpasses* the integrality gap lower bound of 7 after basic preprocessing. This raises the natural open question: could there be preprocessing steps which bring the approximation ratio down to 3? We could also ask: do lift-and-project methods (applied to a potentially different formulation) automatically *capture* these preprocessing steps? We believe that understanding these questions would also shed light on approximating capacitated versions of other problems such as facility location and k-median.

2   Preliminaries

Given an integer k and a metric distance/cost $c : V \times V \to \mathbb{R}_+$ on V with a capacity function $L : V \to \mathbb{Z}_{\geq 0}$, the *capacitated k-center problem* is to choose k vertices to *open*, along with an assignment of every vertex to an open center which minimizes the longest distance between a vertex and the center it is assigned to while honoring the capacity constraints: i.e., no open center v is assigned more vertices than its capacity $L(v)$.

For an undirected graph $G = (V, E)$, $d_G(u, v)$ denotes the distance between $u, v \in V$; $N_G[u]$ denotes the set of vertices in the *closed neighborhood* of u, which includes u itself: i.e., $N_G[u] := \{v \mid (u, v) \in E\} \cup \{u\}$. For $U \subseteq V$, $d_G(v, U)$ denotes the distance from v to U: $d_G(v, U) := \min_{u \in U} d_G(v, u)$. $N_G[U]$ is a shorthand for $\cup_{u \in U} N_G[u]$. When the graph of interest G is clear from the context, we will use d and $N[\cdot]$ instead of d_G and $N_G[\cdot]$, respectively. Let OPT denote the optimal solution value.

Reduction to an unweighted problem using the standard LP relaxation. Our algorithm begins with determining a lower bound τ^* on the optimal solution value: it makes a guess τ at OPT, and tries to decide if $\tau <$ OPT. We simplify this problem by considering an unweighted graph that represents which assignments are "admissible". Let $G_{\leq \tau} = (V, E_{\leq \tau})$ be the unweighted graph on V (with loops on every vertex) where two vertices are adjacent if and only if their distance is at most τ: $E_{\leq \tau} := \{(u, v) \mid c(u, v) \leq \tau\}$. Note that a feasible solution of value τ assigns every vertex to a center that is adjacent in $G_{\leq \tau}$, and conversely, if a solution assigns every vertex to a center that is adjacent in $G_{\leq \tau}$, its value is no greater than τ. For an unweighted graph $G = (V, E)$, the standard LP relaxation $\mathrm{LP}_k(G)$ is the following feasibility LP that fractionally verifies whether there exists a solution that assigns every vertex to an open center that is adjacent in G:

$$\sum_{u \in V} y_u = k;$$
$$x_{uv} \leq y_u, \qquad \forall u, v \in V;$$
$$\sum_{v:(u,v) \in E} x_{uv} \leq L(u) \cdot y_u, \quad \forall u \in V;$$
$$\sum_{u:(u,v) \in E} x_{uv} = 1, \qquad \forall v \in V;$$
$$0 \leq x, y \leq 1.$$

x_{uv} is called an *assignment variable*; y_u is called the *opening variable* of u.

However, the integrality gap of this LP, defined as the maximum ratio $\frac{\mathrm{OPT}}{\tau}$ where $\mathrm{LP}_k(G_{\leq \tau})$ is feasible, is unbounded; hence this LP cannot in general estimate OPT very well. We use the approach of Cygan et al. [8] to address this issue: consider the connected components of $G_{\leq \tau}$; if $\tau \geq$ OPT, a vertex can be assigned only to the vertices in the same connected component. For each connected component G_i of $G_{\leq \tau}$, the algorithm decides the minimum integer

k_i for which $\mathrm{LP}_{k_i}(G_i)$ is feasible; if $\sum_i k_i > k$, this certifies that there exists no solution of value τ or better ($\tau < \mathrm{OPT}$). Now let τ^* be the smallest τ for which the algorithm fails to certify that $\tau < \mathrm{OPT}$; since the algorithm has to fail to provide a certificate for $\tau = \mathrm{OPT}$, we have $\tau^* \leq \mathrm{OPT}$. The algorithm then separately solves the subproblems given by the connected components of $G_{\leq \tau^*}$: given a *connected* graph G for which $\mathrm{LP}_k(G)$ is feasible, our algorithm finds a set of k vertices to open, with an assignment of every vertex to an open center that is within the distance of nine. Note that $d_{G_{\leq \tau^*}}(u, v) \leq 9$ implies $c(u, v) \leq 9\tau^* \leq 9 \cdot \mathrm{OPT}$ from the triangle inequality.

Lemma 2 (Cygan et al. [8]). *Suppose there exists an algorithm that, given a connected graph G, capacity L, and k for which $\mathrm{LP}_k(G)$ is feasible, computes a set of k vertices to open and an assignment of every vertex u to an open center v such that $d(u, v) \leq \rho$ and the capacity constraints are satisfied. Then we can obtain a ρ-approximation algorithm for the capacitated k-center problem.*

3   Distance-r Transfers and Tree Instances

Distance-r transfers. The discussion in Section 2 reduces the task of designing an approximation algorithm for the capacitated k-center problem to that of using a solution (x, y) to $\mathrm{LP}_k(G)$ in order to select k centers so that each vertex in the connected graph G is assigned to a center in a nearby neighborhood. Simple algebraic manipulations show that, for any $U \subseteq V$, the LP solution satisfies $|U| = \sum_{u \in U} \sum_{w:(w,u) \in E} x_{wu} \leq \sum_{w \in N[U]} L(w) \cdot y_w$; note that, if the opening variables y are integral, this exactly corresponds to Hall's condition [11] and hence we can assign every vertex to an adjacent center. However, the LP solution may open each center only by a small fractional amount; in order to obtain an integral solution, it is therefore natural to try to aggregate fractional openings of nearby vertices. As different centers have varying capacities, one difficulty of this approach is that the rounding also needs to ensure that the aggregation does not decrease the available capacity. Consider a center u of capacity $L(u)$ that is open with fraction y_u; we can view it as a center with the *fractional capacity* of $L(u) \cdot y_u$, because in a sense this is the maximum *number* (as a fraction) of vertices this center serves according to the LP. Our rounding procedure will open k centers, while ensuring that we can *transfer* the fractional capacity of each u to one or more of the open centers that are close by (and the performance guarantee is determined by how close these centers are). The following definition formalizes the notion of a distance-r transfer:

Definition 3. *Given a graph $G = (V, E)$ with a capacity function $L : V \to \mathbb{Z}_{\geq 0}$ and $y \in \mathbb{R}_+^V$, a vector $y' \in \mathbb{R}_+^V$ is a distance-r transfer of (G, L, y) if*

(3a): $\sum_{v \in V} y'_v = \sum_{v \in V} y_v$ *and*
(3b): $\sum_{v:d(v,U) \leq r} L(v) y'_v \geq \sum_{u \in U} L(u) y_u$ *for all $U \subseteq V$.*

If y' is the characteristic vector of $S \subseteq V$, we say S is a distance-r transfer of (G, L, y).

The given conditions say that a transfer should not change the total number of open centers, while ensuring that the total fractional capacity in each small neighborhood does not decrease as a result of this transfer. We also remark that multiple transfers can be composed: if y' is a distance-r transfer of (G, L, y) and y'' is a distance-r' transfer of (G, L, y') then y'' is a distance-$(r + r')$ transfer of (G, L, y).

Lemma 4. *For a graph $G = (V, E)$ with a capacity function $L : V \to \mathbb{Z}_{\geq 0}$, let (x, y) be a feasible solution to $\mathrm{LP}_k(G)$. If $S \subseteq V$ is a distance-r transfer of (G, L, y), then every vertex $v \in V$ can be assigned to a center $s \in S$ such that $d_G(v, s) \leq r + 1$, while ensuring no center is assigned more vertices than its capacity. Moreover, $|S| = k$, and this assignment can be found in polynomial time.*

Proof. Consider the natural bipartite matching problem between V and the multiset of open centers that are duplicated to their capacities: i.e, each center $s \in S$ appears in the multiset with multiplicity $L(s)$. Every vertex v in V is connected to every copy of each center $s \in S$ such that $d(v, s) \leq r + 1$. Observe that a matching of cardinality $|V|$ naturally defines an assignment that satisfies the desired properties. We shall now show that there exists such a matching by verifying Hall's condition, i.e., that for all $U \subseteq V$, $|U| \leq \sum_{s \in S : d_G(s, U) \leq r + 1} L(s)$.

As was observed earlier, we have $|U| \leq \sum_{w : d_G(w, U) \leq 1} L(w) \cdot y_w$; from Condition (3b), $|U| \leq \sum_{w : d_G(w, U) \leq 1} L(w) \cdot y_w \leq \sum_{s \in S : d_G(s, U) \leq r + 1} L(s)$. This matching can be found in polynomial time, and $|S| = k$ follows from Condition (3a). $\square$

Tree instances. As was discussed earlier, we solve the general problem via reduction to *tree instances*.

Definition 5. *A tree instance is defined as a tuple (T, L, y), where $T = (V, E)$ is a rooted tree with the capacity function $L : V \to \mathbb{Z}_{\geq 0}$, and opening variables $y \in (0, 1]^V$ satisfy that $\sum_{v \in V} y_v$ is an integer and $y_v = 1$ for every non-leaf node $v \in V$.*

Our reduction uses the standard clustering technique (see e.g. [15]) to partition the given graph into clusters that form a tree with adjacent clusters being close. Then by aggregating the opening within each cluster, we find a distance-2 transfer of the original LP solution, where each cluster now contains at least one fully open vertex. Using these vertices as non-leaf nodes, we transform the tree of clusters into a tree instance. As the adjacent clusters in the tree of clusters were close, it turns out that a distance-r transfer on the tree instance we constructed can be expressed as a distance-$5r$ transfer on the original graph; composing these two transfers gives an integral distance-$(5r + 2)$ transfer. In fact, through a careful choice of non-leaf nodes, we can show that the approximation ratio can be further refined to be $3r + 3$. The complete presentation of this reduction can be found in the full version of this extended abstract.

Lemma 6. *Suppose there exists a polynomial-time algorithm that finds an integral distance-r transfer of a tree instance. Then there exists a $(3r + 3)$-approximation algorithm for the capacitated k-center problem.*

4   Algorithm for Tree Instances

In this section we prove the following.

Lemma 7. *There is a polynomial time algorithm that finds an integral distance-2 transfer of a given tree instance (T, L, y).*

We remark that it is easy to see that some tree-instances do not admit an integral distance-1 transfer and the above lemma is therefore the best possible. One example is the following: the instance consists of a root with six children, where each child is opened with a fraction $2/3$, and all vertices have the same capacity; it is easy to see that any integral solution needs to transfer fractional capacity from one leaf to another (i.e., of distance 2). We now present the algorithm along with the arguments of its correctness.

The algorithm builds up the solution by recursively solving smaller tree instances. The base case is simple: if $|T| \leq 1$ then simply open the vertex in $V(T)$ if any. By the integrality of $\sum_{v \in V(T)} y_v$ this is clearly a distance-2 transfer (actually a distance-0 transfer). Let us now consider the more interesting case when $|T| \geq 2$; then there exists a node r of which every child is a leaf. Let $v_1, \ldots, v_\ell$ be the children of r, in the non-increasing order of capacity: $L(v_1) \geq \cdots \geq L(v_\ell)$. Let T_r denote the subtree rooted at r and $Y := \sum_{i=1}^{\ell} y_{v_i}$. The algorithm considers two separate cases depending on whether Y is an integer.

Let us start with the simpler case when Y is an integer: the algorithm selects the set S_r consisting of the $Y + 1$ vertices of highest capacity in T_r. As every pair of nodes in T_r are within a distance of 2, S_r is a distance-2 transfer of the tree instance induced by T_r. The algorithm then solves the tree instance induced by $\bar{T} := T \setminus T_r$ to obtain a distance-2 transfer $\bar{S}$ of size $\sum_{v \in T} y_v - Y - 1$. It follows that $S := S_r \cup \bar{S}$ is a distance-2 transfer of (T, L, y).

We now consider the final more interesting case when Y is not an integer. In this case, we *cannot* consider T_r and $T \setminus T_r$ as two separate instances because the y-values suggest to either open $\lfloor Y \rfloor + 1$ or $\lceil Y \rceil + 1$ centers in T_r: a choice that depends on the selected centers in $T \setminus T_r$. As at least $\lfloor Y \rfloor + 1$ of the vertices in T_r will be selected as centers in either case, the algorithm will commit itself to open the $\lfloor Y \rfloor + 1$ vertices in T_r of highest capacity. Let S_{commit} denote that set and note that it equals $\{v_1, \ldots, v_{\lfloor Y \rfloor}, r\}$ or $\{v_1, \ldots, v_{\lfloor Y \rfloor}, v_{\lfloor Y \rfloor + 1}\}$ dependent on which node of r and $v_{\lfloor Y \rfloor + 1}$ has the higher capacity ($v_{\lfloor Y \rfloor + 1}$ is well defined since we have that the number of children ℓ is at least $\lceil Y \rceil$ from $y \leq 1$). By the selection of S_{commit}, we have

$$\sum_{u \in V(T_r)} y_u L(u) \leq \sum_{s \in S_{\mathrm{commit}}} L(s) + \bar{y}_p \bar{L}(p), \qquad (1)$$

where $\bar{y}_p = Y - \lfloor Y \rfloor$ and $\bar{L}(p) = \min[L(r), L(v_{\lfloor Y \rfloor + 1})]$. In other words, if the algorithm on the one hand chooses to only open the $\lfloor Y \rfloor + 1$ centers S_{commit} in T_r, then an additional fractional capacity $\bar{y}_p \bar{L}(p)$ needs to be transferred from T_r to an open center in $T \setminus T_r$. On the other hand, if the algorithm chooses to open all the centers $\lceil Y \rceil + 1$ in $S_{\mathrm{commit}} \cup \{v_{\lfloor Y \rfloor + 1}, r\}$ then those centers can accomodate all the fractional capacity in T_r together with $(1 - \bar{y}_p) \bar{L}(p)$ additional capacity.

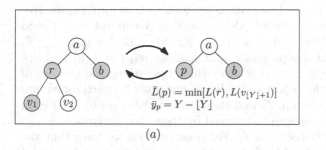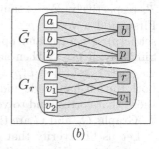

$$\bar{L}(p) = \min[L(r), L(v_{\lfloor Y \rfloor + 1})]$$
$$\bar{y}_p = Y - \lfloor Y \rfloor$$

(a) (b)

Fig. 1. (a) The construction of $\bar{T}$ from T with the subtree T_r rooted at r with children v_1 and v_2; the grey vertices are those selected in potential solutions to $\bar{T}$ and T, respectively. (b) The bipartite graph and the induced subgraphs $\bar{G}$ and G_r that are used in the proof of Claim 8.

We defer this decision to be based on the solution of the smaller tree instance $(\bar{T}, \bar{L}, \bar{y})$ obtained from (T, L, y) as follows (see also Figure 1a): replace T_r by the vertex p that represents the deferred decision and let $\bar{y}, \bar{L}$ be the natural restrictions of y, L on $T \setminus T_r$ with $\bar{y}_p = Y - \lfloor Y \rfloor$ and $\bar{L}(p) = \min[L(r), L(v_{\lfloor Y \rfloor + 1})]$. The algorithm then recursively solves this smaller instance to obtain a distance-2 transfer $\bar{S}$ of $(\bar{T}, \bar{L}, \bar{y})$. From $\bar{S}$ it constructs the solution S to the original problem instance by first replacing p (if p is in $\bar{S}$) by the vertex $v_{\lfloor Y \rfloor + 1}$ or r that was not chosen to be in S_{commit}, and then adding S_{commit} to the set.

We complete the proof of Lemma 7 by arguing that S is a distance-2 transfer of the original tree instance (T, L, y). Note that, as $|\bar{S}| = \sum_{v \in \bar{T}} \bar{y}_v = \sum_{v \in T} y_v - 1 - \lfloor Y \rfloor$, we have $|S| = |\bar{S}| + |S_{\text{commit}}| = \sum_{v \in V} y_v$ as required. It remains to verify Condition (3b) of Definition 3:

Claim 8. *We have* $\displaystyle \sum_{u \in U} y_u L(u) \leq \sum_{s \in S : d(s, U) \leq 2} L(s)$ *for all* $U \subseteq V(T)$.

Proof. Consider the bipartite graph G with left-hand-side $V(T)$, right-hand-side S, and an edge between $v \in V(T)$ and $s \in S$ if $d(s, v) \leq 2$. For simplicity, we slightly abuse notation and think of $V(T)$ and S as disjoint sets. Moreover, let $N(U)$ denote the (open) neighbors of a subset U of vertices in this graph: $N(U) := \{v \mid \exists u \in U \; d_G(u, v) = 1\}$. Let $w : V(T) \cup S \to \mathbb{R}$ be weights on the vertices defined by

$$w(v) = \begin{cases} y_v L(v) & \text{if } v \in V(T) \\ L(v) & \text{if } v \in S \end{cases}.$$

With this notation, we can reformulate the condition of the claim as

$$\sum_{u \in U} w(u) \leq \sum_{s \in N(U)} w(s) \qquad \text{for all } U \subseteq V(T). \tag{2}$$

To prove this, we shall prove a slightly stronger statement by verifying the condition separately on two bipartite graphs G_r and $\bar{G}$ that correspond to T_r and

$\bar{T}$, respectively. We obtain G_r and $\bar{G}$ from G as follows (see also Figure 1b). First, add a vertex p to the left-hand-side by making a copy of $r \in T$ and set $w(p) = \bar{y}_p \bar{L}(p)$ and update $w(r) = y_r L(r) - \bar{y}_p \bar{L}(p) = L(r) - \bar{y}_p \bar{L}(p) \geq 0$. Similarly, if $p \in \bar{S}$ then add a copy p of $r \in S$ and set $w(p) = \bar{L}(p)$ and update $w(r) = L(r) - \bar{L}(p) \geq 0$. Note that after these operations the vertices of both the left-hand-side and the right-hand-side can naturally be partitioned into those that correspond to vertices in T_r and those that correspond to vertices in $\bar{T}$. Graphs G_r and $\bar{G}$ are the subgraphs induced by these two partitions.

Let us first verify that (2) holds for $\bar{G}$. By construction, we have that the total weight $w(U)$ of a subset U of $V(\bar{T})$ is equal to $\sum_{u \in U} \bar{y}_u \bar{L}(u)$ and the total weight $w(N(U))$ of its neighborhood in $\bar{G}$ equals $\sum_{s \in \bar{S}: d(s,U) \leq 2} \bar{L}(s)$. Hence, (2) holds since $\bar{S}$ is a distance-2 transfer of $(\bar{T}, \bar{L}, \bar{y})$.

We conclude the proof of the claim by verifying (2) for G_r. As both the left-hand-side and right-hand-side of G_r correspond to vertices in T_r that all are within distance 2 of each other, we have that G_r is a complete bipartite graph. The total weight of the left-hand-side is by construction $\sum_{u \in T_r} y_u L(u) - \bar{y}_p \bar{L}(p)$ and the total weight of the right-hand-side is $\sum_{s \in T_r \cap S} L(s) - \bar{L}(p) \mathbf{1}_{p \in \bar{S}}$ which equals $\sum_{s \in S_{\text{commit}}} L(s)$. The claim now follows from (1), i.e., that $\sum_{u \in T_r} y_u L(u) - \bar{y}_p \bar{L}(p) \leq \sum_{s \in S_{\text{commit}}} L(s)$.   □

The above claim completed the analysis of the algorithm for finding an integral distance-2 transfer of a given tree instance and Lemma 7 follows.

5   Extensions to Other Problems and Future Directions

Our techniques can be extended to obtain approximation algorithms for other problems. The full version of this extended abstract discusses two problems to which our techniques readily apply; we see this as further evidence that the simplicity of our approach helps in designing better algorithms also for other location problems.

As our 9-approximation algorithm comes close to settling the integrality gap, it is natural to ask if our techniques can be used to obtain a tight result. Recall that our framework consists of first reducing the general problem to tree instances and then solving such instances. Since our algorithm for tree instances is the best possible, any potential improvement must come from the reduction, and we raise this as an open problem.

Finally, as the $\{0, L\}$-case suggests, our preliminary results on additional preprocessing indicate that further investigation is necessary to understand if these techniques can help bring down the integrality gap to the tight factor of 3. More generally, we believe that it is important not only for capacitated k-center but also for other problems, such as facility location and k-median, to understand the power of lift-and-project methods (applied to potentially different formulations). For example, do they automatically capture these preprocessing steps and lead to stronger formulations?

References

1. Arya, V., Garg, N., Khandekar, R., Meyerson, A., Munagala, K., Pandit, V.: Local search heuristics for k-median and facility location problems. SIAM J. Comput. 33(3), 544–562 (2004)
2. Bansal, M., Garg, N., Gupta, N.: A 5-approximation for capacitated facility location. In: Epstein, L., Ferragina, P. (eds.) ESA 2012. LNCS, vol. 7501, pp. 133–144. Springer, Heidelberg (2012)
3. Byrka, J.: An optimal bifactor approximation algorithm for the metric uncapacitated facility location problem. In: APPROX-RANDOM, pp. 29–43 (2007)
4. Charikar, M., Guha, S.: Improved combinatorial algorithms for facility location problems. SIAM J. Comput. 34(4), 803–824 (2005)
5. Charikar, M., Guha, S., Tardos, É., Shmoys, D.B.: A constant-factor approximation algorithm for the k-median problem. J. Comput. Syst. Sci. 65(1), 129–149 (2002)
6. Chudak, F.A., Williamson, D.P.: Improved approximation algorithms for capacitated facility location problems. Math. Program. 102(2), 207–222 (2005)
7. Chuzhoy, J., Rabani, Y.: Approximating k-median with non-uniform capacities. In: SODA, pp. 952–958 (2005)
8. Cygan, M., Hajiaghayi, M., Khuller, S.: LP rounding for k-centers with non-uniform hard capacities. In: FOCS, pp. 273–282 (2012)
9. Gonzalez, T.F.: Clustering to minimize the maximum intercluster distance. Theor. Comput. Sci. 38, 293–306 (1985)
10. Guha, S., Khuller, S.: Greedy strikes back: Improved facility location algorithms. J. Algorithms 31(1), 228–248 (1999)
11. Hall, P.: On representatives of subsets. Journal of the London Mathematical Society 10, 26–30 (1935)
12. Hochbaum, D.S., Shmoys, D.B.: A best possible heuristic for the k-center problem. Mathematics of Operations Research 10, 180–184 (1985)
13. Jain, K., Mahdian, M., Saberi, A.: A new greedy approach for facility location problems. In: STOC, pp. 731–740 (2002)
14. Jain, K., Vazirani, V.V.: Approximation algorithms for metric facility location and k-median problems using the primal-dual schema and lagrangian relaxation. J. ACM 48(2), 274–296 (2001)
15. Khuller, S., Sussmann, Y.J.: The capacitated k-center problem. SIAM J. Discrete Math. 13(3), 403–418 (2000)
16. Korupolu, M.R., Plaxton, C.G., Rajaraman, R.: Analysis of a local search heuristic for facility location problems. J. Algorithms 37(1), 146–188 (2000)
17. Levi, R., Shmoys, D.B., Swamy, C.: LP-based approximation algorithms for capacitated facility location. In: Bienstock, D., Nemhauser, G.L. (eds.) IPCO 2004. LNCS, vol. 3064, pp. 206–218. Springer, Heidelberg (2004)
18. Li, S.: A 1.488 approximation algorithm for the uncapacitated facility location problem. In: Aceto, L., Henzinger, M., Sgall, J. (eds.) ICALP 2011, Part II. LNCS, vol. 6756, pp. 77–88. Springer, Heidelberg (2011)
19. Li, S., Svensson, O.: Approximating k-median problem via pseudo-approximation. In: STOC, pp. 901–910 (2013)
20. Pál, M., Tardos, É., Wexler, T.: Facility location with nonuniform hard capacities. In: FOCS, pp. 329–338 (2001)
21. Shmoys, D.B., Tardos, É., Aardal, K.: Approximation algorithms for facility location problems (extended abstract). In: STOC, pp. 265–274 (1997)
22. Williamson, D.P., Shmoys, D.B.: The Design of Approximation Algorithms. Cambridge University Press (2011)
23. Zhang, J., Chen, B., Ye, Y.: A multiexchange local search algorithm for the capacitated facility location problem. Math. Oper. Res. 30(2), 389–403 (2005)

Sequence Independent, Simultaneous and Multidimensional Lifting of Generalized Flow Covers for the Semi-Continuous Knapsack Problem with Generalized Upper Bounds Constraints

Alejandro Angulo[1], Daniel Espinoza[1], and Rodrigo Palma[2],[*]

[1] Department of Industrial Engineering, Universidad de Chile
[2] Department of Electrical Engineering, Universidad de Chile

Abstract. We consider the semi-continuous knapsack problem with generalized upper bound constraints on binary variables. We prove that generalized flow cover inequalities are valid in this setting. We also prove that, under mild assumptions, they are facet-defining inequalities for the full problem. We then focus on simultaneous lifting of pairs of variables. The associated lifting problem naturally induce multidimensional lifting functions, and we prove that a simple relaxation, in a restricted domain, is a superadditive function. We also prove that in many cases this approximation is actually the optimal lifting function. We then analyze the separation problem, which we separate in two phases: first, find a seed inequality, where we evaluate both exact and heuristic methods; secondly, since the lifting is simultaneous, our class of lifted inequalities might contain an exponential number of them. We choose a strategy of maximizing resulting violation. Finally, we test this class of inequalities on instances arising from electricity planning problems. Our test show that the proposed class of inequalities are strong in the sense that adding a few of these inequalities, they close, on average, 57.70% percent of the root integrality gap, and close 97.70% of the relative gap, while adding very few cuts.

Keywords: Knapsack problem, sequence independent multidimensional lifting, generalized upper bounds.

1 Introduction

Binary knapsack programs are a common model for choosing between discrete alternatives. If the choice is continuous but limited, we can see the model as the classical single-node capacitated network design model [10], if the choice is semi-continuous, then we must consider mixed-binary knapsack programs; this problem is known as the semi-continuous knapsack problem (SCKP). If, in addition, we have that binary variables are partitioned in sets such that we can choose at most one of them from any given set, we have what we call a semi-continuous knapsack problem with generalized

[*] This research was funded by FONDECYT grant 1110024 and Millennium Nucleus Information and Coordination in Networks ICM/FIC P10-024F.

J. Lee and J. Vygen (Eds.): IPCO 2014, LNCS 8494, pp. 64–75, 2014.

upper bound constraints (SCKPGUB). This kind of models are common for representing (possibly non-linear) functions (with only one GUB constraint), or when we are looking at the combined non-linear output of several machines (which is the case on production scheduling problems, as is the case in electricity generation, among many others). In this setting, sequential lifting is too limited, in the sense that whenever we have a constraint $y \leq x$ for $x \in \{0, 1\}$ and $y \in [0, 1]$, lifting must be done first on the integer variable and then on the continuous variable, which precludes finding some facet defining inequalities for the complete problem; thus, simultaneous lifting is essential in this setting.

In this paper we use generalized flow cover inequalities [12] (GFC), show that they are valid in our setting, and in many cases induce facets on faces of the original problem. We then propose a valid sequence-independent multidimensional lifting scheme, to obtain valid inequalities for SCKPGUB; we show that superadditivity on a restricted set of *feasible* right hand sides, and show that this condition is enough to obtain sequence-independent lifting. We also provide sufficient conditions for this lifting to be maximal.

The paper is organized as follows: Section 2 cover some of the known facts on semi-continuous knapsack problems, including some known valid inequalities, and some basic results for the semi-continuous knapsack problem with GUB constraints. Section 3 deals with multidimensional lifting for SCKPGUB, specifically on how to obtain valid sequence-independent lifting functions for them. It also proposes simple algorithms for solving the separation problem; both for the seed inequality, and for selecting maximally violated lifted inequalities. Section 4 present experiments designed to show that the proposed heuristic separation provide good results, and that the lifting step is crucial. Instances are randomly generated and are derived from electricity generation problems. Finally, Section 5 presents our conclusions and further questions on this topic.

2 Definitions and Basic Polyhedral Results

2.1 The Problem

To simplify notation we will write $a \cdot b(A)$ to represent $\sum (a_i b_i : i \in A)$, $a(A)$ to represent $\sum (a_i : i \in A)$ and $[n]$ to represent $\{1, \ldots, n\}$. We consider the semi continuous knapsack problem with generalized upper bound constraints, given by

$$
X_G = \left\{
\begin{array}{ll}
(x, y) \in \{\{0, 1\} \times [0, 1]\}^M : \\
\quad (a \cdot x + m \cdot y)(M) \leq b \\
\quad\quad\quad\quad\quad y_k \leq x_k & \forall k \in M \\
\quad\quad\quad\quad\quad x(M_g) \leq 1 & \forall g \in G
\end{array}
\right\}, \tag{1}
$$

where G is the set of GUB constraints, $M = \{(g_i, j_i)\}_{i=1}^n$, $M_g = \{(g', j') \in M : g' = g\}$. Note that $a_k x_k + m_k y_k$ is a model for a semi-continuous variable with values in $\{0\} \cup [a_k, a_k + m_k]$. The first constraint is the semi-continuous knapsack constraint, the second constraint ensure semi-continuity, and the last constraint imposes the generalized upper bound condition among disjoint sets of binary variables.

2.2 Literature Review

Many special cases of this structure have been studied before. For example, the classical binary knapsack problem was studied by Balas and Jeroslow [3] in a theoretical work where canonical cuts on the unit hypercube were introduced. Based on this work, in 1975, Wolsey [15] and Balas [2] presented facet defining inequalities for the KP using for the first time the notion of cover. In 1978, Balas and Zemel [4] extend previous work, applying lifting procedures to valid inequalities obtained from minimal covers. In 1980, Padberg [11] present $(1, k)$-configurations as a generalization of minimal covers inequalities. Inclusion of GUB constraint in KP (KPGUB), was studied by Johnson and Padberg [8]; they also shown how to transform a general instance of the problem in one with non-negative coefficients only. In [14], Wolsey defined some valid inequalities for the KPGUB, and proved that they are facet defining under certain conditions. Sherali and Lee [13], apply sequential and simultaneous lifting to valid inequalities for KPGUB deduced from minimal covers.

Another special case is when $a_k = 0$ and $|M_g| = 1$. This case is called single node flow sets (SNFS) and their study has been extended from work of Gu et al. [6] from lifting procedures applied to this set. In 2007, Louveaux and Wolsey [9] give an interesting survey of strong valid inequalities for knapsack and single node flow sets.

As can be seen, application of lifting procedures is a fundamental part of cut generation techniques for many specific sets. In 1977, Wolsey [16] presented the first work on this area where the concept of superadditivity was used. In 2000, Gu et al. [7] generalized it and defined sequence independent lifting of general mixed integer programs. In 2004, Atamtürk [1] gave similar results.

All these works can be seen as one-dimensional lifting, since they consider the perturbation of one constraint only. For multidimensional lifting, applications are scarce, being the work of Zeng and Richard [18,19,17] the most interesting. They defined a general framework to derive multidimensional superadditive lifting functions and applied it to the precedence constrained knapsack problem and to the single node flow set. They show that the traditional concept of superadditivity used by Gu et al. [7], can be restricted depending of the problem in which it is applied. We provide a simple proof of this result in the context of SCKPGUB.

2.3 Polyhedral Results

Basic Results for SCKPGUB. From this point onward we will assume that $\bar{a} := \max\{a_k : k \in M\} < b$. With this in place, Proposition 1 follows.

Proposition 1. *1. X_G is full dimensional.*
2. Inequality $y_k \geq 0$ is facet-defining for X_G, $\forall k \in M$.
3. If $a_k + m_k \leq b$, inequality $y_k \leq x_k$ is facet-defining for X_G, $\forall k \in M$.
4. If $\bar{a}_g := \max\limits_{g' \neq g, j \in N_{g'}} \{a_{g'j}\} + \min\limits_{j \in N_g} \{a_{gj}\} < b$, then $x(N_g) \leq 1$ is facet defining for X_G, for $g \in G$.

Generalized Flow Cover Inequalities for SCKP. Consider the set

$$X = \left\{ \begin{array}{c} (x,y) \in \{0,1\}^n \times [0,1]_+^n : \\ (a \cdot x + m \cdot y)(N) \leq b \\ y_j \leq x_j \quad \forall j \in N \end{array} \right\}. \tag{2}$$

Van Roy and Wolsey [12] studied (a generalization of) this polyhedron, and proposed a family of valid inequalities which they called generalized flow cover inequalities (GFC). In our setting, this family of inequalities can be stated as follows:

Given X as defined in (2), a pair (C, C_U) with $C \subset N$, $C_U \subset C$, satisfying $\Gamma := a(C) + m(C_U) - b > 0$ and $m_{C_U} > 0$, is called a *generalized cover*. Defining $\xi_j = a_j$ for $j \in C \setminus C_U$, $\xi_j = a_j + m_j$ for $j \in C_u$, and $\gamma_j = \min\left\{1, \frac{\xi_j}{\Gamma}\right\}$; then

$$\gamma \cdot x(C) + \left(\frac{m}{\Gamma} \cdot (y-x)\right)(C_U) \leq \gamma(C) - 1, \tag{3}$$

is valid for X.

The following theorem give sufficient conditions for (3) to be facet-defining for X.

Theorem 1. *Let (C, C_U) be a generalized flow cover, satisfying* $\sum\limits_{j \in C_U : \xi_j > \Gamma} m_j > \Gamma$. *Then inequality (3), is facet-defining for $X_o := X \cap \{x_i = 0, \forall i \notin C\}$.*

Note that X is a face of X_G, where we choose at most one element from every GUB constraint to be active; from this, the previous theorem give simple conditions for having facet-defining inequalities for this face of X_G. Moreover, since X can also be seen as a relaxation of X_G, we have that (3) also define valid inequalities for X_G.

3   Multidimensional Lifting for SCKPGUB

3.1   Valid Lifting Functions

As was noted before, given a *generalized flow cover* C, C_U in X_G satisfying $|C \cap \{(g,j) : j \in N_g\}| \leq 1$, $\forall g \in G$, then (3) is valid for X_G, and if $m(C_U^+) > \Gamma$, where $C_U^+ = \{k \in C_U : a_k + m_k > \Gamma\}$, then (3) it is a facet-defining inequality for $X_G \cap \{x_i = 0, \forall i \notin C\}$. Following Gu et. al [7], we consider the problem of sequentially lifting pairs of variables[1] (x_k, y_k), $k \notin C$ to obtain

$$\gamma \cdot x(C) + \left(\frac{m}{\Gamma}(y-x)\right)(C_U) + (\alpha \cdot x + \beta \cdot y)(C^c) \leq \gamma(C) - 1. \tag{4}$$

If we index pairs $\{k_i\}_{i=1}^{n-|C|} = \{(g_i, j_i)\}_{i=1}^{n-|C|} = M \setminus C$, and assume that the first $i-1$ pair of variables have been lifted, the i-th lifting functions can be written as

$$h_{k_i}(z, \boldsymbol{v}) = \max\ \alpha_{k_i} x_{k_i} + \beta_{k_i} y_{k_i}$$
$$s.t.\ a_{k_i} x_{k_i} + m_{k_i} y_{k_i} = z$$
$$x_{k_i} = \boldsymbol{v}_{g_i}$$
$$0 \leq y_{k_i} \leq x_{k_i}$$
$$x_{k_i} \in \{0,1\}, \tag{5}$$

---

[1] And then, simultaneously lift them.

and the functions $f_{k_i}(z, \boldsymbol{v})$ are given by

$$f_{k_i}(z, \boldsymbol{v}) = \min \gamma \cdot (1 - x)(C) - \left(\frac{m}{\Gamma}(y - x)\right)(C_U) - (\alpha \cdot x + \beta \cdot y)(K^i) - 1$$
$$\text{s.t. } (a \cdot x + m \cdot y)\left(C \cup K^i\right) \leq b - z$$
$$x\left(M_g \cap \left(C \cup K^i\right)\right) \leq 1 - v_g, \quad \forall g \in G$$
$$0 \leq y_k \leq x_k, \; x_k \in \{0, 1\} \quad \forall k \in C \cup K^i, \tag{6}$$

where $K^i = \{k_1, \ldots, k_{i-1}\}$, $z \in [0, b]$ and $\boldsymbol{v}$ has the dimension of the right hand sides of X_G for GUB constraints and for simple bounds on binary variables, and is defined as $v_{k'} = \delta_{k', k_i}$ for $k \in M$ and $v_{g'} = \delta_{g', g_i}$ for $g \in G$, where $\delta_{a,b} = 1$ if $a = b$ and zero otherwise, i.e. $\boldsymbol{v} = (e_{k_i}, e_{g_i})$. We are interested in finding $\alpha_{k_i}, \beta_{k_i}$ that ensure that $h_{k_i}(z, u\boldsymbol{v}) \leq f_{k_i}(z, u\boldsymbol{v})$ for all $(z, u) \in \{(0, 0)\} \cup \{([a_{k_i}, a_{k_i} + m_{k_i}], 1)\}$ and $\boldsymbol{v} = \{(e_{k_i}, e_{g_i})\}$. This implies that we are not interested in h_{k_i}, f_{k_i} for all possible $(z, \boldsymbol{v}) \in \mathbb{R} \times \{0, 1\}^{G \cup M}$, but only on the true domain of feasible points of X_G.

Although the lifting is multidimensional, at every step, there are only two degrees of freedom in the functions, namely z and $u \in \{0, 1\}$. The analysis for h_{k_i} is easy, for the case where $m_{k_i} > 0$, the optimal value for $h_{k_i}(z, \boldsymbol{v})$ is

$$h_{k_i}(z, \boldsymbol{v}) = \begin{cases} 0 & u = 0, z = 0 \\ \tilde{\alpha} + \tilde{\beta}z & u = 1, a_{k_i} \leq z \leq a_{k_i} + m_{k_i} \end{cases}$$

where $\tilde{\alpha} = \alpha_{k_i} - \frac{a_{k_i}}{m_{k_i}}\beta_{k_i}$ and $\tilde{\beta} = \frac{1}{m_{k_i}}\beta_{k_i}$.

For the case where $m_{k_i} = 0$, the optimal value of the function is

$$h_{k_i}(z, \boldsymbol{v}) = \begin{cases} 0 & u = 0, z = 0 \\ \tilde{\alpha} & u = 1, z = a_{k_i} \end{cases}$$

where $\tilde{\alpha} = \alpha_{k_i}$ and $\tilde{\beta} = \beta_{k_i} = 0$. $\tilde{\alpha}, \tilde{\beta}$ are called *normalized* lifting coefficients.

To study f_{k_i} we start with a simple case in the following propositions:

Proposition 2. *Let* $D = \{(g, j) \in M \setminus C : \exists (g, j') \in C, \xi_{gj'} \geq \Gamma, a_{gj} + m_{gj} \leq \xi_{gj'} - \Gamma\}$. *Then, for all* $k \in D$, *the maximal lifting coefficients* (α_k, β_k) *are* $(0, 0)$.

Proof. Since lifting functions are decreasing in the order, assume that the first elements to lift from the seed inequality are from D. Let k be an element in D. It is known that $f_k(z, \boldsymbol{v}) \geq 0$ and monotone non-decreasing function, and that $f(0, \boldsymbol{0}) = 0$. This implies that is enough to find a feasible point for $z = a_k + m_k$ with objective value equal to zero to prove our result. Let $k_o \in C$ satisfying $\xi_{k_o} > \Gamma$ and $a_k + m_k \leq \xi_{k_o} - \Gamma$, then, setting $(x, y) = (\mathbf{1}_C - e_{k_o}, \mathbf{1}_C - e_{k_o})$, we have that $f_k(z, \boldsymbol{v}) \leq 0$ for $z \in [a_k, a_k + m_k]$, $\boldsymbol{v} = (e_k, e_{g_k})$. ∎

Proposition 3. *If* $k \in M \setminus (C \cup D)$, *then for every optimal solution of the problem* $f_k(z, v)$, *it is always possible to find an optimal solution* x^*, y^* *satisfying* $y_k^* = 0$ *for* $k \in C_L$ *and* $x_k^* = y_k^* = 0$ *for* $k \in D$.

These two propositions allow us to work assuming that $D = \emptyset$ and that $m(C_L) = 0$. Given $\boldsymbol{v} \in \{0, 1\}^G$, and defining $C_{\boldsymbol{v}} = \{(g, j) \in C : v_g = 1\}$, then, we can re-write the first lifting function $f(z_1, v)$ as

$$f_1(z, v) = \gamma(C) - 1 - \max \left(\gamma \cdot x + \frac{m}{\Gamma} \cdot (y - x) \right) (C \setminus C_v)$$
$$s.t. \qquad (\xi \cdot x + m \cdot y)(C \setminus C_v) \leq b - z$$
$$0 \leq y_k \leq x_k, \ x_k \in \{0, 1\} \forall k \in C \setminus C_v. \qquad (7)$$

In general, $f_1(z, v)$ is a complex function; however, in many common cases[2] $f_1(z, v)$ is equivalent to (or is bounded from below by) the function obtained from the following relaxation[3]:

$$\tilde{f}(z, v) = \gamma(C_v) - 1 + \min \ x(C^+ \setminus C_v) + \frac{s}{\Gamma}$$
$$s.t. \ \xi \cdot x(C^+ \setminus C_v) + s \geq z + \Gamma - \xi(C_v)$$
$$0 \leq s, \ x_k \in \{0, 1\}, \ \forall k \in C^+ \setminus C_v, \qquad (8)$$

where $C^+ = \{k \in C : \xi_k > \Gamma\}$. In this form it is easy to prove the following results:

Proposition 4. *By renaming $C^+ \setminus C_v = [r_v]$ while ensuring that $\xi_h^v \geq \xi_{h+1}^v$, defining $\Lambda_h^v = \xi(C_v) + \xi^v([h-1])$, and defining $H(z) = 0$ if $z \leq 0$ and $H(z) = 1$ if $z > 0$, then*

$$\tilde{f}(z, v) = \gamma(C_v) - 1 + \frac{s^*}{\Gamma} + \sum \left(H \left(z - s^* - \Lambda_h^v + \Gamma \right) : h \in [r_v] \right), \qquad (9)$$

where $s^* = (z - \Lambda_{r_v+1}^v + \Gamma)\mathbb{I}(z \geq \Lambda_{r_v+1}^v - \Gamma) + \sum\limits_{h=1}^{r_v} (z - \Lambda_h^v)\mathbb{I}(0 \leq z - \Lambda_h^v \leq \Gamma)$, moreover, the optimal solution for x is given by $x_h^* = H(z - s^* - \Lambda_h^v + \Gamma)$, $\forall h \in [r_v]$.

Theorem 2. *The function $\tilde{f}(z, v)$ is superadditive for $(z, v) \in [0, +\infty) \times \{0, 1\}^G$.*

Corollary 1. *If, for each pair of variables (x_k, y_k), where $k = (g, j)$, we choose the lifting coefficients (α_k, β_k) such that $h_k(z, u) \leq \tilde{f}(z, ue_g)$ for $(z, u) \in \{([a_k, a_k + m_k], 1), (0, 0)\}$, then the lifting process is sequence independent.*

3.2 Algorithmic Separation

The previous results show that we can use $\tilde{f}(z, v)$ to find valid (and in many cases optimal) lifting coefficients for generalized flow cover inequalities (where, for each GUB, at most one binary variable is in the cover) to obtain strong inequalities for X_G.

In this section we deal with the separation problem of such lifted inequalities. More precisely, given x^* a fractional solution in the LP relaxation of (1), try to find a violated lifted constraint. We address this problem in two stages; first, we show how to lift a candidate inequality, then we propose an heuristic to identify a candidate seed inequality. Finally, we keep inequalities that are violated.

---

[2] A simple sufficient condition is that the $m_k \geq \Gamma$ for the two smallest ξ_k coefficients in C

[3] This relaxation is obtained by relaxing integrality of $x \notin C^+$, discarding $y_i \leq x_i$, aggregating all continuous variables into s, and complementing remaining integer variables.

Lifting GUB-Constrained Flow-Cover Inequalities. Although closed form expressions for $\tilde{f}(z, v)$ are possible; it is important to note that given for each pair of lifted variables $k \in M \backslash C$ and its range $[a_k, a_k + m_k]$, there are several maximal pairs of coefficients α_k, β_k satisfying $h_k(z, x_k) \leq \tilde{f}(z, x_k e_{g(k)})$. This implies that the number of possible lifted inequalities derived by this method can be exponential; and a proper method to choose which (set of) inequalities to use is crucial, and should depend on the fractional values of the current fractional point x^*, y^*. Fortunately, a complete description of all pairs of maximal lifting coefficients can be obtained using Algorithm 1, whose complexity is $\mathcal{O}(|C|)$. In our implementation we choose $(\alpha^*, \beta^*) \in \text{argmax}\{x_k^* \alpha + y_k^* \beta : (\alpha, \beta) \in \mathcal{H}\}$, where $\mathcal{H}$ is the set of maximal lifting coefficients.

Algorithm 1. Finding lower envelope of $\tilde{f}(\cdot, v)$.

Require: Breakpoints of $\tilde{f}(\cdot, v)$, $B = \{z_i\}_{i=1}^m$ where $z_i \leq z_{i+1}$ and $|B| \geq 2$. Interval $[a, b]$, actual range for z (we assume $z_1 \leq a \leq b \leq z_n$).
Ensure: $\mathcal{H} = \{\tilde{\alpha}_j, \tilde{\beta}_j\}$, pairs of (normalized) maximal lifting coefficients.
1: $B_{[a,b]} \leftarrow \{a\} \cup \{z_i \in B : a < z_i < b\} \cup \{b\}$ (ordered set)
2: $n \leftarrow |B_{[a,b]}|, \mathcal{H} \leftarrow \emptyset, k_l \leftarrow 1, k_r \leftarrow 2$
3: **if** $n = 1$ **then**
4:     **return** $\{(\tilde{f}(a, v), 0)\}$
5: **loop**
6:     $z_l \leftarrow B_{[a,b]}[k_l], f_l \leftarrow \tilde{f}(z_l, v), z_r \leftarrow B_{[a,b]}[k_r], f_r \leftarrow \tilde{f}(z_r, v)$
7:     $\tilde{\beta} \leftarrow \frac{f_r - f_l}{z_r - z_l}, \tilde{\alpha} \leftarrow f_l - \tilde{\beta} z_l$
8:     **if** $k_r + 1 > n$ **then**
9:         **return** $\mathcal{H} \cup \{(\tilde{\alpha}, \tilde{\beta})\}$
10:     $z_{2r} \leftarrow B_{[a,b]}[k_r + 1], f_{2r} \leftarrow \tilde{f}(z_{2r}, v)$
11:     **if** $\tilde{\alpha} + \tilde{\beta} z_{2r} \leq f_{2r}$ **then**
12:         $k_l \leftarrow k_r, \mathcal{H} \leftarrow \mathcal{H} \cup \{(\tilde{\alpha}, \tilde{\beta})\}$
13:     $k_r \leftarrow k_r + 1$

Algorithm 2. Heuristic to find a GFC.

Require: Fractional point (x^*, y^*).
Ensure: (C, C_U), generalized flow cover.
1: $C \leftarrow \emptyset, C_U \leftarrow \emptyset$
2: **for** $g = 1$ **to** G **do**
3:     $z_g^* \leftarrow \sum_{k \in M_g} (a_k x_k^* + m_k y_k^*)$
4:     Select $\bar{k}_g$ from $\text{argmin}\{k \in M_g : \min\{(a_k - z_g^*)_+, (z_g^* - a_k - m_k)_+\}\}$
5:     **if** $z_g^* > 0$ **then**
6:         $C \leftarrow C \cup \{\bar{k}_g\}$
7:         **if** $|z_g^* - a_{\bar{k}_g}| > |z_g^* - a_{\bar{k}_g} - m_{\bar{k}_g}|$ **then**
8:             $C_U \leftarrow C_U \cup \{\bar{k}_g\}$
9: Apply 1-OPT trying to maximize $\sum_{k \in C} \gamma_k (x_k^* - 1)$
10: **return** (C, C_U)

Finding Generalized Flow Cover Inequalities in Proper Faces of X_G. It is known that finding maximally violated cover inequalities is already $\mathcal{NP}$-hard [6]; and although it is possible to formulate the separation problem of generalized flow cover inequalities as an IP, we propose Algorithm 2[4].

4 Numerical Experiments

4.1 Instances

To evaluate the performance of the inequalities presented in this paper, we consider a set of $3,000$ random instances inspired by the unit commitment problem. Furthermore we assume that the GUB structure and semi-continuous variables are already identified. In these instances $|G| \in \{5, 10, 20, 40, 80\}$, and the number of elements in each GUB constraint where randomly chosen as $|M_g| \sim \{\mathcal{U}[2, 8], \mathcal{U}[7, 13], \mathcal{U}[17, 23]\}$. $a_j \sim \mathcal{U}[10, 150], m_j \sim \mathcal{U}[20, 300], \forall j \in M. b \sim \mathcal{U}[0.25, 0.95]b_{\max}$, where $b_{\max} =$ is the maximum value of the left hand side of the knapsack constraint. The cost coefficients were chosen as $c_k^x \sim 2,500 a_k - \mathcal{U}[370, 1000] - \mathcal{U}[15, 50]a_k, c_k^y \sim 2,500 m_k - \mathcal{U}[15, 50]m_k, \forall k \in M$, to represent typical cost functions in unit commitment instances. To evaluate the effect of cases with $m_k = 0$, half of the instances include GUB constraints where $m_k = 0$ for 40% of the elements in each GUB constraint.

4.2 Quality Measures

We use performance profiles (see [5]) on two quality measures: closed root gap (CG) and closed relative gap (CRG) which we define as

$$CG = 100 \times \frac{z_{LP_n} - z_{LP_o}}{z_{MIP} - z_{LP_o}}, \quad CRG = 100 \times \frac{z_{LP_n}}{z_{MIP}},$$

where z_{MIP} is the optimal objective value of the mixed integer problem, z_{LP_o} is the optimal objective value of the original linear relaxation and z_{LP_n} is the optimal objective value of the final LP relaxation (after several rounds of cuts). Note that for all our instances $z_{LP_o} < z_{MIP}$, and that $z_{MIP} > 0$; thus, we are never dividing by zero. Also, CRG can be seen as what a user will see as the reported gap when using any of the commercial MIP solvers out there (which might be more interesting for practitioners); while CG can be interpreted as the actual improvement in the lower bound due to the given method (which is a proxy on how much we improve the polyhedral representation of the given set for the given objective function).

We do not report running times as the separation process is very quick in all instances and the number of calls of the separation heuristic is always less than fourteen. Moreover, we do not evaluate branch and bound performance, because our instances are exactly instances of a single X_G problem, and are always very easy to solve; whereas actual unit commitment problems can have anywhere between 24 to 336 such substructures, in addition to of other side constraints. This is why we chose to leave this study for a future work.

---

[4] This heuristic is a simple extension of other heuristics to find maximally violated cover inequalities

4.3   The Experiments

The General Cutting Scheme: In each and every case, we apply the cutting scheme described in Algorithm 3.

Algorithm 3. General cutting scheme.

Require: LP^0, initial LP relaxation of X_G.
Ensure: Z, cut generation scheme optimal values.
1: $k \leftarrow 0$, $Z \leftarrow \emptyset$
2: **loop**
3:      Solve current relaxation LP^k.
4:      Obtain optimal value z_k^* and candidate solution (x_k^*, y_k^*).
5:      $Z \leftarrow Z \cup \{z_k^*\}$
6:      **if** $x_k^* \in \{0, 1\}^M$ **then**
7:          **return** Z
8:      From (x_k^*, y_k^*) and using Algorithm 2, find base GFC inequality satisfying $\Gamma \geq 0.1$ (it must be not violated).
9:      Lift seed inequality, expressed as in (4), while maximizing resulting violation v_k.
10:     **if** $\Gamma v_k \geq 0.1$ **then**
11:         $k \leftarrow k + 1$
12:         Add lifted seed inequality to LP^k.
13:     **else**
14:         **return** Z

This scheme can be seen as a basic cutting loop at the root node. We will evaluate the following variations of this scheme:

IP : exact separation of seed inequality and no lifting.
IP+Lift : same as before, but we perform step 9.
Heu : construction heuristic for seed inequality, no lifting.
Heu+1-opt : same as before, but perform 1-opt optimization of seed inequality.
Heu+1-opt+Lift : same as before, but followed by our lifting step.
Heu+Lift : construction heuristic for finding seed inequality, and we perform step 9.

Effectiveness of the Separation Heuristic. Figure 1 shows the performance profile for CG and for CRG on 600 instances with five GUB constraints, where we can solve the IP-separation of the base GFC inequality (top), and shows the performance profile for CG and CRG on all $3,000$ instances (bottom)[5]. Table 1 has a summary of these results.

It is clear that, measured either by CRG or by CG, *Heu+1-opt* performs very close to the IP separation of the base heuristic on the set of small instances; while maintaining its edge over the basic heuristic on all instances. This, plus the fact that the exact separation is far too costly on running time, justify using the proposed method; however, this is not an exhaustive evaluation, and any practical implementation should deal on this matter in much more detail.

---

[5] In our case, each point (x, y) of the plotted curves mean that for the worst $x\%$ of the instances, the given method closes at most $y\%$ of absolute root integrality gap (left), and that the method finished with a lower bound of $y\%$ of the actual integer optimal solution value or less (right).

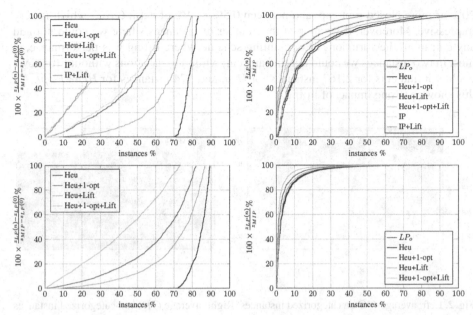

Fig. 1. Left: CG performance profile; Right: CRG performance profile; Top: Instances with $|G| = 5$; Bottom: All instances

Table 1. Summary results of experiments of average CG, CRG and number of cuts, for both, small and all instances for all algorithm variations

	Average CG				Average CRG				Average N_{cuts}			
	All		$|G|=5$		All		$|G|=5$		All		$|G|=5$	
	-Lift	+Lift	-Lift	+Lift	-Lift	+Lift	-Lift	+Lift	-Lift	+Lift	-Lift	+Lift
Heu	15.81	29.63	21.70	35.70	95.01	95.53	82.78	84.56	0.31	1.31	0.31	1.16
Heu+1opt	39.47	57.70	53.73	73.46	96.58	97.70	88.38	92.36	1.64	2.08	1.81	2.14
IP	–	–	55.81	74.13	–	–	89.13	93.11	–	–	3.17	3.36

Robustness of the Results: Another question to ask, is how robust are the results on the size of the instances. To answer this, we categorize our instances according to the number of GUB constraints ($|G|$) and on the number of elements in each GUB constraint ($|M_g|$), and see the average CG and CRG for *Heu+1-opt+Lift*. Figure 2 represent a graphical representation of how the average CR and CRG vary depending on these two criteria. Although it is expected that CG performance deteriorate as we increase both the number of GUB constraints and the number of elements in each GUB, is surprising that this tendency is reversed for CRG. This might be due to the special cost structure used in these instances; but, if true on a larger scale, it can be beneficial that the final relative integrality gap decreases on larger instances.

The Effect of Lifting: As was noted in Section 3, our seed inequality is already valid for X_G; so a natural question is how much we gain by doing the lifting process. Again, Table 1 is clear on this respect. If we measure CG, the effect ranges between 15% to

20% of more closed root gap; and between 0.5% to 4% of extra CRG; which is very impressive. Moreover, if we look at the number of instances where we could not add any cut; in all the variations of our cutting scheme where lifting was performance, at most in two instances we could not find any cut; while for variations without lifting, we could not find cuts for 2,167 instances for *Heu*, and 174 instances for *Heu+1-opt*. All this shows a strong impact of lifting.

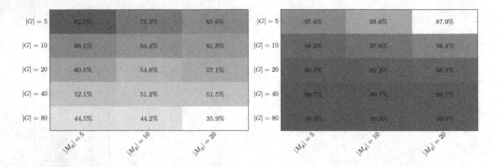

Fig. 2. Left: average CG on categorized instances; Right: average CRG on categorized instances

Number of Added Cuts: A common problem with cutting schemes is that they may require too many cutting rounds to achieve the reported quality. Surprisingly enough, in our experiments, we add, on average 2.14 cuts on all instances; for 92.2% of our instances we add up to three cuts; for 99.33% of our instances we add up to six cuts; and at most 14 cuts in the worst case.

5   Final Comments

In this paper we study sequence-independent multidimensional lifting of generalized flow cover inequalities to obtain strong inequalities for the so-called semi continuous knapsack problem with GUB constraints. We also prove that, under mild assumptions, the starting inequality is facet-defining on a face of our polyhedron. Also, under simple assumptions, we show that the sequence-independent lifting function is indeed the optimal (maximal) lifting function; which together with the previous result, allow us to obtain high-dimensional facets. Unlike one-dimensional lifting, in our setting, our supper additive lifting function define a large *class* of valid inequalities. This introduce the problem of selecting the inequality to be added.

In our computational study, we choose the inequality to be added by maximizing resulting violation. We use a set of $3,000$ randomly generated instances of different sizes to conduct our experiments. Our experiments show that, although the separation problem is $\mathcal{NP}$-hard, by using simple heuristics and our superadditive lifting, it is possible to close, on average, 57.70% root integrality gap, and 97.70% relative gap.

Finally, there are still many open questions: first, can we take advantage of GUB-partitioned binary variables in other classical polyhedral sets to find tighter valid inequalities? Secondly, in our setting, could we extend our analysis for the case where

a_k might be negative? This is an important question in our application, but is also relevant in other applications. Thirdly, Is it possible to efficiently detect the basic GUB and semi-continuous structure in general problems? probably not, but even if we are given the GUB constraints, can we use the proposed methodology in general problems? Other relevant questions are also how to select the seed inequality; or should we be looking at using (at the same time) several seed inequalities that could better complement each other when we add them to the current LP relaxation? We feel that all these questions are relevant points for the practical use of the proposed inequalities, and we hope to tackle them soon.

References

1. Atamtürk, A.: Sequence independent lifting for mixed-integer programming (2004)
2. Balas, E.: Facets of the knapsack polytope. Mathematical Programming 8, 146–164 (1975)
3. Balas, E., Jeroslow, R.: Canonical cuts on the unit hypercube. Mathematical Programming 23, 61–69 (1972)
4. Balas, E., Zemel, E.: Facets of the knapsack polytope from minimal covers. SIAM J. Appl. Math. 34, 119–148 (1978)
5. Dolan, E.D., Moré, J.J.: Benchmarking optimization software with performance profiles. Mathematical Programming 91(2), 201–213 (2002)
6. Gu, Z., Nemhauser, G., Savelsbergh, M.: Lifted flow cover inequalities for mixed 0-1 integer programs. Mathematical Programming 85, 439–468 (1999)
7. Gu, Z., Nemhauser, G., Savelsbergh, M.: Sequence independent lifting in mixed integer programming. Journal of Combinatorial Optimization 4, 109–129 (2000)
8. Johnson, E.L., Padberg, M.W.: A note on the knapsack problem with special ordered sets. Operational Research Letters 1, 18–22 (1981)
9. Louveaux, Q., Wolsey, L.A.: Lifting, superadditivity, mixed integer rounding and single node flow sets revisited. Annals OR 153, 47–77 (2007)
10. Nemhauser, G., Wolsey, L.: Integer and Combinatorial Optimization. Wiley (1988)
11. Padberg, M. (1,k)-configurations and facets for packing problems. Mathematical Programming 18, 94–99 (1980)
12. Roy, T.V., Wolsey, L.: Valid inequalities for mixed 0-1 programs. Discrete Applied Mathematics 14, 199–213 (1986)
13. Sherali, H., Lee, Y.: Sequential and simultaneous lifting of minimal cover inequalities for generalized upper bound constrained knapsack polytopes. SIAM J. Disc. Math. 8, 133–153 (1995)
14. Wolsey, L.: Valid inequalities for 0-1 knapsack and mips with generalized upper bound constraints. Discrete Applied Mathematics 29, 251–261 (1988)
15. Wolsey, L.A.: Facets of linear inequalities in 0-1 variables. Mathematical Programming 8, 165–178 (1975)
16. Wolsey, L.A.: Valid inequalities and superadditivity for 0/1 integer programs. Mathematics of Operations Research 2, 65–77 (1977)
17. Zeng, B.: Efficient Lifting Methods for Unstructured Mixed Integer Programs with Multiple Constraints. Ph.D. thesis, Purdue University. Industrial Engineering Department (2007)
18. Zeng, B., Richard, J.P.P.: Sequence independent lifting for 0-1 knapsack problems with disjoint cardinality constraints (2006)
19. Zeng, B., Richard, J.P.P.: A framework to derive multidimensional superadditive lifting functions and its applications. In: Fischetti, M., Williamson, D.P. (eds.) IPCO 2007. LNCS, vol. 4513, pp. 210–224. Springer, Heidelberg (2007)

On the Unique-Lifting Property

Gennadiy Averkov[1] and Amitabh Basu[2]

[1] Institute of Mathematical Optimization, Faculty of Mathematics,
University of Magdeburg
[2] Department of Applied Mathematics and Statistics, The Johns Hopkins University

Abstract. We study the uniqueness of minimal liftings of cut generating functions obtained from maximal lattice-free polytopes. We prove a basic invariance property of unique minimal liftings for general maximal lattice-free polytopes. This generalizes a previous result by Basu, Cornuéjols and Köppe [3] for *simplicial* maximal lattice-free polytopes, thus completely settling this fundamental question about lifting. We also extend results from [3] for minimal liftings in maximal lattice-free simplices to more general polytopes. These nontrivial generalizations require the use of deep theorems from discrete geometry and geometry of numbers, such as the Venkov-Alexandrov-McMullen theorem on translative tilings, and McMullen's characterization of zonotopes.

1 Introduction

Overview and Motivation. The idea of *cut generating functions* has emerged as a major theme in recent research on cutting planes for mixed-integer linear programming. The main object of study is the following family of mixed-integer sets:

$$X_f(R, P) = \{(s, y) \in \mathbb{R}_+^k \times \mathbb{Z}_+^\ell : f + Rs + Py \in \mathbb{Z}^n\},$$

where $f \in \mathbb{R}^n$, and $R \in \mathbb{R}^{n \times k}$, $P \in \mathbb{R}^{n \times \ell}$. We denote the columns of matrices R and P by r^i, $i = 1, \ldots, k$ and p^j, $j = 1, \ldots, \ell$ respectively. For a fixed $f \in \mathbb{R}^n \setminus \mathbb{Z}^n$, a *cut generating pair* (ψ, π) *for* f is a pair of functions $\psi, \pi : \mathbb{R}^n \to \mathbb{R}$ such that $\sum_{i=1}^k \psi(r^i)s_i + \sum_{j=1}^\ell \pi(p^j)y_j \geq 1$ is a valid inequality (cutting plane) for $X_f(R, P)$, for all matrices R and P. This model and the idea of cut generating pairs arose in the work of Gomory and Johnson from the 70s. We refer the reader to [5] and [6] for surveys of the intense research activity this area has seen in the last 5-6 years.

A very important class of cut generating pairs is obtained using the *gauge function* of *maximal lattice-free polytopes* in $\mathbb{R}^n$. These are convex polytopes $B \subseteq \mathbb{R}^n$ such that $\text{int}(B) \cap \mathbb{Z}^n = \emptyset$ and B is inclusion-wise maximal with this property. Given a maximal lattice-free polytope B such that $f \in \text{int}(B)$, one can express $B = \{x \in \mathbb{R}^n : a^i \cdot (x - f) \leq 1 \quad \forall i \in I\}$. One then obtains a cut generating pair for f by setting $\psi(r) = \max_{i \in I} a^i \cdot r$ for all $r \in \mathbb{R}^n$ (this is known as the gauge of $B - f$), and using any π such that (ψ, π) is a cut generating pair. One commonly used π is defined by $\pi(r) = \min_{w \in \mathbb{Z}^n} \psi(r + w)$ for all $r \in \mathbb{R}^n$. It can be shown that (ψ, π) thus defined forms a valid cut generating pair.

J. Lee and J. Vygen (Eds.): IPCO 2014, LNCS 8494, pp. 76–87, 2014.

Given a particular maximal lattice-free polytope B with $f \in \text{int}(B)$, it is generally possible to find many different functions π such that (ψ, π) is a cut generating pair for f, when ψ is fixed to be the gauge of $B - f$. The different possible π's are called *liftings* and the strongest cutting planes are obtained from *minimal liftings*, i.e., π such that for every lifting π', the inequality $\pi' \leq \pi$ implies $\pi' = \pi$.

Closed-form formulas for cut generating pairs (ψ, π) are highly desirable for computational purposes. This is the main motivation for considering the special class of cut generating pairs obtained from the gauge of maximal lattice-free polytopes and their liftings. The gauge is given by the very simple formula $\psi(r) = \max_{i \in I} a^i \cdot r$, and the hope is that simple formulas can be found for its minimal liftings as well. In this regard, the following results are particularly useful. Dey and Wolsey established the following theorem in [7] for $n = 2$.

Theorem 1. (Theorems 5 and 6 in [4], Theorem 4 in [2].) *Let ψ be the gauge of $B - f$, where B is a maximal lattice-free polytope and $f \in \text{int}(B)$. There exists a compact subset $R'(f, B) \subseteq \mathbb{R}^n$ such that $R'(f, B)$ has nonempty interior, and for every minimal lifting π, $\pi(r) = \psi(r)$ if and only if $r \in R'(f, B)$. Moreover, for all minimal liftings π, $\pi(r) = \pi(r + w)$ for every $w \in \mathbb{Z}^n$, $r \in \mathbb{R}^n$.*

This theorem shows that for a "fat" periodic region $R'(f, B) + \mathbb{Z}^n$, we have a closed-form formula for all minimal liftings (using the formula for ψ). In particular, when all the columns of the matrix P are in $R'(f, B) + \mathbb{Z}^n$, we can efficiently find the cutting plane $\sum_{i=1}^k \psi(r^i) s_i + \sum_{j=1}^\ell \pi(p^j) y_j \geq 1$ from B. Moreover, the above theorem shows that if $R'(f, B) + \mathbb{Z}^n = \mathbb{R}^n$, then there is a unique minimal lifting π. The following theorem from [2] establishes the necessity of this condition.

Theorem 2. (Theorem 5 in [2].) *Let ψ be the gauge of $B - f$, where B is a maximal lattice-free polytope and $f \in \text{int}(B)$. Then ψ has a unique minimal lifting if and only if $R'(f, B) + \mathbb{Z}^n = \mathbb{R}^n$. ($R'(f, B)$ is the region from Theorem 1)*

The above theorems provide a geometric perspective on *sequence independent lifting* and *monoidal strengthening* that started with the work of Balas and Jeroslow [1]. In this context, characterizing pairs f, B with unique minimal liftings becomes an important question in the cut generating function approach to cutting planes. Recent work and related literature can be found in [2–4, 6–8].

Our Contributions. We will denote the convex hull, affine hull, interior and relative interior of a set X using $\text{conv}(X), \text{aff}(X), \text{int}(X)$ and $\text{relint}(X)$ respectively. We call an n-dimensional polytope S in $\mathbb{R}^n$ a *spindle* if $S = (b^1 + C_1) \cap (b^2 + C_2)$ is the intersection of two translated polyhedral cones $b^1 + C_1$ and $b^2 + C_2$, such that the apex $b^1 \in \text{int}(b^2 + C_2)$ and the apex $b^2 \in \text{int}(b^1 + C_1)$.

Let B be a maximal lattice-free polytope in $\mathbb{R}^n$ and let $f \in \text{int}(B)$. By $\text{Fct}(B)$ we denote the set of all facets of B. With each F and f we associate the set $P_F(f) := \text{conv}(\{f\} \cup F)$. With each F and each $z \in F \cap \mathbb{Z}^n$ we associate the spindle $S_{F,z}(f) := P_F(f) \cap (z + f - P_F(f))$. Furthermore, we define $R_F(f) :=$

$\bigcup_{z \in F \cap \mathbb{Z}^n} S_{F,z}(f)$, the union of all spindles arising from the facet F, and the *lifting region* $R(f, B) := \bigcup_{F \in \mathrm{Fct}(B)} R_F(f)$ associated with the point f.

One of the main results in [2] was to establish that $R(f, B) - f$ is precisely the region $R'(f, B)$ described in Theorem 1. In light of Theorem 2, we say B *has the unique-lifting property with respect to* f if $R(f, B) + \mathbb{Z}^n = \mathbb{R}^n$, and B *has the multiple-lifting property with respect to* f if $R(f, B) + \mathbb{Z}^n \neq \mathbb{R}^n$. We summarize our main contributions in this paper.

(i) A natural question arises: is it possible that B has the unique-lifting property with respect to one $f_1 \in \mathrm{int}(B)$, and has the multiple-lifting property with respect to another $f_2 \in \mathrm{int}(B)$? This question was investigated in [3] and the main result was to establish that this cannot happen when B is a *simplicial* polytope. We prove this for general maximal lattice-free polytopes without the simpliciality assumption:

Theorem 3. (Unique-lifting invariance theorem.) *Let B be any maximal lattice-free polytope. For all $f_1, f_2 \in \mathrm{int}(B)$, B has the unique-lifting property with respect to f_1 if and only if B has the unique-lifting property with respect to f_2.*

In view of this result, we can speak about the unique-lifting property of B, without reference to any $f \in \mathrm{int}(B)$.

(ii) To prove Theorem 3, we first show that the volume of $R(f, B)/\mathbb{Z}^n$ (the region $R(f, B)$ sent onto the torus $\mathbb{R}^n/\mathbb{Z}^n$) is an affine function of f (Theorem 4). This is also an extension of the corresponding theorem from [3] for simplicial B. Besides handling the general case, our proof is also significantly shorter and more elegant. We develop a tool for computing volumes on the torus, which enables us to circumvent a complicated inclusion-exclusion argument from [3] (see pages 349-350 in [3]). We view this volume computation tool as an important technical contribution of this paper.

(iii) A major contribution of [3] was to characterize the unique-lifting property for a special class of simplices. We generalize all the results from [3] to a broader class of polytopes, called *pyramids* in Sections 3 and 5 (see Remark 1 and Theorems 5 and 11). For our generalizations, we build tools in Section 4 that invoke non-trivial theorems from the geometry of numbers and discrete geometry, such as the Venkov-Alexandrov-McMullen theorem for translative tilings in $\mathbb{R}^n$, McMullen's characterizations of polytopes with centrally symmetric faces [10] and the combinatorial structure of zonotopes.

(iv) Our techniques give an iterative procedure to construct new families of polytopes with the unique-lifting property in every dimension $n \in \mathbb{N}$. This vastly expands the known list of polytopes with the unique-lifting property. See Remarks 1, 2 and 3.

2   Invariance Theorem on the Uniqueness of Lifting

We consider the torus $\mathbb{T}^n = \mathbb{R}^n/\mathbb{Z}^n$, equipped with the natural Haar measure that assigns volume 1 to $\mathbb{R}^n/\mathbb{Z}^n$. It is clear that a compact set $X \subseteq \mathbb{R}^n$ covers

$\mathbb{R}^n$ by lattice translations, i.e., $X + \mathbb{Z}^n = \mathbb{R}^n$, if and only if $\mathrm{vol}_{\mathbb{T}^n}(X/\mathbb{Z}^n) = 1$, where $\mathrm{vol}_{\mathbb{T}^n}(X/\mathbb{Z}^n)$ denotes the volume of $X/\mathbb{Z}^n$ in $\mathbb{T}^n$.

Theorem 3 will follow as a direct consequence of the following result.

Theorem 4. *The function $f \mapsto \mathrm{vol}_{\mathbb{T}_n}(R(f, B)/\mathbb{Z}^n)$, acting from $\mathrm{int}(B)$ to $\mathbb{R}$, is the restriction of an affine function.*

When B is clear from context, we write $R(f)$ instead of $R(f, B)$.

Lemma 1. *Let $F_1, F_2 \in \mathrm{Fct}(B)$ and let $z_i \in \mathrm{relint}(F_i) \cap \mathbb{Z}^n$ for $i \in \{1, 2\}$. Suppose $x_1 \in \mathrm{int}(S_{F_1, z_1}(f))$ and $x_2 \in \mathrm{int}(S_{F_2, z_2}(f))$ be such that $x_1 - x_2 \in \mathbb{Z}^n$. Then $F_1 = F_2$ and the point $x_1 - x_2$ lies in the linear subspace parallel to the hyperplane $\mathrm{aff}(F_1) = \mathrm{aff}(F_2)$. Thus, $x_1 - x_2 \in \mathrm{aff}(F_1 - F_1)$.*

Proof. For $i \in \{1, 2\}$, if f, x_i and z_i do not lie on a common line, we introduce the two-dimensional affine space $A_i := \mathrm{aff}\{f, x_i, z_i\}$. Otherwise choose A_i to be any two-dimensional affine space containing f, x_i and z_i. The set $T_i := P_{F_i}(f) \cap A_i$ is a triangle, whose one vertex is f. We denote the other two vertices by a_i and b_i. Observe that a_i, b_i are on the boundary of facet F_i such that the open interval $(a_i, b_i) \subseteq \mathrm{relint}(F_i)$. Since z_i lies on the line segment connecting a_i, b_i and $z_i \in \mathrm{relint}(F_i)$, there exists $0 < \lambda_i < 1$ such that $z_i = \lambda_i a_i + (1 - \lambda_i) b_i$. Since $x_i \in \mathrm{int}(S_{F_i, z_i}(f))$, there exist $0 < \mu_i, \alpha_i, \beta_i < 1$ such that $x_i = \mu_i f + \alpha_i a_i + \beta_i b_i$ and $\mu_i + \alpha_i + \beta_i = 1$. Also observe that $x_i \in \mathrm{relint}(T_i \cap (z_i + f - T_i))$. Therefore, $\alpha_i < \lambda_i$ and $\beta_i < 1 - \lambda_i$.

Consider first the case $\mu_1 \geq \mu_2$. In this case, we consider the integral point $z_2 + x_1 - x_2 \in \mathbb{Z}^n$:

$$z_2 + x_1 - x_2 = \lambda_2 a_2 + (1 - \lambda_2) b_2 + \mu_1 f + \alpha_1 a_1 + \beta_1 b_1 - \mu_2 f - \alpha_2 a_2 - \beta_2 b_2$$
$$= (\mu_1 - \mu_2) f + (\lambda_2 - \alpha_2) a_2 + (1 - \lambda_2 - \beta_2) b_2 + \alpha_1 a_1 + \beta_1 b_1.$$

Since $(\mu_1 - \mu_2) + (\lambda_2 - \alpha_2) + (1 - \lambda_2 - \beta_2) + \alpha_1 + \beta_1 = 1$, and each of the terms in the sum are nonnegative, $z_2 + x_1 - x_2$ is a convex combination of points in B. Since B has no point from $\mathbb{Z}^n$ in its interior, and $f \in \mathrm{int}(B)$, the coefficient of f in the above expression must be 0. Thus, $\mu_1 = \mu_2$. So, $z_2 + x_1 - x_2 = (\lambda_2 - \alpha_2) a_2 + (1 - \lambda_2 - \beta_2) b_2 + \alpha_1 a_1 + \beta_1 b_1$, where all coefficients are strictly positive. Since $(a_i, b_i) \subseteq \mathrm{relint}(F_i)$, if $F_1 \neq F_2$, then $z_2 + x_1 - x_2 \in \mathrm{int}(B)$ leading to a contradiction to the fact that B is lattice-free. Thus, $F_1 = F_2$. $\mu_1 = \mu_2$ implies that $x_1 - x_2 \in \mathrm{aff}(F_1 - F_1)$.

The case $\mu_1 \leq \mu_2$ is similar with the same analysis performed on $z_1 + x_2 - x_1$.

An analytical formula for volume on the torus. Let R be a compact subset of $\mathbb{R}^n$. Analytically we can represent R by its indicator function $\mathbf{1}_R$ (which is defined to be equal to 1 on R and equal to 0 outside R). So, the volume of R in $\mathbb{R}^n$ is just the integral $\int_{\mathbb{R}^n} \mathbf{1}_R(x) \, \mathrm{d}\,x$. Of course, $\int_{\mathbb{R}^n} \mathbf{1}_R(x)$ is in general not an appropriate expression for $\mathrm{vol}_{\mathbb{T}^n}(R/\mathbb{Z}^n)$ because in this integral we overcount those points $x \in R$ for which there exists another point $y \in R$ with $x - y \in \mathbb{Z}^n$, i.e., $x - y$ is a point in $(R - R) \cap \mathbb{Z}^n$. Now, the function

$$c_R := \sum_{z \in (R - R) \cap \mathbb{Z}^n} \mathbf{1}_{R - z} \tag{1}$$

is precisely the function which describes whether or not we do overcounting and also how much overcounting we actually do. It can be checked that

$$c_R(x) = |\{y \in R : x \equiv y \,(\mathrm{mod}\,\mathbb{Z}^n)\}| = |(R - x) \cap \mathbb{Z}^n| = |R \cap (x + \mathbb{Z}^n)|.$$

Since $\mathbb{Z}^n$ is discrete and R is compact, $|c_R(x)|$ is a finite natural number for all $x \in R$. If $c_R(x) = 1$ for some $x \in R$, then there is no overcounting. If $c_R(x) > 1$, then there is an overcounting and the number shows how much overcounting has been done for the particular point. To remove the effect of overcounting, we need to divide by $c_R(x)$. Thus, we have

Lemma 2. *Let $R \subseteq \mathbb{R}^n$ be a compact set with nonempty interior. Then*

$$\mathrm{vol}_{\mathbb{T}^n}(R/\mathbb{Z}^n) = \int_R \frac{\mathrm{d}\,x}{c_R(x)}. \tag{2}$$

That is, for every $t \in R/\mathbb{Z}^n$, each element of $X_t := \{x \in R : x \equiv t \,(\mathrm{mod}\,\mathbb{Z}^n)\}$ is counted with the weight $1/|X_t|$ and by this $t \in R/\mathbb{Z}^n$ is counted exactly once (no overcounting!).

We first observe the following property of the function c_R.

Lemma 3. *Let $F \in \mathrm{Fct}(B)$ and let $R = R_F(f)$. Let T be an invertible affine linear map, i.e., $Tx = Lx + t$ for some invertible linear map $L : \mathbb{R}^n \to \mathbb{R}^n$ and $t \in \mathbb{R}^n$. Suppose L leaves the linear subspace parallel to F unchanged, i.e., $Lz = z$ for all $z \in \mathrm{aff}(F - F)$. Then for all $y \in \mathbb{R}^n$,*

$$c_{TR}(Ty) \geq c_R(y).$$

Proof. For any $z \in \mathrm{aff}(F - F)$ and any $y \in \mathbb{R}^n$, $\mathbf{1}_{TR-z}(Ty) = 1 \Leftrightarrow Ty \in TR - z \Leftrightarrow y \in R - L^{-1}z \Leftrightarrow y \in R - z \Leftrightarrow \mathbf{1}_{R-z}(y) = 1$. Therefore,

$$\mathbf{1}_{TR-z}(Ty) = \mathbf{1}_{R-z}(y). \tag{3}$$

Also, by Lemma 1, the set $(R - R) \cap \mathbb{Z}^n$ is contained $\mathrm{aff}(F - F)$. Thus, $(R - R) \cap \mathbb{Z}^n \subseteq (TR - TR) \cap \mathbb{Z}^n$. Therefore,

$$
\begin{aligned}
c_{TR}(Ty) &= \textstyle\sum_{z \in (TR-TR)\cap\mathbb{Z}^n} \mathbf{1}_{TR-z}(Ty) \\
&\geq \textstyle\sum_{z \in (R-R)\cap\mathbb{Z}^n} \mathbf{1}_{TR-z}(Ty) &&\text{We are dropping nonnegative terms} \\
&= \textstyle\sum_{z \in (R-R)\cap\mathbb{Z}^n} \mathbf{1}_{R-z}(y) &&\text{Using (3)} \\
&= c_R(y).
\end{aligned}
$$

An easy technical lemma:

Lemma 4. *Let $g : D \to \mathbb{R}$ be a function defined on a subset $D \subseteq \mathbb{R}^n$. Let $M : D \to \mathbb{R}$ be an affine linear map from D to $\mathbb{R}$ that is not identically 0, i.e., there exists $a \in \mathbb{R}^n, b \in \mathbb{R}$ such that $Mx = a \cdot x + b$ for all $x \in D$. Further,*

$$g(x)Mx' = g(x')Mx$$

for all $x, x' \in D$. Then g is an affine linear map on D.

Proof. Fix $x_0 \in D$ such that $M x_0 \neq 0$. Then for any $x \in D$, $g(x) = \frac{g(x_0)}{M x_0} M x$. Since M is affine, this shows that g is affine.

We finally give the proof of Theorem 4.

Proof (Proof of Theorem 4). First we observe that

$$\mathrm{vol}_{\mathbb{T}^n}(R(f)/\mathbb{Z}^n) = \sum_{F \in \mathrm{Fct}(B)} \mathrm{vol}_{\mathbb{T}^n}(R_F(f)/\mathbb{Z}^n)$$

because for two distinct facets F and F', $\mathrm{int}(R_F(f))$ and $\mathrm{int}(R_{F'}(f))$ do not intersect modulo $\mathbb{Z}^n$ by Lemma 1. Therefore, it suffices to show that for a fixed F, $\mathrm{vol}_{\mathbb{T}^n}(R_F(f)/\mathbb{Z}^n)$ is an affine function in f. Consider $f, f' \in \mathrm{int}(B)$. Fix $v_1, \ldots, v_n \in F$ that are affinely independent. Let A be the matrix formed by the columns $v_1 - f, \ldots, v_n - f$ and A' be the matrix formed by the columns $v_1 - f', \ldots, v_n - f'$. Let $L = A^{-1} A'$ and let $t = -Lf + f'$. Then the invertible affine linear map $T' : \mathbb{R}^n \to \mathbb{R}^n$ defined by $T'x = Lx + t$ maps v_i to v_i for all $i = 1, \ldots, n$, and f to f'. Thus, $T'R_F(f) = R_F(f')$. Moreover, L (therefore L^{-1}) leaves $\mathrm{aff}(F - F)$ unchanged. Note that the Jacobian of T' is given by $|\det(L)|$. Applying Lemma 3 with $R = R_F(f)$ and $T = T'$, and again with $R = R_F(f')$ and $T = T'^{-1}$, we get $c_{T'R_F(f)}(T'y) = c_{R_F(f)}(y)$ for all $y \in \mathbb{R}^n$. Then,

$$
\begin{aligned}
\mathrm{vol}_{\mathbb{T}^n}(R_F(f')/\mathbb{Z}^n) &= \mathrm{vol}_{\mathbb{T}^n}(T'R_F(f)/\mathbb{Z}^n) \\
&= \int_{T'R_F(f)} \frac{\mathrm{d}\,x}{c_{T'R_F(f)}(x)} \\
&= \int_{R_F(f)} |\det(L)| \frac{\mathrm{d}\,y}{c_{T'R_F(f)}(T'y)} \\
&= \int_{R_F(f)} |\det(L)| \frac{\mathrm{d}\,y}{c_{R_F(f)}(y)} \\
&= |\det(L)| \int_{R_F(f)} \frac{\mathrm{d}\,y}{c_{R_F(f)}(y)} \\
&= |\det(L)| \,\mathrm{vol}_{\mathbb{T}^n}(R_F(f)/\mathbb{Z}^n),
\end{aligned}
$$

where the second and last equalities follow from Lemma 2, the third equality follows from the change of variable $y := T'^{-1}x$. Since $|\det(L)| = \frac{|\det(A')|}{|\det(A)|}$, we have

$$\mathrm{vol}_{\mathbb{T}^n}(R_F(f')/\mathbb{Z}^n)|\det(A)| = \mathrm{vol}_{\mathbb{T}^n}(R_F(f)/\mathbb{Z}^n)|\det(A')|. \tag{4}$$

Finally, observe that $\det(A)$ is equal to the determinant of the $(n+1) \times (n+1)$ matrix whose first n rows are formed by the column vectors $v_1, \ldots, v_n, f$ and the last row is the row vector of all 1's. Similarly, $\det(A')$ is given by the determinant of the $(n+1) \times (n+1)$ matrix whose first n rows are formed by the column vectors $v_1, \ldots, v_n, f'$ and the last row is the row vector of all 1's. Hence, there exists a affine linear map $M : \mathbb{R}^n \to \mathbb{R}$ such that $\det(A) = Mf$ and $\det(A') = Mf'$. Thus, (4) and Lemma 4 combine to show that $\mathrm{vol}_{\mathbb{T}^n}(R_F(f)/\mathbb{Z}^n)$ is an affine function of $f \in \mathrm{int}(B)$.

Theorem 4 implies the following.

Corollary 1. *Let B be a maximal lattice-free polytope in $\mathbb{R}^n$. Then the set $\{f \in B : \mathrm{vol}_n(R(f)/\mathbb{Z}^n) = 1\}$ is a face of B.*

Proof. Since $\mathrm{vol}_n(R(f)/\mathbb{Z}^n)$ is always at most 1, the value 1 is a maximum value for the function $\mathrm{vol}_n(R(f)/\mathbb{Z}^n)$. By Theorem 4, optimizing this function over B is a linear program and hence the optimal set is a face of B.

Proof (Proof of Theorem 3). Corollary 1 implies Theorem 3. Indeed, if the set $\{f \in B : \mathrm{vol}_n(R(f)/\mathbb{Z}^n) = 1\}$ is B, then $R(f) + \mathbb{Z}^n = \mathbb{R}^n$ for all $f \in B$ and for all $f \in \mathrm{int}(B)$, B has the unique-lifting property with respect to f. Otherwise, $R(f) + \mathbb{Z}^n \neq \mathbb{R}^n$ for all $f \in \mathrm{int}(B)$ and for all $f \in \mathrm{int}(B)$, B has the multiple-lifting property with respect to f.

3   Unique Lifting in Pyramids

The Construction. We study an iterative procedure for creating higher dimensional maximal lattice-free polytopes. Let $n \geq 1$. Consider any polytope $B \subseteq \mathbb{R}^{n+1}$ such that $B \subseteq \{x \in \mathbb{R}^{n+1} : x_{n+1} = 0\}$, and a point $v^0 \in \mathbb{R}^{n+1}$ such that $v^0_{n+1} > 0$. Let $C(B, v^0)$ be the cone formed with $B - v^0$ as base. We define

$$\mathrm{Pyr}(B, v^0) = (C(B, v^0) + v^0) \cap \{x \in \mathbb{R}^{n+1} : x_{n+1} \geq -1\}.$$

Informally speaking, we put a translated cone through v^0 with B as the base and "cut it off" by the hyperplane $\{x \in \mathbb{R}^{n+1} : x_{n+1} = -1\}$, to create the pyramid $\mathrm{Pyr}(B, v^0)$ (see page 9, Ziegler [13] for a related construction). We will use the terminology that $\mathrm{Pyr}(B, v^0)$ is a *pyramid over* B. The facet of $\mathrm{Pyr}(B, v^0)$ induced by $\{x \in \mathbb{R}^{n+1} : x_{n+1} \geq -1\}$ will be called the *base* of $\mathrm{Pyr}(B, v^0)$. v^0 will be called the *apex* of $\mathrm{Pyr}(B, v^0)$.

Lifting Properties for Pyramids. We say $\mathrm{Pyr}(B, v^0)$ is *2-partitionable* if the integer hull of $\mathrm{Pyr}(B, v^0)$ is contained in $\{x \in \mathbb{R}^{n+1} : -1 \leq x_{n+1} \leq 0\}$. We show that if B is a maximal lattice-free body with the multiple-lifting property, then $\mathrm{Pyr}(B, v^0)$ is also a body with multiple-lifting for any v^0 such that $\mathrm{Pyr}(B, v^0)$ is a 2-partitionable maximal lattice-free pyramid.

Proposition 2. *Let* $B \subseteq \{x \in \mathbb{R}^{n+1} : x_{n+1} = 0\}$ *be a maximal lattice-free polytope (when viewed as an n-dimensional polytope in $\mathbb{R}^n$), and* $v^0 \in \mathbb{R}^{n+1}$ *such that* $\mathrm{Pyr}(B, v^0)$ *is a 2-partitionable maximal lattice-free polytope.*

If B is a body with the multiple-lifting property, then $\mathrm{Pyr}(B)$ *is a body with the multiple-lifting property.*

Proof. Since B is a body with multiple-lifting, there exists a vertex v of B such that the lifting region $R(v)$ satisfies $\mathrm{vol}_{\mathbb{T}^n}(R(v)/\mathbb{Z}^n) < 1$. Consider the edge of $\mathrm{Pyr}(B, v^0)$ passing through v^0 and v and let $\hat{v}$ be the vertex of this edge that lies on the base of $\mathrm{Pyr}(B, v^0)$. The lifting region $R(\hat{v})$ for $\mathrm{Pyr}(B, v^0)$ is a cylinder over $R(v)$ of height 1. Thus, $\mathrm{vol}_{\mathbb{T}^{n+1}}(R(\hat{v})/\mathbb{Z}^{n+1}) < 1$ and therefore, by Theorem 4, $\mathrm{Pyr}(B, v^0)$ is a body with multiple-lifting.

We now show that the unique-lifting property is preserved under the pyramid operation with a special property. This will give us a tool to iteratively construct bodies with unique-lifting in every dimension $n \in \mathbb{N}$.

Proposition 3. *Let* $B \subseteq \{x \in \mathbb{R}^{n+1} : x_{n+1} = 0\}$ *be a maximal lattice-free polytope (when viewed as an n-dimensional polytope in $\mathbb{R}^n$), and $v^0 \in \mathbb{R}^{n+1}$ such that $\mathrm{Pyr}(B, v^0)$ is a maximal lattice-free polytope. Suppose further that the base F_0 of $\mathrm{Pyr}(B, v^0)$ contains an integer translate of B.*

If B is a body with unique-lifting, then $\mathrm{Pyr}(B, v^0)$ is a body with unique-lifting.

Proof. For all the vertices $\hat{v}$ of $\mathrm{Pyr}(B, v^0)$ on F_0, the lifting region $R(\hat{v})$ for $\mathrm{Pyr}(B, v^0)$ contains a cylinder of height 1 over the lifting $R(v)$ for B with respect to the vertex v that lies on the edge connecting v^0 and $\hat{v}$. ($R(\hat{v})$ might contain other spindles that come from integer points in $\mathrm{Pyr}(B, v^0)$ that are not in B, but this can only help with the unique-lifting property.)

So we need to look at the vertex v^0. Let S be the set of integer points in $B \cap \mathbb{Z}^{n+1}$. By our hypothesis, after a unimodular transformation, we can assume that $S - e^{n+1} \subseteq B - e^{n+1} \subseteq F_0$, where $e^{n+1} = (0, 0, \ldots, 0, 1) \in \mathbb{R}^{n+1}$ is the standard unit vector perpendicular to F_0. Let $\bar{v}$ be the projection of v^0 onto the hyperplane $\{x \in \mathbb{R}^{n+1} : x_{n+1} = 0\}$. Since $B - e^{n+1} \subseteq F_0$, we have $\bar{v} \in B$. Let $R(\bar{v})$ be the lifting region in B with respect to $\bar{v}$ (when B is viewed as an n-dimensional polytope in $\mathbb{R}^n$).

We show that for every point $\bar{x} \in R(\bar{v})$, there is an interval of height 1 over $\bar{x}$ that is contained in $R(v^0)$ in $\mathrm{Pyr}(B, v^0)$. By Lemma 1, this will suffice to show that $\mathrm{vol}_{T^{n+1}}(R(v^0)/\mathbb{Z}^{n+1}) = 1$ since $\mathrm{vol}_{T^n}(R(\bar{v})/\mathbb{Z}^n) = 1$ because B is a body with the unique-lifting property.

Let B, when viewed as embedded in $\mathbb{R}^n$, be described by $\{x \in \mathbb{R}^n : a^i \cdot x \leq b_i \ i \in I\}$ where I is the index set for the facets. Then $\mathrm{Pyr}(B, v^0)$ can be described by $\{(x, x_{n+1}) \in \mathbb{R}^{n+1} : a^i \cdot x + \delta^i x_{n+1} \leq b_i \ \forall i \in I, \ x_{n+1} \geq -1\}$. $B - e^{n+1} \subseteq F_0$ implies that $\delta^i \geq 0$ for all $i \in I$.

Consider $\bar{x} \in R(\bar{v})$. Let $\overline{F}$ be the facet of B and $\bar{z} \in \overline{F} \cap \mathbb{Z}^{n+1}$ such that $\bar{x} \in S_{\overline{F}, \bar{z}}(\bar{v})$ (defined with respect to B). Let $z^0 = \bar{z} - e^{n+1} \in F_0$. Now $S_{F_0, z^0}(v^0)$ is given by

$$\begin{aligned} a^i \cdot x + \delta^i x_{n+1} &\leq b_i & \forall i \in I \\ -a^i \cdot x - \delta^i x_{n+1} &\leq -a^i \cdot \bar{z} + \delta^i & \forall i \in I. \end{aligned} \tag{5}$$

Let $j \in I$ be such that $\overline{F}$ is given by $a^j \cdot x \leq b_j$. We now show that the following two points $x_1 = (\bar{x}, \frac{b_j - a^j \cdot \bar{x}}{\delta^j})$ and $x_2 = (\bar{x}, \frac{b_j - a^j \cdot \bar{x}}{\delta^j} - 1)$ are both in $S_{F_0, z^0}(v^0)$. Since $S_{F_0, z^0}(v^0)$ is convex, this will imply that the entire segment of height 1 connecting x_1, x_2 lies inside $S_{F_0, z^0}(v^0) \subseteq R(v^0)$.

Observe that $a^i \cdot \bar{v} + \delta^i v^0_{n+1} = b_i$ for all $i \in I$ and therefore

$$\frac{b_i - a^i \cdot \bar{v}}{\delta^i} = v^0_{n+1} \quad \forall i \in I. \tag{6}$$

Since $\bar{x} \in R(\bar{v})$, there exists $a \in \overline{F}$ such that $\bar{x} = \mu \bar{v} + (1 - \mu)a$ for some $0 \leq \mu \leq 1$. We now check the first set of inequalities in (5) for x_1. For any $i \in I$, the following inequalities are true.

$$a^i \cdot \bar{x} + \delta^i \left(\tfrac{b_j - a^j \cdot \bar{x}}{\delta^j}\right) = a^i \cdot (\mu\bar{v} + (1-\mu)a) + \delta^i \left(\tfrac{b_j - a^j \cdot (\mu\bar{v} + (1-\mu)a)}{\delta^j}\right)$$

$$= \mu(a^i \cdot \bar{v}) + (1-\mu)(a^i \cdot a) + \delta^i \left(\tfrac{\mu b_j - \mu(a^j \cdot \bar{v})}{\delta^j}\right)$$

$$\text{(because } a^j \cdot a = b_j \text{ since } a \in \overline{F})$$

$$= \mu(a^i \cdot \bar{v}) + (1-\mu)(a^i \cdot a) + \mu\delta^i \left(\tfrac{b_i - a^i \cdot \bar{v}}{\delta^i}\right) \quad \text{using (6)}$$

$$= \mu(a^i \cdot \bar{v}) + (1-\mu)(a^i \cdot a) + \mu(b_i - a^i \cdot \bar{v})$$

$$\leq b_i \quad \text{since } a^i \cdot a \leq b^i \text{ because } a \in B.$$

Since the last coordinate of x_2 is less than the last coordinate of x_1 and $\delta^i \geq 0$ for all $i \in I$, we see that the first set of inequalities in (5) are also satisfied for x_2.

We now check the second set of inequalities in (5) for x_2. Consider $\bar{v}, \bar{z}$ and $\bar{x}$ and consider the two dimensional affine hyperplane A passing through all these 3 points. The set $T := A \cap P_{\overline{F}}(\bar{v})$ is a triangle with $\bar{v}$ as one vertex and the other two vertices a and b lying on $\overline{F}$. This is the same construction that was used in the proof of Lemma 1. Since $\bar{z}$ lies on the line segment connecting a, b, there exists $0 \leq \lambda \leq 1$ such that $\bar{z} = \lambda a + (1 - \lambda)b$. Since $\bar{x} \in T$, there exist $0 \leq \mu, \alpha, \beta \leq 1$ such that $\bar{x} = \mu\bar{v} + \alpha a + \beta b$ and $\mu + \alpha + \beta = 1$. Also observe that $\bar{x} \in T \cap (\bar{z} + \bar{v} - T)$. Therefore, $\alpha \leq \lambda$ and $\beta \leq 1 - \lambda$. Now we do the computations. For any $i \in I$,

$$-a^i \cdot \bar{x} - \delta^i \left(\tfrac{b_j - a^j \cdot \bar{x}}{\delta^j} - 1\right) = -a^i \cdot (\mu\bar{v} + \alpha a + \beta b) - \delta^i \left(\tfrac{b_j - a^j \cdot (\mu\bar{v} + \alpha a + \beta b)}{\delta^j}\right) + \delta^i$$

$$= \mu(-a^i \cdot \bar{v}) + \alpha(-a^i \cdot a) + \beta(-a^i \cdot b) - \delta^i \left(\tfrac{\mu b_j - \mu(a^j \cdot \bar{v})}{\delta^j}\right) + \delta^i$$

$$\text{(because } a^j \cdot a = a^j \cdot b = b_j \text{ since } a, b \in \overline{F})$$

$$= \mu(-a^i \cdot \bar{v}) + \alpha(-a^i \cdot a) + \beta(-a^i \cdot b) - \mu\delta^i \left(\tfrac{b_i - a^i \cdot \bar{v}}{\delta^i}\right) + \delta^i$$

$$(\text{using (6)})$$

$$= \alpha(-a^i \cdot a) + \beta(-a^i \cdot b) - \mu b_i + \delta^i$$

$$= -a^i \cdot \bar{z} + a^i \cdot (\lambda a + (1 - \lambda)b) + \alpha(-a^i \cdot a) + \beta(-a^i \cdot b) - \mu b_i + \delta^i$$

$$= -a^i \cdot \bar{z} + (\lambda - \alpha)(a^i \cdot a) + (1 - \lambda - \beta)(a^i \cdot b) - \mu b_i + \delta^i$$

$$\leq -a^i \cdot \bar{z} + (\lambda - \alpha)b_i + (1 - \lambda - \beta)b_i - \mu b_i + \delta^i$$

$$\text{(since } \alpha \leq \lambda, \ \beta \leq 1 - \lambda, \ a^i \cdot a \leq b_i \text{ and } a^i \cdot b \leq b_i \text{ because } a, b \in B)$$

$$= -a^i \cdot \bar{z} + \delta^i.$$

Finally since x_1 has a higher value for the last coordinate than x_2 and $\delta^i \geq 0$, x_1 satisfies the second set of constraints in (5) also.

Remark 1. Propositions 2 and 3 provide strict generalizations of all the results on 2-partitionable simplices from Section 4 in [3]. Furthermore, we can use these propositions to iteratively construct bodies in every dimension $n \in \mathbb{N}$ with or without the unique-lifting property. See Remarks 2 and 3.

Axis-Parallel Simplices. We use Proposition 3 to show that a certain class of simplices has unique-lifting. Let $a = (a_1, \ldots, a_n)$ be an n-tuple of positive reals such that $\frac{1}{a_1} + \ldots + \frac{1}{a_n} = 1$. Then $S(a) := \text{conv}\{0, a_1 e^1, a_2 e^2, \ldots, a_n e^n\}$ is a maximal lattice-free simplex (where $e^1, \ldots, e^n$ form the standard basis for $\mathbb{R}^n$). The following theorem is a generalization of results in [3, 4], where it was proved for the special case when $a_i = n$ for all $i = 1, \ldots, n$.

Theorem 5. $S(a)$ *is a body with unique-lifting for any n-tuple $a = (a_1, \ldots, a_n)$ such that $\frac{1}{a_1} + \ldots + \frac{1}{a_n} = 1$.*

Proof. Observe that $B = S(a) \cap \{x \in \mathbb{R}^n : x_i = 1\}$ can be expressed as $S(a') \subseteq \mathbb{R}^{n-1}$, where $a' = (a'_1, \ldots, a'_{n-1})$ is an $n-1$-tuple where $a'_i = a_i - \frac{a_i}{a_n}$ for $i = 1, \ldots, n-1$. Thus, we can use induction to prove the theorem. The case $n = 1$ is trivial, since $S(a)$ is simply an interval of length 1 and is easily seen to be a body with unique-lifting. For the induction step, we observe that $S(a)$ is a pyramid over $B = S(a) \cap \{x \in \mathbb{R}^n : x_i = 1\}$ with base $S(a) \cap \{x \in \mathbb{R}^n : x_i = 0\}$, and B is an integer translate of $S(a')$: $B = S(a') + e^n$. Further, the base contains $S(a')$ and thus contains an integer translate of B. By the induction hypothesis, $S(a')$ is a body with unique-lifting. Therefore, so is B and by Proposition 3, $S(a)$ is a body with unique-lifting.

Remark 2. We can iteratively build pyramids over $S(a)$ to get more general simplices with the unique-lifting property by repeatedly applying Proposition 3. We simply need to make sure that the base of the pyramid $\text{Pyr}(S(a), v^0)$ we create contains an integer translate of $S(a)$.

4   Spindles That Translatively Tile $\mathbb{R}^n$

In this section, we build some tools from discrete geometry and geometry of numbers. We will apply these tools to gain further insight into pyramids with the unique-lifting property in Section 5.

For any full dimensional polytope $P \subseteq \mathbb{R}^n$, a *ridge* is a face of dimension $n-2$. Let $P \subseteq \mathbb{R}^n$ be a centrally symmetric full dimensional polytope with centrally symmetric facets. Let G be any ridge of P. The *belt* corresponding to G is the set of all facets which contain a translate of G or $-G$. Observe that every full dimensional centrally symmetric polytope P with centrally symmetric facets has belts of even size greater than or equal to 4.

A *zonotope* is a polytope given by a finite set of vectors $V = \{v^1, \ldots, v^k\} \subseteq \mathbb{R}^n$ in the following way:

$$Z(V) = \{\lambda_1 v^1 + \ldots + \lambda_k v^k : -1 \leq \lambda_i \leq 1 \quad \forall i = 1, \ldots, k\}.$$

We begin with a technical lemma about the combinatorial structure of zonotopes in $\mathbb{R}^n$. We omit the proof from this extended abstract.

Lemma 5. *Let Z be a full dimensional zonotope in $\mathbb{R}^n$ such that every belt of Z is of size 4. Then Z is the image of the n-dimensional hypercube under an* *invertible* *affine transformation.*

Theorem 6. *(McMullen [11]) Let $S \subseteq \mathbb{R}^n$ be a full-dimensional centrally symmetric spindle with centrally symmetric facets. Then S is the image of the n-dimensional hypercube under an invertible affine transformation.*

Proof. Let a and $-a$ be the apexes of the spindle S. Consider any belt of S. Since S is centrally symmetric, each belt is even length, i.e., of length k where k is an even natural number. Now there are $k/2$ facets $F_1, \ldots, F_{k/2}$ involved in this belt that are incident on a (and the remaining $k/2$ facets are incident on $-a$). Let G be the $n-2$ dimensional ridge that defines this belt. We project S onto the 2 dimensional space perpendicular to G to get a polygon P. The facets $F_1, \ldots, F_{k/2}$ are all projected onto edges of the polygon. Moreover, observe that a is projected onto all these edges. This implies that $k/2 \leq 2$, otherwise, we have three edges of a polygon incident on the same point in $\mathbb{R}^2$. Thus, $k \leq 4$.

Since every belt has length 4, each $n-2$ ridge in S is centrally symmetric. Therefore, by a theorem of McMullen [10], S is a zonotope. Since S is a zonotope whose belts are length 4, by Lemma 5, S is the image of the n-dimensional hypercube under an invertible affine transformation. $\qquad\blacksquare$

Theorem 6 was communicated to us by Peter McMullen via personal email. We include a complete proof here as the result does not appear explicitly in the literature. The above proof is based on a proof sketch by Prof. McMullen.

We say that a set $S \subseteq \mathbb{R}^n$ *translatively tiles* $\mathbb{R}^n$ if $\mathbb{R}^n$ is the union of translates of S whose interiors are disjoint. We now state the celebrated Venkov-Alexandrov-McMullen theorem on translative tilings.

Theorem 7. *[Venkov-Alexandrov-McMullen (see Theorem 32.2 in [12])] Let P be a compact convex set with non-empty interior that translatively tiles $\mathbb{R}^n$. Then (i) P is a centrally symmetric polytope, (ii) All facets of P are centrally symmetric, and (iii) every belt of P is either length 4 or 6.*

Theorem 8. *Let $S \subseteq \mathbb{R}^n$ be a full-dimensional spindle that translatively tiles space. Then S is the image of the n-dimensional hypercube under an invertible affine transformation.*

Proof. Follows from the Venkov-Alexadrov-McMullen theorem (Theorem 7) and Theorem 6. $\qquad\blacksquare$

5   Maximal Lattice-Free Pyramids with Exactly One Integer Point in the Relative Interior of the Base

Theorem 9. *Let $P \subseteq \mathbb{R}^n$ be a maximal lattice-free pyramid, such that its base contains exactly one integer point in its relative interior. If P is a body with unique-lifting, then P is a simplex.*

The proof of Theorem 9 is a simple application of Theorem 8, by considering the lifting region when f is the apex of P, in which case it is a spindle which transitively tiles space. The details are omitted from this extended abstract.

Remark 3. Using Propositions 2 and 3, one can construct pyramids that are *not* simplices in arbitrarily high dimensions with (or without) the unique-lifting property. For example, start with $B \subseteq \mathbb{R}^2$ as a quadrilateral with (or without) unique-lifting (see [7]) and construct $\text{Pyr}(B, v^0) \subseteq \mathbb{R}^3$; iterate this procedure to get higher dimensional pyramids. The base of such pyramids with unique-lifting will have multiple integer points in its relative interior by Theorem 9.

We recall the following theorem proved in [3].

Theorem 10. *Let Δ be a maximal lattice-free simplex in $\mathbb{R}^n$ $(n \geq 2)$ such that each facet of Δ has exactly one integer point in its relative interior. Then Δ has the unique-lifting property if and only if Δ is an affine unimodular transformation of $\text{conv}\{0, ne^1, \ldots, ne^n\}$.*

We can now generalize this result to pyramids.

Theorem 11. *Let P be a maximal lattice-free pyramid in $\mathbb{R}^n$ $(n \geq 2)$ such that every facet of P contains exactly one integer point in its relative interior. P has the unique-lifting property if and only if P is an affine unimodular transformation of $\text{conv}\{0, ne^1, \ldots, ne^n\}$.*

Proof. Sufficiency follows from Theorem 5. If P has the unique-lifting property, by Theorem 9, P is a simplex. The result then follows from Theorem 10. ∎

References

1. Balas, E., Jeroslow, R.G.: Strengthening cuts for mixed integer programs. European Journal of Operational Research 4, 224–234 (1980)
2. Basu, A., Campelo, M., Conforti, M., Cornuejols, G., Zambelli, G.: Unique lifting of integer variables in minimal inequalities. Mathematical Programming A 141, 561–576 (2013)
3. Basu, A., Cornuéjols, G., Köppe, M.: Unique Minimal Liftings for Simplicial Polytopes. Mathematics of Operations Research 37(2), 346–355 (2012)
4. Conforti, M., Cornuéjols, G., Zambelli, G.: A Geometric Perspective on Lifting. Operations Research 59, 569–577 (2011)
5. Conforti, M., Cornuéjols, G., Zambelli, G.: Corner Polyhedron and Intersection Cuts. Surveys in Operations Research and Management Science 16, 105–120 (2011)
6. Del Pia, A., Weismantel, R.: Relaxations of mixed integer sets from lattice-free polyhedra. 4OR 10(3), 221–244 (2012)
7. Dey, S.S., Wolsey, L.A.: Two Row Mixed Integer Cuts Via Lifting. Mathematical Programming B 124, 143–174 (2010)
8. Dey, S.S., Wolsey, L.A.: Composite Lifting of Group Inequalities and an Application to Two-Row Mixing Inequalities. Discrete Optimization 7, 256–268 (2010)
9. Lovász, L.: Geometry of Numbers and Integer Programming. In: Iri, M., Tanabe, K. (eds.) Mathematical Programming: Recent Developments and Applications, pp. 177–210. Kluwer (1989)
10. McMullen, P.: Polytopes with centrally symmetric faces. Israel Journal of Mathematics 8(2), 194–196 (1970)
11. McMullen, P.: personal communication
12. Gruber, P.M.: Convex and Discrete Geometry, Grundlehren der mathematischen Wissenschaften 336. Springer (2007)
13. Ziegler, G.: Lectures on Polytopes. Springer (1995)

Maximum Weighted Induced Bipartite Subgraphs and Acyclic Subgraphs of Planar Cubic Graphs[*]

Mourad Baïou[1] and Francisco Barahona[2]

[1] CNRS and Université Clermont II, Campus des Cézeaux, BP 125,
63173 Aubière Cedex, France
[2] IBM T. J. Watson research Center, P.O. Box 218, Yorktown Heights,
NY 10589, USA

Abstract. We study the maximum node-weighted induced bipartite subgraph problem in planar graphs with maximum degree three. We show that this is polynomially solvable. It was shown in [6] that it is NP-complete if the maximum degree is four. We extend these ideas to the problem of balancing signed graphs.

We also consider maximum weighted induced acyclic subgraphs of planar directed graphs. If the maximum degree is three, it is easily shown that this is polynomially solvable. We show that for planar graphs with maximum degree four the same problem is NP-complete.

Keywords: Maximum induced bipartite subgraph, balancing signed graphs, maximum induced acyclic subgraph, polynomial algorithm, NP-completeness.

1 Introduction

Given an undirected graph $G = (V, E)$, a graph $H = (W, F)$ is said to be *induced* if $W \subseteq V$ and F is the set of edges in E having both endnodes in W. If every node u has a non-negative weight $w(u)$, the *Maximum Weighted Induced Bipartite Subgraph Problem* (MWBSP) consists of finding an induced bipartite subgraph with maximum total weight. In this paper we show that for planar graphs with degree at most three, this problem is polynomially solvable. We extend this procedure to balancing signed graphs. For planar graphs with degree at most four, it was shown in [6] that this is NP-Complete. This problem was studied in [6], and its connection to via-minimization of integrated circuits and printed circuit boards was discussed. Later in [10] the application to via-minimization and to DNA sequencing has been investigated. The polyhedral approach to this problem has been studied in [2], [3] and [10]. An approximation algorithm for general graphs was given in [16], and an approximation algorithm for planar graphs was presented in [17].

For a directed graph $G = (V, A)$ and induced subgraph is defined in a similar way. If every node u has a non-negative weight $w(u)$, the *Maximum Weighted*

---

[*] This work has been supported by project PICS05891, CNRS-IBM.

J. Lee and J. Vygen (Eds.): IPCO 2014, LNCS 8494, pp. 88–101, 2014.
© Springer International Publishing Switzerland 2014

Induced Acyclic Subgraph Problem (MWASP) consists of finding and induced acyclic subgraph of maximum total weight. If a node-set induces an acyclic subgraph, its complement is called a *Directed Feedback Vertex Set* (DFVS). The DFVS problem is an NP-complete problem that appeared in the first list of NP-complete problems in Karp's seminal paper [19]. It has applications in areas such as operating systems [27], database systems [14], and circuit testing [20]. It was shown that it is NP-Complete for planar directed graphs with in-degree and out-degree at most three, see [15]. A polyhedral approach has been presented in [11], see also [2] and [3]. An approximation algorithm for general directed graphs was given in [8], and for planar directed graphs an approximation algorithm was given in [17]. See [9] for a survey on Feedback Set problems. Here we point out that it is easy to see that for planar graphs with maximum degree three, it is polynomially solvable, then we show that it is NP-complete for planar graphs with in-degree and out-degree at most two, i.e., maximum degree four.

This paper is organized as follows. In Section 2 we present some definitions and recall some classic results that will be used in the sequel. In Section 3 we study maximum weighted induced bipartite subgraphs. In Section 4 we deal with balancing signed graphs. Section 5 is devoted to maximum weighted induced acyclic subgraphs of directed graphs.

2   Preliminaries

In this section we give some definitions and present some classic results to be used in the following sections.

If $G = (V, E)$ is an undirected graph, the *degree* of a node is the number of edges incident to it. We denote by $\Delta(G)$ the maximum among all node degrees of a graph G. If S is a node-set set we denote by $\delta(S)$ the set of edges with exactly one endnode in S. We use $\delta(v)$ instead of $\delta(\{v\})$. If e is an edge with endnodes u and v, we also use uv to denote the edge e. If $F \subseteq E$, the graph $H = (V, F)$ is called a *spanning subgraph* of G.

For a cycle C, its *incidence vector* x^C is defined by $x^C(e) = 1$ if $e \in C$, and $x^C(e) = 0$ otherwise, for each edge $e \in E$. The *cycle space* is obtained by taking sums (mod 2) of incidence vectors of cycles. An element of this space is the incidence vector of a spanning subgraph so that every node has even degree. A *cycle basis* is a basis of this vector space. For a planar graph, its faces minus one, form a cycle basis.

If $D = (V, A)$ is a directed graph, the in-degree (out-degree) of a node is the number of arcs entering (leaving) it. The degree of a node is its in-degree plus its out-degree. Also we denote by $\Delta(D)$ the maximum of all node degrees of a directed graph D. For a node set S we use $\delta^+(S)$ to denote the set $\{(u, v) \mid u \in S, v \notin S\}$. We use $\delta^-(S)$ to denote $\delta^+(V \setminus S)$.

Now we review two classic results in combinatorial optimization.

2.1   The Chinese Postman Problem and Minimum T-joins

Given an undirected connected graph $G = (V, E)$ with nonnegative edge weights $w(e)$ for each edge e, this problem consists of finding a tour of minimum weight, so that every edge is visited at least once. Edmonds & Johnson [7] gave a polynomial algorithm for this. One has to find an edge-set of minimum weight that should be visited twice. This can be formulated as follows.

$$\text{minimize} \sum_{e \in E} w(e)x(e) \tag{1}$$

$$\sum_{e \in \delta(v)} x(e) \equiv \begin{cases} 1 \pmod 2 & \text{if } v \in T, \\ 0 \pmod 2 & \text{if } v \in V \setminus T, \end{cases} \tag{2}$$

$$x(e) \in \{0, 1\} \text{ for all } e \in E. \tag{3}$$

Here T denotes the set of nodes of odd degree. A solution of this corresponds to a set of paths matching the nodes in T. For this Edmonds & Johnson gave a combinatorial algorithm that solves the following linear program.

$$\text{minimize} \sum_{e \in E} w(e)x(e) \tag{4}$$

$$\sum_{e \in \delta(S)} x(e) \geq 1 \quad \text{for each node-set } S \text{ with } |S \cap T| \text{ odd,} \tag{5}$$

$$x(e) \geq 0 \quad \text{for all } e \in E. \tag{6}$$

Their algorithm shows that this linear program always has an optimal solution that is integer valued.

If T is an arbitrary set of nodes with $|T|$ even, the same results hold, and this is called the *Minimum T-join problem*, see [26]. We are going to use this in Section 3.

Let $n = |V|$. If G is a complete graph, this problem can be solved in $O(n^3)$ time, see [12]. If the graph is planar, one can use the planar separator theorem of [22] to solve this in $O(n^{3/2} \log n)$, see [1].

2.2   The Luchessi-Younger Theorem

Let $G = (V, A)$ be a directed graph. An arc-set C is called a *directed cut* if there is a node set $U \subset V$ such that $C = \delta^-(U)$ and $\delta^+(U) = \emptyset$. Suppose that each arc a has a non-negative weight $w(a)$. Lucchesi & Younger [23] proved that the linear program below always has an optimal solution that is integer valued.

$$\text{minimize} \sum_{a \in A} w(a)x(a) \tag{7}$$

$$\sum_{a \in C} x(a) \geq 1 \quad \text{for every directed cut } C, \tag{8}$$

$$x(a) \geq 0 \quad \text{for all } e \in A. \tag{9}$$

Lucchesi [24] gave an $O(n^5 \log n)$ algorithm for this. Later Gabow [13] gave an $O(n^2 m)$ algorithm. Here $n = |V|$ and $m = |A|$.

3   Maximum Weighted Induced Bipartite Subgraphs

In this section we assume that the graph $G = (V, E)$ is planar, with $\Delta(G) = 3$, and with a non-negative weight $w(u)$ for each node $u \in V$. We have to find an induced bipartite subgraph of maximum total weight.

For the case when all node weights are equal to one, i.e., the maximum cardinality version, it was shown in [6] that there a node-set of cardinality k or less whose deletion leaves a bipartite induced subgraph if and only if there is an edge-set of cardinality k or less whose deletion leaves a bipartite spanning subgraph. This shows that the cardinality case reduces to finding a maximum cardinality cut in the same graph; that can be done in polynomial time, see [18]. For general non-negative weights this transformation is not valid, so this case has to be treated in a different way.

This problem is equivalent to look for a node-set of minimum weight that should be deleted to leave a bipartite graph. This can be formulated as the following linear integer program.

$$\text{minimize} \sum_{u \in V} w(u)x(u) \tag{10}$$

$$\sum_{u \in C} x(u) \geq 1, \text{ for each odd cycle } C, \tag{11}$$

$$x(u) \in \{0, 1\}, \text{ for all } u \in V. \tag{12}$$

Consider the linear programming relaxation obtained by replacing (12) by $x(u) \geq 0$, for all $u \in V$. Suppose for instance that G is the graph K_4, and that all weights are equal to one. If we set all variables equal to $1/3$, we have a solution of value $4/3$. On the other hand the optimal value of (10)-(12) is 2. This shows that this linear programming relaxation is not integral. At the end of this section we present a linear programming formulation (17)-(19), that gives the value of a minimum weight node-set to be deleted. This can be seen as an *extended formulation*, since we have three variables for each node.

We assume that G is connected, otherwise each connected component is treated independently. Starting from G, we create a *signed* graph $G' = (V', E')$, where each edge is labeled as *positive* or *negative* as follows.

- If a node u has degree one, let uv be the edge incident to u. Since the node u will appear in every maximum weighted induced bipartite subgraph, we remove the node u and the edge uv. We repeat this until every node has degree at least two.
- For each node u of degree two, let uv_1 and uv_2 be the edges incident to u. We split the node u into u_1 and u_2. We create the following edges.

- u_1u_2, with weight $w(u)$ and labeled positive. This edge is called *artificial*.
- u_1v_1, labeled negative.
- u_2v_2, labeled negative.

– For each node u of degree three, let uv_1, uv_2, uv_3 be the edges incident to u. We split u into u_1, u_2, u_3 and we create the following edges.
  - The edges u_1u_2, u_2u_3, and u_3u_1, all with weight $w(u)/2$ and labeled positive. These edges are called *artificial*.
  - The edges u_1v_1, u_2v_2, u_3v_3, all labeled negative.

A similar transformation has been used for minimizing the number of vias in an integrated circuit, see [25], [5].

Notice that the graph G' is also planar. This construction is illustrated in Figure 1.

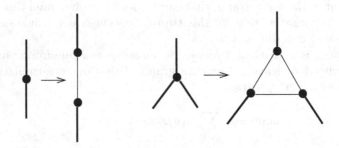

Fig. 1. Construction of the graph G'. Negative edges appear with thick lines. Positives edges with thin lines.

We need the following lemma.

Lemma 1. *Finding a maximum weighted induced bipartite subgraph of G is equivalent to give the labels "a" and "b" to the nodes of G' in such a way that:*

(1) The endnodes of each negative edge have different labels.
(2) The total weight of the positive edges whose endnodes have different labels, is minimum.

Proof. First assume that $U \subseteq V$ induces a bipartite subgraph of G with maximum weight. Let U_1 and U_2 be the bipartition of U. We give the label "a" to each node in G' associated with a node in U_1, and the label "b" to every node in G' associated with a node in U_2. Then for each negative edge that has only one labeled endnode, we give the opposite label to the other endnode. Finally for each negative edge whose endnodes have no label, we give arbitrarily opposite labels to the endnodes.

Let $\bar{U} = V \setminus U$, and $w(\bar{U})$ the total weight of the nodes in $\bar{U}$.

Let λ be the sum of the weights of the positive edges whose endnodes have different label. Since all nodes of G' associated with a node in U have the same label, we have $\lambda \le w(\bar{U})$.

Let $\hat{\lambda}$ be the weight of an optimal labeling satisfying (1) and (2). We have $\hat{\lambda} \leq \lambda$. Let S be the set of nodes of G whose associated nodes in G' have the same label. Clearly S induces a bipartite subgraph. Let $\bar{S} = V \setminus S$ and $w(\bar{S})$ the total weight of the nodes in $\bar{S}$. We have $w(\bar{S}) = \hat{\lambda}$, and

$$w(\bar{U}) \leq w(\bar{S}) = \hat{\lambda} \leq \lambda \leq w(\bar{U}).$$

Now we have to give an algorithm that finds a labeling of G' satisfying (1) and (2) of Lemma 1. For that we need one more definition.

Definition 1. *For a signed graph, and a labeling of the nodes, we say that an edge e is violated if:*

- *e is positive and its endnodes have different labels, or*
- *e is negative and its endnodes have the same label.*

Lemma 2. *If a cycle contains and odd (resp. even) number of negative edges, then for any labeling it has an odd (resp. even) number of violated edges.*

Proof. Consider a cycle with an odd number of negative edges. Start by giving the label "a" to all nodes, then there is an odd number of violated edges. Now if we change the label of a node, either the number of violated edges increases by two, or decreases by two, or remains the same. Then if we keep changing the node-labels the number of violated edges is always odd.

The proof for the other case is similar

Now we have to prove the converse.

Lemma 3. *Consider a signed graph and a set of edges marked as violated, so that for each cycle, if it has an odd (resp. even) number of negative edges then there is an odd (resp.even) number of violated edges. Then there is a set of node-labels according to Definition 1*

Proof. Start with a spanning tree T, pick any node and give it the label "a", then extend the labels through T according to Definition 1.

Then pick an edge $e \notin T$, we have to see that the labels of its endnodes satisfy Definition 1. Let C be the cycle obtained when adding e to T.

Consider the case when C has an odd number of negative edges. Assume that e is positive and marked as violated. We should show that its endnodes have different labels, so assume the opposite. Based on the labels e is not violated. But $C \setminus e$ contains an even number of edges marked as violated, this contradicts Lemma 2. The proof for all other cases is similar.

These two lemmas show that it is equivalent to work with the node labels, or with sets of violated edges satisfying the parity conditions of Lemma 2. From now on we use the second alternative. We associate to each edge e a variable $x(e)$ that should take the value 1 if e is violated and 0 otherwise. Then x should satisfy for each cycle C, the following.

$$\sum_{e \in C} x(e) \equiv \begin{cases} 1 \pmod 2 & \text{if } C \text{ has an odd number of negative edges,} \\ 0 \pmod 2 & \text{if } C \text{ has an even number of negative edges.} \end{cases} \tag{13}$$

Here we have an exponential number of equations over $GF(2)$, but we only need a maximal set that is linearly independent. Thus it is enough to impose equations (13) for the cycles in a cycle basis of G'.

Since G' is a planar graph, we can use the set of faces as a cycle basis. Let $\mathcal{F}$ be the set of faces of G', let $\mathcal{P}$ be the set of positive edges in G', $\mathcal{A}$ the set of artificial edges, and $\mathcal{N}$ the set of negative edges in G'. Here $\mathcal{A}$ coincides with $\mathcal{P}$, in the next section it might not be the case. For each artificial edge e in G', let $\lambda(e)$ be the weight of it. Our problem can be formulated as below.

$$\text{minimize} \sum_{e \in \mathcal{A}} \lambda(e)x(e) \tag{14}$$

$$\sum_{e \in C \cap \mathcal{A}} x(e) \equiv \begin{cases} 1 \pmod 2 & \text{if } C \in \mathcal{F} \text{ and } |C \cap \mathcal{N}| \text{ is odd,} \\ 0 \pmod 2 & \text{if } C \in \mathcal{F} \text{ and } |C \cap \mathcal{N}| \text{ is even,} \end{cases} \tag{15}$$

$$x(e) \in \{0,1\} \text{ for all } e \in \mathcal{A}. \tag{16}$$

Notice that only variables associated with artificial edges have been included in problem (14)-(16). This is because edges associated with the original edges of G, should not be violated. If we work with the dual graph of G', problem (14)-(16) can be solved as a minimum T-join problem, see Section 2. The use of T-joins in the dual graph appears in [26], see also [18], [4].

In Figure 2 we show an example of a graph G and its associated graph G'. If we look at an optimal solution of (14)-(16) in the dual graph, we obtain a set of paths matching pairs of faces that have an odd number of negative edges. We illustrate this in Figure 3, for every violated edge we draw a perpendicular dashed line. This corresponds to an edge of the dual graph. Also we draw a square on each face having an odd number of negative edges. Also in Figure 3 we show the induced bipartite subgraph obtained after removing the nodes associated with the violated edges. It is a simple matter to see that there is an even number of faces with an odd number of negative edges as shown below.

Lemma 4. *A signed planar graph has an even number of faces containing an odd number of negative edges.*

Proof. Start with all edges labeled positive. Pick one positive edge and change its label to negative. Then exactly two faces have an odd number of negative edges. After that, when we change any other label from positive to negative, either the number of faces with an odd number of negative edges increases by two, or decreases by two, or remains the same.

Using planar duality, and the results of Edmonds & Johnson [7], we can see that problem (14)-(16) is equivalent to the linear program below. This is an extended formulation since for each of the nodes in the original graph we might have more than one variable.

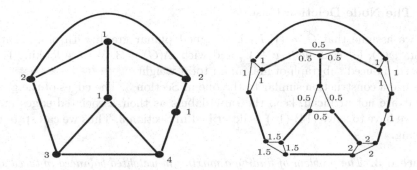

Fig. 2. An example of the graphs G and G'. The numbers near the nodes are their weights. In G' the numbers near the positive edges are their weights.

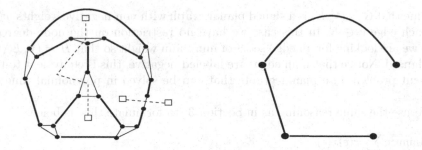

Fig. 3. A solution. The violated edges are crossed with a dashed line. We also show the corresponding induced bipartite subgraph of the original graph.

$$\text{minimize} \sum_{e \in \mathcal{A}} \lambda(e)x(e) \tag{17}$$

$$\sum_{e \in C \cap \mathcal{A}} x(e) \geq 1 \quad \text{for each cycle } C \text{ with } |C \cap \mathcal{N}| \text{ odd,} \tag{18}$$

$$x(e) \geq 0 \quad \text{for all } e \in \mathcal{A}. \tag{19}$$

Now we state the main result of this section.

Theorem 2. *The problem of finding a maximum weighted induced bipartite subgraph of a planar graph G, with $\Delta(G) = 3$, can be solved in $O(n^{3/2} \log n)$ time.*

4   Balancing Signed Graphs

A signed graph is called *balanced* if we can give the labels "a" and "b" to the nodes so that if and edge is positive, its endnodes have the same label; and if an edge is negative, its endnodes have different labels. Here we discuss how to apply the ideas of Section 3 for finding a maximum balanced subgraph.

4.1   The Node Deletion Case

Here we assume that $G = (V, E)$ is a signed planar graph with non-negative weights $w(v)$, for each node $v \in V$, and with $\Delta(G) = 3$. We are looking for a balanced induced subgraph of maximum total weight.

We use a construction similar to the one in Section 3. The edges of the graph G' that are not artificial, keep the same labels as their associated edges in G. Then we have to solve (14)-(16) as described in Section 3. Thus we can state the following.

Theorem 3. *The problem of finding a maximum weighted balanced induced subgraph of a planar graph G, with $\Delta(G) = 3$, can be solved in $O(n^{3/2} \log n)$ time.*

4.2   The Edge Deletion Case

Assume that $G = (V, E)$ is a signed planar graph with non-negative weights $w(e)$ for each edge $e \in E$. In this case we have no restriction on the node-degrees. Here we are looking for an edge set S of minimum weight so that $H = (V, E \setminus S)$ is balanced. Notice that if all edges are labeled negative, this is equivalent to the max-cut problem in a planar graph, that can be solved in polynomial time cf. [18].

We use the same reasoning as in Section 3, to formulate this as below.

$$\text{minimize} \sum_{e \in E} w(e)x(e)$$

$$\sum_{e \in C} x(e) \equiv \begin{cases} 1 & (\text{mod } 2) \quad \text{if } C \text{ is a cycle with an odd number of negative edges,} \\ 0 & (\text{mod } 2) \quad \text{if } C \text{ is a cycle with an even number of negative edges,} \end{cases}$$

$$x(e) \in \{0, 1\} \quad \text{for each edge } e \in E.$$

As before, this can be solved as a minimum T-join problem in the dual graph. Thus we have the following.

Theorem 4. *The problem of finding a maximum weighted balanced spanning subgraph of a planar graph, can be solved in $O(n^{3/2} \log n)$ time.*

5   Maximum Directed Induced Acyclic Subgraphs

Consider a planar directed graph $G = (V, A)$, with node-weights $\lambda(v) \geq 0$, for each node $v \in V$. We study the problem of finding a node set S of maximum weight that induces an acyclic subgraph. The complement of S is a directed feedback set. This problem is also known as the Directed Feedback Vertex Set Problem.

One can try a technique similar to the one in Section 3, splitting nodes, adding extra arcs keeping planarity, and using planar duality. We discuss here the limitations of this technique.

Consider the following simple transformation. Split each node v into v_1 and v_2, add the arc (v_1, v_2), replace every arc (u, v) with (u, v_1), and every arc (v, w) with (v_2, w). Let G' be this new graph. If G' is planar, we work with the dual graph as follows. Let D be the dual graph of G' regardless of the orientations of the arcs. Then for each arc a of G' let $a^\perp$ be the corresponding edge of the dual graph, we give an orientation to $a^\perp$, so that the pair $(a, a^\perp)$ forms a positively oriented basis of $I\!\!R^2$. Let $\vec{D}$ be the directed graph obtained with this orientation. Notice that directed cuts in $\vec{D}$ correspond to directed cycles of G'. Thus it follows from the Theorem of Lucchesi & Younger [23], that the following linear program has an optimal solution that is integer valued, moreover, this can be solved in polynomial time.

$$\text{minimize} \sum_{v \in V} \lambda(v) x(v) \tag{20}$$

$$\sum_{v \in C} x(v) \geq 1 \quad \text{for each directed cycle } C, \tag{21}$$

$$x(v) \geq 0 \quad \text{for all } v \in V. \tag{22}$$

To see this one should start with problem (7)-(9), associated with the graph $\vec{D}$. Then all variables associated with the original arcs are set to zero, and one obtains (20)-(22).

In particular, this transformation works when the degree of every node is at most three. Thus we have the following.

Theorem 5. *A maximum weighted induced acyclic subgraph of a planar directed graph D, with $\Delta(D) = 3$, can be found in $O(n^3)$ time.*

The following theorem shows the limits of this transformation.

Theorem 6. *The minimum feedback vertex set problem is NP-Complete for planar directed graphs D, with $\Delta(D) = 4$, and with in-degree and out-degree at most two.*

Proof. We use a construction similar to the one used in [6] for induced bipartite subgraphs. We start with the following NP-complete problem, see [21].

Planar 3-Satisfiability (P3SAT)
Instance: *A set $U = \{v_i \mid 1 \leq i \leq n\}$ of n boolean variables and a set $C = \{c_j \mid 1 \leq j \leq m\}$ of m clauses over U such that each clause contains exactly three variables or their complements. Furthermore, the following graph is planar:*

$$G_C = (V_C, E_C), \text{ where}$$
$$V_C = \{c_j \mid 1 \leq j \leq m\} \cup \{v_i \mid 1 \leq i \leq n\}, \text{ and}$$
$$E_C = \{c_j v_i \mid v_i \in c_j \text{ or } \bar{v}_i \in c_j\} \cup \{v_i v_{i+1} \mid 1 \leq i \leq n-1\} \cup \{v_n v_1\}.$$

Question: *Is there a truth assignment for U such that each clause in C is true?*

Given a planar embedding of G_C we build a planar directed graph $G = (V, A)$ as follows.

– Each node v_i associated with a variable is replaced by a subgraph called variable component. Its nodes are $\{a^i_{j,1}, \ldots, a^i_{j,8}\}$, for $1 \leq j \leq m$. The arcs are:
- $(a^i_{j,k}, a^i_{j,k+1})$, for $1 \leq k \leq 7$, for $1 \leq j \leq m$.
- $(a^i_{j,8}, a^i_{j+1,1})$, for $1 \leq j \leq m$, with $a^i_{m+1,1} = a^i_{1,1}$.
- $(a^i_{j,k+2}, a^i_{j,k})$, for $k = 1, 3, 5$, $(a^i_{j+1,1}, a^i_{j,7})$; for $1 \leq j \leq m$, with $a^i_{m+1,1} = a^i_{1,1}$.

See Figure 4. There are $4m$ triangles, a directed cycle C_1 with $8m$ nodes and a directed cycle C_2 with $4m$ nodes. The embedding is done so that C_1 is oriented clockwise, and C_2 is oriented counter-clockwise.

– Each node c_j associated with a clause is replaced by three nodes c^1_j, c^2_j, and c^3_j. Assume that v_{i_1}, v_{i_2} and v_{i_3} are the three variables (or their complement) that appear in c_j. Assume that they appear in clockwise order in the embedding of G_C. For each variable v_{i_k} we have two cases:
- If v_{i_k} appears in c_j, we add the arcs $(c^k_j, a^{i_k}_{j,2})$ and $(a^{i_k}_{j,4}, c^{k-1}_j)$, with $c^0_j = c^3_j$.
- If $\bar{v}_{i_k}$ appears in c_j, we add the arcs $(c^k_j, a^{i_k}_{j,4})$ and $(a^{i_k}_{j,6}, c^{k-1}_j)$, with $c^0_j = c^3_j$.

See Figure 5. These arcs are included in only one directed cycle D_j called a clause cycle.

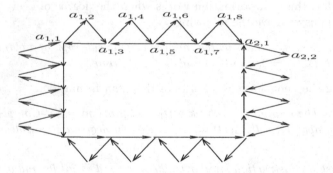

Fig. 4. Subgraph associated with a variable v_i, for $m = 4$. The index i is not shown.

For a variable v_i consider it associated component. We need the following observations.

– A node can cover at most two triangles, and since there are $4m$ triangles, a node-set covering all triangles has cardinality at least $2m$. The triangles can be covered with the nodes $\{a^i_{j,3}, a^i_{j,7}\}$ for $1 \leq j \leq m$, or $\{a^i_{j,1}, a^i_{j,5}\}$ for $1 \leq j \leq m$. Denote by S^i_1 the first set and by S^i_2 the second set. We have $|S^i_1| = |S^i_2| = 2m$. Also any other cycle included in this component is covered by these two sets.

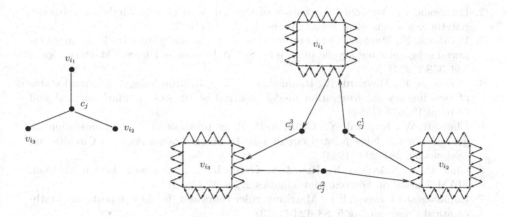

Fig. 5. Nodes and arcs associated with a clause c_j. In this example c_j contains v_{i_1}, v_{i_2} and $\bar{v}_{i_3}$.

- *Consider now any other set S of nodes covering all the triangles. Assume that S contains p nodes of degree two. Each of them covers exactly one triangle, so there are $4m - p$ triangles that should be covered with nodes of degree four. Since each of these nodes covers two triangles, we need at least $2m - p/2$ nodes of degree four. This shows that $|S| \geq 2m + p/2$.*
- *Consider now a node-set T containing only nodes of degree four and covering all triangles. If $|T| = 2m$, each node should cover two distinct triangles, this can only happen if $T = S_1^i$ or $T = S_2^i$. It follows that $2mn$ is a lower bound for the size of a minimum feedback set in G.*

Suppose now that there is an assignment of values to the variables so that each clause is true. If a variable v_i is set to true, we pick the set S_1^i; otherwise v_i is set to false and we pick S_2^i. Thus we obtain a node set F of size $2mn$ that covers every directed cycle contained in each variable component. Now consider a clause cycle D_j corresponding to a clause c_j. Since at least one of the variables included in c_j is set to true, this cycle is covered.

On the other hand if there is a feedback set S of size $2mn$, its restriction to the subgraph associated with a variable v_i is either the set S_1^i or S_2^i. In the first case we set v_i to true, and in the second case we set v_i to false. Since each clause cycle is covered, then each clause is set to true.

Acknowledgements. We are grateful to the referees for helping us to improve the presentation.

References

1. Barahona, F.: Planar multicommodity flows, max cut and the Chinese Postman Problem. In: Polyhedral Combinatorics. DIMACS Series on Discrete Mathematics and Theoretical Computer Science No. 1, pp. 189–202. DIMACS, NJ (1990)

2. Barahona, F., Mahjoub, A.: Facets of the balanced (acyclic) induced subgraph polytope. Mathematical Programming 45, 21–33 (1989)
3. Barahona, F., Mahjoub, A.: Compositions of graphs and polyhedra I: Balanced induced subgraphs and acyclic subgraphs. SIAM Journal on Discrete Mathematics 7, 344–358 (1994)
4. Barahona, F., Maynard, R., Rammal, R., Uhry, J.: Morphology of ground states of two-dimensional frustration model. Journal of Physics A: Mathematical and General 15, 673 (1982)
5. Chen, R.-W., Kajitani, Y., Chan, S.-P.: A graph-theoretic via minimization algorithm for two-layer printed circuit boards. IEEE Transactions on Circuits and Systems 30, 284–299 (1983)
6. Choi, H.-A., Nakajima, K., Rim, C.S.: Graph bipartization and via minimization. SIAM Journal on Discrete Mathematics 2, 38–47 (1989)
7. Edmonds, J., Johnson, E.L.: Matching, euler tours and the chinese postman. Mathematical Programming 5, 88–124 (1973)
8. Even, G., Naor, J.S., Schieber, B., Sudan, M.: Approximating minimum feedback sets and multicuts in directed graphs. Algorithmica 20, 151–174 (1998)
9. Festa, P., Pardalos, P.M., Resende, M.G.: Feedback set problems. In: Handbook of Combinatorial Optimization, pp. 209–258. Springer (1999)
10. Fouilhoux, P., Mahjoub, A.: Solving VLSI design and DNA sequencing problems using bipartization of graphs. Computational Optimization and Applications 51, 749–781 (2012)
11. Funke, M., Reinelt, G.: A polyhedral approach to the feedback vertex set problem. In: Integer Programming and Combinatorial Optimization, pp. 445–459. Springer (1996)
12. Gabow, H.N.: An efficient implementation of edmonds' algorithm for maximum matching on graphs. Journal of the ACM (JACM) 23, 221–234 (1976)
13. Gabow, H.N.: Centroids, representations, and submodular flows. Journal of Algorithms 18, 586–628 (1995)
14. Gardarin, G., Spaccapietra, S.: Integrity of data bases: A general lockout algorithm with deadlock avoidance. In: IFIP Working Conference on Modelling in Data Base Management Systems, pp. 395–412 (1976)
15. Garey, M.R., Johnson, D.S.: Computers and intractability: A guide to the theory of NP-completeness (1979)
16. Garg, N., Vazirani, V.V., Yannakakis, M.: Approximate max-flow min-(multi) cut theorems and their applications. SIAM Journal on Computing 25, 235–251 (1996)
17. Goemans, M.X., Williamson, D.P.: Primal-dual approximation algorithms for feedback problems in planar graphs. Combinatorica 18, 37–59 (1998)
18. Hadlock, F.: Finding a maximum cut of a planar graph in polynomial time. SIAM Journal on Computing 4, 221–225 (1975)
19. Karp, R.: Reducibility among combinatorial problems. In: Miller, R., Thatcher, J., Bohlinger, J. (eds.) Complexity of Computer Computations. The IBM Research Symposia Series, pp. 85–103. Springer US (1972)
20. Leiserson, C.E., Saxe, J.B.: Retiming synchronous circuitry. Algorithmica 6, 5–35 (1991)
21. Lichtenstein, D.: Planar formulae and their uses. SIAM Journal on Computing 11, 329–343 (1982)
22. Lipton, R.J., Tarjan, R.E.: A separator theorem for planar graphs. SIAM Journal on Applied Mathematics 36, 177–189 (1979)
23. Lucchesi, C., Younger, D.: A minimax theorem for directed graphs. J. London Math. Soc. 17(2), 369–374 (1978)

24. Lucchesi, C.L.: A minimax equality for directed graphs, PhD thesis, Thesis (Ph. D.)–University of Waterloo (1976)
25. Pinter, R.Y.: Optimal layer assignment for interconnect. Journal of VLSI and Computer Systems 1, 123–137 (1984)
26. Seymour, P.D.: On odd cuts and plane multicommodity flows. Proceedings of the London Mathematical Society 3, 178–192 (1981)
27. Silberschatz, A., Peterson, J.L., Galvin, P.B.: Operating system concepts. Addison-Wesley Longman Publishing Co., Inc. (1991)

n-Step Cycle Inequalities: Facets for Continuous n-Mixing Set and Strong Cuts for Multi-Module Capacitated Lot-Sizing Problem

Manish Bansal and Kiavash Kianfar

Department of Industrial and Systems Engineering,
Texas A&M University, College Station, TX-77843-3131
{bansal,kianfar}@tamu.edu

Abstract. In this paper, we introduce a generalization of the well-known continuous mixing set (which we refer to as the continuous n-mixing set) $Q^{m,n} := \{(y, v, s) \in \left(\mathbb{Z} \times \mathbb{Z}_+^{n-1}\right)^m \times \mathbb{R}_+^{m+1} : \sum_{t=1}^{n} \alpha_t y_t^i + v_i + s \geq \beta_i, i = 1, \ldots, m\}$. This set is closely related to the feasible set of the *multi-module capacitated lot-sizing* (MML) problem with(out) backlogging. For each $n' \in \{1, \ldots, n\}$, we develop a class of valid inequalities for this set, referred to as the n'-step cycle inequalities, and show that they are facet-defining for $conv(Q^{m,n})$ in many cases. We also present a compact extended formulation for $Q^{m,n}$ and an exact separation algorithm for the n'-step cycle inequalities. We then use these inequalities to generate valid inequalities for the MML problem with(out) backlogging. Our computational results show that our cuts are very effective in solving the MML instances with backlogging, resulting in substantial reduction in the integrality gap, number of nodes, and total solution time.

Keywords: n-step cycle inequalities, n-step MIR, continuous n-mixing, multi-module capacitated lot-sizing with backlogging.

1 Introduction

Polyhedral study of the mixed integer "base" sets which constitute well-structured relaxations of important mixed integer programming (MIP) problems is a promising approach in developing strong cutting planes for these MIP problems. This is because oftentimes one can develop procedures in which the valid inequalities (or facets) developed for the base set are used to generate valid inequalities (or facets) for the original MIPs (see [1, 5–7, 11, 15] for a few examples among many others).

One mixed integer base set studied for this purpose is the continuous mixing set

$$Q := \{(y, v, s) \in \mathbb{Z}^m \times \mathbb{R}_+^{m+1} : y^i + v_i + s \geq \beta_i, i = 1, \ldots, m\},$$

where $\beta_i \in \mathbb{R}$, $i = 1, \ldots, m$ [14]. This set is a generalization of the well-studied mixing set $\{(y, s) \in \mathbb{Z}^m \times \mathbb{R}_+ : y^i + s \geq \beta_i, i = 1, \ldots, m\}$ [6], which itself is a multi-constraint generalization of the base set $\{(y, s) \in \mathbb{Z} \times \mathbb{R}_+ : y + s \geq \beta\}$ that

J. Lee and J. Vygen (Eds.): IPCO 2014, LNCS 8494, pp. 102–113, 2014.

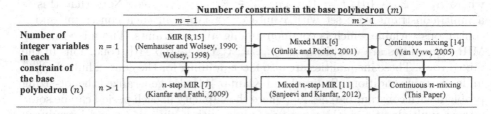

Fig. 1. Generalizations of Mixed Integer Rounding (MIR)

leads to the well-known mixed integer rounding (MIR) inequality (page 127 of [15]). In all these base sets each constraint has only one integer variable. Fig. 1 presents a summary of the relationship between these base sets and other base sets of interest in this paper. The set Q arises as a substructure in relaxations of problems such as lot-sizing (production planning) with backlogging and lot-sizing with stochastic demand. Van Vyve [14] introduced the so-called cycle inequalities for this set and showed that these inequalities along with bound constraints are sufficient to describe the convex hulls of this set. The MIR inequalities (called *1-step* MIR inequalities in this paper) of Nemhauser and Wolsey [8, 15] and the mixed (1-step) MIR inequalities of Günlük and Pochet [6] are special cases of the cycle inequalities for Q (Fig. 1).

In another direction (Fig. 1), Kianfar and Fathi [7] generalized the 1-step MIR inequalities [8] and developed the *n*-step MIR inequalities for the mixed integer knapsack set by studying the base set

$$Q_0^{1,n} = \Big\{(y,s) \in \mathbb{Z} \times \mathbb{Z}_+^{n-1} \times \mathbb{R}_+ : \sum_{t=1}^n \alpha_t y_t + s \geq \beta\Big\},$$

where $\alpha_t \in \mathbb{R}_+ \backslash \{0\}$, $t = 1, \ldots, n$ and $\beta \in \mathbb{R}$. Note that this base set has a single constraint and n integer variables in this constraint. The *n*-step MIR inequalities are valid and facet-defining for the base set $Q_0^{1,n}$ if α_t's and β satisfy the so-called *n*-step MIR conditions (see conditions (4) in Section 2). However, *n*-step MIR inequalities can also be generated for a mixed integer constraint with no conditions imposed on the coefficients. In that case, the external parameters used in generating the inequality are picked such that they satisfy the *n*-step MIR conditions (see [7] for more details). The *n*-step MIR inequalities are facet-defining for the mixed integer knapsack set in many cases [2, 7]. The Gomory mixed integer cut [10] and the 2-step MIR inequalities [5] are the special cases of *n*-step MIR inequalities, corresponding to $n = 1, 2$, respectively.

Recently, Sanjeevi and Kianfar [11] showed that the procedure proposed by Günlük and Pochet [6] to mix 1-step MIR inequalities can be generalized and used to mix the *n*-step MIR inequalities [7] (Fig. 1). As a result, they developed the mixed *n*-step MIR inequalities for a generalization of the mixing set called the *n*-mixing set, i.e.

$$Q_0^{m,n} = \Big\{(y,s) \in (\mathbb{Z} \times \mathbb{Z}_+^{n-1})^m \times \mathbb{R}_+ : \sum_{t=1}^n \alpha_t y_t^i + s \geq \beta_i, i = 1, \ldots, m\Big\},$$

where $\alpha_t \in \mathbb{R}_+ \setminus \{0\}$, $t = 1, \ldots, n$, $\beta_i \in \mathbb{R}$, $i = 1, \ldots, m$. Note that this is a multi-constraint base set with n integer variables in each constraint and a continuous variable which is common among all constraints. The mixed n-step MIR inequalities are valid if α_t and β_i satisfy the n-step MIR conditions in the mixed rows. These inequalities are also facet-defining for the convex hull of $Q_0^{m,n}$ under certain conditions. Sanjeevi and Kianfar [11] also generalized the lot-sizing problem with constant batches [9] (where the capacity in each period can be some integer multiple of a single capacity module with a given size) and introduced the multi-module capacitated lot-sizing (MML) problem. In this generalization, the production capacity in each period can be the summation of some integer multiples of several capacity modules of different sizes. They showed that the mixed n-step MIR inequalities can be used to generate valid inequalities for the MML problem *without backlogging* (which we denote by MML-WB). These inequalities generalize the (k, l, S, I) inequalities which were introduced for the lot-sizing problem with constant batches by Pochet and Wolsey [9].

In this paper, we generalize the concepts of continuous mixing [14] and mixed n-step MIR [11] by introducing a more general base set referred to as the *continuous n-mixing set* which we define as

$$Q^{m,n} := \left\{ (y, v, s) \in (\mathbb{Z} \times \mathbb{Z}_+^{n-1})^m \times \mathbb{R}_+^{m+1} : \sum\nolimits_{t=1}^n \alpha_t y_t^i + v_i + s \geq \beta_i, i = 1, \ldots, m \right\},$$

where $\alpha_t > 0$, $t = 1, \ldots, n$ and $\beta_i \in \mathbb{R}$, $i = 1, \ldots, m$ (see Fig. 1). Note that this set has multiple (m) constraints with multiple (n) integer variables in each constraint; but it is more general than the n-mixing set because in addition to the common continuous variable s, each constraint has a continuous variable v_i of its own. The continuous mixing set Q is the special case of $Q^{m,n}$, where $n = 1$ and $\alpha_1 = 1$, and the n-mixing set of Sanjeevi and Kianfar [11] is the projection of $Q^{m,n} \cap \{v = 0\}$ on (y, s). The continuous n-mixing set arises as a substructure in relaxations of MML-WB and MML *with backlogging* (MML-B). For each $n' \in \{1, \ldots, n\}$, we develop a class of valid inequalities for $Q^{m,n}$ which we refer to as n'-step cycle inequalities (Section 3), and obtain conditions under which these inequalities are facet-defining for $conv(Q^{m,n})$ (Section 4). Note that the n-step MIR inequalities [7] and the mixed n-step MIR inequalities [11] are special cases of the n-step cycle inequalities. In Section 5, we introduce a compact extended formulation for $Q^{m,n}$ and an efficient exact separation algorithm to separate over the set of all n'-step cycle inequalities for set $Q^{m,n}$. Then, in Section 6, we use these inequalities to generate valid inequalities for the MML-WB and MML-B. Our computational results in Section 7 on applying 2-step cycle inequalities using our separation algorithm show that our cuts are very effective in solving MML-B with two capacity modules $(n = 2)$ resulting in considerable reduction in the integrality gap (87.7% in average), the number of nodes (9.9 times in average). Also, the total time (which also includes the cut generation time) taken to solve an instance is in average 4.2 times smaller than the time taken by CPLEX with default settings (except for very easy instances).

2   Necessary Background

In this section, we briefly review the cycle inequalities for the continuous mixing set [14] and the *n*-step MIR inequalities [7] to the extent required as background for the results in this paper.

Van Vyve [14] generated the cycle inequalities for the continuous mixing set Q as follows: Define $\beta_0 := 0$, $f_i := \beta_i - \lfloor \beta_i \rfloor$, $i \in \{0, \ldots, m\}$, and without loss of generality assume that $f_{i-1} \leq f_i$, $i = 1, \ldots, m$. Let $G := (V, A)$ be a directed graph, where $V := \{0, 1, \ldots, m\}$ and $A := \{(i, j) : i, j \in V, f_i \neq f_j\}$. Note that G is a complete graph except for the arcs (i, j) where $f_i = f_j$. An arc $(i, j) \in A$ is called a forward arc if $i < j$ and a backward arc if $i > j$. To each arc $(i, j) \in A$, associate a linear function $\psi_{ij}(y, v, s)$ defined as

$$\psi_{ij}(y, v, s) := \begin{cases} s + v_i + (f_i - f_j + 1)(y^i - \lfloor \beta_i \rfloor) - f_j & \text{if } (i, j) \text{ is a forward arc,} \\ v_i + (f_i - f_j)(y^i - \lfloor \beta_i \rfloor) & \text{if } (i, j) \text{ is a backward arc,} \end{cases}$$

where $v_0 = y^0 = 0$.

Theorem 1 ([14]). *Given an elementary cycle $C = (V_C, A_C)$ in the graph G, the inequality*

$$\sum_{(i,j) \in A_C} \psi_{ij}(y, v, s) \geq 0, \tag{1}$$

referred to as the cycle inequality, is valid for Q.   □

In [14], the validity of the cycle inequality (1) was proved indirectly through the following extended formulation for Q:

$$Q^\delta = \{(y, v, s, \delta) \in \mathbb{R}^m \times \mathbb{R}^m_+ \times \mathbb{R}_+ \times \mathbb{R}^{m+1} :$$
$$\psi_{ij}(y, v, s) \geq \delta_i - \delta_j \text{ for all } (i, j) \in A,$$
$$y^i + v_i + s \geq \beta_i, i = 1, \ldots, m\}.$$

Note that the set of all original inequalities, all cycle inequalities, along with the bound constraints $v, s \geq 0$, define $\text{Proj}_{y,v,s}(Q^\delta)$. Van Vyve [14] showed that $conv(Q) = \text{Proj}_{y,v,s}(Q^\delta)$, which proves Theorem 1. Furthermore, he showed that the separation over $conv(Q)$ can be performed in $O(m^3)$ time by finding a negative weight cycle in G. Similar results were presented for the relaxation of Q to the case where $s \in \mathbb{R}$.

In another direction, Kianfar and Fathi [7] developed the *n*-step MIR inequalities (a generalization of MIR inequalities [8, 15]) for the set $Q_0^{1,n}$. Note that $Q_0^{1,n} = \text{Proj}_{y,s}(Q^{1,n} \cap \{v = 0\})$. The *n*-step MIR inequality for this set is

$$\beta^{(n)} \sum_{t=1}^n \prod_{l=t+1}^n \left\lceil \frac{\beta^{(l-1)}}{\alpha_l} \right\rceil y_t + s \geq \beta^{(n)} \prod_{l=1}^n \left\lceil \frac{\beta^{(l-1)}}{\alpha_l} \right\rceil, \tag{2}$$

where the recursive remainders $\beta^{(t)}$ are defined as

$$\beta^{(t)} := \beta^{(t-1)} - \alpha_t \left\lfloor \beta^{(t-1)}/\alpha_t \right\rfloor, \quad t = 1, \ldots, n, \tag{3}$$

and $\beta^{(0)} := \beta$ (note that $0 \leq \beta^{(t)} < \alpha_t$ for $t = 1, \ldots, n$). By definition if $a > b$, then $\sum_a^b(.) = 0$ and $\prod_a^b(.) = 1$. For inequality (2) to be non-trivial, we assume that $\beta^{(t-1)}/\alpha_t \notin \mathbb{Z}, t = 1, \ldots, n$. Kianfar and Fathi [7] showed that the n-step MIR inequality (2) is valid and facet-defining for the convex hull of $Q_0^{1,n}$ if the so-called n-step MIR conditions, i.e.

$$\alpha_t \left\lceil \beta^{(t-1)}/\alpha_t \right\rceil \leq \alpha_{t-1}, \quad t = 2, \ldots, n, \tag{4}$$

hold. As mentioned in Section 1, they also used inequalities (2) to generate n-step MIR inequalities for single-constraint mixed integer sets with no conditions on the coefficients. Later, Atamtürk and Kianfar [2] showed that those inequalities also have facet-defining properties in several cases (refer to [2, 7] for more details).

3 n-Step Cycle Inequalities for Continuous n-Mixing Set

In this section, we unify the concepts of continuous mixing [14] and n-step MIR [7] by studying the continuous n-mixing set

$$Q^{m,n} := \left\{ (y,v,s) \in (\mathbb{Z} \times \mathbb{Z}_+^{n-1})^m \times \mathbb{R}_+^{m+1} : \sum_{t=1}^n \alpha_t y_t^i + v_i + s \geq \beta_i, i = 1, \ldots, m \right\}$$

introduced in Section 1. $Q^{m,n}$ is a generalization of both the continuous mixing set Q and the n-mixing set $Q_0^{m,n}$. We will show that for each $n' \in \{1, \ldots, n\}$, there exist a family of valid inequalities for $Q^{m,n}$, which we refer to as the n'-step cycle inequalities. In proving the validity of these inequalities, Theorem 1 will become necessary. As mentioned before, Van Vyve [14] proved Theorem 1 indirectly by defining the extended formulation Q^δ and showing that every extreme point (ray) of the set Q has a counterpart in Q^δ (see [14] for details). We have developed a direct proof for Theorem 1, which only uses the original inequalities and the cycle structure. We believe this proof can be insightful in further pursuit of research in this area. Here, we present an alternative form of Theorem 1 and provide the brief sketch of our proof:

Lemma 1. *Let* $C = (V_C, A_C)$ *be a directed Hamiltonian cycle over* q *nodes, where* $V_C = \{1, \ldots, q\}$, $A_C := \{(1, i_2), (i_2, i_3), \ldots, (i_q, 1)\}$, *and* $i_2, \ldots, i_q \in \{2, \ldots, q\}$ *are distinct. Let* $\sigma \in \mathbb{R}, \alpha \in \mathbb{R}_+$, *and to each node* $i \in \{1, \ldots, q\}$ *assign the values* $\omega_i \in \mathbb{R}_+$, $\kappa_i \in \mathbb{Z}$, *and* $\gamma_i \in \mathbb{R}_+$ *such that* $\gamma_i < \alpha, i = 1, \ldots, q$, *and* $\gamma_{i-1} < \gamma_i, i = 2, \ldots, q$. *If*

$$\sigma + \omega_i + \alpha \kappa_i \geq \gamma_i \quad i = 1, \ldots, q, \tag{5}$$

then the cycle inequality

$$\sum_{(i,j) \in F} (\sigma + \omega_i - \gamma_j + (\gamma_i - \gamma_j + \alpha) \kappa_i) + \sum_{(i,j) \in B} (\omega_i + (\gamma_i - \gamma_j) \kappa_i) \geq 0, \tag{6}$$

is valid, where F and B are the sets of forward and backward arcs in A_C, respectively (i.e. $F = \{(i,j) \in A_C : i < j\}$ and $B = \{(i,j) \in A_C : i > j\}$).

Sketch of Proof. For $p \in \{1, \ldots, q\}$, let A_p be the arcs in the path from 1 to i_{p+1} in C, i.e. $A_p := \{(1, i_2), (i_2, i_3), \ldots, (i_p, i_{p+1})\}$ (we define $i_{q+1} := 1$). Denote the set of forward and backward arcs in A_p by F_p and B_p, respectively (note that if $p' < p$, then $A_{p'} \subset A_p$, $F_{p'} \subseteq F_p$, and $B_{p'} \subseteq B_p$). Also, let $T(.)$ be an operator that, when applied on an arc set, returns the set of tail nodes of the arcs in that arc set. Define the index $g_p \in \{i_1, \ldots, i_p\}$ recursively as follows: $g_1 := 1$, and

$$g_p := \begin{cases} g_{p-1} & \text{if } i_p \in T(F_p), g_{p-1} \in T(F_{p-1}), \kappa_{g_{p-1}} \geq \kappa_{i_p}, \\ i_p & \text{if } i_p \in T(F_p), g_{p-1} \in T(F_{p-1}), \kappa_{g_{p-1}} < \kappa_{i_p}, \\ g_{p-1} & \text{if } i_p \in T(F_p), g_{p-1} \in T(B_{p-1}), \kappa_{g_{p-1}} > \kappa_{i_p}, \\ i_p & \text{if } i_p \in T(F_p), g_{p-1} \in T(B_{p-1}), \kappa_{g_{p-1}} \leq \kappa_{i_p}, \\ g_{p-1} & \text{if } i_p \in T(B_p), g_{p-1} \in T(B_{p-1}), \kappa_{g_{p-1}} \leq \kappa_{i_p}, \\ i_p & \text{if } i_p \in T(B_p), g_{p-1} \in T(B_{p-1}), \kappa_{g_{p-1}} > \kappa_{i_p}, \\ g_{p-1} & \text{if } i_p \in T(B_p), g_{p-1} \in T(F_{p-1}), \kappa_{g_{p-1}} < \kappa_{i_p}, \\ i_p & \text{if } i_p \in T(B_p), g_{p-1} \in T(F_{p-1}), \kappa_{g_{p-1}} \geq \kappa_{i_p}, \end{cases}$$

for $p = 2, \ldots, q$ and for $p = 1, \ldots, q$, define

$$\Delta_p := \begin{cases} \gamma_{g_p} - \gamma_{i_{p+1}} & \text{if } g_p \in T(B_p), \\ 0 & \text{if } g_p \in T(F_p). \end{cases}$$

In proving this theorem, we first show that the inequality

$$\sum_{(i,j) \in F_p} (\gamma_i - (\gamma_i - \gamma_j + \alpha)\kappa_i) + \sum_{(i,j) \in B_p} (\gamma_i - \gamma_j)(1 - \kappa_i)$$

$$\leq (|F_p| - 1)\sigma + \sum_{i \in T(A_p) \setminus \{g_p\}} \omega_i - (\gamma_1 - \gamma_{i_{p+1}} + \alpha)\kappa_{g_p} + \gamma_1 + \Delta_p, \tag{7}$$

is valid for $p = 1, \ldots, q$. Then we show that for $p = q$, inequality (7) becomes inequality (6) which completes the proof. □

Now given $n' \in \{1, \ldots, n\}$, we develop the n'-step cycle inequalities for $Q^{m,n}$ as follows: Without loss of generality, we assume $\beta_{i-1}^{(n')} \leq \beta_i^{(n')}$, $i = 2, \ldots, m$, where $\beta_i^{(n')}$ is defined as (3). Also define $\beta_0 := 0$. Now similar to the graph defined for the cycle inequalities (see Section 2), here we define a directed graph $G_{n'} = (V, A)$, where $V := \{0, 1, \ldots, m\}$ and $A := \{(i,j) : i, j \in V, \beta_i^{(n')} \neq \beta_j^{(n')}\}$. $G_{n'}$ is a complete graph except for the arcs (i,j) where $\beta_i^{(n')} = \beta_j^{(n')}$. Here to each arc $(i,j) \in A$, we associate the linear function $\psi_{ij}^{n'}(y, v, s)$ defined as

$$\psi_{ij}^{n'}(y, v, s) := \begin{cases} s + v_i + \sum\limits_{t=n'+1}^{n} \alpha_t y_t^i + \beta_{ij}^{(n')}\left(1 - \phi_i^{n'}(y^i)\right) - \beta_j^{(n')} & \text{if } i < j, \\ v_i + \sum\limits_{t=n'+1}^{n} \alpha_t y_t^i + \left(\beta_i^{(n')} - \beta_j^{(n')}\right)\left(1 - \phi_i^{n'}(y^i)\right) & \text{if } i > j, \end{cases} \tag{8}$$

where $\beta_{ij}^{(n')} := \beta_i^{(n')} - \beta_j^{(n')} + \alpha_{n'}$ for all $(i,j) \in A$, $i < j$, the functions $\phi_i^{n'}(y^i)$, $i = 1, \ldots, m$, are defined as

$$\phi_i^{n'}(y^i) := \prod_{l=1}^{n'} \left\lceil \frac{\beta_i^{(l-1)}}{\alpha_l} \right\rceil - \sum_{t=1}^{n'} \prod_{l=t+1}^{n'} \left\lceil \frac{\beta_i^{(l-1)}}{\alpha_l} \right\rceil y_t^i, \qquad (9)$$

and by definition, $v_0 := 0$, $y^0 := 0$, and $\phi_0^{n'}(y^0) := 1$. We can show that each elementary cycle of graph $G_{n'}$ corresponds to a valid inequality for the set $Q^{m,n}$, which we refer to as the n'-step cycle inequality.

Theorem 2. *Given* $n' \in \{1, \ldots, n\}$ *and an elementary cycle* $C = (V_C, A_C)$ *of graph* $G_{n'}$, *the* n'-*step cycle inequality*

$$\sum_{(i,j) \in A_C} \psi_{ij}^{n'}(y,v,s) \geq 0 \qquad (10)$$

is valid for $Q^{m,n}$ *if the* n'-*step MIR conditions for* $i \in V_C$, *i.e.*

$$\alpha_t \left\lceil \beta_i^{(t-1)}/\alpha_t \right\rceil \leq \alpha_{t-1}, t = 2, \ldots, n', i \in V_C, \qquad (11)$$

hold. □

Remark: For the special case where the parameters $\alpha_1, \ldots, \alpha_{n'}$ are divisible, i.e. $\alpha_t | \alpha_{t-1}$, $t = 2, \ldots, n'$, the n'-step MIR conditions are automatically satisfied no matter what the value of β_i is.

Special Cases: The n-step MIR inequalities of Kianfar and Fathi [7] and the mixed n-step MIR inequalities of Sanjeevi and Kianfar [11] are special cases of the n-step cycle inequalities.

4   Facet-Defining n-Step Cycle Inequalities

In this section, we show that for any $n' \in \{1, \ldots, n\}$, the n'-step cycle inequalities define facets for $conv(Q^{m,n})$ under certain conditions. For $i, j \in \{1, \ldots, m\}$ such that $\beta_i^{(n')} > \beta_j^{(n')}$, we define $\beta_{ij}^{(n',n')} := \beta_i^{(n')} - \beta_j^{(n')}$ and $\beta_{ij}^{(n',t)} := \beta_{ij}^{(n',t-1)} - \alpha_t \left\lfloor \beta_{ij}^{(n',t-1)}/\alpha_t \right\rfloor$, $t = n'+1, \ldots, n$.

Theorem 3. *For* $n' \in \{1, \ldots, n\}$, *the* n'-*step cycle inequality* (10) *for an elementary cycle* $C = (V_C, A_C)$ *of graph* $G_{n'}$ *is facet-defining for* $conv(Q^{m,n})$ *if (in addition to the* n'-*step MIR conditions* (11)*) the following conditions hold*

(a) $\left\lfloor \beta_k^{(d-1)}/\alpha_d \right\rfloor \geq 1$, $d = 2, \ldots, n$, *for all* $(k,l) \in F$,

(b) $\beta_l^{(n')} - \beta_k^{(n')} \geq \max \left\{ \alpha_{d-1} - \alpha_d \left\lceil \beta_k^{(d-1)}/\alpha_d \right\rceil, d = 2, \ldots, n' \right\}$ *for all* $(k,l) \in F$,

(c) $\left\lfloor \beta_{kl}^{(n',d-1)}/\alpha_d \right\rfloor \geq 1$, $d = n'+1, \ldots, n$, *for all* $(k,l) \in B$. □

Theorem 4. *For $n' \in \{1, \ldots, n\}$, the n'-step cycle inequality (10) for an elementary cycle $C = (V_C, A_C)$ of graph $G_{n'}$ is facet-defining for $conv(Q^{m,n})$ if (in addition to the n'-step MIR conditions (11)) the following conditions hold*

(a) $T(F) = \{0\}$,

(b) $\left\lfloor \beta_{kl}^{(n',d-1)}/\alpha_d \right\rfloor \geq 1$, $d = n' + 1, \ldots, n$, *for all* $(k, l) \in B$.   □

Remark: There will be no condition (*c*) in Theorem 3 and no condition (*b*) in Theorem 4 for the case of $n' = n$.

Example 1. Consider a continuous 2-mixing set with 3 rows $Q^{3,2} = \{(y^1, y^2, y^3, v, s) \in (\mathbb{Z} \times \mathbb{Z}_+^1)^3 \times \mathbb{R}_+^3 \times \mathbb{R}_+ : 31y_1^1 + 10y_2^1 + v_1 + s \geq 89, 31y_1^2 + 10y_2^2 + v_2 + s \geq 59, 31y_1^3 + 10y_2^3 + v_3 + s \geq 29\}$. Therefore $\alpha = (\alpha_1, \alpha_2) = (31, 10)$, $\beta_1 = 89$, $\beta_2 = 59$, $\beta_3 = 29$, and we have $\beta_1^{(1)} = 27$, $\beta_1^{(2)} = 7$, $\beta_2^{(1)} = 28$, $\beta_2^{(2)} = 8$, $\beta_3^{(1)} = 29$, and $\beta_3^{(2)} = 9$. Note that $\left\lceil \beta_1^{(1)}/\alpha_2 \right\rceil = \left\lceil \beta_2^{(1)}/\alpha_2 \right\rceil = \left\lceil \beta_3^{(1)}/\alpha_2 \right\rceil = 3$ and $\beta_1^{(2)} < \beta_2^{(2)} < \beta_3^{(2)}$. Now, consider a complete directed graph $G_2 = (V, A)$, where $V = \{0, \ldots, 3\}$. The linear function $\psi_{ij}^2(y, v, s)$ associated with each arc $(i, j) \in A$ is defined by (8) where $n' = 2$ and $\phi_i^2(y^i) = \left\lceil \beta_i^{(1)}/\alpha_2 \right\rceil \lceil \beta_i/\alpha_1 \rceil - \left\lceil \beta_i^{(1)}/\alpha_2 \right\rceil y_1^i - y_2^i$, for $i = 1, \ldots, 3$. The 2-step MIR conditions (11) are satisfied. Therefore, the 2-step cycle inequalities corresponding to the cycles in graph G_2 are valid for $Q^{3,2}$. For $n' = 2$, the additional conditions required in Theorem 3 are also satisfied, i.e. (*a*) $\left\lfloor \beta_k^{(1)}/\alpha_2 \right\rfloor = 2 \geq 1$, for $k \in \{1, 2, 3\}$, (*b*) $\beta_l^{(2)} - \beta_k^{(2)} \geq 1 = \alpha_1 - \alpha_2 \left\lceil \beta_k^{(1)}/\alpha_2 \right\rceil$ for all $(k, l) \in A$, $1 \leq k < l \leq 3$, and there is no condition (*c*) for $n' = n = 2$. Therefore, the 2-step cycle inequality (10) corresponding to each cycle $C = (V_C, A_C)$ in graph G_2, where $V_C \subseteq \{1, 2, 3\}$, defines a facet for $conv(Q^{3,2})$. Moreover, based on Theorem 4, the 2-step cycle inequality (10) corresponding to each cycle $C = (V_C, A_C)$ in graph G_2, where $T(F) = \{0\}$, also defines a facet for $conv(Q^{3,2})$ because there is no condition (*b*) for $n' = n = 2$.   □

5   Extended Formulation and Separation Algorithm

Theorem 5. *The following linear program is a compact extended formulation for $Q^{m,n}$, if conditions (11) hold.*

$$\psi_{ij}^{n'}(y, v, s) \geq \delta_i^{n'} - \delta_j^{n'} \text{ for all } (i, j) \in A, n' \in \{1, \ldots, n\}$$

$$\sum_{t=1}^n \alpha_t y_t^i + v_i + s \geq \beta_i, i = 1, \ldots, m$$

$$y \in (\mathbb{R} \times \mathbb{R}_+^{n-1})^m, v \in \mathbb{R}_+^m, s \in \mathbb{R}_+, \delta \in \mathbb{R}^{n(m+1)}.$$ □

Separation Algorithm. Given a point $(\hat{y}, \hat{v}, \hat{s})$ and $n' \in \{1, \ldots, n\}$, it is possible to solve the exact separation problem over all the n'-step cycle inequalities for the set $Q^{m,n}$. The goal is to find an n'-step cycle inequality (10) that is violated by $(\hat{y}, \hat{v}, \hat{s})$, if any. This can be done by detecting a negative weight cycle (if any) in the directed graph $G_{n'} = (V, A)$ with weights $\psi_{ij}^{n'}(\hat{y}, \hat{v}, \hat{s})$ for each arc $(i, j) \in A$. This means that the most negative cycle in $G_{n'}$ (if it exists) corresponds to the n'-step cycle inequality that is most violated by $(\hat{y}, \hat{v}, \hat{s})$. However, the problem of finding the most negative cycle in a graph is strongly NP-hard [12]. A method proposed by Cherkassy and Goldberg [3] (which is a combination of the cycle detection strategy of Tarjan [13] and the Bellman-Ford-Moore's labeling algorithm [4]), denoted by BFCT, is one of the fastest known algorithms to detect a negative cycle. BFCT terminates when it finds the first negative cycle; however, there may be cycles with smaller weight in the graph which would lead to stronger inequalities. Therefore, we devised a modified version of BFCT, denoted by MBFCT, which does not stop after finding the first negative cycle and continues the search for other negative cycles (if any) until a certain termination condition is satisfied. Out of all the cycles found by MBFCT, the one with the most negative weight is used to generate the n'-step cycle inequality (10) that separates $(\hat{y}, \hat{v}, \hat{s})$ with the largest violation among all generated cycles.

6   Cuts for Multi-Module Capacitated Lot-Sizing Problem

In this section, we use n-step cycle inequalities to develop cutting planes for MML-(W)B problem. We define MML-B as follows. Let $P := \{1, \ldots, m\}$ be the set of time periods and $\{\alpha_1, \ldots, \alpha_n\}$ be the set of sizes of the n available capacity modules. The setup cost per module of size $\alpha_t, t = 1, \ldots, n$ in period p is denoted by f_p^t. Given the demand, the production per unit cost, the inventory per unit cost, and the per unit shortage (backlog) cost in period p, denoted by d_p, c_p, h_p, and b_p, respectively, the MML-B problem can be formulated as:

$$\min \sum_{p \in P} c_p x_p + \sum_{p \in P} h_p s_p + \sum_{p \in P} b_p r_p + \sum_{p \in P} \sum_{t=1}^{n} f_p^t z_p^t \tag{12}$$

$$s_{p-1} - r_{p-1} + x_p = d_p + s_p - r_p, p \in P \tag{13}$$

$$x_p \leq \sum_{t=1}^{n} \alpha_t z_p^t, p \in P \tag{14}$$

$$(z, x, r, s) \in \mathbb{Z}_+^{m \times n} \times \mathbb{R}_+^m \times \mathbb{R}_+^{m+1} \times \mathbb{R}_+^{m+1} \tag{15}$$

where x_p is the production in period p, s_p and r_p are the inventory and backlog, respectively, at the end of period p, $s_0 = r_m = 0$, and z_p^t is the number of capacity modules of size $\alpha_t, t = 1, \ldots, n$, used in period p. Let X^{MML-B} denote the set of feasible solutions to constraints (13)-(15). Note that every valid inequality for X^{MML-B} also gives a valid inequality for the set of feasible solutions to the MML-WB problem which is the projection of $X^{MML-B} \cap \{r = 0\}$ on (z, x, s).

In order to generate valid inequalities for X^{MML-B}, we consider periods $k, \ldots, l$, for any $k, l \in P$ where $k < l$. Let $S \subseteq \{k, \ldots, l\}$ such that $k \in S$. For $i \in S$, let $S_i := S \cap \{k, \ldots, i\}$, $m_i = \min\{p : p \in S \setminus S_i\}$ with $m_i = l + 1$ if $S \setminus S_i = \emptyset$, and $b_i = \sum_{p=k}^{m_i - 1} d_p$. Now, by adding equalities (13) from period

k to period $m_i - 1$, relaxing $x_p, p \in S_i$, to its upper bound based on (14) and dropping $r_{k-1}, s_{m_i-1} (\geq 0)$, we get the following valid inequality:

$$s_{k-1} + r_{m_i-1} + \sum_{p \in \{k,...,m_i-1\} \backslash S_i} x_p + \sum_{t=1}^{n} \alpha_t \sum_{p \in S_i} z_p^t \geq b_i. \qquad (16)$$

Notice that inequality (16) is of the same form as the defining inequalities of the continuous n-mixing set $Q^{m,n}$ where $s = s_{k-1}, v_i = r_{m_i-1} + \sum_{p \in \{k,...,m_i-1\} \backslash S_i} x_p$, $y_t^i = \sum_{p \in S_i} z_p^t$, and $\beta_i = b_i$ (notice that $s, v_i \in \mathbb{R}_+, y_t^i \in \mathbb{Z}_+, t = 1, \ldots, n$). Therefore we can form a set of base inequalities consisting of inequalities (16) for all $i \in S$ such that the n-step MIR conditions, i.e. $\alpha_t \left\lceil b_i^{(t-1)} / \alpha_t \right\rceil \leq \alpha_{t-1}, t = 2, \ldots, n$, hold. We construct a directed graph for these base inequalities in the same fashion as we did for the continuous n-mixing set $Q^{m,n}$ in Section 3. The n-step cycle inequalities corresponding to each elementary cycle C in this graph will be valid for X^{MML-B}. We refer to these inequalities as the n-step (k, l, S, C) cycle inequalities. The same procedure also provides a new class of valid inequalities for MML-WB which subsume the valid inequalities generated using the mixed n-step MIR inequalities [11] for MML-WB.

Note that a procedure similar to what was presented above for n can also be used to develop n'-step (k, l, S, C) cycle inequalities for MML-(W)B problem for any $n' \in \{1, \ldots, n\}$ in general.

7 Computational Results

In this section, we computationally evaluate the effectiveness of the n-step cycle inequalities for the MML-B problem using our separation algorithm (discussed in Section 5). We chose $n = 2$ for our experiments in this paper and refer to the MML-B problem with two capacity modules ($n = 2$) as 2ML-B. We created random 2ML-B instances with 60 time periods, i.e. $P = \{1, \ldots, 60\}$, and varying cost and capacity characteristics. The demand d_p, production cost c_p, and holding cost h_p in each period were drawn from integer $uniform[10, 190]$, integer $uniform[81, 119]$, and real $uniform[1, 19]$, respectively. For each instance of 2ML-B, the backlog cost b_p in each period equals h_p plus a real number drawn from $uniform[1, 10]$. We used three sets of capacity modules $\alpha = (\alpha_1, \alpha_2)$: (70, 34), (100, 35), and (180, 80). Two sets of setup costs $(f_p^1, f_p^2), p \in P$ were used for these modules: (1000, 600), leading to easy instances, and (5000, 2600), leading to hard instances. Note that some of the instance generation ideas we used here are inspired by the ideas used in [11] for 2ML without backlogging.

For each 2ML-B instance, we first solved the problem (defined by (12)-(15) for $n = 2$) without adding any of our own cuts using CPLEX 11.0 with its default settings (2ML-B-DEF). In a separate run, we used our cut generation algorithm to add 2-step (k, l, S, C) cycle inequalities to the problem at the root node. This algorithm calls our separation algorithm for several choices of (k, l, S) to generate 2-step (k, l, S, C) cycle inequalities that are violated by the LP relaxation optimal solution, which is updated after adding each cut. Note that each

Table 1. Results of computational experiments on 2ML-B instances (each entry in this table corresponds to the average for 10 instances)

Instance		2ML-B-DEF		2ML-B-CUTS					
(f_p^1, f_p^2)	(α_1, α_2)	T_{Def}	Nodes	Cuts	T_{Cut}	T_{Opt}	T_{Total}	Nodes	Gap%
(1000, 600)	(70,34)	0.34	582	109	4.06	0.18	4.24	197	91.54
	(100, 35)	0.31	691	91	2.89	0.10	2.99	116	88.23
	(180, 80)	0.13	277	125	3.68	0.03	3.71	23	92.52
(5000, 2600)	(70,34)	1133.81	5271562	86	4.88	135.73	140.61	274503	83.20
	(100, 35)	11.54	31909	81	3.71	6.13	9.84	12065	83.27
	(180, 80)	28.66	117942	101	4.63	3.76	8.39	6986	87.43

choice of (k, l, S) provides one set of base inequalities (16) (where $n = 2$) and we solve an exact separation problem over the set of all 2-step (k, l, S, C) cycle inequalities corresponding to the base inequalities which satisfy the n-step MIR conditions (discussed in Section 6). We then removed the inactive cuts and used CPLEX 11.0 with its default settings to solve the problem (2ML-B-CUTS). We implemented our codes in Microsoft Visual C++ 2010.

The results of our computational experiments are shown in Table 1. Each row of this table reports the average results for 10 instances of the corresponding category. We report the percentage of the integrality gap closed by our cuts, i.e. $Gap\% = 100 \times (zcut - zlp)/(zmip - zlp)$, where zlp, $zcut$, and $zmip$ are the optimal objective values of the LP relaxation without our cuts, LP relaxation with our cuts, and MIP, respectively. We also report the number of active 2-step (k, l, S, C) cycle cuts added at the root node (Cuts), the number of branch-and-bound nodes (Nodes), and the time (in seconds) to solve 2ML-B-DEF to optimality (T_{Def}), the time (in seconds) to generate 2-step (k, l, S, C) cycle cuts (T_{Cut}), and the time (in seconds) to solve 2ML-B-CUTS to optimality (excluding the cut generation time) (T_{Opt}). The total time to solve 2ML-B-CUTS including the cut generation time is also reported $(T_{Total} = T_{Cut} + T_{Opt})$.

Comparing the time to optimize the problem before and after adding the cuts (i.e. T_{Opt} vs. T_{Def}), we see significant improvement obtained by adding the cuts in both easy instances (on average 3.1 times) and hard instances (on average 6 times). There is also a substantial reduction in the number of branch-and-bound nodes (on average 6.9 times for easy instances and 12.9 times for hard instances). The percentage of integrality gap closed by our cuts is between 83.1% and 92.5% (the average is 87.7%). These results show the strength of 2-step (k, l, S, C) cycle inequalities. Also, observe that for the hard instances, the cut generation time (T_{Cut}) is negligible compared to (T_{Def}). This combined with the highly improved optimization time after adding the cuts has resulted in a total solution time (T_{Total}) which is on average 4.2 times smaller than the total time to solve 2ML-B-DEF (T_{Def}). The collection of these observations show that the 2-step (k, l, S, C) cycle inequalities are very effective in solving the 2ML-B problems.

8   Concluding Remarks

We unified the concepts of the continuous mixing and the n-step MIR by developing a class of valid inequalities (n-step cycle inequalities) for continuous n-mixing set (a generalization of the continuous mixing set and the n-mixing set) where the coefficients satisfy the n-step MIR conditions. We provided the facet-defining properties of the n'-step cycle inequalities, $n' \in \{1, \ldots, n\}$, for the continuous n-mixing set. We also presented a compact extended formulation for the continuous n-mixing set and an exact separation algorithm over the set of all n'-step cycle inequalities, $n' \in \{1, \ldots, n\}$. We showed that these inequalities can be used to generate cuts for the multi-module capacitated lot-sizing problems with(out) backlogging. Our computational results showed that the n-step cycle inequalities and our separation algorithm are very effective in solving these problems.

References

1. Atamtürk, A., Günlük, O.: Mingling: mixed-integer rounding with bounds. Mathematical Programming 123(2), 315–338 (2010)
2. Atamtürk, A., Kianfar, K.: n-step mingling inequalities: new facets for the mixed-integer knapsack set. Mathematical Programming 132(1), 79–98 (2012)
3. Cherkassky, B.V., Goldberg, A.V.: Negative-cycle detection algorithms. Mathematical Programming 85(2), 277–311 (1999)
4. Cormen, T.H., Stein, C., Rivest, R.L., Leiserson, C.E.: Introduction to Algorithms, 3rd edn. McGraw-Hill Higher Education (2009)
5. Dash, S., Günlük, O.: Valid inequalities based on simple mixed-integer sets. Mathematical Programming 105(1), 29–53 (2006)
6. Günlük, O., Pochet, Y.: Mixing mixed-integer inequalities. Mathematical Programming 90(3), 429–457 (2001)
7. Kianfar, K., Fathi, Y.: Generalized mixed integer rounding inequalities: facets for infinite group polyhedra. Mathematical Programming 120(2), 313–346 (2009)
8. Nemhauser, G.L., Wolsey, L.A.: A recursive procedure to generate all cuts for 0–1 mixed integer programs. Mathematical Programming 46(1-3), 379–390 (1990)
9. Pochet, Y., Wolsey, L.A.: Lot-Sizing with constant batches: Formulation and valid inequalities. Mathematics of Operations Research 18, 767–785 (1993)
10. Ralph, E.: Gomory: An algorithm for the mixed integer problem. Tech. Rep. RM-2597, RAND Corporation (1960)
11. Sanjeevi, S., Kianfar, K.: Mixed n-step MIR inequalities: Facets for the n-mixing set. Discrete Optimization 9(4), 216–235 (2012)
12. Shigeno, M., Iwata, S., McCormick, S.T.: Relaxed most negative cycle and most positive cut canceling algorithms for minimum cost flow. Mathematics of Operations Research 25(1), 76–104 (2000)
13. Tarjan, R.E.: Data structures and network algorithms. Society for Industrial and Applied Mathematics, Philadelphia, PA, USA (1983)
14. Vyve, M.V.: The continuous mixing polyhedron. Mathematics of Operations Research 30(2), 441–452 (2005)
15. Wolsey, L.A.: Integer Programming. Wiley, New York (1998)

On the Adaptivity Gap
of Stochastic Orienteering[*]

Nikhil Bansal[1],[**] and Viswanath Nagarajan[2]

[1] Eindhoven University of Technology, The Netherlands
bansal@gmail.com
[2] IBM T.J. Watson Research Center, New York
viswanath@us.ibm.com

Abstract. The input to the *stochastic orienteering* problem [14] consists of a budget B and metric (V, d) where each vertex $v \in V$ has a job with a deterministic reward and a *random* processing time (drawn from a known distribution). The processing times are independent across vertices. The goal is to obtain a non-anticipatory policy (originating from a given root vertex) to run jobs at different vertices, that maximizes expected reward, subject to the total distance traveled plus processing times being at most B. An *adaptive* policy is one that can choose the next vertex to visit based on observed random instantiations. Whereas, a *non-adaptive* policy is just given by a fixed ordering of vertices. The *adaptivity gap* is the worst-case ratio of the expected rewards of the optimal adaptive and non-adaptive policies.

We prove an $\Omega\left((\log \log B)^{1/2}\right)$ lower bound on the adaptivity gap of stochastic orienteering. This provides a negative answer to the $O(1)$-adaptivity gap conjectured in [14], and comes close to the $O(\log \log B)$ upper bound proved there. This result holds even on a line metric.

We also show an $O(\log \log B)$ upper bound on the adaptivity gap for the *correlated* stochastic orienteering problem, where the reward of each job is random and possibly correlated to its processing time. Using this, we obtain an improved quasi-polynomial time $\min\{\log n, \log B\} \cdot \tilde{O}(\log^2 \log B)$-approximation algorithm for correlated stochastic orienteering.

1 Introduction

In the *orienteering* problem [11], we are given a metric (V, d) with a starting vertex $\rho \in V$ and a budget B on length. The objective is to compute a path originating from ρ having length at most B, that maximizes the number of vertices visited. This is a basic vehicle routing problem (VRP) that arises as a subroutine in algorithms for a number of more complex variants, such as VRP with time-windows, discounted reward TSP and distance constrained VRP.

The stochastic variants of orienteering and related problems such as traveling salesperson and vehicle routing have also been extensively studied. In particular, several dozen variants have been considered depending on which parameters are

---

[*] The full version of this paper can be found on the Arxiv [3].
[**] Supported by the Dutch NWO Grant 639.022.211.

J. Lee and J. Vygen (Eds.): IPCO 2014, LNCS 8494, pp. 114–125, 2014.
© Springer International Publishing Switzerland 2014

stochastic, the choice of the objective function, the probability distributions, and optimization models such as *a priori* optimization, stochastic optimization with recourse, probabilistic settings and so on. For more details we refer to a recent survey [18] and references therein.

Here, we consider the following stochastic version of the orienteering problem defined by [14]. Each vertex has a job with a deterministic reward and random processing time (also referred to as size); these processing times are independent across vertices. The processing times model the random delays encountered at the node, say due to long queues or activities such as filling out a form, before the reward can be collected. The distances in the metric correspond to travel times between vertices, which are deterministic. The goal is to compute a *policy*, which describes a path originating from the root ρ that visits vertices and runs the respective jobs, so as to maximize the total expected reward subject to the total time (for travel plus processing) being at most B. Stochastic orienteering also generalizes the well-studied stochastic knapsack problem [9,5,4] (when all distances are zero). We also consider a further generalization, where the reward at each vertex is also random and possibly *correlated* to its processing time.

A feasible solution (policy) for the stochastic orienteering problem is represented by a decision tree, where nodes encode the "state" of the solution (previously visited vertices and the residual budget), and branches denote random instantiations. Such solutions are called *adaptive* policies, to emphasize the fact that their actions may depend on previously observed random outcomes. Often, adaptive policies can be very complex and hard to reason about. For example, even for the stochastic knapsack problem an optimal adaptive strategy may have exponential size (and several related problems are PSPACE-hard) [9].

Thus a natural approach for designing algorithms in the stochastic setting is to: (i) restrict the solution space to the simpler class of *non adaptive* policies (eg. in our stochastic orienteering setting, such a policy is described by a fixed permutation to visit vertices in, until the budget B is exhausted), and (ii) design an efficient algorithm to find a (close to) optimum non-adaptive policy.

While non-adaptive policies are often easier to optimize over, the drawback is that they could be much worse than the optimum adaptive policy. Thus, a key issue is to bound the *adaptivity gap*, introduced by [9] in their seminal paper, which is the worst-case ratio (over all problem instances) of the optimal adaptive value to the optimal non-adaptive value.

In recent years, increasingly sophisticated techniques have been developed for designing good non-adaptive policies and for proving small adaptivity gaps [9,12,8,2,13,14]. For stochastic orienteering, [14] gave an $O(\log \log B)$ bound on the adaptivity gap, using an elegant probabilistic argument (previous approaches only gave a $\Theta(\log B)$ bound). More precisely, they considered certain $O(\log B)$ correlated probabilistic events and used martingale tails bounds on suitably defined stopping times to bound the probability that none of these events happen. In fact, [14] conjectured that the adaptivity gap for stochastic orienteering was $O(1)$, suggesting that the $O(\log \log B)$ factor was an artifact of their analysis.

1.1 Our Results and Techniques

Adaptivity Gap for Stochastic Orienteering. Our main result is:

Theorem 1. *The adaptivity gap of stochastic orienteering is* $\Omega\left((\log\log B)^{1/2}\right)$, *even on a line metric.*

This answers negatively the $O(1)$-adaptivity gap conjectured in [14], and comes close to the $O(\log\log B)$ upper bound proved there. To the best of our knowledge, this gives the first non-trivial $\omega(1)$ adaptivity gap for a natural problem.

The lower bound proceeds in three steps and is based on a somewhat intricate construction. We begin with a basic instance described by a directed binary tree of height $\log\log B$ that essentially represents the optimal adaptive policy. Each processing time is a Bernoulli random variable: it is either zero, in which case the optimal policy goes to its left child, or a carefully set positive value, in which case the optimal policy goes to its right child. The edge distances and processing times are chosen so that when a non-zero size instantiates, it is always possible to take a right edge, while the left edges can only be taken a few times. On the other hand, if the non-adaptive policy chooses a path with mostly right edges, then it cannot collect too much reward.

In the first step of the proof, we show that any non-adaptive policy in this directed tree has an $\Omega((\log\log B)^{1/2})$ adaptivity gap. The main technical difficulty here is to show that every fixed path (which may possibly skip vertices, and gain advantage over the adaptive policy) either runs out of budget B or collects low expected reward. In the second step, we drop the directions on the edges and show that the adaptivity gap continues to hold (up to constant factors). The optimum adaptive policy that we compare against remains the same as in the directed case, and the key issue here is to show that the non-adaptive policy cannot gain too much by backtracking along the edges. To this end, we use some properties of the distances on edges in our instance. In the final step, we embed the undirected tree onto a line at the expense of losing another $O(1)$ factor in the adaptivity gap. The problem here is that pairs of nodes that are far apart on the tree may be very close on the line. To get around this, we exploit the asymmetry of the tree distances and some other structural properties to show that this has limited effect.

Correlated Stochastic Orienteering. Next, we consider the correlated stochastic orienteering problem, where the reward at each vertex is also random and possibly correlated with its processing time (the distributions are still independent across vertices). In this setting, we prove the following.

Theorem 2. *The adaptivity gap of correlated stochastic orienteering is* $O(\log\log B)$.

This improves upon the $O(\log B)$-factor adaptivity gap that is implicit in [14], and matches the adaptivity gap upper bound known for uncorrelated stochastic orienteering. The proof makes use of a martingale concentration inequality [19] (as [14] did for the uncorrelated problem), but dealing with the reward-size correlations requires a different definition of the stopping time. For the uncorrelated case, the stopping time [14] used a single "truncation threshold" (equal to B minus the travel time) to compare the instantiated sizes and their expectation. In the correlated setting, we use $\log B$ different truncation thresholds (all powers of 2), irrespective of the travel time, to determine the stopping criteria.

Algorithm for Correlated Stochastic Orienteering. Using some structural properties in the proof of the adaptivity gap upper bound above, we obtain an improved *quasi-polynomial* time algorithm for correlated stochastic orienteering. (A quasi-polynomial time algorithm is one that runs in $2^{\log^c N}$ time on inputs of size N, where c is some constant.)

Theorem 3. *There is an $O\left(\alpha \cdot \log^2 \log B / \log \log \log B\right)$-approximation algorithm for correlated stochastic orienteering, running in time $(n + \log B)^{O(\log B)}$. Here $\alpha \le \min\{O(\log n), O(\log B)\}$ denotes the best approximation ratio for the orienteering with deadlines problem.*

The *orienteering with deadlines* problem is defined formally in Section 1.3. Previously, [14] gave a polynomial time $O(\alpha \cdot \log B)$-approximation algorithm for correlated stochastic orienteering. They also showed that this problem is at least as hard to approximate as the deadline orienteering problem, i.e. an $\Omega(\alpha)$-hardness of approximation (this result also holds for quasi-polynomial time algorithms). Our algorithm improves the approximation ratio to $O(\alpha \cdot \log^2 \log B)$, but at the expense of quasi-polynomial running time. We note that the running time in Theorem 3 is quasi-polynomial for general inputs where probability distributions are described *explicitly*, since the input size is $n \cdot B$. If probability distributions are specified implicitly, the runtime is quasi-polynomial only for $B \le 2^{poly(\log n)}$.

As a corollary of Theorem 3, we obtain a polynomial-time *bicriteria* approximation algorithm, that for any fixed $\epsilon > 0$, computes an $O\left(\frac{\log(1/\epsilon)\cdot\log^2\log(1/\epsilon)}{\log\log\log(1/\epsilon)}\right)$-approximate solution, which violates the budget B by a $1 + \epsilon$ factor.

The algorithm in Theorem 3 is based on finding an approximate non-adaptive policy, and losing an $O(\log \log B)$-factor by Theorem 2. There are three main steps in the algorithm: (i) we enumerate over $\log B$ many "portal" vertices (suitably defined) on the optimal policy; (ii) using these portal vertices, we approximately solve a *configuration LP relaxation* for paths between portal vertices; (iii) we randomly round the LP solution. The quasi-polynomial time is only due to the enumeration. In formulating and solving the configuration LP relaxation, we also use some ideas from the earlier $O(\alpha \cdot \log B)$-approximation algorithm [14]. Solving the configuration LP requires an algorithm for deadline orienteering (as the dual separation oracle), and incurs an α-factor loss in the approximation ratio. This configuration LP is a "packing linear program", for which we can use fast combinatorial algorithms [16,10]. The final rounding step uses randomized rounding with alteration, and loses an extra $O(\frac{\log \log B}{\log \log \log B})$ factor.

1.2 Related Work

The deterministic orienteering problem was introduced by Golden et al. [11]. It has several applications, and many exact approaches and heuristics have been applied to this problem, see eg. the survey [17]. The first constant-factor approximation algorithm was due to Blum et al. [6]. The approximation ratio has been improved [1,7] to the current best $2 + \epsilon$.

Dean et al. [9] were the first to consider stochastic packing problems in this adaptive optimization framework: they obtained a constant-factor approximation algorithm and adaptivity gap for the *stochastic knapsack* problem (where items have random sizes). The approximation ratio has subsequently been improved to $2+\epsilon$, due to [5,4]. The stochastic orienteering problem [14] is a common generalization of both deterministic orienteering and stochastic knapsack.

Gupta et al. [13] studied a generalization of the stochastic knapsack problem, to the setting where the reward and size of each item may be correlated, and gave an $O(1)$-approximation algorithm and adaptivity gap for this problem. Recently, Ma [15] improved the approximation ratio to $2 + \epsilon$.

The correlated stochastic orienteering problem was studied in [14], where the authors obtained an $O(\log n \cdot \log B)$-approximation algorithm and an $O(\log B)$ adaptivity gap. They also showed the problem to be at least as hard to approximate as the deadline orienteering problem, for which the best approximation ratio known is $O(\log n)$ [1].

A related problem to stochastic orienteering was considered by Guha and Munagala [12] in the context of the *multi-armed bandit* problem. As observed in [14], the approach in [12] yields an $O(1)$-approximation algorithm (and adaptivity gap) for the variant of stochastic orienteering with two *separate* budgets for the travel and processing times. In contrast, our result shows that stochastic orienteering (with a single budget) has super-constant adaptivity gap.

1.3   Problem Definition

An instance of stochastic orienteering (StocOrient) consists of a metric space (V, d) with vertex-set $|V| = n$ and symmetric integer distances $d : V \times V \to \mathbb{Z}^+$ (satisfying the triangle inequality) that represent travel times. Each vertex $v \in V$ is associated with a stochastic job, with a deterministic reward $r_v \geq 0$ and a random processing time (also called size) $S_v \in \mathbb{Z}^+$ distributed according to a known probability distribution. The processing times are independent across vertices. We are also given a starting "root" vertex $\rho \in V$, and a budget $B \in \mathbb{Z}^+$ on the total time available. A solution (policy) must start from ρ, and visit a sequence of vertices (possibly adaptively). Each job is executed non-preemptively, and the solution knows the precise processing time only upon completion of the job. The objective is to maximize the expected reward from jobs that are completed before the horizon B; note that there is no reward for partially completing a job. The approximation ratio of an algorithm is the ratio of the expected reward of an optimal policy to that of the algorithm's policy.

We assume that all times (travel and processing) are integer valued and lie in $\{0, 1, \cdots, B\}$. In the *correlated* stochastic orienteering problem (CorrOrient), the job sizes and rewards are both random, and correlated with each other. The distributions across different vertices are still independent. For each vertex $v \in V$, we use S_v and R_v to denote its random size and reward, respectively. We assume an explicit representation of the distribution of each job $v \in V$: for each $s \in \{0, 1, \cdots, B\}$, job v has size $S_v = s$ and reward $r_v(s)$ with probability $\Pr[S_v = s] = \pi_v(s)$. Note that the input size is nB.

An *adaptive policy* is a decision tree where each node is labeled by a job/vertex of V, with the outgoing arcs from a node labeled by u corresponding to the possible sizes in the support of S_u. A *non-adaptive policy* is simply given by a path P starting at ρ: we just traverse this path, processing the jobs that we encounter, until the total (random) size of the jobs plus the distance traveled reaches B. A randomized non-adaptive policy may pick a path P at random from some distribution before it knows any of the size instantiations, and then follows this path as above. Note that in a non-adaptive policy, the order in which jobs are processed is independent of their processing time instantiations.

In our algorithm for CorrOrient, we use the *deadline orienteering* problem as a subroutine. The input to this problem is a metric (U, d) denoting travel times, a reward and deadline at each vertex, start (s) and end (t) vertices, and length bound D. The objective is to compute an $s - t$ path of length at most D that maximizes the reward from vertices visited before their deadlines. The best approximation ratio for this problem is $\alpha = \min\{O(\log n), O(\log B)\}$ [1,7].

1.4   Organization

Due to space limitations in the main body, we only describe the lower bound on the adaptivity gap in the directed tree case. We note that the $O(\log \log B)$ upper bound [14] holds even for directed metrics. The other results can be found in the full version of the paper [3].

2   Lower Bound on the Adaptivity Gap

Here we describe our lower bound instance which shows that the adaptivity gap is $\Omega(\sqrt{\log \log B})$ even for an undirected line metric. The proof and the description of the instance is divided into three steps. First we describe an instance where the underlying graph is a directed complete binary tree, and prove the lower bound for it. The directedness ensures that all policies follow a path from root to the leaf (possibly with some nodes skipped) without any backtracking. Second, we show that the directed assumption can be removed at the expense of an additional $O(1)$ factor in the adaptivity gap. In particular this means that the nodes on the tree can be visited in any order starting from the root. Finally, we "embed" the undirected tree into a line metric, and show that the adaptivity gap stays the same up to a constant factor.

2.1   Directed Binary Tree

Let $L \geq 2$ be an integer and $p := 1/\sqrt{L}$. We define a complete binary tree $\mathcal{T}$ of height L with root ρ. All the edges are directed from the root towards the leaves. The *level* $\ell(v)$ of any node v is the number of nodes on the shortest path from v to any leaf. So all the leaves are at *level* one and the root ρ is at level L. We refer to the two children of each internal node as the left and right child, respectively. Each node v of the tree has a job with some deterministic reward r_v

and a random size S_v. Each random variable S_v is Bernoulli, taking value zero with probability $1 - p$ and some positive value s_v with the remaining probability p. The budget for the instance is $B = 2^{2^{L+1}}$.

To complete the description of the instance, we need to define the values of the rewards r_v, the job sizes S_v, and the distances $d(u, v)$ on edges $e = (u, v)$.

Defining Rewards. For any node v, let $\tau(v)$ denote the number of right-branches taken on the path from the root to v. We define the reward of each node v to be $r_v := (1 - p)^{\tau(v)}$.

Defining Sizes. Let $e(x) := 2^x$ for any $x \in \mathbb{R}$. The size at the root, $s_\rho := e(2^L) = 2^{2^L}$. The rest of the sizes are defined recursively. For any non-root node v at level $\ell(v)$ with u denoting its parent, the size is:

$$s_v := \begin{cases} s_u \cdot e\left(2^{\ell(v)}\right) & \text{if } v \text{ is the right child of } u \\ s_u \cdot e\left(-2^{\ell(v)}\right) & \text{if } v \text{ is the left child of } u \end{cases}$$

In other words, for a node v at level ℓ, consider the path from ρ to v, $P = (\rho = u_L, u_{L-1}, \ldots, u_{\ell+1}, u_\ell = v)$. Let $k = \sum_{j=L}^{\ell}(-1)^{i(u_j)} 2^j$ where $i(u_j) = 1$ if u_j is the left child of its parent u_{j+1}, and 0 otherwise (we assume $i(\rho) = 0$). Then $s_v = e(k)$.

Observe that for a node v, each node u in its left (resp. right) subtree has $s_u < s_v$ (resp. $s_u > s_v$).

It remains to define distances on the edges. This will be done in an indirect way, and it is instructive to first consider the adaptive policy that we will work with. In particular, the distances will be defined in such a way that the adaptive policy can always continue till it reaches a leaf node.

Adaptive policy $\mathcal{A}$. Consider the policy $\mathcal{A}$ that goes left at node u whenever it observes size zero at u, and goes right otherwise.

Clearly, the *residual budget* $b(v)$ at node v under $\mathcal{A}$ will satisfy the following: $b(\rho) = B = e\left(2^{L+1}\right) = 2^{2^{L+1}}$, and

$$b(v) := \begin{cases} b(u) - s_u - d(u, v) & \text{if } v \text{ is the right child of } u \\ b(u) - d(u, v) & \text{if } v \text{ is the left child of } u \end{cases}$$

Defining Distances. We define the distances so that the residual budgets $b(\cdot)$ under $\mathcal{A}$ satisfy the following: $b(\rho) = B$, and for any node v with parent u,

$$b(v) := \begin{cases} b(u) - s_u & \text{if } v \text{ is the right child of } u \\ s_u & \text{if } v \text{ is the left child of } u \end{cases}$$

This implies the following lengths on edges. For any node v with parent u,

$$d(u, v) := \begin{cases} 0 & \text{if } v \text{ is the right child of } u \\ b(u) - s_u = b(u) - b(v) & \text{if } v \text{ is the left child of } u \end{cases}$$

In Claim 2 below we will show that the distances are non-negative, and hence well-defined. Figure 1 gives a pictorial view of the instance.

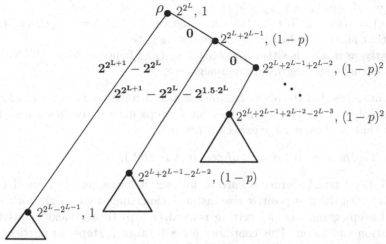

The distances are shown (in bold) on the edges.
The sizes s_v and rewards r_v are shown next to the nodes.

Fig. 1. The binary tree $\mathcal{T}$

Claim 1. *If a node w is a left child of its parent, then $b(w) = s_w \cdot e\left(2^{\ell(w)}\right)$.*

Proof. Let u be the parent of w. By definition of sizes, $s_w = s_u \cdot e\left(-2^{\ell(w)}\right)$. As $b(w) = s_u$ by the definition of residual budgets, the claim follows. □

Claim 2. *For any node u, we have $3 \cdot s_u \leq b(u)$. This implies that all the residual budgets and distances are non-negative.*

Proof. Let w denote the lowest level node on the path from ρ to u that is the left child of its parent (if u is the left child of its parent, then $w = u$); if there is no such node, set $w = \rho$. Note that by Claim 1 and the definition of s_ρ and $b(\rho)$, in either case it holds that $b(w) = s_w \cdot e\left(2^{\ell(w)}\right)$.

Let π denote the path from w to u (including w but not u; so $\pi = \emptyset$ if $w = u$). Since π contains only right-branches, $b(u) = b(w) - \sum_{y \in \pi} s_y$ and hence $b(u) \geq b(w) - 3\sum_{y \in \pi} s_y$. Thus to prove $3 \cdot s_u \leq b(u)$ it suffices to show $3(s_u + \sum_{y \in \pi} s_y) \leq b(w)$. For brevity, let $s := s_w$ and $\ell = \ell(w)$. By definition of sizes,

$$s_u + \sum_{y \in \pi} s_y \leq \sum_{i=1}^{\ell} s \cdot e\left(2^{\ell-1} + 2^{\ell-2} + \cdots 2^i\right) = s \cdot \sum_{i=1}^{\ell} e\left(2^\ell - 2^i\right)$$

$$\leq s \cdot e\left(2^\ell\right) \cdot \sum_{i \geq 1} 4^{-i} \leq \frac{1}{3} \cdot s \cdot e\left(2^\ell\right) = \frac{b(w)}{3},$$

as desired. Here the right hand side of the first inequality is simply the total size of nodes in the w to leaf path using all right branches. The inequality in the second line follows as $e\left(-2^i\right) = 2^{-2^i} \leq 2^{-2i} = 4^{-i}$ for all $i \geq 1$.

Thus we always have $3 \cdot s_u \leq b(u)$.

As $b(v) = b(u) - s_u$ if v is the right child of u, or $b(v) = s_u$ otherwise, this implies that all the residual-budgets are non-negative.

Similarly, as $d(u,v)$ is either 0 or $b(u) - s_u$ (and hence at least $2/3b(u)$), this implies that all edge lengths are non-negative.   □

This claim shows that the above instance is well defined, and that $\mathcal{A}$ is a feasible adaptive policy that always continues for L steps until it reaches a leaf. Next, we show that $\mathcal{A}$ obtains large expected reward.

Lemma 1. *The expected reward of policy $\mathcal{A}$ is $\Omega(L)$.*

Proof. Notice that $\mathcal{A}$ accrues reward as follows: it keeps getting reward 1 (and going left) until the first positive size instantiation, then it goes right for a single step and keeps going left and getting reward $(1 - p)$ till the next positive size instantiation and so on. This continues for a total of L steps. In particular, at any time t it collects reward $(1-p)^i$, if exactly i nodes have positive sizes among the t nodes seen.

Let X_i denote the Bernoulli random variable that is 1 if the ith node in $\mathcal{A}$ has a positive size instantiation, and 0 otherwise. So $E[X_i] = p$, and $E[X_1 + \ldots + X_L] = Lp = \sqrt{L}$. By Markov's inequality, the probability that more than $2\sqrt{L}$ nodes in $\mathcal{A}$ have positive sizes is at most half. Hence, with probability at least $\frac{1}{2}$ the reward collected in the last node of $\mathcal{A}$ is at least $(1-p)^{2\sqrt{L}}$. That is, the total expected reward of $\mathcal{A}$ is at least $\frac{1}{2} \cdot L \cdot (1-p)^{2\sqrt{L}} \approx L/2 \cdot e^{-2} = \Omega(L)$.   □

2.2   Bounding Directed Non-adaptive Policies

We now show that any non-adaptive policy $\mathcal{N}$ that is constrained to visit vertices according to the partial order given by the tree $\mathcal{T}$ gets reward $O(\sqrt{L})$. Notice that these correspond precisely to non-adaptive policies on the directed tree $\mathcal{T}$.

The key property we need from the size construction is the following.

Lemma 2. *For any node v, the total size instantiation observed under $\mathcal{A}$ before v is strictly less than s_v.*

Proof. Consider the path π from the root to v, and let $k_1 < k_2 < \cdots < k_t$ denote the levels at which π "turns left". That is, for each i, the node u_i at level k_i in path π satisfies (a) u_i is the right child of its parent, and (b) π contains the left child of u_i if it goes below level k_i. (If v is the right child of its parent then $u_1 = v$ and $k_1 = \ell(v)$.) Let s_i denote the size of u_i, the level k_i node in π. Also, set $k_{t+1} = L$ corresponding to the root. Below we use $[t] := \{1, 2, \cdots, t\}$.

Observe that a positive size instantiation is seen in $\mathcal{A}$ only along right branches. So for any $i \in [t]$, the total size instantiation seen in π between levels k_i and k_{i+1} is at most:

$$s_i \cdot \left[e\left(-2^{k_i}\right) + e\left(-2^{k_i} - 2^{k_i+1}\right) + e\left(-2^{k_i} - 2^{k_i+1} - 2^{k_i+2}\right) \cdots \right]$$
$$\leq s_i \cdot e\left(-2^{k_i}\right) \cdot (1 + 1/2 + 1/4 + \cdots) \quad \leq \quad 2\, s_i \cdot e\left(-2^{k_i}\right) \qquad (1)$$

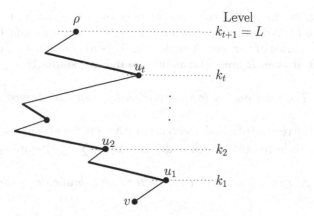

Path π from ρ to v, with "left turning" nodes $u_1, \cdots, u_t$.
The thick lines denote right-branches (where positive sizes are seen).

Fig. 2. The path π in proof of Lemma 2

Now, note that for any $i \in [t]$, the sizes s_{i+1} and s_i are related as follows:

$$s_{i+1} \le s_i \cdot e\left(-2^{k_i} + 2^{k_i+1} + 2^{k_i+2} + \cdots + 2^{k_{i+1}-1}\right) = s_i \cdot e\left(-2^{k_i} - 2^{k_i+1} + 2^{k_{i+1}}\right)$$
$$\le \frac{s_i}{4} \cdot e\left(-2^{k_i} + 2^{k_{i+1}}\right) \tag{2}$$

The first inequality uses the fact that the path from u_{i+1} to u_i is a sequence of (at least one) left-branches followed by a sequence of (at least one) right-branches: so s_i/s_{i+1} is minimized for the path with a sequence of left branches followed by a single right branch (at level k_i).

Using (2), we obtain inductively that:

$$s_{i+1} \cdot e\left(-2^{k_{i+1}}\right) \le \frac{1}{4} \cdot s_i \cdot e\left(-2^{k_i}\right) \le \frac{1}{4^i} \cdot s_1 \cdot e\left(-2^{k_1}\right), \qquad \forall i \in [t]. \tag{3}$$

Using (1) and (3), the total size instantiation seen in π (this does not include the size at v) is at most:

$$\sum_{i=1}^t 2\, s_i \cdot e\left(-2^{k_i}\right) \le 2 \sum_{i=1}^t \frac{1}{4^{i-1}} \cdot s_1 \cdot e\left(-2^{k_1}\right) < 4\, s_1 \cdot e\left(-2^{k_1}\right). \tag{4}$$

Finally, observe that the size at the level k_1 node is

$$s_1 \le s_v \cdot e\left(2^{k_1-1} + 2^{k_1-2} + \cdots + 2^1\right) = s_v \cdot e\left(2^{k_1} - 2\right),$$

since k_1 is the lowest level at which π turns left (i.e. π keeps going left below level k_1 until v). Together with (4), it follows that the total size instantiation seen before v is strictly less than

$$4\, s_1 \cdot e\left(-2^{k_1}\right) \le 4\, e\left(-2^{k_1}\right) \cdot s_v \cdot e\left(2^{k_1} - 2\right) = 4\, e\left(-2\right) s_v = s_v.$$

This completes the proof of Lemma 2.   □

We now show that any non-adaptive policy on the directed tree $\mathcal{T}$ achieves reward $O(\sqrt{L})$. Note that any such solution $\mathcal{N}$ is just a root-leaf path in $\mathcal{T}$ that skips some subset of vertices. A node v in $\mathcal{N}$ is an *L-branching* node if the path $\mathcal{N}$ goes left after v. *R-branching* nodes are defined similarly.

Claim 3. *The total reward from R-branching nodes is at most $\sqrt{L}$.*

Proof. As the reward of a node decreases by a factor of $(1-p)$ upon taking a right branch, the total reward of such nodes is at most $\sum_{i=0}^{L}(1-p)^i \leq \frac{1}{p} = \sqrt{L}$. $\quad\square$

Claim 4. $\mathcal{N}$ *can not get any reward after two L-branching nodes instantiate to positive sizes.*

Proof. For any node v in tree $\mathcal{T}$, let $A_d(v)$ (resp. $A_s(v)$) denote the distance traveled (resp. size instantiated) in the adaptive policy $\mathcal{A}$ until v; here $A_s(v)$ does not include the size of v. Observe that Lemma 2 implies that $A_s(v) < s_v$ for all nodes v.

In the non-adaptive solution $\mathcal{N}$, let u and v be any two L-branching nodes that instantiate to positive sizes s_u and s_v; say u appears before v. Under this outcome, we will show that $\mathcal{N}$ exhausts its budget after v. Note that the distance traveled to node v in $\mathcal{N}$ is exactly $A_d(v)$, the same as that under $\mathcal{A}$. So the total distance plus size in $\mathcal{N}$ is at least $A_d(v) + s_v + s_u$, which (as we show next) is more than the budget B.

By definition of the residual budgets, $b(v) = B - A_d(v) - A_s(v)$. Moreover, the residual budget $b(u')$ at the left child u' of u equals s_u. Since the residual budgets are non-increasing down the tree $\mathcal{T}$, we have $B - A_d(v) - A_s(v) = b(v) \leq b(u') = s_u$, i.e. $A_d(v) \geq B - A_s(v) - s_u$. Hence, the total distance plus size in $\mathcal{N}$ is at least

$$A_d(v) + s_v + s_u \quad \geq \quad B - A_s(v) + s_v \quad > \quad B,$$

where the last inequality follows from Lemma 2. So $\mathcal{N}$ can not obtain reward from any node after v. $\quad\square$

Combining the above two claims, we obtain:

Claim 5. *The expected reward of any directed non-adaptive policy is at most $3\sqrt{L}$.*

Proof. Using Claim 4, the expected reward from L-branching nodes is at most the expected number of L-branching nodes until two positive sizes instantiate, i.e. at most $\frac{2}{p} = 2\sqrt{L}$. Claim 3 implies that the expected reward from R-branching nodes is at most $\sqrt{L}$. Adding the two types of rewards, we obtain the claim. $\quad\square$

This proves an $\Omega(\sqrt{\log \log B})$ adaptivity gap on directed metrics. As noted earlier, the $O(\log \log B)$ upper bound [14] also holds for directed metrics.

References

1. Bansal, N., Blum, A., Chawla, S., Meyerson, A.: Approximation algorithms for deadline-TSP and vehicle routing with time-windows. In: STOC, pp. 166–174 (2004)
2. Bansal, N., Gupta, A., Li, J., Mestre, J., Nagarajan, V., Rudra, A.: When LP is the cure for your matching woes: Improved bounds for stochastic matchings. Algorithmica 63(4), 733–762 (2012)
3. Bansal, N., Nagarajan, V.: On the adaptivity gap of stochastic orienteering. CoRR, abs/1311.3623 (2013)
4. Bhalgat, A.: A $(2+\epsilon)$-approximation algorithm for the stochastic knapsack problem (2011) (unpublished manuscript)
5. Bhalgat, A., Goel, A., Khanna, S.: Improved approximation results for stochastic knapsack problems. In: SODA, pp. 1647–1665 (2011)
6. Blum, A., Chawla, S., Karger, D.R., Lane, T., Meyerson, A., Minkoff, M.: Approximation algorithms for orienteering and discounted-reward TSP. SIAM J. Comput. 37(2), 653–670 (2007)
7. Chekuri, C., Korula, N., Pál, M.: Improved algorithms for orienteering and related problems. ACM TALG 8(3) (2012)
8. Chen, N., Immorlica, N., Karlin, A.R., Mahdian, M., Rudra, A.: Approximating matches made in heaven. In: Albers, S., Marchetti-Spaccamela, A., Matias, Y., Nikoletseas, S., Thomas, W. (eds.) ICALP 2009, Part I. LNCS, vol. 5555, pp. 266–278. Springer, Heidelberg (2009)
9. Dean, B.C., Goemans, M.X., Vondrák, J.: Approximating the stochastic knapsack problem: The benefit of adaptivity. Math. Oper. Res. 33(4), 945–964 (2008)
10. Garg, N., Könemann, J.: Faster and simpler algorithms for multicommodity flow and other fractional packing problems. SIAM J. Comput. 37(2), 630–652 (2007)
11. Golden, B.L., Levy, L., Vohra, R.: The orienteering problem. Naval Research Logistics 34(3), 307–318 (1987)
12. Guha, S., Munagala, K.: Multi-armed bandits with metric switching costs. In: Albers, S., Marchetti-Spaccamela, A., Matias, Y., Nikoletseas, S., Thomas, W. (eds.) ICALP 2009, Part II. LNCS, vol. 5556, pp. 496–507. Springer, Heidelberg (2009)
13. Gupta, A., Krishnaswamy, R., Molinaro, M., Ravi, R.: Approximation algorithms for correlated knapsacks and non-martingale bandits. In: FOCS, pp. 827–836 (2011)
14. Gupta, A., Krishnaswamy, R., Nagarajan, V., Ravi, R.: Approximation algorithms for stochastic orienteering. In: SODA, pp. 1522–1538 (2012)
15. Ma, W.: Improvements and generalizations of stochastic knapsack and multi-armed bandit approximation algorithms: Extended abstract. In: SODA, pp. 1154–1163 (2014)
16. Plotkin, S.A., Shmoys, D.B., Tardos, E.: Fast approximation algorithms for fractional packing and covering problems. In: FOCS, pp. 495–504 (1991)
17. Vansteenwegena, P., Souffriaua, W., Oudheusdena, D.V.: The orienteering problem: A survey. Eur. J. Oper. Res. 209(1), 1–10 (2011)
18. Weyland, D.: Stochastic Vehicle Routing - From Theory to Practice. PhD thesis, University of Lugano, Switzerland (2013)
19. Zhang, T.: Data dependent concentration bounds for sequential prediction algorithms. In: COLT, pp. 173–187 (2005)

A Utility Equivalence Theorem
for Concave Functions

Anand Bhalgat[1,*] and Sanjeev Khanna[2,**]

[1] Facebook Inc.
bhalgat@fb.com
[2] University of Pennsylvania
sanjeev@cis.upenn.edu

Abstract. Given any two sets of independent non-negative random variables and a non-decreasing concave utility function, we identify sufficient conditions under which the expected utility of sum of these two sets of variables is (almost) equal. We use this result to design a polynomial-time approximation scheme (PTAS) for utility maximization in a variety of risk-averse settings where the risk is modeled by a concave utility function. In particular, we obtain a PTAS for the asset allocation problem for a risk-averse investor as well as the risk-averse portfolio allocation problem.

1 Introduction

Given an arbitrary non-decreasing concave function $\mathbf{U} : \mathbf{R}^+ \to \mathbf{R}^+$ (with $\mathbf{U}(0) = 0$) and two sets of independent non-negative random variables $\mathbf{X} = \{X_1, X_2, \ldots, X_n\}$ and $\mathbf{X}' = \{X_1', X_2', \ldots, X_m'\}$, we consider the following question: what similarity between $\mathbf{X}$ and $\mathbf{X}'$ suffices to ensure that $\mathbf{E}\left[\mathbf{U}(\sum_i X_i)\right] \approx \mathbf{E}\left[\mathbf{U}(\sum_i X_i')\right]$? This question is naturally motivated by the problem of utility maximization in a risk-averse setting when the risk is modeled by a concave utility function [1,14]. As a concrete example, consider the *asset allocation problem* faced by a risk averse investor. The investor has a set of assets to choose from where each asset is associated with a cost. The investor has a prior on the future value of each asset; the future value of each asset is independent and the investor would like to buy a subset of assets (subject to his budget constraint) that maximizes the expected utility of his total future wealth. As the reader might notice, this problem is equivalent to the stochastic knapsack problem with random profits. While the stochastic knapsack problem with random profits has previously been studied with an objective to maximize the probability of achieving a certain target wealth (see e.g. [4,9,15,16]), these results does not translate to any guarantees when the utility function is concave. Our main result is a utility equivalence theorem that identifies sufficient conditions to ensure that $\mathbf{E}\left[\mathbf{U}(\sum_i X_i)\right] \approx \mathbf{E}\left[\mathbf{U}(\sum_i X_i')\right]$. Using this theorem, we design polynomial-time approximation schemes to solve a

---

* Work done when at University of Pennsylvania.
** Corresponding author.

J. Lee and J. Vygen (Eds.): IPCO 2014, LNCS 8494, pp. 126–137, 2014.

variety of concave utility maximization problems including the stochastic knapsack problem and a general portfolio allocation problem.

A work that is similar in flavor to our results is the recent work of Li and Deshpande [13] that addressed utility maximization for an arbitrary utility function in presence of general combinatorial feasibility constraints. A key component of their approach is to represent the utility function using a sum of exponential functions. They give an additive ϵ-approximation to the optimal utility (for the given instance) assuming that the *utility function is always bounded from above by* 1. This gives a (multiplicative) PTAS only when the optimal expected utility is within a constant factor (or some function of ϵ) of the maximum possible utility in any realization for an allocation. While this approach is useful for settings where the objective is to minimize (or maximize) the probability of hitting a certain target, it does not yield useful guarantees when the utility function is an arbitrary monotone concave function, as is the case for the asset allocation problem for a risk-averse investor.

Our Result: Our main result is a utility equivalence theorem (Theorem 1) that identifies a set of sufficient conditions under which any two sets of independent random variables have similar utility. Informally speaking, we show that the possible values realized by the variables can be partitioned into three regions, *small*, *large* and *huge*, and for approximate utility equivalence (within a $(1 \pm \epsilon)$ factor for some given $\epsilon > 0$), it suffices that the two sets match in

(a) the total support for each large value,

(b) the total expectation from small values,

(c) the total expected utility from huge values, and

(d) the set of random variables that have significant support in the large value region.

Given an instance of the utility maximization problem, the values of the above four properties in an optimal solution defines its *features*. The utility equivalence theorem allows us to conclude that the number of distinct features of interest is polynomially bounded (in the size of the ground set of elements). One can then guess the features of the variables in an optimal solution, and find a feasible solution that is required to match the optimal set of variables only on these features. Whenever it is possible to find a feasible solution with matching features in polynomial time, this approach then allows us to obtain a PTAS for the underlying optimization problem. We use this approach to obtain PTAS results for many important applications, that include the asset allocation problem for a risk-averse investor [4,9,15,16], risk-averse portfolio allocation problem [8,12,11,5,10], and the stochastic spanning tree problem [7,6].

In order to find a solution that matches the optimal solution on the feature set, similar to the approach used by Li and Deshpande [13], we establish a reduction from utility maximization to the feasibility of a deterministic profit problem; whenever the latter problem permits an exact pseudo-polynomial time algorithm, there exists a PTAS for the utility maximization problem. Our reduction yields a PTAS for monotone concave utility functions that was not possible to obtain using the earlier approach. We refer the reader to Section 4 for details.

In a recent work, Bhalgat *et al* [3] identified a set of conditions under which the expected utility of an independent set of variables $\mathbf{X}'$ lower bounds (to within a factor of $(1 - 1/e)$) the expected utility of a correlated set of variable $\mathbf{X}$. We note important differences between these two works:

a) Our result is about utility equivalence $((1 \pm \epsilon)$-approximation), whereas [3] establishes an approximate lower bound of $(1 - 1/e)$, and

b) To establish utility equivalence, we require random variables in both sets to be independent; in comparison [3], an independent set of random variables is used to approximate an *arbitrarily correlated set* of random variables.

Organization: We provide basic definitions and prove useful properties of concave functions in Section 2. We establish the utility equivalence theorem in Section 3 and illustrate its applications in Section 4.

2   Preliminaries

We note, the results in this paper (structural properties of random variables as well as algorithmic results) easily extend to a setting where each random variable is associated with a continuous distribution on its value and we are given an oracle to draw samples from this distribution. For expositional simplicity, we assume that distributions are discrete.

For notational conciseness, given any set $\mathbf{X} = \{X_1, X_2, \ldots, X_n\}$ of random variables, we use $\mathbf{U}(\mathbf{X})$ to denote $\mathbf{U}(\sum_i X_i)$. We next define a useful operation on a pair of random variables and establish its properties.

Merging and Splitting of Random Variables: Let $\mathbf{X} = \{X_1, X_2, \ldots, X_n\}$ be a set of independent non-negative random variables. For any two variables X_i, X_j in $\mathbf{X}$ such that $\mathbf{Pr}\,[X_i \neq 0] + \mathbf{Pr}\,[X_j \neq 0] \leq 1$, the *merge* operation on the variables X_i, X_j replaces them by a new variable Y such that, for each value $p > 0$, $\mathbf{Pr}\,[Y = p] = \mathbf{Pr}\,[X_i = p] + \mathbf{Pr}\,[X_j = p]$, and Y is independent of variables in $\mathbf{X} \backslash \{X_i, X_j\}$. The next lemma shows that the merge operation can only increase expected utility; its proof is deferred to the full version of the paper.

Lemma 1. *Let* $\mathbf{X} = \{X_1, X_2, \ldots, X_n\}$ *be a set of independent non-negative random variables, and let* Y *be the variable formed by merging* $X_1, X_2 \in \mathbf{X}$. *Then*

$$\mathbf{E}\,[\mathbf{U}(\mathbf{X})] \leq \mathbf{E}\,[\mathbf{U}((\mathbf{X} \setminus \{X_1, X_2\}) \cup \{Y\})]$$

Splitting: We will later define a converse operation, called the *splitting* operation, where we split a random variable into a set of independent random variables. The precise definition is tailored to our needs in Section 3.

Stochastic Dominance: Given two non-negative distributions D_1 and D_2, we say D_1 *stochastically dominates* D_2, denoted by $\mathsf{D}_1 \succeq \mathsf{D}_2$, if $\forall a \geq 0, \mathbf{Pr}_{X \sim \mathsf{D}_1}(X \geq a) \geq \mathbf{Pr}_{X \sim \mathsf{D}_2}(X \geq a)$. We note below an important (folklore) property of concave functions in the following lemma.

Lemma 2. *Let* $\mathbf{U}$ *be a non-decreasing concave function, and let* X, Y_1, Y_2 *be independent non-negative random variables such that* Y_i *is distributed according to* D_i, $i \in \{1, 2\}$. *If* $\mathsf{D}_1 \succeq \mathsf{D}_2$, *then*

(a) $\mathbf{E}_{X,Y_1 \sim \mathsf{D}_1} [\mathbf{U}(X + Y_1) - \mathbf{U}(Y_1)] \leq \mathbf{E}_{X,Y_2 \sim \mathsf{D}_2} [\mathbf{U}(X + Y_2) - \mathbf{U}(Y_2)],$ *and*

(b) $\mathbf{E}_{X,Y_1 \sim \mathsf{D}_1} [\mathbf{U}(X + Y_1) - \mathbf{U}(X))] \geq \mathbf{E}_{X,Y_2 \sim \mathsf{D}_2} [\mathbf{U}(X + Y_2) - \mathbf{U}(X)]$

3   The Utility Equivalence Theorem

In this section, we state and prove the utility equivalence theorem for any two sets of independent non-negative random variables $\mathbf{X} = \{X_1, X_2, \ldots, X_n\}$ and $\mathbf{X}' = \{X_1', X_2', \ldots, X_n'\}$ (see Theorem 1 for the exact conditions required for utility equivalence). The power of this theorem lies in the fact that it shows that as long as $\mathbf{X}$ and $\mathbf{X}'$ agree only on a coarse footprint of attributes, $\mathbf{E}[\mathbf{U}(\mathbf{X})]$ and $\mathbf{E}[\mathbf{U}(\mathbf{X}')]$ are comparable.

In the rest of the section, we will use OPT to denote $\mathbf{E}[\mathbf{U}(\mathbf{X})]$. Let $\mathcal{P}$ be the set of values realized by the random variables in $\mathbf{X}$, and let $\epsilon > 0$ be the accuracy parameter. For each value $p_j \in \mathcal{P}$, let $q_j = \sum_i \mathbf{Pr}[X_i = p_j]$, i.e. q_j denotes the total support for value p_j in set $\mathbf{X}$. To define our notion of a coarse footprint of $\mathbf{X}$, we start by partitioning $\mathcal{P}$ into three sets of values $\mathcal{P}_{\mathrm{hg}}$ (*huge* values), $\mathcal{P}_{\mathrm{lg}}$ (*large* values) and $\mathcal{P}_{\mathrm{sm}}$ (*small* values) as follows:

(a) $p_j \in \mathcal{P}_{\mathrm{hg}}$ if $p_j \geq \mathbf{U}^{-1}(\mathsf{OPT}/\epsilon)$.

(b) Let p^* be the largest element of $\mathcal{P}$ such that either $\sum_{p_j \notin \mathcal{P}_{\mathrm{hg}}, p_j \geq p^*} q_j \geq 1/\epsilon^4$ or $p^* = \mathbf{U}^{-1}(\epsilon^4 \mathsf{OPT})$. Then, $p_j \in P_{\mathrm{lg}}$ if $p_j \geq p^*$ and $p_j \notin \mathcal{P}_{\mathrm{hg}}$. Note that expected number of large value realizations is $O(\frac{1}{\epsilon^4})$. Note, once OPT and $\mathbf{X}$ are specified, p^* is uniquely defined.

(c) p_j is in $\mathcal{P}_{\mathrm{sm}}$ otherwise.

Informally speaking, we will exploit the following properties of this partitioning: *huge values occur occasionally* (and the rest of the realization can be ignored in such scenarios), *small values occur with high support* and hence admit concentration bounds, and the *total support for large values is bounded*.

Discretization of Large Values: With a small loss in the utility, we can reduce the number of distinct large values to be considered to $\mathrm{poly}(1/\epsilon)$. Specifically, we will assume that large values come only from the set $\mathbf{U}^{-1}(\epsilon^5 t \mathsf{OPT})$ where t is an integer between 0 and $\frac{1}{\epsilon^6}$. Any large value not in this set can be rounded down to the nearest large value – the loss in utility due to this discretization is at most $\epsilon^5 \mathsf{OPT}$ per realization of a large value. By concavity, the total loss in expected utility due to this discretization is at most $\epsilon^5 \mathsf{OPT}$ times the expected number of large value realizations. Since there are $O(\frac{1}{\epsilon^4})$ large value realizations in expectation, the expected loss in the utility due to discretization is $O(\epsilon \mathsf{OPT})$. Next, we split the set of random variables based on their support for large values.

Essential and Optimal Random Variables: We classify the set of random variables in $\mathbf{X}$ into two sets, based on their support for large values: X_i is *essential* if for some large (discretized) value p_j, $\mathbf{Pr}(X_i = p_j) \geq \epsilon^7$, and it is *optional* otherwise. As the total support over all large values is bounded by $1/\epsilon^4$, we get the following property:

Observation 1. *The number of essential random variables is at most $1/\epsilon^{11}$.*

We are now ready to state the utility equivalence theorem.

Theorem 1. *Fix an $\epsilon > 0$. Given any two sets of independent non-negative random variables, $\mathbf{X} = \{X_1, X_2, \cdots, X_n\}$ and $\mathbf{X}' = \{X_1', X_2', \cdots, X_m'\}$, and any non-decreasing concave function $\mathbf{U} : \mathbf{R}^+ \to \mathbf{R}^+$, if $\mathbf{X}$ and $\mathbf{X}'$ satisfy the following set of properties:*

(a) $\mathbf{X}$ and $\mathbf{X}'$ have the same set of essential random variables,
(b) for each $p_j \in \mathcal{P}_{lg}$, $\sum_i \mathbf{Pr}\,[X_i = p_j] = \sum_i \mathbf{Pr}\,[X_i' = p_j]$,
(c) $\sum_{i,p_j \in \mathcal{P}_{sm}} p_j \times \mathbf{Pr}\,[X_i = p_j] = \sum_{i,p_j \in \mathcal{P}_{sm}} p_j \times \mathbf{Pr}\,[X_i' = p_j]$, and
(d) $\sum_{i,p_j \in \mathcal{P}_{hg}} \mathbf{U}(p_j) \times \mathbf{Pr}\,[X_i = p_j] = \sum_{i,p_j \in \mathcal{P}_{hg}} \mathbf{U}(p_j) \times \mathbf{Pr}\,[X_i' = p_j]$,
then $\mathbf{E}\,[\mathbf{U}(\mathbf{X}')] \in (1 \pm O(\epsilon))\,(\mathbf{E}\,[\mathbf{U}(\mathbf{X})])$.

In the rest of the section, we prove Theorem 1. We define a set of independent random variables, $\mathbf{X}^{lg} = \{X_1^{lg}, X_2^{lg}, \ldots, X_n^{lg}\}$ where for each $i, p_j \in \mathcal{P}_{lg}$, $\mathbf{Pr}\,(X_i = p_j) = \mathbf{Pr}\,(X_i^{lg} = p_j)$ and X_i^{lg} is 0 otherwise; i.e. X_i^{lg} corresponds to the marginal distributions of X_i in the large value region. Further, let $R = \mathbf{E}\left[\sum_{i,p_j \in \mathcal{P}_{sm}} p_j \times \mathbf{Pr}\,[X_i = p_j]\right]$ and $H = \sum_{i,p_j \in \mathcal{P}_{hg}} \mathbf{U}(p_j) \times \mathbf{Pr}\,[X_i = p_j]$.

We now show that $\mathbf{E}\,[\mathbf{U}(\mathbf{X})]$ can be approximately expressed in terms of R, H and $\sum_i X_i^{lg}$, essentially linearizing the contribution of small values and huge values.

Theorem 2. $\mathbf{E}\,[\mathbf{U}(\mathbf{X})] \in (1 \pm O(\epsilon))\,(\mathbf{E}\,[\mathbf{U}\,(R + \sum_i X_i^{lg})] + H)$.

Proof. For each variable X_i, define variable X_i^{lg+sm} that has same distribution as X_i in the small and the large value regions; and it has 0 support in the huge value region; X_i^{lg+sm}s are independent. Using concavity and monotonicity of $\mathbf{U}$, we get,

$$\mathbf{E}\,[\mathbf{U}(\mathbf{X})] \leq \mathbf{E}\,[\mathbf{U}(\textstyle\sum_i X_i^{lg+sm})] + \textstyle\sum_{i,p_j \in \mathcal{P}_{hg}} \mathbf{Pr}\,[X_i = p_j] \times \mathbf{U}(p_j)$$

We next consider an experiment to lower bound the value of $\mathbf{E}\,[\mathbf{U}(\mathbf{X})]$. Arrange variables in order X_1 through X_n. For each $1 \leq i \leq n$, let E_i be an event that some variable in X_1 through X_i has realized to a huge value. Clearly, $\mathbf{Pr}\,[E_i] \leq \epsilon$ for each i. We measure the utility of first huge realization, if any. It is lower bounded by,

$$\textstyle\sum_{i,p_j \in \mathcal{P}_{hg}} \mathbf{Pr}\,[X_i = p_j] \times \mathbf{U}(p_j) \times (1 - \mathbf{Pr}\,[E_{i-1}]) \geq (1 - \epsilon)H$$

Next, we lower bound the utility from large and small value realizations; we measure their contribution only in the event of $\overline{E_n}$. We note, *for any given variable X_i, the information that it does not realize to a huge value, only increases its probability for other (small and large) values.* Thus, the utility from small and large value realizations can be lower bounded by

$$(1 - \mathbf{Pr}\,[E_n]) \times \mathbf{E}\,[\mathbf{U}(\textstyle\sum_i X_i^{lg+sm})]$$

This establishes

$$\mathbf{E}\left[\mathbf{U}(\mathbf{X})\right] \in (1 \pm O(\epsilon)) \left(\mathbf{E}\left[\mathbf{U}(\sum_i X_i^{\text{lg+sm}})\right] + \sum_{i, p_j \in \mathcal{P}_{\text{hg}}} \mathbf{Pr}\left[X_i = p_j\right] \times \mathbf{U}(p_j)\right)$$

It remains to separate the small value realizations; there are two cases to consider. If $\mathbf{U}(R) \leq \epsilon \mathbf{E}\left[\mathbf{U}(\mathbf{X})\right]$, then it can be ignored with only $\epsilon \mathbf{E}\left[\mathbf{U}(\mathbf{X})\right]$ loss in the utility. Otherwise, using Chernoff bounds, the total realized value of small values is $(1 \pm \epsilon)R$ w.h.p. $(1 - \epsilon)$. However, this does not immediately yield the required bound. To establish the required bound, we compare the values of $\mathbf{U}(\sum_i X_i^{\text{lg+sm}})$ and $\mathbf{U}(R + \sum_i X_i^{\text{lg}})$ in each realization using a coupling argument. In a realization of $\mathbf{X}^{\text{lg+sm}}$, let E_i' be an event in which the sum of small value realizations of variables in $\mathbf{X}_{-\mathbf{i}}^{\text{lg+sm}}$ sum to at least $(1 - \epsilon)R$. Clearly, $\mathbf{Pr}\left[E_i'\right] \geq 1 - \epsilon$. We note,

(a) $X_i^{\text{lg+sm}}$ is independent of E_i', and

(b) Conditioned on $E_i' = \text{TRUE}$, the distribution of $X_i^{\text{lg+sm}}$, treating small values as 0, is same as the distribution of X_i^{lg}.

Thus, we get

$$\mathbf{E}\left[\mathbf{U}(\sum_i X_i^{\text{lg+sm}})\right] \geq (1 - \epsilon) \times \mathbf{E}\left[\mathbf{U}\left((1 - \epsilon)R + \sum_i [X_i^{\text{lg}} | E_i']\right)\right]$$

The right hand side is minimized when E_i's are perfectly correlated; in which case we get,

$$\mathbf{E}\left[\mathbf{U}(\sum_i X_i^{\text{lg+sm}})\right] \geq (1 - \epsilon) \times (1 - \epsilon)\mathbf{E}\left[\mathbf{U}\left(R + \sum_i X_i^{\text{lg}}\right)\right]$$

The same coupling argument can also be used to establish the upper bound. This completes the proof of the theorem. $\qquad\square$

Thus it suffices to bound the value of $\mathbf{E}\left[\mathbf{U}\left(R + \sum_i X_i^{\text{lg}}\right)\right]$ where R is an arbitrary positive constant corresponding to the total expected value from small values and $\mathbf{X}^{\text{lg}} = \{X_1^{\text{lg}}, X_2^{\text{lg}}, \dots, X_n^{\text{lg}}\}$ are independent. We next define an operation *split* on variables in $\mathbf{X}^{\text{lg}}$.

Split: In this step, we replace each *optional* variable X_i^{lg} by a set of independent random variables $\{Z_{ijk} | p_j \in \mathcal{P}_{\text{lg}}, 1 \leq l \leq \ell \to \infty\}$, such that, for each j, k, $\mathbf{Pr}\left(Z_{ijk} = p_j\right) = \frac{\mathbf{Pr}(X_i^{\text{lg}} = p_j)}{\ell}$ and it is 0 otherwise. In Lemmas 3 and 4, we establish that the split operation can be performed on optional random variables in $\mathbf{X}^{\text{lg}}$ simultaneously without loosing the expected utility by much.

Lemma 3. *Let X_i^{lg} be any optional random variable and let Y be any non-negative random variable independent of X_i^{lg}. Consider a set of independent variables $\{Z_{ijk} | 1 \leq k \leq \ell \to \infty, p_j \in \mathcal{P}_{\text{lg}}\}$ that are generated by the split operation on X_i^{lg}. Then we have*

(a) $\mathbf{E}\left[\mathbf{U}\left(Y + X_i^{\text{lg}}\right)\right] - \mathbf{E}[\mathbf{U}(Y)] \geq \mathbf{E}\left[\mathbf{U}\left(Y + \sum_{j,k} Z_{ijk}\right)\right] - \mathbf{E}[\mathbf{U}(Y)]$

(b) $\mathbf{E}\left[\mathbf{U}\left(Y + X_i^{\text{lg}}\right)\right] - \mathbf{E}[\mathbf{U}(Y)] \leq (1 + O(\epsilon))\left(\mathbf{E}\left[\mathbf{U}\left(Y + \sum_{j,k} Z_{ijk}\right)\right] - \mathbf{E}[\mathbf{U}(Y)]\right)$

Proof. It suffices to establish the lemma for each realization R of Y. Let E_{ij} be an event in which at least one variable in set $\{Z_{ijk} | 1 \leq k \leq \ell\}$ realizes to a non-zero value (which is always p_j). Next, we establish an important property of these events.

Observation 2

$$\mathbf{Pr}\left[E_{ij} \cap_{j' \neq j} \overline{E_{ij'}}\right] \geq (1 - O(\epsilon))\mathbf{Pr}\left(X_i^{lg} = p_j\right) \quad and$$
$$\mathbf{Pr}\left[E_{ij} \cap_{j' \neq j} \overline{E_{ij'}}\right] \leq \mathbf{Pr}\left(X_i^{lg} = p_j\right)$$

Proof. We first compute the value of $\mathbf{Pr}[E_{ij}]$. Trivially, $\mathbf{Pr}[E_{ij}] \leq \mathbf{Pr}\left(X_i^{lg} = p_j\right)$. Further,

$$\mathbf{Pr}[E_{ij}] = \ell \times \left(1 - \frac{\mathbf{Pr}(X_i^{lg} = p_j)}{\ell}\right)^\ell \times \frac{\mathbf{Pr}(X_i^{lg} = p_j)}{\ell}$$
$$\text{using } \mathbf{Pr}\left(X_i^{lg} = p_j\right) \leq \epsilon^7, \ell \to \infty \text{ and for small } \epsilon$$
$$\geq (1 - 2\epsilon^7) \times \ell \times \frac{\mathbf{Pr}(X_i^{lg} = p_j)}{\ell}$$
$$= (1 - 2\epsilon^7)\mathbf{Pr}\left(X_i^{lg} = p_j\right)$$

Further, for each $p_{j'} \in \mathcal{P}_{lg}$,

$$\mathbf{Pr}[\overline{E_{ij'}}] \geq 1 - \mathbf{Pr}\left(X_i^{lg} = p_{j'}\right) \geq 1 - \epsilon^7$$

Using union bound,

$$\mathbf{Pr}[\cap_{j' \neq j}\overline{E_{ij'}}] \geq 1 - \epsilon^7 \times \frac{1}{\epsilon^6} = 1 - \epsilon$$

The lemma follows, as E_{ij} is independent of $\cap_{j' \neq j}\overline{E_{ij'}}$.   □

We now prove,

$$\mathbf{E}\left[\mathbf{U}\left(R + \sum_{j,k} Z_{ijk}\right)\right] - \mathbf{U}(R) \geq (1 - \epsilon)\mathbf{E}\left[\mathbf{U}\left(R + X_i^{lg}\right)\right] - \mathbf{U}(R)$$

Using Observation 2, we get

$$\mathbf{E}\left[\mathbf{U}\left(R + \sum_{j,k} Z_{ijk}\right)\right] - \mathbf{U}(R) \geq \sum_j (\mathbf{U}(R + p_j) - \mathbf{U}(R)) \times \mathbf{Pr}\left[E_{ij} \cap_{j' \neq j} \overline{E_{ij'}}\right]$$
$$\geq (1 - O(\epsilon))(\mathbf{U}(R + p_j) - \mathbf{U}(R))\mathbf{Pr}\left(X_i^{lg} = p_j\right)$$
$$= (1 - O(\epsilon))\mathbf{E}\left[\mathbf{U}\left(R + X_i^{lg}\right) - \mathbf{U}(R)\right]$$

It remains to prove

$$\mathbf{E}\left[\mathbf{U}\left(R + \sum_{j,k} Z_{ijk}\right)\right] - \mathbf{U}(R) \leq \mathbf{E}\left[\mathbf{U}\left(R + X_i^{lg}\right)\right] - \mathbf{U}(R)$$

This follows as X_i^{lg} can be seen as merged from $\{Z_{ijk}|p_j \in \mathcal{P}_{lg}, 1 \leq k \leq \ell\}$ and using Lemma 1. This completes the proof.   □

The proof of Lemma 3 in fact establishes a stronger property: for each optional random variable X_i^{lg}, define a set of random variables $\{T_{ij}|p_j \in \mathcal{P}_{lg}\}$; $T_{ij} = p_j$ if $[E_{ij} \cap_{j' \neq j} \overline{E_{ij'}}] = \mathbf{true}$, and 0 otherwise.

Corollary 1. *Let $X_i^{l_g}$ be any optional random variable and let Y be any non-negative random variable independent of $X_i^{l_g}$. Consider another set of independent variables $\{Z_{ijk}|1 \le k \le \ell \to \infty, p_j \in \mathcal{P}_{l_g}\}$ that are generated by the split operation on $X_i^{l_g}$. Then we have*

(a) $Y + X_i^{l_g} \succeq Y + \sum_j T_{ij}$,

(b) $\mathbf{E}\left[\mathbf{U}\left(Y + X_i^{l_g}\right)\right] - \mathbf{E}[\mathbf{U}(Y)] \ge \mathbf{E}\left[\mathbf{U}\left(Y + \sum_j T_{ij}\right)\right] - \mathbf{E}[\mathbf{U}(Y)]$,

(c) $\mathbf{E}\left[\mathbf{U}\left(Y + X_i^{l_g}\right)\right] - \mathbf{E}[\mathbf{U}(Y)] \le (1 + O(\epsilon))\left(\mathbf{E}\left[\mathbf{U}\left(Y + \sum_j T_{ij}\right)\right] - \mathbf{E}[\mathbf{U}(Y)]\right)$.

We now simultaneously split the set of all optimal random variables. Let S_1 and S_2 be the set of essential and optional random variables in $\mathbf{X}^{l_g}$ respectively. Define a random variable $Y = R + \sum_{i \in S_1} X_i^{l_g}$. Note, each random variable $X_i^{l_g} \in S_2$ is independent of Y. We next establish that all optional random variables can be simultaneously split without significantly affecting the expected utility.

Lemma 4. *(a)* $\mathbf{E}\left[\mathbf{U}\left(Y + \sum_{i \in S_2} X_i^{l_g}\right)\right] \le (1 + O(\epsilon))\mathbf{E}\left[\mathbf{U}\left(Y + \sum_{i \in S_2, j, k} Z_{ijk}\right)\right]$, *and*

(b) $\mathbf{E}\left[\mathbf{U}\left(Y + \sum_{i \in S_2} X_i^{l_g}\right)\right] \ge \mathbf{E}\left[\mathbf{U}\left(Y + \sum_{i \in S_2, j, k} Z_{ijk}\right)\right]$.

Proof. For expositional simplicity, we drop S_2 from the summation sign. For the remainder of proof, $\sum_i$ always refers to $\sum_{i \in S_2}$. We will first show that

$$\mathbf{E}\left[\mathbf{U}\left(Y + \sum_i X_i^{l_g}\right)\right] \le (1 + O(\epsilon))\mathbf{E}\left[\mathbf{U}\left(Y + \sum_{i,j} T_{ij}\right)\right]$$

which implies, (using monotonicity of $\mathbf{U}$, and as $\sum_k Z_{ijk} \ge T_{ij}$ per realization)

$$\mathbf{E}\left[\mathbf{U}\left(Y + \sum_i X_i^{l_g}\right)\right] \le (1 + O(\epsilon))\mathbf{E}\left[\mathbf{U}\left(Y + \sum_{i,j,k} Z_{ijk}\right)\right]$$

Define a random variable $\mathtt{XSUM}_i = Y + \sum_{i' \le i} X_{i'}^{l_g}$, and $\mathtt{TSUM}_i = Y + \sum_{i' \le i, j} T_{i'j}$. Using induction on Corollary 1, we get that, for each i, $\mathtt{XSUM}_i$ stochastically dominates $\mathtt{TSUM}_i$. We note,

$$\mathbf{E}\left[\mathbf{U}\left(Y + \sum_i X_i^{l_g}\right)\right] = \mathbf{E}\left[U(Y) + \left(\sum_i \mathbf{U}(X_i^{l_g} + \mathtt{XSUM}_{i-1}) - \mathbf{U}(\mathtt{XSUM}_{i-1})\right)\right]$$

Using Corollary 1

$$\le (1 + O(\epsilon))\mathbf{E}\left[\mathbf{U}(Y) + \left(\sum_i \mathbf{U}(\sum_j T_{ij} + \mathtt{XSUM}_{i-1}) - \mathbf{U}(\mathtt{XSUM}_{i-1})\right)\right]$$

Using stochastic dominance of $\mathtt{XSUM}_{i-1}$ over $\mathtt{TSUM}_{i-1}$

$$\le (1 + O(\epsilon))\mathbf{E}\left[\mathbf{U}(Y) + \left(\sum_i \mathbf{U}(\sum_j T_{ij} + \mathtt{TSUM}_{i-1}) - \mathbf{U}(\mathtt{TSUM}_{i-1})\right)\right]$$

$$= (1 + O(\epsilon))\mathbf{E}\left[\mathbf{U}\left(Y + \sum_{i,j} T_{ij}\right)\right]$$

It remains to establish that,

$$\mathbf{E}\left[\mathbf{U}\left(Y + \sum_i X_i^{l_g}\right)\right] \ge \mathbf{E}\left[\mathbf{U}\left(Y + \sum_{i,j,k} Z_{ijk}\right)\right]$$

This follows from Lemma 1, as each $X_i^{l_g}$ can be seen as merged from $\{Z_{ijk}|1 \le k \le \ell\}$. This completes the proof. $\square$

Theorem 1 follows as both $\mathbf{E}[\mathbf{U}(\mathbf{X})]$ and $\mathbf{E}[\mathbf{U}(\mathbf{X'})]$ can be approximated within $(1 \pm \epsilon)$ factor by the sum of same set of random variables. This completes the proof.

4   Applications to Stochastic Optimization Problems

We now illustrate how Theorem 1 can be used to to design a polynomial-time approximation scheme (PTAS) for concave utility maximization in a variety of stochastic optimization settings. We illustrate this in an abstract setting first, and then derive as a corollary a PTAS result for several stochastic optimization problems including stochastic knapsack with random profits and the stochastic spanning tree problem. We then show another interesting application, namely the problem of portfolio allocation for a risk-averse investor which cannot be directly captured in our abstract framework. However, we show that a PTAS can still be designed using a direct application of Theorem 1.

4.1   Framework for Concave Utility Maximization

An Abstract Utility Maximization Problem: We are given a set $E = \{e_1, e_2, \ldots, e_n\}$ of n elements; each element $e_i \in E$ is associated with a random non-negative profit X_i that is drawn from a distribution D_i, independently of profit of other elements. Let $F \subseteq 2^E$ be the set of feasible subsets of E. We are given an arbitrary non-decreasing concave utility function $U : R^+ \to R^+$ with $U(0) = 0$. The objective is to find a feasible set with maximum utility, i.e.

$$\text{Max}_{S \in F} \mathbf{E} \left[U(\textstyle\sum_{e_i \in S, X_i \sim D_i} X_i) \right]$$

We use $\text{RAND}(E, F, U, D)$ to refer to the above concave utility maximization problem, where $D = \{D_1, D_2, \ldots, D_n\}$ is the set of profit distributions for elements in E. We now relate this stochastic utility maximization problem to a deterministic feasibility problem with an identical space of feasible subsets.

A Deterministic Feasibility Problem: As above, we are given a set $E = \{e_1, e_2, \ldots, e_n\}$ of n elements. However, each element $e_i \in E$ is now associated with a *deterministic* profit p_i, and we are given a target value T. The goal is to solve the following feasibility problem:

$$\text{Does } \exists S \in F \text{ such that } \textstyle\sum_{e_i \in S} p_i = T?$$

We use $\text{DET}(E, F, \mathcal{P}, T)$ to refer to the above feasibility problem, where $\mathcal{P} = \{p_1, p_2, \ldots, p_n\}$ is the set of profit values associate with elements in E. The following theorem summarizes our result:

Theorem 3. *If there is an algorithm that solves the problem* $\text{DET}(E, F, \mathcal{P}, T)$ *with running time polynomial in n and the value of the profit target T, then there is a polynomial-time approximation scheme for the problem* $\text{RAND}(E, F, U, D)$.

Proof. Let X_i be the random variable corresponding to profit from element e_i. Let $\mathbb{A}$ be an optimal feasible set, and let OPT be its expected utility. We can assume w.l.o.g. that the value OPT is known to within a factor of $(1 \pm \epsilon/n)$. This can be done by a simple enumeration scheme that does not rely on any knowledge of the space of feasible subsets F: Let $e_i \in \mathbb{A}$ be such that $\mathbf{U}(X_i) = \max_{e_j \in S} \mathbf{E}\left[\mathbf{U}(X_j)\right]$. Then we know that OPT is guaranteed to be in the range $\mathbf{E}\left[\mathbf{U}(X_i)\right]$ and $n \times \mathbf{E}\left[\mathbf{U}(X_i)\right]$. We can try all choices of element X_i (and the ranges induced by them); within each range, it suffice to guess the value of the OPT within an additive error of $\frac{\epsilon}{n}\mathbf{E}\left[\mathbf{U}(X_i)\right]$.

We define $\mathbb{A}$'s feature set as

(a) the total support q_j for each large value p_j, i.e. $\sum_{e_i \in \mathbb{A}} \mathbf{Pr}\left[X_i = p_j\right]$,

(b) the expectation from small values R, i.e. $\sum_{e_i \in \mathbb{A}, p_j \in \mathcal{P}_{sm}} p_j \times \mathbf{Pr}\left[X_i = p_j\right]$,

(c) the utility from huge values H: $\sum_{e_i \in \mathbb{A}, p_j \in \mathcal{P}_{hg}} \mathbf{U}(p_j) \times \mathbf{Pr}\left[X_i = p_j\right]$, and

(d) the set of essential items corresponding to essential random variables.

We now guess $\mathbb{A}$'s feature set; we refer to each guess as a *configuration*. A careful reader will notice that the boundary between small and large values (p^*) is also dependent upon set $\mathbb{A}$: however, it does not need to be explicitly guessed. Each configuration implicitly defines a small-large boundary. Given the optimal (but unknown) set $\mathbb{A}$, there exists a configuration that matches the optimal set, hence the implicit definition of small-large boundary in the configuration also matches the small-large boundary for set $\mathbb{A}$.

We next bound the total number of distinct configurations:

(a) For each $p_j \in \mathcal{P}_{lg}$, the value of q_j is guessed to the nearest multiple of ϵ^9/n. The total error over all value buckets is less than ϵ^2/n, and so the loss in the utility due to rounding of the probability masses is bounded $(\epsilon \mathsf{OPT})/n$.

(b) R is guessed to the nearest multiple of $\mathbf{U}^{-1}\left((\epsilon \mathsf{OPT})/n\right)$.

(c) H is guessed to the nearest multiple of $(\epsilon \mathsf{OPT})/n$.

(d) There are at most $1/\epsilon^{11}$ essential random variables (by Observation 1); they are guessed explicitly from the set E.

Observation 3. *The number of distinct configurations is bounded by $n^{poly(1/\epsilon)}$.*

Similarly, we define a feature set for each element $e_i \in E$ by considering the distribution associated with it:

(a) its support for each large value $p_j \in \mathcal{P}_{lg}$,

(b) its expectation from small values: $\sum_{p_j \in \mathcal{P}_{sm}} p_j \times \mathbf{Pr}\left[X_i = p_j\right]$,

(c) its expected utility from huge values: $\sum_{p_j \in \mathcal{P}_{hg}} \mathbf{U}(p_j) \times \mathbf{Pr}\left[X_i = p_j\right]$, and

(d) a $\{0,1\}$ value that indicates whether it is a member of the (guessed) set of essential random variables.

We discretize the feature set of each element in the same manner as we discretized the feature set of the optimal solution.

We next use the technique developed in [13] where the entire feature set of an element (as well as the guessed optimal allocation) is represented using an integer and each feature takes the position of a specific digit; the value of this digit indicates the value of the corresponding feature. For any feature of type (a), (b) or (c), we use the discretized value. For the feature (d), the value for an element is either 1 (when it is an essential random variable) or 0 (otherwise); this value for the guessed optimal allocation is the number of essential random variables.

Given any set and any feature, it is easy to see that, if we add up the feature's value over all members of the set, then it matches the corresponding value for the entire set. We then choose as the base of the integral representation any integer that is at least $n + 1$ times the maximum value of an element for any feature. The maximum value of this integer is bounded by $n^{poly(1/\epsilon)}$. Let INT_i and $\text{INT}_\mathbb{A}$ be the integral representations of element e_i and the optimal set $\mathbb{A}$ respectively. We next invoke the pseudo-polynomial time algorithm for solving the feasibility for the deterministic version as follows. We assign a profit of INT_i to element e_i, and ask,

$$\text{Does } \exists S \subseteq F \text{ such that } \sum\nolimits_{e_i \in S} \text{INT}_i = \text{INT}_\mathbb{A}$$

This completes the proof of the Theorem 3. □

Next, we list two important stochastic combinatorial optimization problems for which a PTAS is implied using this reduction. We note, the earlier works on these problems (stochastic knapsack with random profits [4,9,15,16] and stochastic spanning tree problem [7,6]) have primarily focussed on the *value-at-risk* objective, i.e. either maximizing (or minimizing) the probability of crossing a given threshold value, and no prior results were known (to the best of our knowledge) for monotone utility functions.

Stochastic Knapsack with Random Profits: As discussed in Section 1, this problem models the asset allocation problem faced by a risk averse investor. For the knapsack problem, dynamic programming can be used to check whether there exists a set that has profit exactly equal to the target value.

Stochastic Spanning Tree: In this problem, the objective is to pick up a spanning tree with maximum (or similarly minimum) expected utility. It can be checked in pseudo-polynomial time (using techniques in [2]) whether there is a spanning tree with given weight.

Risk Averse Portfolio Allocation Problem: This problem is similar to the stochastic knapsack problem with random profits; we note the differences. Each item corresponds to an investment option and an arbitrary nonnegative amount of money can be invested in each investment option. The investor has a total budget of B dollars. Let λ_i be the his investment in option i. The rate of return (per unit investment) of option i is drawn from distribution D_i, independent of other options. The investor's objective is to maximize the expected utility of his future wealth; i.e. $\mathbf{E}_{\forall i, X_i \sim D_i} (\mathbf{U}(\sum_i \lambda_i \times X_i))$. While this problem has been extensively studied in the literature [8,12,11,5,10], no approximation schemes are known for this problem.

Theorem 3 cannot be applied directly here to get a polynomial time implementation as the ground element set is of unbounded size. However an explicit dynamic program can be designed using a direct application of Theorem 1. We guess the feature set of the optimal portfolio allocation along with investments (and their corresponding values) corresponding to essential random variables. The feasibility of each configuration can be checked using a dynamic program as follows: $A(i, B', \text{sm}, \{q_j|p_j \in \mathcal{P}_{\text{sm}}\}, \text{hg})$ indicates whether there is a feasible allocation that uses investments i through n, invests *at most* B' dollars, and has (i) sm as the total expected value in the small region, (ii) hg as the expected utility in the huge value region, and (iii) has a support of q_j for $p_j \in \mathcal{P}_{\text{lg}}$. The state space of the dynamic program is $n^{poly(1/\epsilon)}$, since the value for each parameter needs only to be stored in the increments of ϵ/n^2 (relative to the maximum value for this parameter).

Note that a direct application of this approach may violate the budget constraint; and the final solution may use a budget up to $(1 + \epsilon/n)B$. This is easy to fix: simply scale down the allocation to each investment (by a common factor) so that the total investment satisfies the budget constraint. Using concavity of $\mathbf{U}$, this will reduce the expected utility only by $\epsilon \text{OPT}/n$. Finally, we emphasize that the running time of the algorithm is *polynomial* in n and $\log B$.

References

1. Arrow, K.J.: The theory of risk aversion. Aspects of the Theory of Risk Bearing (1965)
2. Barahona, F., Pulleyblank, W.: Exact arborescences, matchings and cycles. Discrete Applied Mathematics (1987)
3. Bhalgat, A., Chakraborty, T., Khanna, S.: Mechanism design for a risk-averse seller. In: Goldberg, P.W. (ed.) WINE 2012. LNCS, vol. 7695, pp. 198–211. Springer, Heidelberg (2012)
4. Carraway, R.L., Schmidt, R.L., Weatherford, L.R.: An algorithm for maximizing target achievement in the stochastic knapsack problem with normal returns. Naval Res. Logist. 40, 161–173 (1993)
5. Fishburn, P., Porter, B.: Optimal portfolios with one safe and one risky asset: Effects of changes in rate of return and risk. Management Science (1976)
6. Geetha, S., Nair, K.: On stochastic spanning tree problem. Networks (1993)
7. Shiode, S., Ishii, H., Nishida Yoshikazu, T.: Stochastic spanning tree problem. Discrete Applied Mathematics (1981)
8. Hadar, J., Seo, T.: Asset proportions in optimal portfolios. Review of Economics Studies (1988)
9. Henig, M.: Risk criteria in stochastic knapsack. Oper. Res. 38(5), 820–825 (1990)
10. Kijima, M., Ohnishi, M.: Portfolio selection problems via the bivariate characterization of stochastic dominance relations. Journal Mathematical Finance (1996)
11. Landsberger, M., Meilijson, I.: Demand for risky financial assets: A portfolio analysis. Journal of Economic Theory (1990)
12. Lappan, H., Hennessey, D.: Symmetry and order in the portfolio allocation problem. Economic Theory (2002)
13. Li, J., Deshpande, A.: Maximizing expected utility for stochastic combinatorial optimization problems. In: FOCS (2011)
14. Pratt, J.W.: Risk aversion in the small and in the large. Econometrica 32(1-2), 122–136 (1964)
15. Sneidovik, M.: Preference order stochastic knapsack problems: Methodological issues. J. Oper. Res Soc. 31(11), 1025–1032 (1980)
16. Steinberg, E., Parks, M.S.: Preference order dynamic program for a knapsack problem with stochastic rewards. J. Oper. Res Soc. 30(2), 141–147 (1979)

Network Improvement for Equilibrium Routing*

Umang Bhaskar, Katrina Ligett, and Leonard J. Schulman

California Institute of Technology, Pasadena, CA 91125 USA
{umang,katrina,schulman}@caltech.edu

Abstract. In routing games, agents pick routes through a network to minimize their own delay. A primary concern for the network designer in routing games is the average agent delay at equilibrium. A number of methods to control this average delay have received substantial attention, including network tolls, Stackelberg routing, and edge removal.

A related approach with arguably greater practical relevance is that of making investments in improvements to the edges of the network, so that, for a given investment budget, the average delay at equilibrium in the improved network is minimized. This problem has received considerable attention in the literature on transportation research. We study a model for this problem introduced in transportation research literature, and present both hardness results and algorithms that obtain tight performance guarantees.

- In general graphs, we show that a simple algorithm obtains a 4/3-approximation for affine delay functions and an $O(p/\log p)$-approximation for polynomial delay functions of degree p. For affine delays, we show that it is NP-hard to improve upon the 4/3 approximation.
- Motivated by the practical relevance of the problem, we consider restricted topologies to obtain better bounds. In series-parallel graphs, we show that the problem is still NP-hard. However, we show that there is an FPTAS in this case.
- Finally, for graphs consisting of parallel paths, we show that an optimal allocation can be obtained in polynomial time.

1 Introduction

Routing games are widely used to model and analyze networks where traffic is routed by multiple users, who typically pick their route to minimize their delay [22]. Routing games capture the uncoordinated nature of traffic routing. A prominent concern in the study of these games is the overall social cost, which is usually taken to be the average delay suffered by the players at equilibrium. It is well known that equilibria are generally suboptimal in terms of social cost. The ratio of the average delay of the worst equilibrium routing to the optimal routing

---

* Supported in part by NSF Awards 1038578 and 1319745, an NSF CAREER Award (1254169), the Charles Lee Powell Foundation, and a Microsoft Research Faculty Fellowship.

J. Lee and J. Vygen (Eds.): IPCO 2014, LNCS 8494, pp. 138–149, 2014.

that minimizes the average delay is called the price of anarchy; tight bounds on the price of anarchy are well-studied and are known for various classes of delay functions [18, Chapter 18].

However, the notion of price of anarchy assumes a fixed network. In reality, of course, networks change, and such changes may intentionally be implemented by the network designer to improve quality of service. This raises the question of how to identify cost-effective network improvements. Our work addresses this fundamental design problem. Specifically, given a budget for improving the network, how should the designer allocate the budget among edges of the network to minimize the average delay at equilibrium in the resulting, improved network? This crucial question arises frequently in network planning and expansion, and yet seems to have received no attention in the algorithmic game theory literature. This is surprising considering the attention given to other methods of improving equilibria, e.g., edge tolls and Stackelberg routing.

Our model of network improvement is adopted from a widely studied problem in transportation research [27] called the Continuous Network Design Problem (CNDP) [1]. In this model, each edge in the network has a delay function that gives the delay on the edge as a function of the traffic carried by the edge. Specifically, the delay function on each edge consists of a free-flow term (a constant), plus a congestion term that is the ratio of the traffic on the edge to the conductance of the edge, raised to a fixed power. The cost to the network designer of increasing the conductance of an edge by one unit is an edge-specific constant. Our objective is to select an allocation of the improvement budget to the edges that minimizes the social cost of equilibria in the improved network.

The continuous network design problem, along with the discrete network design problem that deals with the creation (rather than improvement) of edges, has been referred to as "one of the most difficult and challenging problems facing transport" [27]. The CNDP is generally formulated as a mathematical program with the budget allocated to each edge and the traffic at equilibrium as variables. Since the traffic is constrained to be at equilibrium, such a formulation is also called a Mathematical Program with Equilibrium Constraints (MPEC). Further, since the traffic at equilibrium is itself obtained as a solution to a optimization problem, this is also a bilevel optimization problem. Both bilevel optimization problems and MPECs have a number of other applications and have been studied independent of the CNDP as well (e.g., [6]).

Owing both to the rich structure of the problem and its practical relevance, the CNDP has received considerable attention in transportation research. Because of the nonconvexity and the complex nature of the constraints, the bulk of the literature focuses on heuristics, and proposed algorithms are evaluated by performance on test data rather than formal analysis. Many of these algorithms are surveyed in [27]. More recent papers give algorithms that obtain global optima [15,16,25], but make no guarantees on the quality of solutions that can be obtained in polynomial time.

In this paper, we consider a model with fixed demands, separable polynomial delay functions on the edges and constant improvement costs. This particular model, and further restrictions of it, have been the focus of considerable attention, e.g., [10,12,17], and is frequently used for test instances. The model captures many of the essential characteristics of the more general problems, such as the bilevel and nonconvex nature of the problem and the equilibrium constraints. Our work gives the first algorithmic results with proven output quality and runtime for the network improvement problem.

Our Contributions. We first focus on general graphs, and show that an algorithm that relaxes equilibrium constraints on the flow gives an approximation guarantee that is tight for linear delays.

- We show that for general networks with multiple sources and sinks and polynomial delays, the algorithm described gives an $O(p/\log p)$-approximation to the optimal allocation, where p is the maximum degree of the polynomial delay functions. If $p = 1$, this gives a 4/3-approximation algorithm.
- We show that the approximation ratio for linear delays is tight, even for the single-commodity case: by a reduction similar to that used by Roughgarden [23], we show that it is NP-hard to obtain an approximation ratio better than 4/3.

The hardness of approximation crucially depends on the generality of the network topology. The practical relevance of the network improvement problem then motivates us to consider restricted topologies of networks. We first consider series-parallel graphs and give tight approximation guarantees for single-commodity instances.

- We show that obtaining the optimal allocation even in instances with affine delays on *series-dipole graphs*, a very limited subset of series-parallel graphs, is NP-hard.
- However, in series-parallel networks with polynomial delays, we can obtain a nearly-optimal allocation in polynomial time, i.e., we give an FPTAS[1].

Finally, we give efficient exact optimization algorithms for even more restricted instances: parallel s-t paths with linear delays. Even this is a non-convex optimization problem, but we nevertheless show that first-order conditions are sufficient for optimality. However, due to the nonconvexity, gradient descent methods may not converge in poly-time. We use the structure of the first-order conditions to give a poly-time algorithm to find Opt. (And an even simpler algorithm when each path is a single edge.) Note that finding the optimal Stackelberg routing is NP-hard even in parallel links [21].

Our work thus presents a fairly comprehensive set of approximation guarantees for network improvement. Both the lower bounds and approximation guarantees we obtain are tight for the topologies we consider. Our results thus complement the work in transportation research on the problem, by formalizing the (existing) intuition that the problem is hard, and giving tight approximation algorithms for restricted instances.

---

[1] A fully polynomial-time approximation scheme (FPTAS) is a sequence of algorithms $\{A_\epsilon\}$ so that, for any $\epsilon > 0$, A_ϵ runs in time polynomial in the input and $1/\epsilon$ and outputs a solution that is at most a $(1 + \epsilon)$ factor worse than the optimal solution.

2 Related Work

Routing games as a model of traffic on roads were introduced by Wardrop in 1952 [26]. The equilibrium in these games is thus known as Wardrop equilibrium. Beckmann et al. [3] showed that equilibria in routing games are obtained as the solution to a strictly convex optimization problem if all delay functions are increasing, thus establishing the existence and uniqueness of equilibria. The price of anarchy — the ratio of the social cost at the worst equilibrium to the minimum social cost — was introduced by Papadimitriou [19] as a formal measure of inefficiency. For routing games with the social cost given by the average delay, the price of anarchy is known to be $4/3$ for linear delays [24], and $\Theta(p/\log p)$ for delay functions that are polynomials of degree p [20].

Significant research has gone into the use of tolls to improve the efficiency of routing games. Tolls to induce any minimal routing including the routing of minimum total delay can be obtained as the solution to a linear program [3,8,13,28]. Another method studied for improving the efficiency of routing is Stackelberg routing, which assumes that a fraction of the traffic is centrally controlled and is routed to improve efficiency. Obtaining the optimal Stackelberg routing is NP-hard even in parallel links [21], although a fully-polynomial time approximation scheme is known for this case [14]. Roughgarden [23] studies the problem of removing edges from a network to minimize the delay at equilibrium in the resulting network. The problem is strongly NP-hard, and there is no algorithm with an approximation ratio better than $n/2$ for general delay functions.

The importance of the network improvement problem has caused it to receive significant attention in transportation research, where the version we are considering is known as the continuous network design problem. Early research focused on heuristics that did not give any guarantees about the quality of the solution obtained. These results are surveyed in [27].

More recent work in the transportation literature has also tried to obtain algorithms that obtain global minima for the continuous network design problem. Early approaches include the use of simulated annealing [10] and genetic algorithms [30]. Li et al. [15] reduce the problem to a sequence of mathematical programs with concave objectives and convex constraints, and show that the accumulation point of the sequence of solutions is a global optimum. If the sequence is terminated early, they show weak bounds on the quality of the solution that are consequential only under strong assumptions on the delay function and agents' demands. Wang and Lo [25] reformulate the problem as a mixed integer linear program (MILP) by replacing the equilibrium constraints by constraints containing binary variables for each path, and using a number of linear segments to approximate the delay functions. This approach was further developed by Luathep et al. [16] who replaced the possibly exponentially many path variables by edge variables and gave a cutting constraint algorithm for the resulting MILP. The last two methods converge to a global optimum of the linearized approximation in finite time, but require solving a MILP with a possibly exponential number of variables and constraints.

A variant of the problem where the initial conductance of every edge in the network is zero, and the budget is part of the objective rather than a hard constraint, is studied by Marcotte [17] and, independent of our work, by Gairing et al. [11]. Unlike the work cited earlier, these papers give provable guarantees on the performance of polynomial-time algorithms. Marcotte gives an algorithm that is a 2-approximation for monomial delay functions and a 5/4-approximation for linear delay functions. Gairing et al. present an algorithm that improves upon these upper bounds, give an optimal polynomial-time algorithm for single-commodity instances, and show that the problem is APX-hard in general. In our problem, the budget is a hard constraint, and edges may have arbitrary initial capacities. Our problem is demonstrably harder than this variant: e.g., in contrast to the polynomial-time algorithm for single-commodity instances given by Gairing et al. [11], we show that in our problem no approximation better than 4/3 is possible even in single-commodity instances.

3   Notation and Preliminaries

$G = (V, E)$ is a directed graph with $|E| = m$ and $|V| = n$. If G is a two-terminal graph, then it has two special vertices s and t called the source and the sink, collectively called terminals. A u-v path $p = ((v_0, v_1), (v_1, v_2), \ldots, (v_{k-1}, v_k))$ is a sequence of edges with $v_0 = u$, $v_k = v$ and edges $(v_i, v_{i+1}) \in E$. In a two-terminal graph, each edge is on an s-t path. We use $\mathcal{P}$ to denote the set of all s-t paths. Given vertices s', t' in graph G, vector $(f_e)_{e \in E}$ is an s'-t' flow of value d if:

$$\sum_{(u,w) \in E} f_{uw} - \sum_{(w,u) \in E} f_{wu} = 0, \ \forall u \in V \setminus \{s', t'\}$$

$$\sum_{(s,w) \in E} f_{sw} - \sum_{(w,s) \in E} f_{ws} = d$$

$$f_e \geq 0, \ \forall e \in E.$$

We use $|f|$ to denote the value of flow f. A path decomposition of an s'-t' flow f is a set of flows $\{f_p\}$ along s'-t' paths p that satisfies $f_e = \sum_{p:e \in p} f_p$, $\forall e$. A path decomposition for flow f so that $f_p > 0$ for at most m paths can be obtained in polynomial time [2]. Without reference to a path decomposition, we use $f_p > 0$ to indicate that $f_e > 0$ for all $e \in p$.

Each edge $e \in E$ has an increasing delay function $l_e(x)$ that gives the delay on the edge as a function of the flow on the edge. For flow f and path p, $l_p(f) := \sum_{e \in p} l_e(f_e)$ is the delay on path p. Further, $f_e l_e(f_e)$ is the total delay on edge e, and the total delay of flow f is $\sum_{e \in E} f_e l_e(f_e)$.

Routing games. A routing game is a tuple $\Gamma = (G, l, K)$ where G is a directed graph, l is a vector of delay functions on edges, and $K = \{s_i, t_i, d_i\}_{i \in I}$ is a set of triples where d_i is the total traffic routed by players of commodity i from s_i to t_i. Each player of commodity i in a routing game controls infinitesimal traffic and picks an s_i-t_i path p on which to route her flow, as her strategy. The strategies induce a flow f^i. Let $f = \sum_i f^i$, then the delay of a player that selects path p as her strategy is $l_p(f)$. In the single-commodity case, $|I| = 1$. We say a flow f is a valid flow for routing game Γ if $f = \sum_{i \in I} f^i$ where each f^i is an s_i-t_i flow of value d_i.

At equilibrium in a routing game, each player minimizes her delay, subject to the strategies of the other players. This is also called a Wardrop equilibrium.

Definition 1. *A set of flows $\{f^i\}_{i \in I}$ where f^i is an s_i-t_i flow of value d_i is a Wardrop equilibrium if for all $i \in I$, for any s_i-t_i paths p, q such that $f^i_p > 0$, $l_p(f) \le l_q(f)$.*

The *equilibrium flow*, i.e., the collection of flows $\{f^i\}_{i \in I}$ that form a Wardrop equilibrium, is also the solution to the following mathematical program:

$$\min \sum_{e \in E} \int_0^{f_e} l_e(x)\, dx, \text{ s.t. } f = \sum_{i \in I} f^i \text{ and } f^i \text{ is an } s_i\text{-}t_i \text{ flow of value } d_i.$$

Since the delay functions are increasing, the program has a strictly convex objective with linear constraints, and hence the first-order conditions are necessary and sufficient for optimality. Further, because of strict convexity, the equilibrium flow is unique. Definition 1 then corresponds to the first-order conditions for optimality of the convex program. By Definition 1, each s_i-t_i path p with $f^i_p > 0$ has the same delay at equilibrium. Let L^i be this common path delay. Then the total delay $\sum_e f_e l_e(f_e) = \sum_i d_i L^i$, where $f = \sum_{i \in I} f^i$ and $\{f^i\}_{i \in I}$ is the equilibrium flow. The average delay is $\sum_i d_i L^i / \sum_i d_i$.

Network Improvement. In the network improvement problem, we are given a routing game Γ, where the delay function on each edge e is of the form $l_e(x) = (x/c_e)^{n_e} + b_e$. We call c_e the *conductance*, $1/c_e$ the *resistance*, and b_e the *length* of edge e. We assume $c_e \ge 0$ and $n_e > 0$, and hence the delay is an increasing function of the flow on the edge. The delay function on an edge is affine if $n_e = 1$. Each edge has a marginal cost of improvement, μ_e. Upon spending β_e to improve edge e, the conductance of the edge increases to $c_e + \mu_e \beta_e$. For a given budget B, a valid allocation is a vector $\beta = (\beta_e)_{e \in E}$ so that $\sum_e \beta_e \le B$ and $\beta_e \ge 0$ for each $e \in E$. The objective is to determine a valid allocation of the budget B to the edges to minimize the average delay obtained at equilibrium with the modified delay functions $l_e(x, \beta_e) = (x/(c_e + \mu_e \beta_e))^{n_e} + b_e$. Delay functions are affine if $n_e = 1$ on all edges.

Let $\beta = (\beta_e)_{e \in E}$ be the vector of edge allocations. Since the flow at equilibrium is unique, for any β, the average delay at equilibrium is unique. $L(\beta)$ is this unique average delay as a function of the edge allocations. When considering a flow f other than the equilibrium flow, we use $L(f, \beta)$ to denote the average delay of flow f with the modified delay functions. We will also have occasion to allocate budget to units other than edges, e.g., paths, and will slightly abuse notation to express the average delay in terms of these units.

Our problem corresponds to the following (non-linear, possibly non-convex) optimization problem:

$$\min_{\beta} L(\beta), \text{ s.t. } \sum_e \beta_e \leq B, \; \beta_e \geq 0 \; \forall e \in E. \tag{1}$$

We use β^* to denote an optimal solution for this problem, and define $L^* := L(\beta^*)$. As is common in nonlinear optimization, instead of an exact solution we will obtain a solution that is within a specified additive tolerance of ϵ of the exact solution, i.e., a valid allocation $\hat{\beta}$ so that $L(\hat{\beta}) - \epsilon \leq L(\beta^*)$. An algorithm is polynomial-time if it obtains such a solution in time polynomial in the input size and $\log(1/\epsilon)$. Since the problem has linear constraints, the first-order conditions are necessary for optimality (e.g., [29]). By the first-order conditions for optimality, for any edges e and e',

$$\beta_e > 0 \Rightarrow \frac{\partial L(\beta)}{\partial \beta_e} \leq \frac{\partial L(\beta)}{\partial \beta_{e'}}. \tag{2}$$

For any edge e and allocation β, define $c_e(\beta) = c_e + \mu_e \beta_e$. For a path p, $b_p = \sum_{e \in p} b_e$ is the length of path p. For affine delay functions, define $c_p(\beta) = 1/\sum_{e \in p} \frac{1}{c_e(\beta)}$ as the conductance of path p, and the resistance of path p as the reciprocal of the conductance: $r_p(\beta) = 1/c_p(\beta)$. For $k \in \mathbb{Z}_+$, $[k] := \{1, 2, \ldots, k\}$.

All missing proofs appear in the full version of the paper [4].

4   General Graphs

We present upper bounds for a classical approach introduced in [7,17], and show that it gives a good approximation for the general network improvement problem: with multiple commodities, in general graphs, and with polynomial delay functions. This approach relaxes the equilibrium constraints on the flow and solves the resulting convex program. The analysis relies on an unusual application of the well-known price of anarchy bounds for routing games [20,24].

Theorem 2. *There is a polynomial-time algorithm that gives a 4/3-approximation for the network improvement problem with affine delay functions, and an $O(p/\log p)$-approximation if all delay functions have degree at most p.*

The upper bounds in the theorem are tight for affine delays, even for single-commodity routing games. To show this, we give a reduction from the problem of 2-Directed Disjoint Paths, which is known to be NP-complete [9]. Our reduction is similar to that given by [23] for the problem of removing edges from a network to improve the total delay at equilibrium in the resulting network.

Theorem 3. *It is NP-hard to obtain an approximation ratio better than 4/3 even in single-commodity instances of network improvement with affine delays.*

5   Series-Parallel Graphs

5.1   NP-Hardness for Series-Dipole Graphs

We show that even in fairly simple *series-dipole networks*, with affine delay functions, the network improvement problem is NP-hard. Series-dipole graphs are a special case of series-parallel graphs, and consist of a series of *dipole graphs*—subgraphs consisting of parallel edges. In fact, we show that even when each dipole consists of just two edges, computing the optimal allocation is NP-hard.

Theorem 4. *The network improvement problem in series-dipole graphs with two edges in each dipole and affine delay functions is weakly NP-hard.*

In the proof in the full version [4], we show a reduction from the Partition problem. An instance of Partition consists of n items, each with value v_i and $\sum_{i \in [n]} v_i = 2V$. The problem is to select a subset S of the items so that $\sum_{i \in S} v_i = V$. In our reduction, there is a dipole for each item. Let $L_i(x)$ be the delay at equilibrium across the terminals of dipole i for an optimal allocation of x to the edges of dipole i. The crucial part of our reduction is to construct dipoles where the *sum* $L_i(x) + x$ is minimized at exactly two points (Figure 1). These points intuitively correspond to the inclusion and exclusion of item i in set S. Further, the reduction must ensure that (i) the difference in the allocation between these two points is exactly v_i, and (ii) the difference in the delay at equilibrium between these two points is exactly v_i. The construction of such dipoles with just affine delays draws upon structural properties of dipoles.

5.2   An FPTAS for Series-Parallel Graphs

For single-commodity instances of network improvement in series-parallel networks with polynomial delays, we show in this section that we can obtain near-optimal algorithms that run in polynomial time, i.e., we obtain an FPTAS. The standard approach for an FPTAS is to discretize the space of possible solutions, and use an efficient algorithm such as dynamic programming to find the optimal solution in this discretized space. In our problem, this approach encounters a

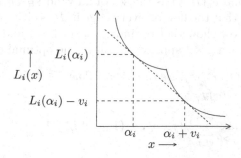

Fig. 1. The $L_i(x)$ curve has slope -1 at exactly two points, $x = \alpha_i$ and $x = \alpha_i + v_i$, and hence $L_i(x) + x$ is minimized at exactly these points

number of difficulties. Firstly, since the equilibrium flow varies as the allocation to edges changes, it is not sufficient to discretize the space of allocations. We need to discretize the space of flows as well, and consider both flows and allocations as variables in our optimization problem. Secondly, we cannot use the total delay as our minimization objective. Since both the flow and allocation are variables in our optimization problem, minimizing the total delay can only yield an $O(p/\log p)$-approximation as obtained earlier. To obtain a better approximation, we will minimize a different objective. Thirdly, since we need multiplicative guarantees, on edges where the optimal allocation β_e^* is small the standard discretization will be insufficient. We handle this last case in the full version, and discuss how we solve the first two problems here.

We first discuss our objective. Instead of the total delay, our objective will be to find the allocation and flow that minimize the maximum delay on paths with positive flow. There are two reasons why this is a good objective. Firstly, for single-commodity equilibrium flows, this is exactly the average delay, since at equilibrium any path with positive flow has the same delay. Lemma 5 shows the second reason why this is a good objective: in series-parallel graphs, the equilibrium flow minimizes this value. The lemma holds only for series-parallel graphs, and the standard examples of inefficiency in the Braess graphs show that the lemma does not hold in general.

Lemma 5. *Let f be the equilibrium flow in routing game Γ on series-parallel graph G, and g be any s-t flow of value d. Then $\max_{p:f_p>0} l_p(f) \leq \max_{p:g_p>0} l_p(g)$.*

As usual, β^* is the optimal allocation for the network improvement instance, and f^* and L^* are the equilibrium flow and average delay for allocation β^*. We now construct discretized spaces F_ϵ and A_ϵ of possible flows and allocations respectively. Define $\nu := \max_e n_e$ as the maximum exponent of the delay function on any edge. Given a parameter $\epsilon > 0$, $\lambda := \epsilon^2/m^2$ is our unit of discretization, where as before $m = |E|$. For clarity of presentation, we assume that $1/\lambda$ is integral. Further, as mentioned earlier, we assume that $\beta_e^* \geq \lambda B/\epsilon$ on every edge, and remove this assumption in the full version. For any subgraph H of G and $k \in \mathbb{Z}_+$, define $A_\epsilon(H,k)$ as the set of all valid allocations of budget $kB\lambda$ to edges in H, so that the allocation to each edge is either 0 or an integral multiple of $B\lambda$. Similarly, define $F_\epsilon(H,k)$ as the set of all valid s_H-t_H flows on the edges of H of value $kd\lambda$, so that the flow on every edge in H is either zero or an integral multiple of $d\lambda$. We now show that optimizing over flows and allocations in the discretized space gives a good approximation to the optimal delay.

Lemma 6. *In time $O(m^3/\epsilon^2)$, we can obtain flow $\hat{f} \in F_\epsilon(G,1/\lambda)$ and allocation $\hat{\beta} \in A_\epsilon(G,1/\lambda)$ that satisfy*

$$\max_{p:\hat{f}_p>0} \sum_{e\in p} l_e(\hat{f}_e,\hat{\beta}_e) \leq (1+\epsilon)^\nu L^*/(1-\epsilon)^\nu.$$

Here, we only prove the existence of $\hat{f}$ and $\hat{\beta}$. The algorithm to obtain $\hat{f}$ and $\hat{\beta}$ uses dynamic programming and takes advantage of the recursive construction of series-parallel graphs, and we leave this to the full version. For the existence, we first show there exist $\hat{f}$ and $\hat{\beta}$ that are near f^* and β^* on every edge.

Claim 7. *There exists flow $\hat{f} \in F_\epsilon(G, 1/\lambda)$ that satisfies $\hat{f}_e \leq (1 + \epsilon)f_e^*$ for all $e \in E$, and allocation $\hat{\beta} \in A_\epsilon(G, 1/\lambda)$ that satisfies $\hat{\beta}_e \geq \beta_e^*(1 - \epsilon)$.*

Proof. For allocation $\hat{\beta}$, round β_e^* down to the nearest multiple of λB to obtain $\hat{\beta}_e$, and on an abitrary edge e_1, allocate $\hat{\beta}_{e_1} = B - \sum_{e \neq e_1} \hat{\beta}_e$. Since the allocation to every other edge is an integral multiple of $B\lambda$, so is the allocation to edge e_1. Allocation $\hat{\beta}$ is obviously a valid allocation of budget B, and thus $\hat{\beta} \in A_\epsilon(G, 1/\lambda)$. Further, on every edge, $\hat{\beta}_e \geq \beta_e^* - B\lambda \geq \beta_e^* - \epsilon\beta_e^* = \beta_e^*(1-\epsilon)$, where the inequality is since by assumption $\beta_e^* \geq B\lambda/\epsilon$.

For $\hat{f}$, let $\{f_p^*\}_{p \in P}$ be a flow decomposition of f^* with at most m paths in P with $f_p^* > 0$. There is a path q with $f_q^* \geq md\lambda/\epsilon$ in this decomposition. For $p \neq q$, round f_p^* down to the nearest multiple of $d\lambda$ to obtain $\hat{f}_p$. Assign the remaining flow $d - \sum_{p \neq q} \hat{f}_p$ to path q. The flow on every edge is then an integral multiple of $d\lambda$, and $\hat{f}$ is a flow of value v. Hence $\hat{f} \in F_\epsilon(G, 1/\lambda)$. Further, for $p \neq q$, $\hat{f}_p \leq f_p^*$. For path q, $\hat{f}_q \leq f_q^* + md\lambda \leq f_q^* + \epsilon f_q^* = (1 + \epsilon)f_q^*$. Hence for every edge, $\hat{f}_e \leq (1 + \epsilon)f_e^*$. □

For flow $\hat{f}$ and allocation $\hat{\beta}$ obtained in Claim 7, if $\hat{f}_e > 0$, then $f_e^* > 0$. Hence for path p with $\hat{f}_e > 0$ for all $e \in p$, by Claim 7 and since the delay functions are polynomials of degree at most ν,

$$\sum_{e \in p} l_e(\hat{f}_e, \hat{\beta}_e) \leq \left(\frac{1 + \epsilon}{1 - \epsilon}\right)^\nu \sum_{e \in p} l_e(f_e^*, \beta_e^*) = \left(\frac{1 + \epsilon}{1 - \epsilon}\right)^\nu L^* .$$

The proof of the lemma follows immediately.   □

Note that the flow $\hat{f}$ obtained is not an equilibrium flow for the allocation $\hat{\beta}$, but by Lemma 5 it must be close to an equilibrium flow. We now show the existence of an FPTAS for series-parallel graphs.

Theorem 8. *For any $\epsilon' > 0$, for single-commodity network improvement in series-parallel graphs there is a $(1 + \epsilon')$-approximate algorithm that runs in time polynomial in $1/\epsilon'$, ν, and the size of the input, where ν is the maximum degree of any delay function.*

Proof. Let $\epsilon > 0$ be a parameter we fix later. Let $\hat{f}$ and $\hat{\beta}$ be the flow and allocation from Lemma 6, and f be the equilibrium flow for allocation $\hat{\beta}$. By Lemma 6 and 5,

$$L(f, \hat{\beta}) = \max_{p:f_p > 0} \sum_{e \in p} l_e(f_e, \hat{\beta}_e) \leq \max_{p:\hat{f}_p > 0} \sum_{e \in p} l_e(\hat{f}_e, \hat{\beta}_e) \leq (1 + \epsilon)^\nu L^*/(1 - \epsilon)^\nu .$$

The theorem is satisfied by choosing $\epsilon = \epsilon'/6\nu$.   □

6   Parallel Paths

If G is an s-t path, then the delay at equilibrium $L(\beta)$ is a convex function. Thus, obtaining the optimal allocation requires minimizing a convex function subject to linear constraints, which is polynomial-time solvable by, e.g., interior-point methods [5].

When G consists of parallel paths between s and t, $L(\beta)$ may not be a convex function of β. We prove however that in the single-commodity case, with all delay functions affine, the first-order conditions for optimality are sufficient. Hence, any solution that satisfies the first-order optimality conditions is a global minimum. The nonconvexity however implies that algorithms based on gradient descent may not converge in polynomial time. Instead, we give an algorithm that solves a particular convex relaxation within a binary search framework to obtain the optimal allocation in polynomial time.

Theorem 9. *There is a polynomial-time algorithm for single-commodity network improvement in graphs consisting of parallel s-t paths with affine delay functions.*

Further, if each s-t path is a single edge, we prove that there is an optimal solution where the entire budget is spent on a single edge. This characterization yields a simple optimal algorithm for the case of parallel edges — try all edges, and allocate the entire budget to the edge for which the delay obtained is minimum.

References

1. Abdulaal, M., LeBlanc, L.J.: Continuous equilibrium network design models. Transportation Research Part B: Methodological 13(1), 19–32 (1979)
2. Ahuja, R.K., Magnanti, T.L., Orlin, J.B.: Network flows: theory, algorithms, and applications. Prentice-Hall, Inc., Upper Saddle River (1993)
3. Beckmann, M., McGuire, C.B., Winsten, C.B.: Studies in the economics of transportation. Yale University Press (1956)
4. Bhaskar, U., Ligett, K., Schulman, L.J.: The network improvement problem for equilibrium routing. CoRR abs/1307.3794 (2013)
5. Boyd, S., Vandenberghe, L.: Convex optimization. Cambridge University Press (2004)
6. Colson, B., Marcotte, P., Savard, G.: An overview of bilevel optimization. Annals of Operations Research 153(1), 235–256 (2007)
7. Dantzig, G.B., Harvey, R.P., Lansdowne, Z.F., Robinson, D.W., Maier, S.F.: Formulating and solving the network design problem by decomposition. Transportation Research Part B: Methodological 13(1), 5–17 (1979)
8. Fleischer, L., Jain, K., Mahdian, M.: Tolls for heterogeneous selfish users in multicommodity networks and generalized congestion games. In: FOCS, pp. 277–285 (2004)
9. Fortune, S., Hopcroft, J.E., Wyllie, J.: The directed subgraph homeomorphism problem. Theor. Comput. Sci. 10, 111–121 (1980)
10. Friesz, T.L., Cho, H.J., Mehta, N.J., Tobin, R.L., Anandalingam, G.: A simulated annealing approach to the network design problem with variational inequality constraints. Transportation Science 26(1), 18–26 (1992)

11. Gairing, M., Harks, T., Klimm, M.: Complexity and approximation of the continuous network design problem. CoRR abs/1307.4258 (2013)
12. Harker, P.T., Friesz, T.L.: Bounding the solution of the continuous equilibrium network design problem. In: Proceedings of the Ninth International Symposium on Transportation and Traffic Theory, pp. 233–252 (1984)
13. Karakostas, G., Kolliopoulos, S.G.: Edge pricing of multicommodity networks for heterogeneous selfish users. In: FOCS, pp. 268–276 (2004)
14. Kumar, V.S.A., Marathe, M.V.: Improved results for stackelberg scheduling strategies. In: Widmayer, P., Triguero, F., Morales, R., Hennessy, M., Eidenbenz, S., Conejo, R. (eds.) ICALP 2002. LNCS, vol. 2380, pp. 776–787. Springer, Heidelberg (2002)
15. Li, C., Yang, H., Zhu, D., Meng, Q.: A global optimization method for continuous network design problems. Transportation Research Part B: Methodological 46(9), 1144–1158 (2012)
16. Luathep, P., Sumalee, A., Lam, W.H.K., Li, Z.C., Lo, H.K.: Global optimization method for mixed transportation network design problem: a mixed-integer linear programming approach. Transportation Research Part B: Methodological 45(5), 808–827 (2011)
17. Marcotte, P.: Network design problem with congestion effects: A case of bilevel programming. Mathematical Programming 34(2), 142–162 (1986)
18. Nisan, N., Roughgarden, T., Tardos, E., Vazirani, V.V.: Algorithmic Game Theory. Cambridge University Press, New York (2007)
19. Papadimitriou, C.H.: Algorithms, games, and the internet. In: STOC, pp. 749–753 (2001)
20. Roughgarden, T.: The price of anarchy is independent of the network topology. J. Comput. Syst. Sci. 67(2), 341–364 (2003)
21. Roughgarden, T.: Stackelberg scheduling strategies. SIAM Journal on Computing 33(2), 332–350 (2004)
22. Roughgarden, T.: Selfish Routing and the Price of Anarchy. The MIT Press (2005)
23. Roughgarden, T.: On the severity of Braess's paradox: designing networks for selfish users is hard. Journal of Computer and System Sciences 72(5), 922–953 (2006)
24. Roughgarden, T., Tardos, E.: How bad is selfish routing? Journal of the ACM 49(2), 236–259 (2002)
25. Wang, D.Z.W., Lo, H.K.: Global optimum of the linearized network design problem with equilibrium flows. Transportation Research Part B: Methodological 44(4), 482–492 (2010)
26. Wardrop, J.G.: Some theoretical aspects of road traffic research. In: Proc. Institute of Civil Engineers, Pt. II, vol. 1, pp. 325–378 (1952)
27. Yang, H., Bell, M.G.H.: Models and algorithms for road network design: a review and some new developments. Transport Reviews 18(3), 257–278 (1998)
28. Yang, H., Huang, H.J.: The multi-class, multi-criteria traffic network equilibrium and systems optimum problem. Transportation Research Part B: Methodological 38(1), 1–15 (2004)
29. Ye, J.J.: Necessary and sufficient optimality conditions for mathematical programs with equilibrium constraints. Journal of Mathematical Analysis and Applications 307(1), 350–369 (2005)
30. Yin, Y.: Genetic-algorithms-based approach for bilevel programming models. Journal of Transportation Engineering 126(2), 115–120 (2000)

Finding Small Stabilizers for Unstable Graphs

Adrian Bock[1], Karthekeyan Chandrasekaran[2], Jochen Könemann[3],
Britta Peis[4], and Laura Sanità[3]

[1] EPFL, Switzerland
adrianaloysius.bock@epfl.ch
[2] Harvard University, USA
karthe@seas.harvard.edu
[3] University of Waterloo, Canada
{jochen,laura.sanita}@uwaterloo.ca
[4] RWTH Aachen University, Germany
britta.peis@oms.rwth-aachen.de

Abstract. An undirected graph $G = (V, E)$ is *stable* if its inessential
vertices (those that are exposed by at least one maximum matching) form
a stable set. We call a set of edges $F \subseteq E$ a *stabilizer* if its removal from G
yields a stable graph. In this paper we study the following natural edge-
deletion question: given a graph $G = (V, E)$, can we find a minimum-
cardinality stabilizer?

Stable graphs play an important role in cooperative game theory. In
the classic *matching game* introduced by Shapley and Shubik [19] we are
given an undirected graph $G = (V, E)$ where vertices represent players,
and we define the *value* of each subset $S \subseteq V$ as the cardinality of a
maximum matching in the subgraph induced by S. The *core* of such a
game contains all *fair* allocations of the *value* of V among the players,
and is well-known to be non-empty iff graph G is *stable*. The stabilizer
problem addresses the question of how to modify the graph to ensure
that the core is non-empty.

We show that this problem is vertex-cover hard. We then prove that
there is a minimum-cardinality stabilizer that avoids some maximum
matching of G. We use this insight to give efficient approximation algorithms for sparse graphs and for regular graphs.

1   Introduction

Given an undirected graph $G = (V, E)$, a subset of edges $M \subseteq E$ is a *matching*
if every vertex $v \in V$ is incident to at most one edge in M. Dually, a subset of
vertices $U \subseteq V$ is called *vertex cover* if every edge has at least one endpoint in
U. The corresponding optimization problems of finding a matching and vertex
cover of largest and smallest size, respectively, have a rich history in the field of
Combinatorial Optimization. Relaxing canonical integer programming formulations for these problems yields the following primal-dual pair of linear programs:

$$\nu_f(G) := \max\{\mathbf{1}^T x : x(\delta(v)) \le 1 \; \forall v \in V, \; x \ge 0\} \tag{P}$$

J. Lee and J. Vygen (Eds.): IPCO 2014, LNCS 8494, pp. 150–161, 2014.
© Springer International Publishing Switzerland 2014

where $\delta(v)$ denotes the set of edges incident to v, and

$$\tau_f(G) := \min\{\mathbf{1}^T y : y_u + y_v \geq 1 \ \forall uv \in E, \ y \geq 0\}. \tag{D}$$

We will henceforth refer to feasible solutions of (P) and (D) as *fractional* matchings and vertex covers, respectively. An application of duality theory easily yields

$$\nu(G) \leq \nu_f(G) = \tau_f(G) \leq \tau(G)$$

where $\nu(G)$ and $\tau(G)$ denote the size of a maximum matching and a minimum vertex cover, respectively.

In this paper, we study graphs G with the property $\nu(G) = \tau_f(G)$. We denote the family of graphs satisfying this property to be *stable* graphs. Stable graphs subsume the well-studied class of *König-Egerváry* (KEG) graphs (e.g., see [20, 13, 14, 15]) for which $\nu(G) = \tau(G)$. Stable graphs arise quite naturally in the study of cooperative *matching games* introduced by Shapley and Shubik in their seminal paper [19]. An instance of this game is associated with an undirected graph $G = (V, E)$ where vertices represent players. We define the *value* of each subset $S \subseteq V$ as the cardinality of a maximum matching in the subgraph $G[S]$ induced by S, and the *core* of the game consists of all *stable* allocations of total value $\nu(G)$ among the vertices in V in which no coalition of vertices has an incentive to deviate. This is formally defined as

$$\mathrm{core}(G) := \left\{ y \in \mathbb{R}_+^V : \sum_{v \in S} y_v \geq \nu(G[S]) \ \forall S \subseteq V, \ \sum_{v \in V} y_v = \nu(G) \right\}.$$

It is well-known (e.g., see [8]) that $\mathrm{core}(G)$ is non-empty iff G is stable.

Matching games in turn are closely related to *network bargaining*, a natural, recent generalization of Nash's famous bargaining solution [16] to networks due to Kleinberg and Tardos [11]. Here, we are given an undirected graph $G = (V, E)$ whose vertices correspond to players, and whose edges correspond to potential unit-value deals between the incident players. Each player is allowed to engage in at most one deal with one of its neighbors. Hence, a permissible outcome is naturally associated with a matching M among the vertices of G, as well as an allocation $y \in \mathbb{R}_+^V$ of $|M|$ among M's endpoints. Kleinberg and Tardos define an allocation to be *stable* if $y_u + y_v \geq 1$ for all $uv \in E$. The authors further define an *outside option* α_u for each vertex $u \in V$ as

$$\alpha_u := \max\{1 - y_v : uv \in \delta(u) \setminus M\},$$

and say that an outcome (M, y) is *balanced* if for every edge $uv \in M$, the surplus $1 - \alpha_u - \alpha_v$ is split evenly among u and v. The main result in [11] is that an instance of network bargaining has a stable outcome iff it has a balanced one. One now realizes (see also [5]) that a stable outcome exists iff the core of the underlying matching game instance is non-empty and hence iff G is stable.

In this paper, we focus on *unstable* instances of the matching game, where the core is empty. Our motivating goal is to establish strategies for *stabilizing* such instances in the *least intrusive way*; i.e., we would like to alter the input graph in few places and ideally maintain the value of the grand coalition formed by the set of vertices V in the process. The following natural edge-deletion *stabilizer problem* formalizes this: given a graph $G = (V, E)$, find the smallest edge set $F \subseteq E$ such that the subgraph $G \setminus F := (V, E \setminus F)$ is stable.

Stable graphs form a proper superclass of KEGs which in turn form a superclass of bipartite graphs. Readers familiar with the literature of bipartite graphs would immediately recognize that the stabilizer problem closely resembles the optimization problems of deleting the minimum number of edges to convert a given graph into a KEG or a bipartite graph, both of which have been well-studied in the literature (e.g., see [1, 15]). The investigation of structural properties of unstable graphs has a long history (e.g., see [21, 3, 17]), but there are few algorithmic results on how to convert an unstable graph to a stable graph. Biró et al. [6] recently studied the minimum stabilizer problem in the weighted setting, where maximum-weight matchings are considered instead of maximum-cardinality matchings. The authors showed that the problem is NP-hard in this case, and leave the complexity of the question in the unweighted setting open.

1.1   Our Results

We first show that removing a minimum stabilizer from a given graph G does not reduce the cardinality of the maximum matching. Hence the value of the grand coalition of the associated matching game remains the same.

Theorem 1. *For every minimum stabilizer F, we have $\nu(G \setminus F) = \nu(G)$.*

The proof of this theorem is algorithmic: given any stabilizer F, we can efficiently find a maximum matching M in G and a stabilizer F' such that $F' \subseteq F$ and $M \cap F' = \emptyset$. The last equality implies that M is still a maximum matching in $G \setminus F'$. The result motivates the following intermediate M-*stabilizer* problem: given a maximum matching M, find a minimum-cardinality stabilizer F_M that is disjoint from M. In the network bargaining setting, this question asks how to convert a specific maximum matching into one with a stable allocation through minimal edge deletions in the underlying network. Biró et al. [6] previously showed that this problem is NP-hard. We strengthen the hardness result and complement it with a tight algorithmic counterpart.

Theorem 2. *The M-stabilizer problem is NP-hard, and no efficient $(2 - \varepsilon)$-approximation algorithm exists for any $\varepsilon > 0$ assuming the Unique Games Conjecture [10]. Furthermore, the M-stabilizer problem admits an efficient 2-approximation algorithm.*

The hardness proof employs an approximation preserving reduction from vertex cover. The approximation algorithm uses linear programming, and one shows that a suitable linear programming relaxation for the problem has a half-integral optimal solution. Turning to the stabilizer problem, we first extend the hardness result obtained for M-stabilizers answering the open question in [6]. Interestingly, our hardness result holds for *factor-critical* graphs (see next subsection for the definition).

Theorem 3. *The stabilizer problem is NP-hard. Furthermore, no efficient $(2 - \varepsilon)$-approximation algorithm exists for any $\varepsilon > 0$ assuming the Unique Games Conjecture [10].*

Theorems 1 and 2 suggest that the crux of the hardness of the stabilizer problem lies in finding the *right* maximum matching that survives the removal of a minimum stabilizer. Once such a matching is found one could indeed *simply* apply our 2-approximation for the M-stabilizer problem. However, not every maximum matching survives the removal of a minimum stabilizer. In fact, for two different maximum matchings M and M', the cardinality of F_M and $F_{M'}$ can differ by a factor of $\Omega(|V|)$ even on a planar factor-critical graph [7]. In Section 3.1, we present an approximation algorithm whose approximation factor depends on the *sparsity* of the graph. We say that a graph $G = (V, E)$ is ω-*sparse* if $|E(S)| \leq \omega |S|$ for all vertex subsets $S \subseteq V$.

Theorem 4. *There exists an efficient $O(\omega)$-approximation algorithm for the stabilizer problem, where ω is the sparsity of the input graph.*

We note that the above result implies a constant factor approximation algorithm for graphs with constant sparsity, e.g., *planar graphs*. We do not know whether a constant factor approximation algorithm can be developed for arbitrary graphs. However, we give a 2-approximation algorithm for regular graphs (graphs where all vertex degrees are equal). In the network bargaining setting, this gives a 2-approximation algorithm to stabilize networks in which every player has the same number of potential deals to make.

Theorem 5. *There exists an efficient 2-approximation algorithm for the stabilizer problem in regular graphs.*

The analysis of our algorithm combines some classic results about matchings and vertex covers such as the structure of basic solutions of (P) and (D) and the Edmonds-Gallai decomposition.

The proof of Theorem 1 is presented in Section 2, and that of Theorems 4 and 5 are presented in Section 3. The proofs of Theorems 2 and 3 are deferred to the full version of the paper [7].

1.2   Related Work

The problem of removing vertices or edges from a graph in order to attain a certain graph property is natural, and thus not surprisingly, its variants have been studied extensively. Much of the work on deletion problems addresses *monotone* graph properties (e.g., see [22, 2]) that are invariant under edge-removal or vertex-removal. Crucially, graph stability is not a monotone property as one easily verifies: the triangle is not stable, and adding a single pendant edge to one of its vertices yields a stable graph.

Our work is closely related to that of Mishra et al. [15] on vertex-removal and edge-removal problems to attain the König-Egerváry graph property. Similar to stability, KEG is not a monotone property. Mishra et al. showed that it is NP-hard to approximate the corresponding edge-deletion problem to within a factor of 2.88. Assuming the Unique Games Conjecture, no constant-factor approximation may exist for the problem. We note that the reductions used in [15] will likely not be helpful for proving hardness for the stabilizer problem as the graphs constructed in the reduction are stable. On the positive side, the authors show that, for a given graph $G = (V, E)$ one can efficiently find a KEG (and hence stable) subgraph with at least $3|E|/5$ edges.

The recent paper by Könemann et al. [12] addressed the related, NP-hard problem of finding a minimum-cardinality *blocking set* in an input graph $G = (V, E)$. Here one wants to find a set of edges $F \subseteq E$ such that $G \setminus F$ has a fractional vertex cover of size at most $\nu(G)$. Importantly, the resulting graph $G \setminus F$ is not required to be stable; indeed, the cardinality of a minimum blocking set can differ from the cardinality of a minimum stabilizer by a factor of $\Omega(|V|)$.

1.3   Preliminaries

Given an undirected graph G and a matching M in G, a path is called M-*alternating* if it alternates edges from M and those from $E \setminus M$. An odd cycle of length $2k + 1$ in which exactly k edges are in M is called an M-*blossom*. An M-*flower* is an even M-alternating path from an exposed vertex to a vertex u such that there exists a blossom through u. For a subset of vertices $S \subseteq V$, we use $E(S)$ to denote the set of edges in the graph induced by S and $G[S]$ to denote the subgraph induced by S. A graph $G = (V, E)$ is called *factor-critical* if for all $v \in V$, $G[V \setminus \{v\}]$ has a perfect matching; i.e., a matching that does not expose any vertex. A vertex v is called *inessential* for G if there exists a maximum matching M that exposes v, and *essential* otherwise. In this paper, we will also use the following characterization of stable graphs.

Theorem 6 ([11]). *The following are equivalent: (i) G is stable, (ii) The set of inessential vertices of G forms a stable set, (iii) G contains no M-flower for any maximum matching M. Moreover, if G is not stable, then G contains an M-flower for every maximum matching M.*

Given a graph G, the Edmonds-Gallai decomposition is a partition of its vertex set into three parts $B(G)$, $C(G)$, $D(G)$, where $B(G)$ is the set of inessential vertices, the set $C(G)$ consists of the neighbors of $B(G)$ and $D(G) = V \setminus (B(G) \cup C(G))$. We list several standard but useful properties.

Theorem 7 ([18]). *Given a graph G, the Edmonds-Gallai decomposition of the graph $B(G), C(G), D(G)$ can be computed in polynomial time. Further, we have the following properties.*

1. *Each component of $G[B(G)]$ is factor-critical.*
2. *Every maximum matching M in G exposes at most one vertex in each component K of $G[B(G)]$.*
3. *If U is a non-trivial factor-critical component in $G[B(G)]$ (i.e., a factor-critical component with more than one vertex), then $\nu(G \setminus E(U)) < \nu(G)$.*

The following proposition is a consequence of the Edmonds-Gallai decomposition theorem, which follows from classic results by Balas [3] and Pulleyblank [17]. We include its proof in the full version of this paper [7].

Proposition 1. *Let M be a maximum matching in G that also matches the maximum possible number of isolated vertices in $G[B(G)]$. Let k be the number of non-trivial factor-critical components with at least one vertex exposed by M. Then $k = 2(\nu_f(G) - \nu(G))$.*

2   Maximum Matchings and Minimum Stabilizers

We first show that the deletion of any minimum stabilizer does not alter the cardinality of the maximum matching.

Proof (of Theorem 1). Let F be a minimum stabilizer. Find a maximum matching M in G such that $|M \cap F|$ is minimum. Suppose $|M \cap F| \neq 0$.

Consider $G' := G \setminus (F \setminus M)$, the graph obtained by removing all the edges of $F \setminus M$ from G. Clearly M is still a maximum matching in G'. However, since F is minimum, G' is not stable. By Theorem 6, this implies that there exists an M-flower in G' starting at an M-exposed vertex w.

Suppose the M-flower contains an edge $uv \in F$. Then, $uv \in M$, since all other edges from F have been removed in G'. Therefore, we can find an even M-alternating path P from w to either u or v. Switching along the edges of this path, we obtain another maximum matching $M' = M \Delta P$ in G with $|F \cap M'| < |F \cap M|$, a contradiction.

It follows that the M-flower does not contain any edge from F, and therefore the M-flower also exists in $G \setminus F$. However, since $G \setminus F$ is stable, this implies that $M \setminus F$ is not a maximum matching in $G \setminus F$. Apply Edmonds' maximum matching algorithm on the graph $G \setminus F$ initialized with the matching $M \setminus F$, and construct an $M \setminus F$-alternating tree starting with the exposed vertex w. There are two possibilities: either we find an augmenting path P or a frustrated tree rooted at w. In the first case, the path P starts with w and ends with a $M \setminus F$-exposed vertex, say w'. However, such a path cannot exist in G because M is a maximum matching, and therefore w' must have been incident to an edge $f \in M \cap F$. Also, note that the path P is in $G \setminus F$. Hence, $P + f$ is an even M-alternating path in G containing exactly one edge in $M \cap F$. Switching along the edges of this path, we obtain another maximum matching $M' = M \Delta P$ in G with $|F \cap M'| < |F \cap M|$, a contradiction.

The only remaining possibility is that we find a frustrated tree T rooted at w. Let $G[T] = (V_T, E_T)$ be the graph induced by all vertices in the frustrated tree T

(after expanding pseudonodes). In this case, $M \cap E_T$ is a maximum matching in $G[T]$, and the M-flower is contained in E_T. However, if we continue Edmonds' algorithm, it would remove the vertices of the frustrated tree, and continue running in the resulting subgraph to find a maximum matching. Therefore it ends by computing a maximum matching M^* in $G \setminus F$ with $M^* \cap E_T = M \cap E_T$. Therefore, we have a M^*-flower in $G \setminus F$, again a contradiction.   □

We remark here that the above proof is algorithmic, therefore given a stabilizer F, we can find in polynomial time a maximum matching M in G and another stabilizer $F' \subseteq F$ such that $M \cap F' = \emptyset$. The first step of computing a maximum matching M in G with minimum intersection with F can be done by assigning a cost of one to the edges in F, zero to the rest of the edges, and computing a min-cost matching in G of cardinality $\nu(G)$.

We next prove a lower bound on the cardinality of a stabilizer.

Theorem 8. *For every minimum stabilizer F, we have $|F| \geq 2(\nu_f(G) - \nu(G))$.*

Proof. Let $B(G), C(G), D(G)$ denote the Edmonds-Gallai decomposition and let M be a maximum matching in G that also matches the maximum possible number of isolated vertices in $G[B(G)]$. Let $U_1, \ldots, U_k$ denote the non-trivial components in $G[B(G)]$ with at least one vertex exposed by M. Let F be a minimum stabilizer and $H = G \setminus F$. For each component $U_1, \ldots, U_k$, at least one vertex $v_i \in U_i$ becomes essential in H. Suppose not, then all vertices of some U_i are inessential in H. This implies that F contains all edges in $G[U_i]$. Thus, by Theorem 7, we have that $\nu(H) < \nu(G)$, a contradiction to Theorem 1.

Pick a maximum matching N in H. Then, N will cover all these vertices $v_1, \ldots, v_k$ that are essential in H. Since $G[U_i]$ is factor-critical and M matches all but one vertex in U_i using edges in $G[U_i]$, we may assume without loss of generality, that M misses all these vertices. The graph $M \triangle N$ is a disjoint union of even cycles and even paths since $|M| = |N| = \nu(G)$. Consider the k disjoint paths starting at the vertices $v_1, \ldots, v_k$ in $M \triangle N$. We observe that at least one of the M edges in each of these paths should belong to F, otherwise we can obtain a maximum matching in H that exposes the starting vertex v_i, thus contradicting $v_i \notin B(H)$. Hence $|F| \geq k$. The result follows by Proposition 1.   □

3   Finding Small Stabilizers

In this section, we return to the problem of finding small stabilizers. The following two sections present algorithms for the problem in sparse, and regular graphs, respectively.

3.1   An $O(\omega)$-approximation Algorithm for Sparse Graphs

Before proving Theorem 4, we state and prove the following lemma that is the main ingredient of our algorithm.

Lemma 1. *Let G be a graph with $\nu_f(G) > \nu(G)$. There exists an efficient algorithm to find a set of edges L with $|L| = O(\omega)$, such that*

(i) $\nu(G \setminus L) = \nu(G)$,
(ii) $\nu_f(G \setminus L) \leq \nu_f(G) - \frac{1}{2}$.

In other words, Lemma 1 shows that we can find a small subset of edges to remove from G without decreasing the size of the maximum matching but reducing the size of the minimum fractional vertex cover. The proof of Lemma 1 will use two classic results on the structure of fractional and integral matchings.

Theorem 9. *[4] Every basic feasible solution to (P) has components equal to $0, 1$ or $\frac{1}{2}$, and the edges with half integral components induce vertex disjoint cycles.*

Theorem 10. *[3, 21] Let $\hat{x}$ be a maximum fractional matching in a graph G having half integral fractional components for a minimum number of odd cycles $C_1, \ldots, C_q$. Let $\hat{M} := \{e \in E : \hat{x}_e = 1\}$ and M_i be a maximum matching in C_i. Then $M = \hat{M} \cup M_1 \cup \cdots \cup M_q$ is a maximum matching in G. Moreover, such $\hat{x}$ and M can be found in time polynomial in the number of vertices.*

We are now ready to prove Lemma 1.

Proof (Proof of Lemma 1). Consider $\hat{x}$ and M as in Theorem 10 for the graph G. By duality theory, there exists a fractional vertex cover y with $\mathbf{1}^T y = \mathbf{1}^T \hat{x}$ satisfying complementary slackness conditions with $\hat{x}$. Moreover, we can always find such a vector y with half integral components (e.g., see [9]). We will give an efficient algorithm to find a vertex u with the following properties:

(a) $y_u = \frac{1}{2}$,
(b) $L_u := \{uw : y_w = \frac{1}{2}\}$ satisfies $\nu(G \setminus L_u) = \nu(G)$ and $|L_u| \leq 4\omega$.

First, let us argue that $(a) + (b)$ implies the result. Assume we can find such a vertex u. The only non-trivial conclusion that needs to be verified is that $\nu_f(G \setminus L_u) \leq \nu_f(G) - 1/2$. Consider the vector y' defined as $y'_v = y_v$ for all $v \neq u$ and $y'_u = 0$ otherwise. Then y' is a fractional vertex cover for $G \setminus L_u$ (vertex u cannot be adjacent to vertices with y-value zero because y is a fractional vertex cover for G).

Now let us prove that a vertex u satisfying $(a) + (b)$ can be found efficiently. Consider an arbitrary cycle in $\hat{x}$, e.g., C_1. Since $\hat{x}_e > 0$ for every edge $e = uv$ in C_1, it follows that the vertex cover constraint is tight (i.e., $y_u + y_v = 1$ holds) for all edges in C_1, and therefore $y_v = \frac{1}{2}$ for all vertices in C_1.

Set $H := C_1$, and *mark* all vertices in C_1. Note that C_1 is an odd cycle, therefore if we remove any subset of edges incident to one marked vertex in H, then we do not decrease the size of a maximum integral matching in the resulting graph. Repeat the following process, which will maintain a collection of four invariants for the graph H: (i) Every vertex in H has y-value $\frac{1}{2}$, (ii) removing any subset of edges incident to one marked vertex of H does not decrease the size of a maximum matching, (iii) from any marked vertex, there is an even-length M-alternating path to a vertex in C_1, (iv) at least half of the vertices of H are marked. All properties clearly hold initially when H consists of C_1 only.

1. If there is a marked vertex in H with $|L_u| \leq 4\omega$, then u satisfies properties (a) and (b). STOP.
2. Otherwise, consider an arbitrary marked vertex u in H that is adjacent to a vertex $w \notin H$ with $y_w = \frac{1}{2}$. Such a w must be matched in M as otherwise, we could obtain an M-augmenting path in G by concatenating wu, the even-length M-alternating path from u to C_1 guaranteed by property (iii) and an appropriate even-length M-alternating path along C_1 to the M-exposed vertex on C_1.
3. Let z be the vertex matched to w by M. By complementary slackness, $y_z = \frac{1}{2}$. Add w and z to H and mark z. Go to 1.

It is straightforward to verify that properties (i)–(iv) continue to hold throughout the execution of the above process. Thus, it only remains to show that we can always find a vertex w as specified in Step 2 above; i.e., if all marked vertices u have $|L_u| > 4\omega$, then there exists a marked vertex in H that is adjacent to a vertex $w \notin H$ with $y_w = 1/2$. Suppose not. Consider the subgraph $G[H]$ induced by the vertices in H. This subgraph has the property that the degree of every marked vertex u in $G[H]$ is at least $|L_u| > 4\omega$. However, by (iv), the number of marked vertices is more than half the total number of vertices in $G[H]$. This contradicts the ω-sparsity of $V(H)$ in G. Finally, it is easy to see that the above process runs in polynomial time. $\square$

With this Lemma at hand, we are now ready to prove our main theorem. We will use the following algorithm:

Algorithm 1.
INITIALIZE $G' = G$.
FOR $i = 1, \ldots, 2(\nu_f(G) - \nu(G))$:

1. Let L be the set of edges returned by the algorithm in Lemma 1 when its input is the current graph G'.
2. Set $G' \leftarrow G' \setminus L$.
3. If G' is stable, STOP.

Proof (Proof of Theorem 4)

Let G be an unstable graph. We use Algorithm 1. We will now prove that (a) whenever the above algorithm stops, the current graph G' is stable, and (b) the total number of edges removed during the complete execution of the algorithm is $O(\omega) \cdot |F^*|$, where F^* is a minimum stabilizer. Clearly (a) + (b) implies the result.

First, let us argue about stability. If the algorithm stops in step (iii) for some iteration $i < 2(\nu_f(G) - \nu(G))$, this is clear. So we may assume that the algorithm stops after performing all $2(\nu_f(G) - \nu(G))$ iterations. The graph G' output at this point has $\nu_f(G') \leq \nu_f(G) - \frac{1}{2}(2(\nu_f(G) - \nu(G))) = \nu(G) = \nu(G')$. This is because, by Lemma 1, in each iteration the size of a minimum fractional vertex cover decreases by at least $\frac{1}{2}$ while the size of the maximum matching is maintained. Hence, by definition of stability, G' is stable.

By Lemma 1, in each iteration we remove $O(\omega)$ edges and the total number of iterations is at most $2(\nu_f(G) - \nu(G))$. The bound on the approximation factor follows from Theorem 8. The running time bound also follows since the number of applications of the algorithm in Lemma 1 is at most $2(\nu_f(G) - \nu(G)) \leq |F^*| \leq |E|$ times.   □

We end the section with an observation about our algorithm that will be useful for our approximation results on regular graphs.

Proposition 2. *The stabilizer output by Algorithm 1 has size at most $2(\nu_f(G) - \nu(G)) \cdot \Delta(G)$, where $\Delta(G)$ is the maximum degree of a vertex in G.*

Proof. In each iteration of the algorithm, we remove a subset of edges incident to some vertex. Therefore we remove at most $\Delta(G)$ edges in each iteration. Further, the number of iterations is at most $2(\nu_f(G) - \nu(G))$.   □

3.2   A 2-approximation Algorithm for Regular Graphs

In this section, we give a 2-approximation algorithm for solving the stabilizer problem in regular graphs.

Proof. (Proof of Theorem 5) We use Algorithm 1. Consider a d-regular graph G, i.e., a graph where every vertex has degree d. Let $k := 2(\nu_f(G) - \nu(G))$. By Proposition 2, the size of F output by the algorithm is at most kd. We complete the proof by showing that every stabilizer in G is of size at least $kd/2$.

Consider the Edmonds-Gallai decomposition of G, namely $B(G)$, $C(G)$, $D(G)$. Let S denote the isolated vertices in $G[B]$. Consider a maximum matching M in G that also matches the maximum possible number of vertices in S. By Proposition 1, the number of non-trivial factor-critical components in $G[B(G)]$ with at least one vertex exposed by M is equal to k.

Let S_u denote the vertices in S that are exposed by M. We first observe that the size $\nu(G)$ of the maximum matching in G is $(|V| - k - |S_u|)/2$. Consider the following primal and dual linear programs.

$$\min \sum_{e \in E} z_e \qquad (\mathcal{P})$$
$$y_u + y_v + z_{uv} \geq 1 \ \forall \ uv \in E$$
$$\sum_{u \in V} y_u = \nu(G)$$
$$y, z \geq 0$$

$$\max \sum_{e \in E} \alpha_e - \gamma \nu(G) \qquad (\mathcal{D})$$
$$\alpha(\delta(u)) \leq \gamma \ \forall \ u \in V$$
$$0 \leq \alpha \leq 1$$

By setting z to be the indicator vector of the minimum stabilizer, we can obtain y such that (y, z) is a feasible solution to the primal program. This is because, if z is the indicator vector of a stabilizer in G, then by definition there exists a fractional vertex cover y in $G \setminus \text{Support}(z)$ with size equal to $\nu(G \setminus \text{Support}(z))$. We also know by Theorem 1 that for every minimum stabilizer F, $\nu(G \setminus F) = \nu(G)$.

Thus, the primal program is a relaxation of the stabilizer problem. Consequently, the objective value of any feasible solution to the dual program is a lower bound on the size of a minimum stabilizer. We will provide a dual feasible solution with objective value at least $kd/2$.

Consider the dual solution $(\gamma = d, \ \alpha_e = 1 \ \forall \ e \in E)$. Since the graph is d-regular we have that $\alpha(\delta(u)) = d$. Thus, all dual constraints are satisfied and hence, it is a dual feasible solution. The objective value is

$$\sum_{e \in E} \alpha_e - \gamma \nu(G) = \frac{d|V|}{2} - d \left(\frac{|V| - k - |S_u|}{2} \right) = d \left(\frac{k + |S_u|}{2} \right) \geq \frac{kd}{2}.$$

$\square$

Concluding Remarks. We conclude the paper with a remark about the linear program $(\mathcal{P})$. If we add the integrality constraints on the z variables, we obtain an integer program (IP) and it follows by our result that the integrality gap of the resulting IP is at most 2 for d-regular graphs. Könemann et al. [12] proved a $\Theta(n)$-bound on the integrality gap of the IP for general graphs. However, the resulting IP is *not* a formulation for our minimum stabilizer problem, since the integral optimum solution of the IP could be $\Omega(n)$ away from the size of a minimum stabilizer for arbitrary graphs (not necessarily regular). In order to obtain a formulation for our stabilizer problem, we could introduce additional variables x and impose the existence of a matching in $G \setminus \text{Support}(z)$ of size $\nu(G)$:

$$\min \sum_{e \in E} z_e$$

$$y_u + y_v + z_{uv} \geq 1 \ \forall \ uv \in E,$$

$$\sum_{u \in V} y_u = \nu(G),$$

$$x(\delta(v)) \leq 1 \ \forall \ v \in V, \ \sum_{e \in E} x_e = \nu(G), \ x(E[S]) \leq \frac{|S| - 1}{2} \ \forall \ S \subseteq V, \ |S| \ \text{odd},$$

$$x_e + z_e \leq 1 \ \forall \ e \in E,$$

$$x, y, z \geq 0, \ x, z \ \text{integral}.$$

However, we can show a lower bound of $\Omega(n)$ on the integrality gap of the above formulation. We refer the reader to the full version of the paper [7] for the example that exhibits the integrality gap.

References

[1] Agarwal, A., Charikar, M., Makarychev, K., Makarychev, Y.: $O(\sqrt{\log n})$ approximation algorithms for min UnCut, min 2CNF deletion, and directed cut problems. In: Proceedings of the ACM Symposium on Theory of Computing, pp. 573–581 (2005)

[2] Alon, N., Shapira, A., Sudakov, B.: Additive approximation for edge-deletion problems. Annals of Mathematics 170, 371–411 (2009)

[3] Balas, E.: Integer and fractional matchings. Annals of Discrete Mathematics 11, 1–13 (1981)

[4] Balinski, M.L.: On maximum matching, minimum covering and their connections. In: Proceedings of the Princeton Symposium on Mathematical Programming (1970)

[5] Bateni, M., Hajiaghayi, M., Immorlica, N., Mahini, H.: The cooperative game theory foundations of network bargaining games. In: Proc., International Colloquium on Automata, Languages and Processing, pp. 67–78 (2010)

[6] Biró, P., Bomhoff, M., Golovach, P.A., Kern, W., Paulusma, D.: Solutions for the stable roommates problem with payments. In: Golumbic, M.C., Stern, M., Levy, A., Morgenstern, G. (eds.) WG 2012. LNCS, vol. 7551, pp. 69–80. Springer, Heidelberg (2012)

[7] Bock, A., Chandrasekaran, K., Könemann, J., Peis, B., Sanità, L.: Finding small stabilizers for unstable graphs, Full version in preparation (2014)

[8] Chalkiadakis, G., Elkind, E., Wooldridge, M.: Computational aspects of cooperative game theory. In: Synthesis Lectures on Artificial Intelligence and Machine Learning. Morgan & Claypool Publishers (2011)

[9] Hochbaum, D.: Approximation algorithms for the set covering and vertex cover problems. SIAM Journal on Computing 11(3), 555–556 (1982)

[10] Khot, S.: On the power of unique 2-Prover 1-Round games. In: Proceedings of the ACM Symposium on Theory of Computing, pp. 767–775 (2002)

[11] Kleinberg, J.M., Tardos, É.: Balanced outcomes in social exchange networks. In: Proceedings of the ACM Symposium on Theory of Computing, pp. 295–304 (2008)

[12] Könemann, J., Larson, K., Steiner, D.: Network bargaining: Using approximate blocking sets to stabilize unstable instances. In: Proceedings of the Symposium on Algorithmic Game Theory, pp. 216–226 (2012)

[13] Korach, E.: Flowers and Trees in a Ballet of K_4, or a Finite Basis Characterization of Non-König-Egerváry Graphs, Tech. Report 115, IBM Israel Scientific Center (1982)

[14] Korach, E., Nguyen, T., Peis, B.: Subgraph characterization of Red/Blue-Split Graph and König Egerváry Graphs. In: Proceedings of the ACM-SIAM Symposium on Discrete Algorithms, pp. 842–850 (2006)

[15] Mishra, S., Raman, V., Saurabh, S., Sikdar, S., Subramanian, C.R.: The complexity of König subgraph problems and above-guarantee vertex cover. Algorithmica 61(4), 857–881 (2011)

[16] Nash, J.: The bargaining problem. Econometrica 18, 155–162 (1950)

[17] Pulleyblank, W.R.: Fractional matchings and the edmonds-gallai theorem. Discrete Appl. Math. 16, 51–58 (1987)

[18] Schrijver, A.: Combinatorial optimization. Springer, New York (2003)

[19] Shapley, L.S., Shubik, M.: The assignment game: the core. International Journal of Game Theory 1(1), 111–130 (1971)

[20] Sterboul, F.: A characterization of the graphs in which the transversal number equals the matching number. J. Combin. Theory Ser. B, 228–229 (1979)

[21] Uhry, J.P.: Sur le problème du couplage maximal. RAIRO 3, 13–20 (1975)

[22] Yannakakis, M.: Node- and edge-deletion NP-complete problems. In: Proceedings of the ACM Symposium on Theory of Computing, pp. 253–264 (1978)

The Triangle Splitting Method
for Biobjective Mixed Integer Programming

Natashia Boland, Hadi Charkhgard, and Martin Savelsbergh

School of Mathematical and Physical Sciences,
The University of Newcastle, NSW 2308, Australia
{Natashia.Boland,Martin.Savelsbergh}@newcastle.edu.au,
Hadi.Charkhgard@uon.edu.au

Abstract. We present the first criterion space search algorithm, the triangle splitting method, for finding the efficient frontier of a biobjective mixed integer program. The algorithm is relatively easy to implement and converges quickly to the complete set of nondominated points. A computational study demonstrates the efficacy of the triangle splitting method.

Keywords: biobjective mixed integer program, triangle splitting method, efficient frontier.

1 Introduction

Multiobjective optimization, one of the earliest fields of study in operations research, has been experiencing a resurgence of interest in the last decade. This is due, in part, to the ever increasing adoption of optimization-based decision support tools in industry. Since most real-world problems involve multiple, often conflicting, goals, the want for multiobjective optimization decision support tools is not surprising. The development of effective, and relatively easy to use, evolutionary algorithms for multiobjective optimization is another contributing factor. Finally, the availability of cheap computing power has played a role. Solving multiobjective optimization problems is highly computationally intensive (more so than solving single-objective optimization problems) and thus the availability of cheap computing power has acted as an enabler.

Exact algorithms for multiobjective optimization can be divided into solution space search algorithms, i.e., methods that search in the space of feasible solutions, and criterion space search algorithms, i.e., methods that search in the space of objective function values. It has long been argued (see for example [3]) that criterion space search algorithms have advantages over solution space search algorithms and are likely to be more successful. Our motivation for focusing on criterion space search algorithms is that we want to exploit the power of commercial (single-objective) integer programming solvers. Several extremely powerful commercial integer programming solvers exist, e.g., the IBM ILOG CPLEX Optimizer, the FICO Xpress Optimizer, and the Gurobi Optimizer, and criterion space search algorithms can take full advantage of their features. Embedding

J. Lee and J. Vygen (Eds.): IPCO 2014, LNCS 8494, pp. 162–173, 2014.
© Springer International Publishing Switzerland 2014

a commercial integer programming solver in any algorithm has the additional advantage that enhancements made to the commercial solver immediately result in improved performance of the algorithm in which it is embedded.

In this study, we focus on biobjective mixed integer programs (BOMIPs). Computing the efficient frontier of a BOMIP is challenging because of two reasons. First, the existence of unsupported nondominated points, i.e., nondominated points that cannot be obtained by optimizing a convex combination of the objective functions. Secondly, the existence of continuous parts in the efficient frontier, i.e., parts where all points of a line segment are nondominated. Figure 1 shows an example of a nondominated frontier of a BOMIP. Observe that this nondominated frontier contains isolated points as well as closed, half open, and open line segments.

There are only a few studies that present algorithms for finding the efficient frontier of a BOMIP (several of these algorithms were later shown to be incomplete or incorrect). All these algorithms are solution space search algorithms and are based on the following observation (which is made more precise in Section 2). If S_I is the projection of the set of feasible solutions to a BOMIP on to the space of the integer variables, then fixing the integer variables to the values of s for any $s \in S_I$ changes the BOMIP to a biobjective linear program (BOLP). Furthermore, if $\mathcal{Y}_N(s)$ for $s \in S_I$ denotes the nondominated frontier of the resulting BOLP, then the nondominated frontier of the BOMIP is the set of nondominated points in $\bigcup_{s \in S_I} \mathcal{Y}_N(s)$. This observation suggests a natural algorithm for computing the nondominated frontier: enumerate the solutions in S_I, for each of these solutions find the nondominated frontier of the associated BOLP, take the union of these nondominated frontiers, and, finally, eliminate any dominated points from this set.

Unfortunately, this natural algorithm has a number of drawbacks in practice:

- The set $\bigcup_{s \in S_I} \mathcal{Y}_N(s)$ can become prohibitively large, which implies that storing and maintaining this set of points may require an excessive amount of memory;

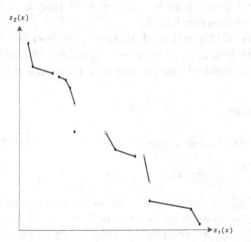

Fig. 1. Example of a nondominated frontier of a BOMIP

- Eliminating dominated points from the set $\bigcup_{s \in S_I} \mathcal{Y}_N(s)$ can become prohibitively time-consuming; and
- The nondominated frontier is only available upon completion, i.e., during the course of the algorithm the set of points maintained contains both dominated and nondominated points.

Most of the research on algorithms for BOMIPs has focused on addressing the first two drawbacks, either by developing specialized data structures and methods for efficiently storing and maintaining the set $\bigcup_{s \in S_I} \mathcal{Y}_N(s)$, or by curtailing the enumeration of $s \in S_I$, i.e., by recognizing or determining that for a given $s \in S_I$ all the points in $\mathcal{Y}_N(s)$ will (eventually) be eliminated. The third drawback, unfortunately, is an inherent feature of this algorithm and cannot be avoided.

We have developed a completely different algorithm, the triangle splitting algorithm, which does not suffer from any of these drawbacks. To the best of our knowledge, it is the first criterion space search algorithm for finding the efficient frontier of a BOMIP. The algorithm recursively explores smaller and smaller rectangles and right triangles that may contain as yet unknown nondominated points. The triangle splitting algorithm has the following advantages:

- It is a criterion space search method that relies on a small number of ideas and techniques, which makes it both easy to understand and easy to implement.
- It has minimal requirements in terms of information storage.
- It maintains at any time during its execution a set of points which are guaranteed to be part of the nondominated frontier.
- It relies on solving single objective mixed integer programs (MIPs) and benefits automatically from any advances in single objective MIP solvers.

A computational study demonstrates the efficacy of the triangle splitting algorithm. Its performance is as good or better than the best known algorithm for BOMIPs [2] and it can handle instance sizes that far exceed the size that solution space algorithms can handle.

In the remainder of this extended abstract, we describe the basic version of the algorithm. In Section 2, we give preliminaries, and in Section 3, we introduce the triangle splitting method and present computational results.

2   Preliminaries

A multiobjective mixed integer programming problem (MOMIP) can be stated as follows

$$\min_{x \in \mathcal{X}} \; z(x) := \{z_1(x), ..., z_p(x)\},$$

where $\mathcal{X} \subseteq \mathbb{Z}^n \times \mathbb{R}^m$ is defined by a set of linear constraints and represents the *feasible set in the decision space* and the image $\mathcal{Y}$ of $\mathcal{X}$ under vector-valued function $z = \{z_1, ..., z_p\}$ represents the *feasible set in the criterion space*, i.e., $\mathcal{Y} := z(\mathcal{X}) := \{y \in \mathbb{R}^p : y = z(x) \text{ for some } x \in \mathcal{X}\}$. We will sometimes use

$x = (x_I, x_C)$, for $x \in \mathcal{X}$, to distinguish the integer and continuous variables in a feasible solution. For convenience, we also use the notation $\mathbb{R}^p_\geq := \{y \in \mathbb{R}^p : y \geq 0\}$ for the nonnegative orthant of $\mathbb{R}^p$, and $\mathbb{R}^p_> := \{y \in \mathbb{R}^p : y > 0\}$ for the positive orthant of $\mathbb{R}^p$.

Definition 1. *A feasible solution $x' \in \mathcal{X}$ is called efficient or Pareto optimal, if there is no other $x \in \mathcal{X}$ such that $z_k(x) \leq z_k(x')$ for $k = 1, ..., p$ and $z(x) \neq z(x')$. If x' is efficient, then $z(x')$ is called a nondominated point. The set of all efficient solutions $x' \in \mathcal{X}$ is denoted by $\mathcal{X}_E$. The set of all nondominated points $y' = z(x') \in \mathcal{Y}$ for some $x' \in \mathcal{X}_E$ is denoted by $\mathcal{Y}_N$ and referred to as the nondominated frontier or the efficient frontier.*

Definition 2. *Let $x' \in \mathcal{X}_E$. If there is a $\lambda \in \mathbb{R}^{n+m}$ such that x' is an optimal solution to $\min_{x \in \mathcal{X}} \lambda^T z(x)$, then x' is called a supported efficient solution and $y = z(x')$ is called a supported nondominated point.*

Definition 3. *Let $\mathcal{Y}^e$ be the set of extreme points of $Conv(\mathcal{Y})$. A point $y \in \mathcal{Y}$ is called an extreme supported nondominated point if $y \in \mathcal{Y}^e \cap \mathcal{Y}_N$.*

Theorem 1. *For a multiobjective linear program (MOLP), we have that $\mathcal{Y}$ is closed and convex if $\mathcal{X}$ is closed.*

Corollary 1. *For a MOLP, the nondominated points in $\mathcal{Y}_N$ are supported and connected, i.e., between any pair of nondominated points there exists a sequence of nondominated points with the property that all points on the line segment between consecutive points in the sequence are also nondominated.*

Let $\mathcal{X}_I = proj_I(\mathcal{X})$, where $proj_I(\mathcal{X}) := \{x \in \mathbb{Z}^n : \text{there exists a } x_C \in \mathbb{R}^m \text{ such that } (x, x_C) \in \mathcal{X}\}$, and for $\bar{x} \in \mathcal{X}_I$ let $\mathcal{Y}_N(\bar{x})$ denote the nondominated frontier of the MOLP defined by

$$\min_{(x_I, x_C) \in \mathcal{X} : x_I = \bar{x}} z(x) := \{z_1(x), ..., z_p(x)\}.$$

Finally, let the function $\mathcal{ND} : \mathbb{R}^p \to \mathbb{R}^p$ be one that takes a set of points $\mathcal{P}$ in the criterion space and removes any point $p \in \mathcal{P}$ that is dominated by any other point $p' \in \mathcal{P}$. We have the following theorem (see for example Gardenghi et al. [4]).

Theorem 2. *The nondominated frontier $\mathcal{Y}_N = \mathcal{ND}(\bigcup_{\bar{x} \in \mathcal{X}_I} \mathcal{Y}_N(\bar{x}))$.*

Theorem 2 forms the basis of most solution space search algorithms for MOMIPs. These algorithms essentially enumerate $\mathcal{X}_I$, compute $\mathcal{Y}_N(\bar{x})$ for all $\bar{x} \in \mathcal{X}_I$, form the union of the resulting nondominated frontiers, and eliminate any dominated points, i.e., set $\mathcal{Y}_N = \mathcal{ND}(\bigcup_{\bar{x} \in \mathcal{X}_I} \mathcal{Y}_N(\bar{x}))$.

The first "branch-and-bound" algorithm for solving multiobjective mixed 0-1 integer programs was proposed by Mavrotas and Diakoulaki [6]. The enumeration algorithm ensures that at the leaf nodes of the search tree, the values of the 0-1 variables are fixed at either zero or one. The algorithm maintains a list of potential nondominated points, which is updated at each leaf node of the search tree, i.e., potential nondominated points are added to the list and dominated points are removed from the list. In a follow-up paper, Mavrotas and Diakoulaki [7] show that their initial scheme for updating the list was incomplete in the sense that some dominated points might erroneously remain in the list. They propose an a posteriori filtering method to remedy the situation. Vincent et al. [8] show another deficiency of the branch-and-bound algorithm of Mavrotas and Diakoulaki, namely that it may not identify all nondominated points, i.e., some nondominated points are missed. They show that this issue can be corrected for biobjective mixed 0-1 integer programs. More recently, Belotti et al. [2] propose a different branch-and-bound algorithm to compute the nondominated frontier of a biobjective mixed integer program. It more closely resembles the traditional branch-and-bound for single objective mixed integer programs, in the sense that bounding strategies are employed to fathom nodes during the search.

Next, we introduce concepts and notation that will facilitate the presentation and discussion of the triangle splitting method. For the remainder of the paper, we restrict ourselves to BOMIPs. Let $z^1 = (z_1^1, z_2^1)$ and $z^2 = (z_1^2, z_2^2)$ be two points in the criterion space with $z_1^1 \leqslant z_1^2$ and $z_2^2 \leqslant z_2^1$. We denote with $R(z^1, z^2)$ the rectangle in the criterion space defined by the points z^1 and z^2. Furthermore, we denote with $T(z^1, z^2)$ the right triangle in the criterion space defined by the points z^1, (z_1^2, z_2^1), and z^2. Finally, we denote with $H(z^1, z^2)$ the line segment in the criterion space defined by the points z^1 and z^2, i.e., the hypotenuse of triangle $T(z^1, z^2)$.

A point $\bar{z}$ in criterion space corresponding to a solution with smallest value for $z_2(x)$ among all solutions with smallest value for $z_1(x)$ among all feasible solutions with objective function values in $T(z^1, z^2)$, if one exists, can be found by solving two mixed integer programs in sequence, namely

$$\bar{z}_1 = \min_{x \in \mathcal{X}} z_1(x)$$
$$\text{subject to } z(x) \in T(z^1, z^2),$$

followed by

$$\bar{z}_2 = \min_{x \in \mathcal{X}} z_2(x)$$
$$\text{subject to } z(x) \in T(z^1, z^2) \text{ and } z_1(x) \leq \bar{z}_1.$$

As this is an operation that will be performed frequently in our criterion space search algorithm, we introduce the following notation to represent the process:

$$\bar{z} = \operatorname*{lex\,min}_{x \in \mathcal{X}}\{z_1(x), z_2(x) : z(x) \in T(z^1, z^2)\}.$$

Finding a point $\bar{z}$ in criterion space corresponding to a solution with smallest value for $z_1(x)$ among all solutions with smallest value for $z_2(x)$ among all feasible solutions with objective function values in $T(z^1, z^2)$, if one exists, can be done similarly, and we introduce the following notation to represent that process:

$$\bar{z} = \mathop{\text{lex min}}_{x \in \mathcal{X}}\{z_2(x), z_1(x) : z(x) \in T(z^1, z^2)\}.$$

It is often convenient to assume that the points of the efficient frontier are listed in order of nondecreasing value of the first objective function. In that case, the first and last point of the efficient frontier can be found by solving

$$z^T := \mathop{\text{lex min}}_{x \in \mathcal{X}}\{z_1(x), z_2(x) : z(x) \in R((-\infty, \infty), (-\infty, \infty))\}.$$

and

$$z^B := \mathop{\text{lex min}}_{x \in \mathcal{X}}\{z_2(x), z_1(x) : z(x) \in R((-\infty, \infty), (-\infty, \infty))\},$$

respectively (where the feasible region is defined by a rectangle instead of a triangle).

The next propositions and their corollaries provide the basis for the development of the triangle splitting method.

Proposition 1. *Let z^1 and z^2 be two points in the criterion space with $z_2^2 < z_2^1$ and let v be such that $z_2^2 < v < z_2^1$. If $\{(z - \mathbb{R}_{>}^2) \cap \mathcal{Y}_N : z \in H(z^1, z^2)\} = \emptyset$, then $\text{lex min}_{x \in \mathcal{X}}\{z_1(x), z_2(x) : z_2(x) \leq v, \; z(x) \in T(z^1, z^2)\}$ returns a nondominated point if it has a solution.*

Corollary 2. *Let z^1 and z^2 be two points in the criterion space with $z_2^2 < z_2^1$ and let v be such that $z_2^2 < v < z_2^1$. If $\{(z - \mathbb{R}_{>}^2) \cap \mathcal{Y}_N : z \in H(z^1, z^2)\} = \emptyset$ and $z^2 \in \mathcal{Y}_N$, then $\bar{z}^1 = \text{lex min}_{x \in \mathcal{X}}\{z_1(x), z_2(x) : z_2(x) \leq v, \; z(x) \in T(z^1, z^2)\} \in \mathcal{Y}_N$. Furthermore, if $\bar{z}^1 = z^2$, then z^2 is the only nondominated point in $T(z^1, z^2)$ with $z_2(x) \leq v$.*

Proposition 2. *Let z^1 and z^2 be two points in the criterion space with $z_1^1 < z_1^2$ and let v be such that $z_1^1 < v < z_1^2$. If $\{(z - \mathbb{R}_{>}^2) \cap \mathcal{Y}_N : z \in H(z^1, z^2)\} = \emptyset$, then $\text{lex min}_{x \in \mathcal{X}}\{z_2(x), z_1(x) : z_1(x) \leq v, \; z(x) \in T(z^1, z^2)\}$ returns a nondominated point if it has a solution.*

Corollary 3. *Let z^1 and z^2 be two points in the criterion space with $z_1^1 < z_1^2$ and let v be such that $z_1^1 < v < z_1^2$. If $\{(z - \mathbb{R}_{>}^2) \cap \mathcal{Y}_N : z \in H(z^1, z^2)\} = \emptyset$ and $z^1 \in \mathcal{Y}_N$, then $\bar{z}^2 = \text{lex min}_{x \in \mathcal{X}}\{z_2(x), z_1(x) : z_1(x) \leq v, \; z(x) \in T(z^1, z^2)\} \in \mathcal{Y}_N$. Furthermore, if $\bar{z}^2 = z^1$, then z^1 is the only nondominated point in $T(z^1, z^2)$ with $z_1(x) \leq v$.*

Theorem 3. *Let z^1 and z^2 be two nondominated points in the criterion space. If $\{(z - \mathbb{R}_{>}^2) \cap \mathcal{Y}_N : z \in H(z^1, z^2)\} = \emptyset$ and there exists an $x \in \mathcal{X}_I$ and x_C^1 and $x_C^2 \in \mathbb{R}^m$ such that $(x, x_C^1), (x, x_C^2) \in \mathcal{X}$, $z_1((x, x_C^1)) \leq z_1^1$, $z_2((x, x_C^1)) \leq z_2^2$, $z_1((x, x_C^2)) \leq z_1^2$, and $z_2((x, x_C^2)) \leq z_2^2$, then $H(z^1, z^2) \subset \mathcal{Y}_N$.*

Theorem 3 implies that for a triangle $T(z^1, z^2)$ such that $\{(z - \mathbb{R}^2_\geq) \cap \mathcal{Y}_N : z \in H(z^1, z^2)\} = \emptyset$, the following *line detection* mixed integer program (MIP) establishes whether $H(z^1, z^2) \subset \mathcal{Y}_N$:

$$\max z_1((x_I, x_C^2))$$

subject to

$$z_1((x_I, x_C^1)) \leq z_1^1$$
$$z_2((x_I, x_C^1)) \leq z_2^1$$
$$\lambda_1 z_1((x_I, x_C^2)) + \lambda_2 z_2((x_I, x_C^2)) = \lambda_1 z_1^1 + \lambda_2 z_2^1$$
$$(x_I, x_C^1) \in \mathcal{X}, (x_I, x_C^2) \in \mathcal{X}$$

where $\lambda_1 = z_2^1 - z_2^2$ and $\lambda_2 = z_1^2 - z_1^1$. The constraint $\lambda_1 z_1((x_I, x_C^2)) + \lambda_2 z_2((x_I, x_C^2)) = \lambda_1 z_1^1 + \lambda_2 z_2^1$ expresses that the ratio of the horizontal distance and the vertical distance is the same as $\frac{z_1^2 - z_1^1}{z_2^1 - z_2^2}$, which implies that the point is on the imaginary line connecting z^1 and z^2. Note that if the optimal value is greater than or equal to z_1^2, then z^1 and z^2 satisfy the conditions of Theorem 3.

Determining whether the points on the hypotenuse of a triangle are all nondominated, i.e., whether the hypotenuse of the triangle is part of the nondominated frontier, is a core component of the triangle splitting algorithm. Another core component of the triangle splitting method is the weighted sum method [1]. The weighted sum method is used to find all locally extreme supported nondominated points in a rectangle defined by two nondominated points z^1 and z^2. The weighted sum method uses the following optimization problem to search for extreme supported nondominated points in rectangle $R(z^1, z^2)$:

$$z^* = \min_{x \in \mathcal{X}} \lambda_1 z_1(x) + \lambda_2 z_2(x)$$

subject to $z(x) \in R(z^1, z^2)$

with $\lambda_1 = z_2^1 - z_2^2$ and $\lambda_2 = z_1^2 - z_1^1$, i.e., the objective function is parallel to the line that connects z^1 and z^2 in the criterion space. Figure 2 shows an example with $z^1 = z^T$ and $z^2 = z^B$. It is easy to see that the optimum point z^* is an as yet unknown locally supported nondominated point if $\lambda_1 z_1^* + \lambda_2 z_2^* < \lambda_1 z_1^1 + \lambda_2 z_2^1$. That is, the optimization either returns a new locally supported nondominated point z^* or a convex combination of z^1 and z^2. When an as yet unknown locally supported nondominated point z^* is returned, the method is applied recursively to search $R(z^1, z^*)$ and $R(z^*, z^2)$ for additional as yet unknown locally supported nondominated points. Note that the set of locally supported nondominated points returned by the weighted sum method is guaranteed to include *all* locally extreme supported nondominated points (but it may also include locally supported nondominated points that are not extreme).

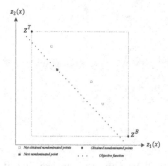

Fig. 2. Searching for a nondominated point using the weighted sum optimization problem

3   The Triangle Splitting Method

The triangle splitting method maintains a priority queue with rectangles and triangles, each of which still has to be explored, i.e., may still contain as yet unknown nondominated points. Each element of the priority queue is characterized by two nondominated points z^1 and z^2, a shape, **rectangle** or **triangle**, and a splitting direction, **horizontal** or **vertical**. The algorithm also maintains an ordered list of nondominated points, which is updated after finding a new nondominated point or after detecting that all points on the hypotenuse of a triangle are nondominated. The nondominated points are maintained in order of nondecreasing value of their first objective value. In addition to the nondominated point itself, there is an indicator that specifies whether the nondominated point is connected to the next nondominated point in the list (indicator value 1) or not (indicator value 0), i.e., whether all points on the line segment defined by the two nondominated points are also nondominated. The list is initialized with $(z^T, 0)$ and $(z^B, 0)$. The priority queue is initialized with $(z^T, z^B, \textbf{rectangle}, \textbf{horizontal})$.

Next, we discuss how rectangles and triangles are explored. A rectangle is explored by applying the weighted sum method to find locally extreme supported nondominated points and divide the rectangle into one or more triangles. The locally supported nondominated points are added to the list of nondominated points and the triangles are added to the priority queue (with the same splitting direction). See Figure 3 for an example of the exploration of a rectangle. Exploring a triangle $T(z^1, z^2)$ starts by determining whether all the points on the hypotenuse of the triangle are nondominated. (Note that by construction, there are no nondominated points "below" the hypotenuse, i.e., $\{(z - \mathbb{R}^2_{\geqq}) \cap \mathcal{Y}_N : z \in H(z^1, z^2)\} = \emptyset$.) If so, the list of nondominated points is updated accordingly, i.e., element $(z^1, 0)$ is changed to $(z^1, 1)$. (Note that z^1 and z^2 appear consecutively in the list of nondominated points). Otherwise, the triangle will be split into two rectangles, either by splitting the triangle horizontally at $\frac{z_2^1 + z_2^2}{2}$ or by splitting the triangle vertically at $\frac{z_1^1 + z_1^2}{2}$, which will then be added to the priority queue unless they cannot contain as yet unknown nondominated points.

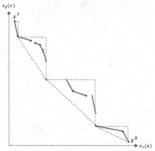

Fig. 3. The triangles determined by the exploration of initial rectangle $R(z^T, z^B)$ for the BOMIP that gives rise to the nondominated frontier of Figure 1

More specifically, when splitting triangle $T(z^1, z^2)$ horizontally at height $\frac{z_2^1 + z_2^2}{2}$, we start by computing

$$\bar{z}^1 = \operatorname*{lex\,min}_{x \in \mathcal{X}} \{z_1(x), z_2(x) : z_2(x) \leq \frac{z_2^1 + z_2^2}{2}, \, z(x) \in T(z^1, z^2)\}.$$

If $\bar{z}_2^1 = \frac{z_2^1 + z_2^2}{2}$, i.e., if the resulting point is on the cut, then we set $\bar{z}^2 = \bar{z}^1$. If not, then we compute

$$\bar{z}^2 = \operatorname*{lex\,min}_{x \in \mathcal{X}} \{z_2(x), z_1(x) : z_1(x) \leq \bar{z}_1^1, \, z(x) \in T(z^1, z^2)\}.$$

By construction of $\bar{z}^1$ and $\bar{z}^2$, all as yet unknown nondominated points in triangle $T(z^1, z^2)$ must be in rectangles $R(z^1, \bar{z}^2)$ and $R(\bar{z}^1, z^2)$. Note that when $\bar{z}^2 = z^1$, rectangle $R(z^1, \bar{z}^2)$ consists of a single nondominated point and does not need to be explored further (and, similarly, when $\bar{z}^1 = z^2$ rectangle $R(\bar{z}^1, z^2)$ consists of a single nondominated point and does not need to be explored further). The two situations that can occur during horizontal splitting at height $\frac{z_2^1 + z_2^2}{2}$ are illustrated in Figures 4 and 5. Note that if $\bar{z}^1$ is on the cut, then it is easy to see that $\bar{z}^2 = \bar{z}^1$ and we do not need to compute $\bar{z}^2$. (In this case, $\bar{z}^1 = \bar{z}^2$ is almost surely on a line segment of the nondominated frontier.) Splitting a triangle vertically at height $\frac{z_1^1 + z_1^2}{2}$ proceeds analogously.

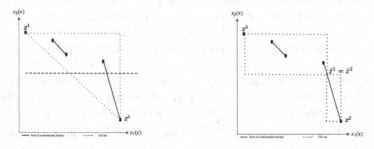

Fig. 4. Horizontal splitting of triangle $T(z^1, z^2)$ when $\bar{z}^1$ is on the cut

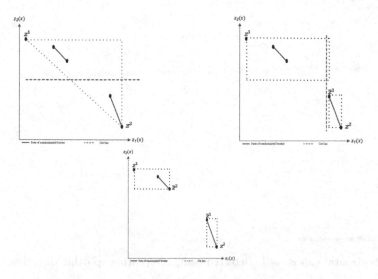

Fig. 5. Horizontal splitting of triangle $T(z^1, z^2)$ when $\bar{z}^1$ is not on the cut

The splitting direction plays an important role in the triangle splitting method. Recall that a nondominated frontier may contain horizontal and vertical discontinuities or gaps. If the same splitting direction is used throughout the algorithm, finding these gaps can be unnecessarily time-consuming or even impossible.

Figure 6 shows the progression of the triangle splitting method for a nondominated frontier with a horizontal gap when only horizontal cutting is employed to split triangles. In this situation, the triangle splitting methods continuous to split the continuous segment of the nondominated frontier until the area of the remaining rectangles is less than a pre-specified tolerance. To avoid these situations, an *alternating splitting direction* strategy is employed, i.e., the splitting

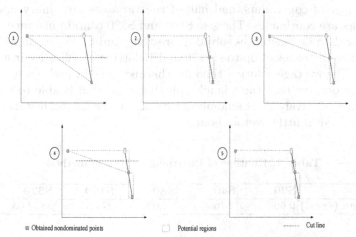

Fig. 6. A horizontal gap may not be detected when splitting horizontally only

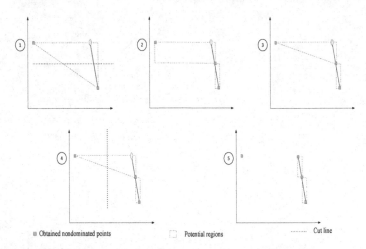

■ Obtained nondominated points   ☐ Potential regions   ·········· Cut line

Fig. 7. A horizontal gap is easily detected using alternating splitting directions

direction for newly generated rectangles is set to the opposite of the direction that was used to create these rectangles. Figure 7 shows the progression of the triangle splitting method for the same nondominated frontier when an alternating splitting direction strategy is employed. We see that the horizontal gap is detected easily.

Thus, rectangles are added to the priority queue with the opposite splitting direction.

To investigate the efficacy of the triangle splitting method, we used the class of biobjective 0-1 mixed integer programs introduced by Mavrotas et al. [6]. This class has been used in all previous computational studies of algorithms for BOMIPs. We generated five sets of instances for our computational study, each consisting of five instances and identified by the number of constraints m in the instance, i.e., S20, S40, S80, S160, and S320. The number of variables is equal to the number of constraints, and half of the variables are binary and half of the variables are continuous. The sets S160 and S320 contain instances that are much larger than any instances solved in previous studies. For example, Vincent et al. [8] solve instances of up to size 70 and Belotti et al. [2] solve instances up to size 80. The average solution time for the instances in each set is shown in Table 1. We observe that the triangle splitting algorithm is able to solve large instances in a relatively short amount of time. Even the largest instances (in Set S320) are solved in little over an hour.

Table 1. Runtime of the triangle splitting method

Set	S20	S40	S80	S160	S320
Time (secs.)	0.60	2.76	29.08	274.54	3852.63

References

1. Aneja, Y.P., Nair, K.P.K.: Bicriteria transportation problem. Management Science 27, 73–78 (1979)
2. Belotti, P., Soylu, B., Wiecek, M.M.: A branch-and-bound algorithm for biobjective mixed-integer programs,
   http://www.optimization-online.org/DB_HTML/2013/01/3719.html
3. Benson, H.P., Sun, E.: A weight set decomposition algorithm for finding all efficient extreme points in the outcome set of multiple objective linear program. European Journal of Operational Research 139, 26–41 (2002)
4. Gardenghi, M., Gómez, T., Miguel, F., Wiecek, M.M.: Algebra of efficient sets for multiobjective complex systems. Journal of Optimization Theory and Applications 149, 385–410 (2011)
5. Isermann, H.: The enumeration of the set of all efficient solutions for a linear multiple objective program. Operational Research Quarterly 28(3), 711–725 (1977)
6. Mavrotas, G., Diakoulaki, D.: A branch and bound algorithm for mixed zero-one multiple objective linear programming. European Journal of Operational Research 107, 530–541 (1998)
7. Mavrotas, G., Diakoulaki, D.: Multi-criteria branch and bound: A vector maximization algorithm for mixed 0-1 multiple objective linear programming. Applied Mathematics and Computation 171, 53–71 (2005)
8. Vincent, T., Seipp, F., Ruzika, S., Przybylski, A., Gandibleux, X.: Multiple objective branch and bound for mixed 0-1 linear programming: Corrections and improvements for biobjective case. Computers & Operations Research 40(1), 498–509 (2013)

Cut Generation through Binarization

Pierre Bonami[1] and François Margot[2]

[1] IBM ILOG CPLEX, Madrid (Spain) and
LIF, Aix-Marseille Université, Marseille (France)
`pierre.bonami@es.ibm.com`
[2] Tepper School of Business, Carnegie Mellon University, Pittsburgh, USA
`fmargot@andrew.cmu.edu`

Abstract. For a mixed integer linear program where all integer variables are bounded, we study a reformulation introduced by Roy that maps general integer variables to a collection of binary variables. We study theoretical properties and empirical strength of rank-2 simple split cuts of the reformulation. We show that for a pure integer problem with two integer variables, these cuts are sufficient to obtain the integer hull of the problem, but that this does not generalize to problems in higher dimensions. We also give an algorithm to compute an approximation of the rank-2 simple split cut closure. We report empirical results on 22 benchmark instances showing that the bounds obtained compare favorably with those obtained with other approximate methods to compute the split closure or lattice-free cut closure. It gives a better bound than the split closure on 6 instances while it is weaker on 9 instances, for an average gap closed 3.8% smaller than the one for the split closure.

Keywords: Split cuts, closure, binarization.

1   Introduction

Cut generation is an essential part of any Branch-and-Cut software solving mixed-integer linear programs (MILPs). Many cut families and related cut generation algorithms have been developed such as Gomory Mixed-Integer (GMI) cuts [16], split cuts [9], lift-and-project cuts [3], and intersection cuts from lattice-free sets (ILF) [2,10]. Theoretical comparisons show that GMI cuts are a proper subset of split cuts which, in turn are a subset of ILF cuts. Generating GMI cuts is very fast, while generating split cuts or ILF cuts fast enough to provide a significant improvement over GMI cuts when solving an instance is quite challenging.

It is known [5] that the split closure (i.e., adding all split cuts to the linear relaxation of the problem) of usual benchmark instances often gives a tight relaxation of the feasible set. Using a cutting plane approach, most of this strength can be obtained in a reasonable time [6,14]. Iterating these procedures, one can get higher rank cuts and stronger relaxations. However, split cuts are not enough to generate the integer hull for all MILPs [9]. In this paper, we investigate an indirect way to generate stronger cuts introduced by Roy [20], lifting the linear

J. Lee and J. Vygen (Eds.): IPCO 2014, LNCS 8494, pp. 174–185, 2014.
© Springer International Publishing Switzerland 2014

relaxation of the problem to a higher space and generating *simple* rank-2 split cuts, where simple means that the disjunctions used are of the form $x_j \leq b - 1$ or $x_j \geq b$. Using simple disjunctions makes the generation of the cuts easier, but could potentially lead to cuts much weaker than rank-2 split cuts. We show, however, that the lifting used before generating the cuts gives cuts stronger than simple rank-2 split cuts in the original space. In particular, for pure integer instances with two variables, we show that the proposed cuts give the convex hull of the feasible set, a feat not matched by rank-2 simple split cuts in the original space. Empirical results show that the bounds obtained on 22 MIPLIB3 [7] and MIPLIB2003 [1] compare favorably with those obtained with other approximate methods to compute the split closure or ILF closure.

2   Binarization, t-Splits, and t-Simple Splits

We consider mixed integer linear programs of the form:

$$\max c^\top x$$
$$\text{s.t.}$$
$$Ax \leq b, \tag{MILP}$$
$$0 \leq x_j \leq u_j, \qquad \text{for } j = 1, 2, \ldots, p$$
$$x \in \mathbb{Z}_+^p \times \mathbb{R}^{n-p},$$

where A is an $m \times n$ rational matrix of full row rank, $c \in \mathbb{Q}^n$ and $b \in \mathbb{Q}^m$ and $u_j \in \mathbb{Z}_+$ is finite for all $j = 1, 2, \ldots, p$. We denote by P the linear relaxation of (MILP) obtained by replacing the constraints $x_j \in \mathbb{Z}_+$ by $x_j \in \mathbb{R}_+$ for $j = 1, 2, \ldots, p$ and by P_I the convex hull of the feasible set of (MILP).

The definition of (MILP) is quite general, the only actual restriction being that the projection of P onto the space of the integer variables is bounded.

A *cut* for (MILP) is an inequality valid for P_I but not for P. In this paper, we propose a cut generation procedure based on binarization of general integer variables and disjunctive cuts using simple disjunctions.

The *full binary reformulation* of (MILP), introduced by Roy [20], is obtained by adding to (MILP), for each $j = 1, \ldots, p$, a set of u_j binary variables $z_{1j}, \ldots, z_{u_j, j}$ and the following ordering constraints.

$$x_j = \sum_{i=1}^{u_j} z_{ij} \tag{1}$$
$$z_{ij} \geq z_{i+1,j} \quad i = 1, \ldots, u_j - 1 \tag{2}$$
$$z_{ij} \in \{0, 1\} \quad i = 1, \ldots, u_j . \tag{3}$$

We denote by $\mathrm{B}(P)_I$ the full binary reformulation of (MILP) and by $\mathrm{B}(P)$ the continuous relaxation of the latter. The distinctive feature of this reformulation is the presence of the ordering constraints (2). Alternative binary reformulations are discussed in [19,20]. The results in [19] indicate that trying to solve directly the reformulated problem using these alternative reformulations is likely to result in higher computational time. This is why Roy proposes the full binary reformulation to generate cuts that can easily be expressed in the space of the variables x [20]. The results of Roy indicate that combining the full binary reformulation and lift-and-project cuts can lead to significantly better cuts than those obtained using the binary reformulations of [19]. In this paper, we investigate the theoretical and empirical strengths of this reformulation when using simple split disjunctions of rank 1 or 2 to generate cuts.

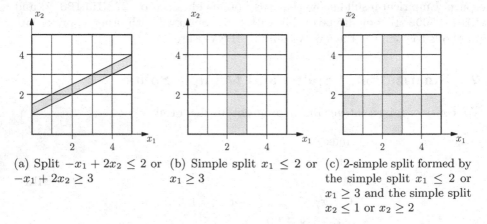

(a) Split $-x_1 + 2x_2 \leq 2$ or $-x_1 + 2x_2 \geq 3$

(b) Simple split $x_1 \leq 2$ or $x_1 \geq 3$

(c) 2-simple split formed by the simple split $x_1 \leq 2$ or $x_1 \geq 3$ and the simple split $x_2 \leq 1$ or $x_2 \geq 2$

Fig. 1. Points not satisfying the split or 2-simple split are shaded

An illustration of the following definitions is provided in Figure 1. A *split* for (MILP) is a disjunction of the form $a^T x \leq b - 1$ or $a^T x \geq b$ for a nonzero $a \in \mathbb{Z}^n$ with $a_j = 0$ for $j = p + 1, \ldots, n$ and $b \in \mathbb{Z}$. A *simple split* for (MILP) is a disjunction of the form $x_j \leq b - 1$ or $x_j \geq b$ for some $b \in \mathbb{Z}$ and some $1 \leq j \leq p$. Notice that if x_j is a binary variable, a simple split on x_j is the disjunction $x_j \in \{0, 1\}$. For any integer $t \geq 1$, a t-split (resp. t-simple split) for (MILP) is the intersection of t splits (resp. t simple splits) for (MILP). (The term t-branch split has been used previously [17] in place of t-split, but since we are using this term often in the sequel, we prefer the short hand.) Note that a 1-split is just a split.

For a t-split (resp. t-simple split) S for (MILP), let P_S be the set of points in P satisfying all splits in S. Observe that as $P_I \subseteq P_S$, all valid inequalities for conv(P_S) are valid for P_I. As B(P) is of the form (MILP), the definitions of split, simple splits, and related inequalities given above apply.

The t-*simple split closure of* P (resp. B(P)), denoted by Si$_t(P)$ (resp. Si$_t(\text{B}(P))$), is the polyhedron obtained as the intersection of conv(P_S) (resp. conv(B$(P)_S$)) for all possible t-simple splits S for (MILP) (resp. B$(P)_I$). The projection of Si$_t(\text{B}(P))$ onto the space of the x variables is denoted by pSi$_t(\text{B}(P))$. As we assume that all integer variables are bounded in (MILP), the number of possible t-simple splits is finite and thus all three sets are obviously polyhedra. A valid inequality for S$_t(P)$ (resp. Si$_t(P)$) is a t-*split* (resp. t-*simple split*) cut for (MILP). A valid inequality for pSi$_t(\text{B}(P))$ is a *projected t-simple split* cut for (MILP).

Notice that the definition of split cut depends on P as well as on which variables are constrained to be integer in the corresponding (MILP). For ease of notation, we will talk of split inequalities for P leaving out the dependence on integer variables. This should not create any confusion, as the integer variables are defined in (MILP) or its binarization and never change.

In Section 3, we will show that $\text{pSi}_t(\text{B}(P)) \subseteq \text{Si}_t(P)$ and that this inclusion is sometimes strict, meaning that binarization allows us to generate stronger cuts than t-simple split inequalities for P.

We can use the closure operators iteratively, defining various rank-k closure of a polyhedron for any positive integer k. For example, the *rank-k t-split closure* of P, denoted by $\text{S}_t^k(P)$, is defined recursively by $\text{S}_t^k(P) = \text{S}_t(\text{S}_t^{k-1}(P))$ with $\text{S}_t^1(P) = \text{S}_t(P)$. The *rank-$k$ t-simple split closure* of P (or $\text{B}(P)$), denoted by $\text{Si}_t^k(P)$ (or $\text{Si}_t^k(\text{B}(P))$) is defined similarly.

Note that in the remainder of the paper, for the various notations defined above, we omit the superscript k or the subscript t when its value is 1.

In this paper, we focus on $\text{B}(P)$, its rank-2 simple split closure $\text{Si}^2(\text{B}(P))$, and its 2-simple split closure $\text{Si}_2(\text{B}(P))$. We first show in Section 3 that the rank-2 simple split closure of $\text{B}(P)$ is never weaker than its 2-simple split closure. In the computational experiments of Section 4, we thus use $\text{Si}^2(\text{B}(P))$, but we show in Section 3 that the projection of $\text{Si}_2(\text{B}(P))$ on the x-variables is P_I for an (MILP) when $n = p = 2$, i.e., with only two integer variables. This result cannot be generalized for problems in three dimensions (either pure or mixed-integer). In Section 4, we compare the lower bounds obtained using rank-2 projected simple split inequalities on problems from MIPLIB3 and MIPLIB2003 with published bounds obtained using split or lattice-free set inequalities.

3 Relations between Closures and Pure Case in $\mathbb{R}^2$

We start by the trivial observation that the t-(simple) split closure contains the $(t + 1)$-(simple) split closure.

Lemma 1. *For $t \geq 1$, we have $\text{S}_{t+1}(P) \subseteq \text{S}_t(P)$ and $\text{Si}_{t+1}(P) \subseteq \text{Si}_t(P)$.*

It is possible to show that, when $n = p = 2$, then $\text{S}^2(P) = \text{S}_2(P) = P_I$, but there are instances for which $\text{S}(P) \neq P_I$. An example of the latter is when P is the square with corners $(0.8, 2.5)$, $(2.5, 0.8)$, $(4.2, 2.5)$, and $(2.5, 4.2)$. It is of course easy to construct examples where $\text{Si}^t(P) \neq P_I$ and $\text{Si}_t(P) \neq P_I$ for any fixed integer $t \geq 1$, such as when P is the triangle with vertices $(0, 0)$, $(1, 0)$, and $(t + 0.5, t + 0.3)$.

The next lemma shows that the projection of the t-simple split closure of $\text{B}(P)$ on the x-variables is not weaker than the t-simple split closure of P.

Lemma 2. *For $t \geq 1$, we have $\text{pSi}_t(\text{B}(P)) \subseteq \text{Si}_t(P)$.*

The main result of Section 3 is that $\text{pSi}_2(\text{B}(P)) = P_I$ for any (MILP) with $n = p = 2$, showing that the inclusion given in Lemma 2 is sometimes strict.

Lemma 3. *Let S be a t-simple disjunction for $\mathrm{B}(P)$. Then there exists a t-simple disjunction S' using only simple disjunctions on some z_{ij} variables such that $\mathrm{B}(P)_{S'} \subseteq \mathrm{B}(P)_S$.*

Lemma 3 shows that any t-simple split cut for $\mathrm{B}(P)$ can be obtained using t-simple splits on the z variables only. As a result, when we use simple splits for $\mathrm{B}(P)$ in the remainder of the paper, we only discuss simple splits on the binary variables z_{ij} and ignore those on the x_j variables.

Let us now turn to the comparison between rank-t simple split closure and t-simple split closure for $\mathrm{B}(P)$. Dash et al. [12] have shown that, in general, 2-split closure and rank-2 split closure are incomparable. However, for simple t-splits involving only binary variables, we have the following.

Lemma 4. *Any t-simple split inequality obtained using only simple splits on binary variables can be generated as a rank-t simple split inequalities.*

Putting lemmas 3 and 4 together, we get the following.

Corollary 1. *For $\mathrm{B}(P)$, any t-simple split inequality is a rank-t simple split inequality, i.e., $\mathrm{Si}^t(\mathrm{B}(P)) \subseteq \mathrm{Si}_t(\mathrm{B}(P))$.*

There are cases where the inclusion given in Corollary 1 is strict. In the remainder of this section, we consider the case where P is a polytope in $\mathbb{R}^2$ and both variables are integer. The main result of this paper is the following.

Theorem 1. *Let $P = \{x \in \mathbb{R}^2 \mid Ax \leq b, 0 \leq x_j \leq u_j \text{ for } j = 1, 2\}$ be the linear relaxation of a problem of the form* (MILP) *with $n = p = 2$. Then the projection of the 2-simple split closure of $\mathrm{B}(P)$ is P_I, i.e. $\mathrm{pSi}_2(\mathrm{B}(P)) = P_I$.*

As mentioned above, rank-2 split inequalities or 2-split inequalities are necessary to obtain P_I when $n = p = 2$. Theorem 1 shows that, for (MILP) with $n = p = 2$, projected 2-simple split inequalities for P are as strong as rank-2 split inequalities as well as 2-simple split inequalities for P and they are stronger than rank-k simple split inequalities for P, for any $k \geq 1$; this is interesting, as 2-simple split disjunctions for $\mathrm{B}(P)$ are in finite number and much easier to generate than general split inequalities for P. Note that Chen et al. [8] show that in the general case $n \geq p \geq 1$, there exists a finite value for t such that $\mathrm{Si}_t(P) = P_I$, but this bound depends on P itself (a trivial upper bound on t is $\prod_{j=1}^{p} u_j$, but their analysis is a little bit stronger than that).

Unfortunately, generalizations of Theorem 1 to $\mathbb{R}^3$ cannot be obtained. It is possible to construct examples where $\mathrm{pSi}_p(\mathrm{B}(P)) \neq P_I$ when $P \subseteq \mathbb{R}^3$ and $p = 2$ or $p = 3$.

4   Computational Experiments

In this section, we propose a new recursive procedure to optimize over $\mathrm{Si}^2(\mathrm{B}(P))$ as well as a practical construction of partial binary reformulations for problems with integer variables that can take large values. Finally, we present a series of experiments using these two procedures on problems from MIPLIB3 and MIPLIB2003 having general integer variables.

4.1   Closure Optimization Algorithms

The procedures we use to optimize over simple split closures are based on the algorithm for optimizing over $\mathrm{Si}(P)$ given in [6]. The algorithms described in this section can be used to compute rank-1 or rank-2 simple split closures of any MILP, but to keep notation and exposition as simple as possible, we focus on our specific formulation and task, namely computing the rank-2 simple split closure of $\mathrm{B}(P)$, using disjunctions on binary variables z_{ij} only.

To simplify the presentation, let Q be the MILP obtained by a binary reformulation of (MILP). The reformulation might be either the full reformulation described in Section 2 or the partial one we describe in Section 4.2. We assume that all variables in Q are renamed, such that the binary variables z_{ij} of the reformulation are now variables x_j for $j = 1, 2, \ldots, q$. For any index $j = 1, \ldots, q$, we denote by Q_j the polyhedron $Q_j = \mathrm{conv}\{x \in Q \mid x_j \in \{0, 1\}\}$. Following the notation introduced in Section 2, for $k \geq 1$ we denote by $\mathrm{Si}^k(Q)$ the rank-k simple split closure of Q. Lemma 3 shows that $\mathrm{Si}^k(Q)$ can be obtained using only simple splits on variables x_j for $j = 1, 2, \ldots, q$.

Rank-1 simple-split closure optimization. To optimize over $\mathrm{Si}(Q)$, we use a simple cutting plane algorithm. At each iteration, let $\overline{x}$ be the optimal solution over the current approximation of $\mathrm{Si}(Q)$. The feasibility of $\overline{x}$ for $\mathrm{Si}(Q)$ is verified by solving, for each $j = 1, \ldots, q$ with $0 < \overline{x}_j < 1$, the LP described in [6]:

$$\max \ y_j$$
$$\text{s.t.}$$
$$A\overline{x} + b(\overline{x}_j - 1) \leq Ay \leq b\overline{x}_j \qquad (\mathrm{MLP}_j)$$
$$0 \leq y \leq \overline{x},$$
$$y \in \mathbb{R}^n.$$

If the optimal solution $\overline{y}$ to (MLP_j) has $\overline{y}_j = \overline{x}_j$, then $\overline{x} \in Q_j$; otherwise, a cut separating $\overline{x}$ from Q_j can be generated using the dual solution of (MLP_j) as shown in Lemma 1 in [6].

Rank-2 simple-split closure optimization. The procedure for optimizing over $\mathrm{Si}^2(Q)$ consists in a recursive application of the algorithm for optimizing over $\mathrm{Si}(Q)$ described above. Suppose that we have a complete linear description of $\mathrm{Si}(Q)$ as $\{x \in \mathbb{R}^n_+ : Ax \leq b,\ Dx \leq d\}$. Then, optimizing over $\mathrm{Si}^2(Q)$ can be achieved by taking into consideration the cuts $Dx \leq d$ in the separation problem (MLP_j) and using the algorithm of the previous paragraph. Of course, the main roadblock for this approach is obtaining the complete description of $\mathrm{Si}(Q)$. This can, however, be sidestepped by using a cut generation procedure within the solution of the separation problem (MLP_j). The aim of this secondary cut generation step is to find the rank-1 inequalities that are useful to generate a rank-2 cut cutting $\overline{x}$, if one exists.

First, we optimize over $\mathrm{Si}(Q)$. Let $\overline{x}$ be the optimal point in $\mathrm{Si}(Q)$. Then, for all $j = 1, \ldots, q$ such that $\overline{x}_j \notin \{0, 1\}$, there exist $\overline{x}^{j,1}$ and $\overline{x}^{j,0} \in Q$ such that

$\bar{x} = \bar{x}_j \, \bar{x}^{j,1} + (1 - \bar{x}_j) \, \bar{x}^{j,0}$, with $\bar{x}_j^{j,1} = 1$ and $\bar{x}_j^{j,0} = 0$. Note that these points can be obtained from the solution $\bar{y}$ of (MLP$_j$), namely $\bar{x}^{j,1} = \frac{\bar{y}}{\bar{x}_j}$ and $\bar{x}^{j,0} = \frac{\bar{x} - \bar{y}}{1 - \bar{x}_j}$.

To try to cut $\bar{x}$, we use a generalization of (MLP$_j$):

$$\max \ y_j$$
$$\text{s.t.}$$
$$A\bar{x} + b(\bar{x}_j - 1) \le Ay \le b\bar{x}_j$$
$$D^1\bar{x} + d^1(\bar{x}_j - 1) \le D^1 y \le d^1 \, \bar{x}_j \qquad (MLP_j^2(D^1, d^1))$$
$$0 \le y \le \bar{x},$$
$$y \in \mathbb{R}^n \ ,$$

where the $D^1 x \le d^1$ are rank-1 cuts that may possibly be combined to cut $\bar{x}$. The pseudo-code of the algorithm is given below.

Algorithm 1. Separation of $\bar{x}$ from $\text{Si}^2(Q)$

---

0. **Initialization.**
   Let $\bar{x}$ be a point in $\text{Si}(Q)$, and $\tilde{D}^1 x \le \tilde{d}^1$ be a partial linear description of $\text{Si}(Q)$.
   Let $D^1 x \le d^1$ be an empty system.
1. For each $j \in \{1, 2, \dots, q\}$ with $\bar{x}_j$ fractional, repeat steps 2.-5.
2. **Solution of rank-2 separator**
   Let $\bar{y}$ be an optimal solution of $(MLP_j^2(D^1, d^1))$
3. **Generate a cut?**
   If $\bar{y}_j < 1$, a cut can be generated from the dual solution of $(MLP_j^2(D^1, d^1))$.
   STOP.
   Otherwise let $\bar{x}^1 = \frac{\bar{y}}{\bar{x}_j}$ and $\bar{x}^0 = \frac{\bar{x} - \bar{y}}{1 - \bar{x}_j}$.
4. **Find existing rank-1 cuts**
   If there is a cut in $\tilde{D}^1 x \le \tilde{d}^1$ not satisfied by $\bar{x}^1$ or $\bar{x}^0$, add it to $D^1 x \le d^1$.
   Go to 1.
5 **Find new rank-1 cuts**
   For all fractional components j' of $\bar{x}^1$ (resp. $\bar{x}^0$), solve (MLP$_{j'}$) to separate $\bar{x}^1$ (resp. $\bar{x}^0$) from $\text{Si}(Q)$. If violated rank-1 cuts are found, add them to $D^1 x \le d^1$.
   Go to 1. Otherwise, $\bar{x} \in \text{Si}^2(Q)$, **STOP**

The following lemma shows the correctness of the algorithm.

Lemma 5. *If Algorithm 1 terminates without generating a cut then $\bar{x} \in \text{Si}^2(Q)$.*

In Section 4.3, we use the separation procedures presented here both to separate rank-2 cuts for the original instances with general integers and their binary reformulations. Extending the procedure to general integer variables can be obtained using bound translations to get $0 \le \bar{x}_j \le 1$ for $j = 1, \dots, p$.

Finally, note that monoidal strengthening [4] can be used on all cuts found by the separation algorithms. Each strengthened cut dominates its non-strengthened counterpart. The resulting cuts are not simple split cuts anymore but split cuts. Therefore, the algorithms no longer compute $Si(Q)$ or $Si^2(Q)$, but an approximation of $S(Q)$ or $S^2(Q)$.

4.2   Partial Binary Reformulation

The full binary reformulation presented in Section 2 is unattractive in practice when some of the upper bounds u_j are large. We thus devised a practical binarization scheme, using at most K binary variables for each integer variable in (MILP), where $K \geq 1$ is a parameter of the method.

For each variable x_j of (MILP) with $j \in \{1, 2, \ldots, p\}$, either (i) $K \geq u_j$ and we use the full binary reformulation for variable x_j using u_j binary variables and corresponding inequalities (1)-(3); or (ii) $K < u_j$, and we use a *partial binary reformulation* by adding to the problem K binary variables, $z_{1j}, \ldots, z_{K,j}$, and two general integer variables z_j^-, z_j^+. More precisely, we define

$$M_1 = \min\{\max\{0, \lfloor \overline{x}_j \rfloor - K/2\}, \, u_j - K\}$$
$$M_2 = \max\{\min\{u_j, \lfloor \overline{x}_j \rfloor + K/2\}, \, K\} \ .$$

and use the following constraints

$$x_j = z_j^- + z_j^+ + \sum_{i=1}^{K} z_{ij}$$
$$z_{ij} \geq z_{i+1,j} \qquad\qquad \text{for } i = 1, \ldots, K-1$$
$$M_1 \geq z_j^- \geq M_1 \, z_{1,j}$$
$$(u_j - M_2)z_{Kj} \geq z_j^+$$
$$z_{ij} \in \{0,1\} \qquad\qquad \text{for } i = 1, \ldots, K$$
$$z_j^-, z_j^+ \in \mathbb{Z}_+ \ .$$

4.3   Results

The closure optimization algorithms are implemented in C++. The implementation uses CPLEX 12.4 in its default settings to solve all linear programs. The closure optimization algorithms are generic in that they work indifferently on (MILP) or on instances reformulated using a binary reformulation.

The tolerances of the algorithm are set as follows. Let the fractionally of $\overline{x}_j$ be $f_j := |\overline{x}_j - \lfloor \overline{x}_j + 0.5 \rfloor|$. We try to generate a cut using the simple split on x_j if $f_j < 10^{-2}$. Whenever a separation problem (MLP$_j$) or ($MLP_j^2(D^1, d^1)$) is solved, a cut is generated only if the optimal value of the problem is smaller than $\overline{x}_j - 10^{-2}$. A generated inequality is considered as cutting $\overline{x}$ if the violation of $\overline{x}$ scaled by the norm of the left hand side of the cut is larger than 10^{-5}. A maximum time of 10,800 CPU seconds for the whole execution is used.

We consider the 22 instances from MIPLIB3 and MIPLIB2003 having general integer variables. All instances are preprocessed in order to find tight bounds for the integer variables by minimizing and maximizing the value of each general integer variables over the continuous relaxation. After this tightening, all integer variables have finite upper and lower bounds on the considered instances.

In order to assess the additional strength of binary formulations, we compare the fraction of the gap closed by optimizing over $Si(P)$, $Si(Q)$, $Si^2(P)$ and $Si^2(Q)$. We also compare the strength of the bounds obtained when monoidal strengthening is used on all cuts.

To generate the partial binary formulations Q, we use the procedure described in Section 4.2 with parameter $K = 10$. Only one partial binary formulation is constructed for each instance. Since our goal is to assess the additional gap that can be closed by using a binary formulation, we choose as the reference point for constructing the partial binary formulation an approximation x' of the optimal solution over $Si(P)$ on the original instance. The time taken to compute x' is included in the reported computation time and is at most half of the maximum time allocated to solve an instance. Note that when optimizing over $Si(Q)$ or $Si^2(Q)$, we only generate cuts using simple splits on binary variables, ignoring the integrality of the variables z_j^+ and z_j^-.

Table 1 presents the percent gap closed by optimizing over $Si(P)$, $Si(Q)$, $Si^2(P)$, and $Si^2(Q)$. For each instance and each closure, we report the percent gap closed by optimizing over the closure with respect to the gap between the optimal solution and LP relaxation values. We also report in the columns labeled Si^2/Si the percent gap closed by Si^2 of the gap remaining for Si. Whenever the optimal value over a closure could not be computed within the time limit, we put a ">" sign in front of the corresponding number.

In the discussion of the results, by abuse of language, we simply refer to the name of each closure to refer to the result obtained by optimizing over it. There are 3 instances for which we can not compute $Si(P)$ within the time limit. For these, we cannot report values for the improvement obtained by the other closures. There are 5 additional instances for which we can not compute $Si^2(P)$ exactly and 8 more on which we can not compute $Si^2(Q)$. Even with these incomplete results, we see that $Si(Q)$ and $Si^2(Q)$ close more gap than their counterparts $Si(P)$ and $Si^2(P)$. Another interesting finding is that the gap closed by Si^2 with respect to Si is very significant: $Si^2(P)$ closes on average 40.8% of the gap remaining for $Si(P)$. Moreover, the effect of binarization does not seem to fade with increasing rank. On average, $Si^2(Q)$ closes 2.8% more fraction gap over $Si(Q)$ than $Si^2(P)$ over $Si(P)$.

Table 2 gives the gap closed when monoidal strengthening is applied on all cuts. We denote by Si^* and Si^{*2} the methods where strengthening is applied. Note that for this experiment, contrary to what is done for Table 1, we generate cuts using simple splits on the variables z_j^- and z_j^+. For reference, we also include in this table the gap closed by alternate approximations of the split closure from the literature: The approximation of the split closure found in [14,15] (we report the best value among the 6 methods proposed in the paper), the procedure for separating 2 dimensional lattice free cuts proposed in [18] (we report only the results for exact separation reported in Table 2 of [18]), the procedure for

Table 1. Gap closed when optimizing over Si, Si^2 for the original formulation and the partial binary reformulation

name	P			Q		
	Si	Si^2	Si^2/Si	Si	Si^2	Si^2/Si
arki001	20.0	27.9	9.9	24.5	> 37.2	> 16.8
bell3a	64.6	68.7	11.7	64.6	68.7	11.7
bell5	86.2	89.8	25.7	87.5	92.9	43.3
blend2	26.8	35.5	11.9	26.8	35.5	12.0
flugpl	12.5	14.9	2.7	14.5	23.9	11.1
gen	68.7	87.7	60.6	68.7	87.6	60.4
gesa2	58.0	76.1	43.2	59.6	> 78.8	> 47.5
gesa2_o	59.4	76.7	42.5	61.0	> 78.9	> 46.0
gesa3	78.9	99.2	96.0	88.1	> 99.5	> 95.4
gesa3_o	81.6	99.3	96.3	89.4	> 99.6	> 95.9
gt2	92.4	99.6	95.3	92.4	> 99.6	> 95.3
momentum2	> 48.8	> 48.8	> —	> 48.8	> 48.8	> —
msc98-ip	> 42.3	> 42.3	> —	> 42.3	> 42.3	> —
mzzv11	> 77.0	> 77.0	> —	> 76.0	> 76.0	> —
mzzv42z	84.2	> 93.3	> 57.4	85.0	> 91.1	> 40.7
noswot	0.0	0.0	0.0	0.0	0.0	0.0
qnet1	93.9	> 96.9	> 48.6	93.0	> 97.1	> 58.3
qnet1_o	87.0	> 98.2	> 86.2	86.5	> 98.1	> 86.2
roll3000	14.7	> 31.1	> 19.3	14.9	> 33.3	> 21.6
rout	27.2	> 45.1	> 24.7	27.2	> 45.1	> 24.6
timtab1	27.1	42.8	21.6	30.2	> 55.5	> 36.2
timtab2	20.8	38.0	21.7	22.6	> 41.6	> 24.5
Average	53.3	63.1	40.8	54.7	65.1	43.6

generating 2-split cuts (Table 1 in [13]), and finally in the column labeled "split" the best known approximation of the split closure taken from [5,11,14,15].

The use of monoidal strengthening results in fewer instances where computations can not be completed within the time limit. The results are similar to those of Table 1 but with larger gaps closed by all methods and larger differences between gap closed using P or Q. The difference between Si$^*(P)$ and Si$^*(Q)$ is now 3.2% on average while the difference between Si$^{*2}(P)$ and Si$^{*2}(Q)$ is 5.4% on average. On average, Si$^{*2}(Q)$ closes 7% more fraction gap over Si$^*(Q)$ than Si$^{*2}(P)$ over Si$^*(P)$.

Comparing with results from the literature, we first observe that the approximation of Si$^{*2}(Q)$ is always stronger than those obtained in [18] or [13]. Second, while our approximation of Si$^{*2}(Q)$ is not as strong as the best known values for the split closure on average, there are 6 instances on which it gives a better bound. There are 7 instances on which the values are identical (either no gap is closed or the problem is solved), and on the remaining 9 instances the split

Table 2. Gap closed when optimizing over Si^*, Si^{*2} (i.e, using monoidal strengthening) for the original formulation and the partial binary reformulation. Problems for which the final point was within the integrality tolerance are marked with a star. For each instance, the value of the largest gap closed is in bold.

name	P			Q			[14]	[18]	[13]	split
	Si^*	Si^{*2}	Si^{*2}/Si^*	Si^*	Si^{*2}	Si^{*2}/Si^*				
arki001	33.3	51.9	27.9	44.8	> 64.9	> 36.5	43.7		55.1	**83.1**
bell3a	64.6	68.7	11.7	68.3	72.3	12.4	69.1	59.0	67.2	**99.6**
bell5	86.5	95.4	66.1	90.2	**99.4**	94.0	90.7	91.2	17.8	92.9
blend2	28.6	37.8	12.9	28.6	38.5	13.9	23.1	28.2	21.6	**46.5**
flugpl	12.5	16.3	4.3	18.0	*100	100	98.5	44.5	14.1	**100**
gen	83.0	100	100	91.6	*100	100	96.5		79.1	**100**
gesa2	61.7	86.1	63.7	64.5	> 89.1	> 69.2	80.8		65.0	**99.7**
gesa2_o	67.7	88.6	64.8	70.5	> 91.0	> 69.3	81.9	45.8	63.2	**100**
gesa3	93.6	99.7	94.8	96.6	**100**	100	95.9	51.3	83.0	95.8
gesa3_o	93.8	99.7	95.5	97.2	**100**	100	95.7	59.6	85.3	95.2
gt2	94.4	100	99.3	94.4	> 100	> 99.6	98.7	58.3	77.1	98.4
momentum2	48.4	48.4	0.0	48.4	> **48.7**	> 0.5	48.3	—	—	48.3
msc98-ip	> 50.0	> 50.0	> —	> 50.0	> 50.0	> —	54.4	—	—	**54.4**
mzzv11	99.9	> 100	> 100	**100**	*100	100	99.9	—	—	**100**
mzzv42z	100	100	—	*100	*100	—	100	—	—	**100**
noswot	0.0	0.0	0.0	0.0	0.0	0.0	0.0			0.0
qnet1	94.4	100	99.8	97.5	* 100	100	99.6	7.2	29.4	**100**
qnet1_o	88.2	100	99.7	89.1	* 100	99.5	100	26.5	36.5	**100**
roll3000	54.8	> 65.8	> 24.4	55.3	> 66.7	> 25.6	92.1	—	—	**92.1**
rout	51.9	> 72.2	> 42.2	52.8	> **72.2**	> 41.0	50.8	1.4	4.0	70.7
timtab1	49.6	> 72.0	> 44.5	63.0	> 74.2	> 30.3	81.8	—	—	**81.8**
timtab2	43.0	> 54.4	> 20.0	48.6	> 58.4	> 19.1	72.7	—	—	**72.7**
Average	63.6	73.0	53.6	66.8	78.4	60.6	76.1			82.2

closure gives a better bound (for 7 out of these 9 instances the computation over $Si^{*2}(Q)$ could not be completed within the set time limit).

Acknowledgments. F. Margot was supported by ONR grant N00014-12-10032 and did part of this work while visiting the Ecole Polytechnique Fédérale de Lausanne (Switzerland).

References

1. Achterberg, T., Koch, T., Martin, A.: MIPLIB 2003. Oper. Res. Lett. 34, 361–372 (2006)
2. Andersen, K., Louveaux, Q., Weismantel, R., Wolsey, L.: Cutting planes from two rows of a simplex tableau. In: Fischetti, M., Williamson, D.P. (eds.) IPCO 2007. LNCS, vol. 4513, pp. 1–15. Springer, Heidelberg (2007)

3. Balas, E., Ceria, S., Cornuéjols, G.: A lift-and-project cutting plane algorithm for mixed 0-1 programs. Math. Program. 58, 295–323 (1993)
4. Balas, E., Jeroslow, R.G.: Strengthening cuts for mixed integer programs. European J. Oper. Res. 4(4), 224–234 (1980)
5. Balas, E., Saxena, A.: Optimizing over the split closure. Math. Program. 113, 219–240 (2008)
6. Bonami, P.: On optimizing over lift-and-project closures. Math. Program. Computation 4, 151–179 (2012)
7. Bixby, R.E., Ceria, S., McZeal, C.M., Savelsbergh, M.W.P.: An updated mixed integer programming library: MIPLIB 3.0. Optima 58, 12–15 (1998)
8. Chen, B., Küçükyavuz, S., Sen, S.: Finite disjunctive programming characterizations for general mixed-integer linear programs. Oper. Res. 59, 202–210 (2011)
9. Cook, W., Kannan, R., Schrijver, A.: Chvátal closures for mixed integer programming problems. Math. Program. 47, 155–174 (1990)
10. Cornuéjols, G., Margot, F.: On the facets of mixed integer programs with two integer variables and two constraints. Math. Program. 120, 419–456 (2009)
11. Dash, S., Günlük, O., Lodi, A.: MIR closures of polyhedral sets. Math. Program. 121, 33–60 (2010)
12. Dash, S., Günlük, O., Molinaro, M.: On the relative strength of different generalizations of split cuts. Working paper (2012)
13. Dash, S., Günlük, O., Vielma, J.P.: Computational experiments with cross and crooked cross cuts. Working paper (2011)
14. Fischetti, M., Salvagnin, D.: Approximating the split closure. INFORMS Journal on Computing 25, 808–819 (2013)
15. Fischetti, M., Salvagnin, D.: Personnal communication
16. Gomory, R.E.: An Algorithm for Integer Solutions to Linear Programs. In: Graves, R.L., Wolfe, P. (eds.) Recent Advances in Mathematical Programming, pp. 269–302. McGraw-Hill, New York (1963)
17. Li, Y., Richard, J.-P.P.: Cook, Kannan and Schrijver's example revisited. Discrete Optim. 5, 724–734 (2008)
18. Louveaux, Q., Poirrier, L., Salvagnin, D.: The strength of multi-row models. Working paper (2012)
19. Owen, H., Mehrotra, S.: On the value of binary expansions for general mixed-integer linear programs. Oper. Res. 50, 810–819 (2002)
20. Roy, J.-S.: "Binarize and Project" to generate cuts for general mixed-integer programs. Algorithmic Oper. Res. 2 (2007)

A $\frac{5}{4}$-Approximation for Subcubic 2EC Using Circulations*

Sylvia Boyd, Yao Fu, and Yu Sun

School of Electric Engineering and Computer Science (EECS), University of Ottawa
Ottawa, Ontario K1N 6N5, Canada
{sboyd,yfu099,ysun007}@uottawa.ca

Abstract. In this paper we study the NP-hard problem of finding a minimum size 2-edge-connected spanning subgraph (henceforth 2EC) in cubic and subcubic multigraphs. We present a new $\frac{5}{4}$-approximation algorithm for 2EC for subcubic bridgeless graphs, improving upon the current best approximation ratio of $\frac{5}{4} + \varepsilon$. Our algorithm involves an elegant new method based on circulations which we feel has potential to be more broadly applied. We also study the closely related integrality gap problem, i.e. the worst case ratio between the integer linear program for 2EC and its linear programming relaxation, both theoretically and computationally. We show this gap is at most $\frac{9}{8}$ for all subcubic bridgeless graphs with up to 16 nodes. Moreover, we present a family of graphs that demonstrate the integrality gap is at least $\frac{8}{7}$, even when restricted to subcubic bridgeless graphs. This represents an improvement over the previous best known bound of $\frac{9}{8}$.

Keywords: minimum 2-edge-connected subgraph problem, approximation algorithm, circulations, integrality gap, subcubic graphs.

1 Introduction

Given an unweighted bridgeless multigraph $G = (V, E)$, $|V| = n$, the *minimum size 2-edge-connected spanning subgraph problem* (henceforth 2EC) consists of finding a 2-edge-connected spanning subgraph H of G with the minimum number of edges. Note that a *2-edge-connected* graph $G = (V, E)$ is one that remains connected after the removal of any edge. An edge removal disconnects a graph into two components is called a *bridge*. In a solution for 2EC, multiple copies of an edge $e \in E$ are not allowed (and also not necessary).

The problem 2EC is one of the most extensively studied problems in network design. It relates to the optimal design of a network that can survive the loss of a link, and thus has many real world applications. However, it is known to be NP-hard and also MAX SNP-hard even for subcubic graphs [1], where a graph is cubic if every node has degree 3, and subcubic if every node has degree at

* This research was partially supported by the Natural Science and Engineering Research Council of Canada (NSERC).

J. Lee and J. Vygen (Eds.): IPCO 2014, LNCS 8494, pp. 186–197, 2014.
© Springer International Publishing Switzerland 2014

most 3. Thus research has focused on finding good approximation algorithms. Unfortunately, finding improved approximation algorithms seems to be difficult for the more general weighted version of 2EC where, as with the closely related travelling salesman problem (TSP), the best known approximation ratio for metric weights has remained at $\frac{3}{2}$ [2] without any improvement for over 30 years.

Given the difficulty of this problem, people have turned to the study of approximation algorithms for special cases, which has proven to be a more successful approach for 2EC than studying its more general weighted form. In such studies, not only improved results were obtained but also new innovative methods which may lead to more general results.

In this paper we focus on the simplest form of 2EC that still remains NP-hard, i.e. 2EC for subcubic graphs. In Section 2 we describe the framework for a new innovative method for designing approximation algorithms for 2EC based on circulations. Similar types of circulations were used in [3] in the approximation of graph TSP, however the goal was quite different in that context. In fact, to the best of our knowledge, circulations have not previously been used in the way we describe to approximate 2EC. We demonstrate the usefulness of our method by using it to develop a new $\frac{5}{4}$-approximation algorithm for 2EC on bridgeless subcubic graphs. This algorithm improves upon the previous best approximation ratio of $\frac{5}{4} + \varepsilon$ given by Csaba, Karpinski and Krysta [1] for 2EC on such graphs. We feel this algorithm not only provides a modest improvement in the approximation ratio, but also, and perhaps more importantly, provides an improvement in the simplicity and elegance of the method and proof.

A related approach for finding approximated 2EC solutions is to study the integrality gap α^{2EC}, which is the worst case ratio between the optimal value for 2EC and the optimal value for its linear programming (henceforth LP) relaxation [2]. As a critical topic throughout this paper, we study α^{2EC} intensively. There are two main reasons this is useful. First, the integrality gap itself serves as an indicator of the quality of the lower bound given by the LP relaxation. This is important for methods, such as branch and bound and approximation, that depend on good lower bounds for their success. Secondly, an algorithmic proof for $\alpha^{2EC} = k$ yields a k-approximation algorithm for 2EC [2]. In this paper, we give an upper bound on the value of α^{2EC} on bridgeless cubic graphs with an algorithmic proof, while lower bounds on the integrality gap of 2EC are investigated through computational studies. We show that the integrality gap of 2EC is strictly less than $\frac{5}{4}$ for bridgeless cubic graphs, improving on the previous best known bound of $\frac{5}{4} + \varepsilon$ [1]. We also conduct a computational study by designing a program that calculates α^{2EC} exactly for all simple graphs $G \in \mathcal{G}$, where $\mathcal{G}$ contains all test cases in three categories: (1) General bridgeless graphs for $3 \leq n \leq 10$; (2) Cubic bridgeless graphs for $6 \leq n \leq 16$; (3) Subcubic bridgeless graphs for $3 \leq n \leq 16$. Using the knowledge gained through the data analysis for the computational study, we obtain a family of subcubic bridgeless graphs G which shows $\alpha^{2EC} \geq \frac{8}{7}$ asymptotically, which improves upon the previous best lower bound of $\frac{9}{8}$ [4].

1.1   Literature Review on 2EC

Constant factor approximation algorithms for 2EC have been intensively studied. In 1994, Khuller and Vishkin [5] found a $\frac{3}{2}$-approximation, which was improved by Cheriyan, Sebő and Szigeti [6] to $\frac{17}{12}$. The ratio was later improved to $\frac{4}{3}$ in 2000 by Vempala and Vetta [7]. One year later, Krysta and Kumar [8] improved the approximation ratio to $\frac{4}{3} - \epsilon$ where $\epsilon = \frac{1}{1344}$. Recently, Sebő and Vygen [4] designed a simpler and more elegant $\frac{4}{3}$-approximation algorithm for 2EC.

In the meantime, research on 2EC has also been conducted for special classes of graphs, especially on cubic bridgeless graphs, on which 2EC still remains NP-hard. In 2001, along with their $(\frac{4}{3} - \epsilon)$-approximation algorithm for 2EC on general graphs, Krysta and Kumar [8] also presented an approximation algorithm for 2EC on cubic graphs with the approximation ratio of $\frac{21}{16} + \epsilon$. One year later, Csaba, Karpinski and Krysta [1] designed a $(\frac{5}{4} + \epsilon)$-approximation algorithm for 2EC on subcubic graphs. In 2004, Huh [9] presented an algorithm yielding a $\frac{5}{4}$-approximation on cubic 3-edge-connected graphs. A more recent improvement came from Boyd, Iwata and Takazawa [10] with a $\frac{6}{5}$-approximation algorithm for 2EC on cubic 3-edge-connected graphs.

Concerning the integrality gap α^{2EC} of 2EC, Csaba, Karpinski and Krysta [1] proved that for maximum degree 3 graphs, the integrality gap of the LP relaxation for 2EC is at most $\frac{5}{4} + \epsilon$ for any fixed $\epsilon > 0$. It was also stated in [1] that the best known lower bound on α^{2EC} is $\frac{10}{9}$ for maximum degree 3 graphs (and thus subcubic graphs). In 2013, Boyd, Iwata and Takazawa [10] show $\alpha^{2EC} \leq \frac{6}{5}$ for 3-edge-connected cubic graphs. Around the same time, Sebő and Vygen [4] proved that $\frac{9}{8} \leq \alpha^{2EC} \leq \frac{4}{3}$ in general.

1.2   Notation and Background

For the purpose of this paper, any graph $G = (V, E)$ is considered to be a multigraph without loops. We use n to denote $|V|$. Denoted by $ILP(G)$, the integer linear program of 2EC for a given graph $G = (V, E)$ is given below. Note that for any $S \subset V$, $\delta(S)$ is the set of edges with one end in S and the other end not in S, and for any $F \subseteq E$, we use the notation $x(F)$ to denote $\sum_{e \in F} x_e$.

$$\text{Minimize} \quad \sum_{e \in E} x_e \tag{1}$$

$$\text{subject to} \quad x(\delta(S)) \geq 2 \quad \text{for all} \quad \emptyset \subset S \subset V, \tag{2}$$

$$0 \leq x_e \leq 1 \quad \text{for all} \quad e \in E, \tag{3}$$

$$x_e \text{ integer} \quad \text{for all} \quad e \in E. \tag{4}$$

By relaxing the integrality constraints (4) of $ILP(G)$, the LP relaxation of $ILP(G)$, denoted by $LP(G)$, is obtained. We use the notation $OPT(G)$ and $OPT_{LP}(G)$ to denote the optimal objective value for $ILP(G)$ and $LP(G)$ respectively.

2 Using Circulations to Obtain 2-Edge-Connected Spanning Subgraphs

In this section we outline a new method for finding approximation algorithms for 2EC which is based on circulations and depth first search (DFS) trees. Similar ideas were used in [3] for approximation for graph TSP, but not in the same way or for the same purpose as they are used here. To the best of our knowledge, these ideas represent a new framework for 2EC approximation.

Given a digraph $D = (V, A)$, $f \in \mathbb{R}^A$ is called a *circulation* for D if $f(\delta^{in}(v)) = f(\delta^{out}(v))$ for all $v \in V$. For an arc $e \in A$, f_e is called the *flow* of e. Given arc demands $d \in \mathbb{R}^A$ and arc capacities $u \in \mathbb{R}^A$, a circulation is called *feasible* if $d_e \leq f_e \leq u_e$ for all $e \in A$. The *support graph* of a circulation f is the subgraph $D_f = (V, A_f)$, where A_f is the set of arcs a in A for which $f_a > 0$. Finally, given arc costs $c \in \mathbb{R}^A$, the *minimum cost circulation problem* is as follows:

$$\begin{aligned}
\text{minimize} \quad & cf \\
\text{subject to} \quad & f(\delta^{in}(v)) = f(\delta^{out}(v)) \quad \text{for all } v \in V, \\
& d_e \leq f_e \leq u_e \quad \quad \text{for all } e \in A.
\end{aligned}$$

The following is well known for circulations ([11], also see [12]).

Theorem 1. *Given a minimum cost circulation problem for which d and u are integer-valued and for which there exists a feasible circulation f, there exists an optimal circulation f^* which is integer-valued and can be found in polynomial time.*

Given a 2-edge-connected graph $G = (V, E)$, we now define a minimum cost circulation problem $P(G)$ based on G. To begin, give G an orientation by growing a spanning tree from an arbitrary root $r \in V$ using DFS. Call the edges in the tree *tree edges* and the rest of the edges in E *back edges*. Let the directed graph $D = (V, A)$ be the orientation of G obtained in the usual way using the DFS tree, i.e. by directing all the tree edges away from r and all the back edges towards r. Let $A = T \cup B$ where T is the set of directed tree edges and B is the set of directed back edges. Note that the arc set T forms a spanning arborescence of D, and that the edges uv of G are in one to one correspondence with the arcs (u, v) of D, a fact that we exploit by referring to edges and arcs interchangeably. Define the minimum cost circulation problem $P(G)$ as follows:

$$d_e = \begin{cases} 1 \text{ for } e \in T \\ 0 \text{ otherwise,} \end{cases} \quad c_e = \begin{cases} 1 \text{ for } e \in B \\ 0 \text{ otherwise,} \end{cases} \quad u_e = \begin{cases} 1 \text{ for } e \in B \\ \infty \text{ otherwise.} \end{cases}$$

Let f be any feasible circulation for $P(G)$. By Theorem 1, there exists an integer feasible circulation f^* of cost at most cf. The support graph $D_{f^*} = (V, A_{f^*})$ of f^* will consist of the edges of T plus the edges of $e \in B$ with $f_e^* = 1$. Thus $|A_{f^*}| \leq (n - 1) + cf$. Moreover, the corresponding edges in G form a 2-edge-connected spanning subgraph of G.

Given the circulation problem $P(G)$ and a lower bound β for $OPT_{LP}(G)$, the above suggests the following scheme for finding a k-approximation for 2EC:

(1) Find a feasible (perhaps fractional) circulation f for $P(G)$ such that $(n - 1) + cf \leq k\beta$.
(2) Find an optimal integer circulation f^* for $P(G)$. The support graph of f^* provides a 2-edge-connected spanning subgraph $H = (V, J)$ of G with $|J| \leq (n - 1) + cf^* \leq (n - 1) + cf \leq k\beta$. Since $\beta \leq OPT_{LP}(G) \leq OPT(G)$, we have a k-approximation algorithm.

Next we provide a very useful lower bound for $OPT_{LP}(G)$ to be used in the above framework. Given any 2-edge-connected graph $G = (V, E)$, we let F represent the set of edges in G which are in 2-edge cuts. We can use F to obtain a lower bound on $OPT_{LP}(G)$ (and thus on $OPT(G)$ for 2EC) as follows.

Lemma 1. *We have the following lower bound for $OPT(G)$:*

$$OPT_{LP}(G) \geq \frac{1}{2} \sum_{v \in V} max(2, |\delta(v) \cap F|).$$

Proof. Let x' be any feasible solution for $2EC_{LP}$. For any edge in F we must have $x'_e = 1$, thus for any node $v \in V$, $x'(\delta(v)) \geq max(2, |\delta(v) \cap F|)$.
The result follows. □

3   A $\frac{5}{4}$-Approximation Algorithm

In this section we use the ideas presented in the previous section to provide a $\frac{5}{4}$-approximation algorithm for 2EC for subcubic bridgeless multigraphs and also show that $\alpha^{2EC} < \frac{5}{4}$ for such graphs. We begin with some preliminaries.

Given a graph $G = (V, E)$, a cut $\delta(S)$ for $S \subset V$ is called *proper* if $2 \leq |S| \leq n - 2$. For $S \subset V$, let $\bar{S} = V \setminus S$, and let $G \downarrow S$ be the graph obtained by shrinking S into a single pseudonode v_s.

For the remainder of this section, let $G = (V, E)$ be a subcubic bridgeless 2-edge-connected multigraph, and as before let F be the set of edges in E which are in 2-edge cuts. Let $V = V_2 \cup V_3$, where V_i is the set of nodes in V of degree i, $i = 2, 3$. We consider the number of edges in F incident with each node in V_3, which partitions V_3 into four sets which we denote by $V_{3F_i}(G) = \{v \in V_3 : |F \cap \delta(v)| = i\}$ for $i = 0, 1, 2, 3$. Note that by Lemma 1 we have

$$OPT_{LP}(G) \geq n + \frac{|V_{3F_3}(G)|}{2}. \tag{5}$$

We now construct the minimum cost circulation problem $P(G)$ as described in Section 2, using the same notation. Recall T is the set of tree edges formed by the DFS and B is the set of back edges. For each edge $e \in T$, let $\delta_{T_e} \subset B$ be the set of back edges in the unique cut in D for which e is the only tree edge in the cut. Each edge b in δ_{T_e} forms a unique directed cycle C_b in $T \cup \{b\}$ which contains tree edge e. Let $f \in \mathbb{R}^A$ be the circulation for $P(G)$ obtained by setting the flow to $\frac{1}{2}$ around each cycle C_b for $b \in B \setminus F$ and to 1 around each cycle

C_b for $b \in B \cap F$, and then summing these cycle flows. More specifically, f is defined as follows:

$$f_a = \begin{cases} \frac{1}{2} \text{ for } a \in B \setminus F, \\ 1 \text{ for } a \in B \cap F, \\ \sum (f_b : b \in \delta_{T_a}) \text{ for } a \in T. \end{cases}$$

Lemma 2. *The circulation f is a feasible circulation for $P(G)$ and has cost* $cf = \frac{n - |V_2|}{4} + \frac{|B \cap F|}{2} + \frac{1}{2}$.

Proof. Since every tree edge not in F must be in the cycle C_e for at least 2 back edges, f is clearly a feasible circulation for $P(G)$. Moreover, since $c_a = 0$ for $a \in T$ and $c_a = 1$ for $a \in B$, $cf = \frac{1}{2}|B \setminus F| + |B \cap F| = \frac{|B|}{2} + \frac{|B \cap F|}{2}$. Since $|E| = \frac{1}{2}(2|V_2| + 3|V_3|)$, and $n = |V_2| + |V_3|$, we have $|B| = |E| - |T| = |E| - (n - 1) = \frac{n - |V_2|}{2} + 1$ and the result follows. $\qquad \square$

We now describe our recursive approximation algorithm, which is based on the ideas from Section 2 along with a careful specification of how we grow the DFS tree. Note that we assume $|V_3| \neq 0$, as otherwise the problem is trivial.

Algorithm Approx 5_4
Input: A bridgeless subcubic graph $G = (V, E)$.
Output: A 2-edge-connected spanning subgraph $H = (V, J)$ of G.

Grow a DFS tree T in G starting at any node $r \in V_3$. We grow the DFS tree according to the following two rules: **Rule 1** If we have a choice for the next edge to add to the tree, we always add one in F, if possible. **Rule 2** If we have more than one choice of F edge to add, we add one that goes to a degree 2 node, if possible.

Case (a): {Base Case} For all $e = (u, v) \in B \cap F$ we have $u \in V_2$.
Construct the minimum cost circulation problem $P(G)$ using the DFS tree T above, and find the optimal solution f^* for $P(G)$ which is integer-valued. Let J be the edges in G corresponding to the edges of the support graph of f^*. Return graph $H = (V, J)$.

Case (b): {Recursive Step} There is an edge $e = (u, v) \in B \cap F$ with $u \in V_3$. In this case we will show that e is in a proper 2-edge cut $\delta(S) = \{e, e'\}$ in G. Apply Algorithm Approx 5_4 recursively to $G \downarrow S$ and $G \downarrow \overline{S}$ to obtain graphs $H' = (V', J')$ and $H'' = (V'', J'')$ respectively. Combine these graphs into a graph $H = (V, J)$ by removing the pseudonodes v_S and $v_{\overline{S}}$ and adding the edges e and e'. Return graph $H = (V, J)$.

Theorem 2. *The graph $H = (V, J)$ from Algorithm Approx 5_4 is a 2-edge-connected spanning subgraph of G such that $|J| \leq \frac{5}{4}n + \frac{|V_{3F_3}(G)|}{2} - \frac{1}{2}$.*

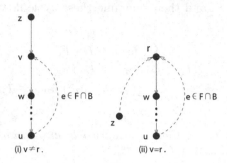

Fig. 1. The configuration of nodes u, v, w and z

Proof

In the digraph $D = (V, A)$ resulting from the DFS, consider any back arc $e = (u, v) \in B \cap F$, i.e. e is in a 2-edge cut in G. Since v has a back arc directed into it, it cannot be a leaf of the DFS tree, thus there exists a tree edge (v, w) directed out of v. If $v \neq r$ there is also a tree edge (z, v) directed into v and v has degree 3, so no other arcs are incident with it. If $v = r$, r was chosen to have degree 3 thus there is another arc zv incident with v. This arc must be in B, because if it was in T it would form a cut edge in G. (See Figure 1). Finally, we cannot have $u = w$ as this would imply the pair of multi-edges uv form a 2-edge cut in G, since this would then be the only 2-edge cut in which e can be contained. However, this can only occur when $n = 2$ and $|V_3| = 0$, which is not the case here.

First we show that $v \in V_{3F_2}(G)$ or $v \in V_{3F_3}(G)$. Clearly $v \notin V_{3F_0}(G)$ as $uv \in F$. Moreover, when we grow from v in the DFS, we would have had a choice to grow along vw or vu. So by our construction Rule 1, we must have $vw \in F$ also, and thus $v \in V_{3F_2}(G)$ or $v \in V_{3F_3}(G)$. Note that this implies that if $v = r$ and the other edge zv in B incident with r is also in F, then $r \in V_{3F_3}(G)$.

Case (a): For all $e = (u, v) \in B \cap F$, $u \in V_2$.

We begin by proving the following claim.

Claim 1: $|V_2| \geq 2(|B \cap F| - t)$, where $t = 1$ if $r \in V_{3F_3}(G)$ and 0 otherwise.

Proof of Claim 1. For any edge $e = (u, v) \in B \cap F$ consider node w. When we grow from node v in the DFS we had a choice of two F edges to grow along next, namely vw and vu. We chose vw, which means by Rule 2 of our DFS construction that node $w \in V_2$. So for each edge $e \in B \cap F$, we have $u \in V_2$ and $w \in V_2$. If we can show that all of these $u's$ and $w's$ are distinct from each other whenever $r \notin V_{3F_3}(G)$, the claim is proved.

Since w is on the unique cycle formed by e with the DFS tree and $u \neq w$, there must be a tree edge leaving w. So w cannot be the tail of any back arc. Thus the $w's$ are distinct from the $u's$. The $w's$ are also distinct from each other if $r \notin V_{3F_3}(G)$, as there is only one unique back arc into node v. Finally, it is clear the $u's$ are distinct from each other since any degree 2 node is the tail of at most one back arc. The claim follows.   □

Now consider the cost of our feasible circulation f. By Lemma 2 and Claim 1 we have

$$cf = \frac{n - |V_2|}{4} + \frac{|B \cap F|}{2} + \frac{1}{2} \leq \frac{n}{4} - \frac{2(|B \cap F| - t)}{4} + \frac{|B \cap F|}{2} + \frac{1}{2}. \qquad (6)$$

Moreover, by using $|V_{3F_3}(G)| \geq t$ and simplifying (6) we have

$$cf \leq \frac{n}{4} + \frac{|V_{3F_3}(G)|}{2} + \frac{1}{2}. \qquad (7)$$

By Theorem 1 we know the optimal circulation f^* from the algorithm will have cost at most cf, thus by (7) it follows that

$$|J| \leq (n-1) + \frac{n}{4} + \frac{|V_{3F_3}(G)|}{2} + \frac{1}{2} = \frac{5}{4}n + \frac{|V_{3F_3}(G)|}{2} - \frac{1}{2}$$

as required. This completes Case (a).

Case (b): For some $e = (u, v) \in B \cap F$, $u \in V_3$.
To begin we show that there exists a proper 2-edge cut in G.

Since $uv \in F$, it must be in a 2-edge cut in G with another edge e'. If $e' = vw$, then this would imply zv is a bridge in G, contradicting the fact that G is bridgeless. Thus $e' \neq vw$.

Let S be the set of nodes defining the 2-edge cut $\{uv, e'\}$ (i.e. $\delta(S) = \{uv, e'\}$) such that $u \in S$, and let C_e be the unique cycle formed by $e = (u, v)$ and T. Clearly $(v, w) \in C_e$. Since vw cannot be in $\delta(S)$, we have $w \notin S$, $v \notin S$. Thus $\delta(S)$ crosses C_e in an edge of T, and $e' \in (T \cap C_e) \setminus \{(v, w)\}$. Moreover, all back edges from nodes in S have both ends contained inside S.

Finally, consider node u. Since $u \in V_3$, there must be another edge qu incident with u with $q \in S$. Thus $\delta(S)$ is a proper 2-edge cut in G.

Now consider the proper 2-edge cut $\delta(S) = \{e, e'\}$. Both $G \downarrow S$ and $G \downarrow \bar{S}$ are also subcubic 2-edge connected graphs with fewer nodes than G, thus we can apply Algorithm 5_4 recursively to them to obtain 2-edge connected spanning subgraphs $H' = (V', J')$ and $H'' = (V'', J'')$ respectively such that

$$|J'| \leq \frac{5|V'|}{4} + \frac{|V_{3F_3}(G \downarrow S)|}{2} - \frac{1}{2} \text{ and } |J''| \leq \frac{5|V''|}{4} + \frac{|V_{3F_3}(G \downarrow \bar{S})|}{2} - \frac{1}{2}.$$

Now consider the graph $H = (V, J)$ for G we obtain by removing v_S from H', $v_{\bar{S}}$ from H'' and adding the edges e and e'. Graph H is a 2-edge connected spanning subgraph of G. Moreover, $n = |V'| + |V''| - 2$, $|J| = |J'| + |J''| - 2$, and $|V_{3F_3}(G)| = |V_{3F_3}(G \downarrow S)| + |V_{3F_3}(G \downarrow \bar{S})|$. Hence $|J| = |J'| + |J''| - 2 \leq \frac{5}{4}(n+2) + \frac{|V_{3F_3}(G)|}{2} - 1 - 2 = \frac{5}{4}n + \frac{|V_{3F_3}(G)|}{2} - \frac{1}{2}$. $\qquad \square$

Using (5) and Theorem 1, we have the following corollaries to Theorem 2:

Corollary 1. *Algorithm 5_4 is a $\frac{5}{4}$-approximation algorithm for 2EC for subcubic bridgeless graphs.*

Corollary 2. *The integrality gap α^{2EC} is less than $\frac{5}{4}$ for subcubic bridgeless graphs.*

4   Computational Study on the Integrality Gap of 2EC

In this section we report on a computational study where we investigate the worst-case ratio between $ILP(G)$ and $LP(G)$ for graphs with a small number of nodes, thus obtains a lower bound on α^{2EC} for those types of graphs. Here we give a brief summary of the methods used and results obtained. For more details, see [13].

4.1   Methodology

It is known that the computational complexity for solving $ILP(G)$ is NP-hard. However, it is practically possible to solve $ILP(G)$ in reasonable time for graphs G of small size. Therefore the graphs in this experimental study were limited to the following three sets:

- General simple graphs $\mathbb{G} = \bigcup_{k=3}^{10} \mathbb{G}_k$, where $\mathbb{G}_k$ denotes the set of all non-isomorphic 2-edge-connected simple graphs with k nodes;
- Cubic simple graphs $\mathbb{C} = \bigcup_{k=6}^{16} \mathbb{C}_k$, where $\mathbb{C}_k$ denotes the set of all non-isomorphic 2-edge-connected cubic simple graphs with k nodes; and
- Subcubic simple graphs $\mathbb{S} = \bigcup_{k=3}^{16} \mathbb{S}_k$, where $\mathbb{S}_k$ denotes the set of all non-isomorphic 2-edge-connected subcubic simple graphs with k nodes.

With the purpose to find out more about the lower bound for the integrality gap α^{2EC} of the LP relaxation for 2EC, we calculated the ratio, denoted by $\alpha(G)$, between the optimal objective value $OPT(G)$ and $OPT_{LP}(G)$ for all graphs $G \in \mathbb{G}$, $G \in \mathbb{C}$ and $G \in \mathbb{S}$. Denote the complete set of all graphs studied in this experiment as $\mathcal{G} = \mathbb{G} \cup \mathbb{C} \cup \mathbb{S}$, the maximum ratio $\alpha(G)$ among all $G \in \mathcal{G}$ provides a lower bound for the value of α^{2EC} for graphs of that type.

By using the **nauty** package (Version 2.4), developed by Brendan D. McKay [14], we were able to obtain all non-isomorphic connected graphs of a certain category (i.e. general, cubic, subcubic), and then eliminate all the graphs with bridges. We then formulated $ILP(G)$ and $LP(G)$ for each graph in our set. Finally we used Gurobi™ Optimizer (Version 5.0) to obtain solutions to $ILP(G)$ and $LP(G)$ for each G.

The program designed for our experiments was developed using the C programming language, on a 64-bit system running Mircosoft®Windows 7 Professional, with a Lenovo®Thinkpad X201 laptop equipped with Intel®Core™i5 M480 @ 2.67GHz, and 4.00 GB installed memory (RAM).

4.2   Analysis of Results

Facing a large amount of data, it became difficult for us to analyze all results with the limited resources. For example, the number of all non-isomorphic 2-edge-connected graphs on 10 nodes is 9,804,368, and it took the program approximately 11 days to finish the experiment process for all graphs $G \in \mathbb{G}_{10}$. In order to learn more about the lower bound for the value of α^{2EC} in general and the upper bound for particular classes and sizes of graphs, more attention was given to the data that resulted in a higher ratio between $OPT(G)$ and $OPT_{LP}(G)$. Figure 2 demonstrates the trend in the changes of the maximum ratios between $OPT(G)$ and $OPT_{LP}(G)$ for $G \in \mathbb{G}_k$, $G \in \mathbb{C}_k$ and $G \in \mathbb{S}_k$. Table 1 gives a summary of the maximum value of the ratio between $OPT(G)$ and $OPT_{LP}(G)$ for graphs in each of the three categories.

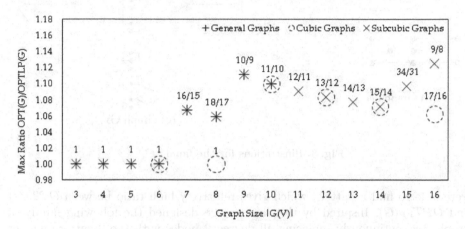

Fig. 2. Experimental result data analysis

Table 1. Summary of the experimental study

Graph Category	Max $\alpha(G)$	Corresponding $|V(G)|$
$\mathbb{G}_k$ $(3 \le k \le 10)$	10/9	9
$\mathbb{C}_k$ $(6 \le k \le 16)$	11/10	10
$\mathbb{S}_k$ $(3 \le k \le 16)$	9/8	16

Let $\alpha(\mathbb{G}_k)$, $\alpha(\mathbb{C}_k)$ and $\alpha(\mathbb{S}_k)$ denote the maximum ratios between $OPT(G)$ and $OPT_{LP}(G)$ for $G \in \mathbb{G}_k$, $G \in \mathbb{C}_k$ and $G \in \mathbb{S}_k$ respectively. It is noted from Figure 2 that our result on $\alpha(\mathbb{S}_{16})$ reached the highest value we obtained (i.e. $\frac{9}{8}$) with only 16 nodes, which was the previous best known lower bound on the integrality gap of the LP relaxation for 2EC [4]. In addition, it is noted that for each value of k $(3 \le k \le 10)$, the set of graphs with k nodes that gave the worst ratio among all graphs with the same size always included subcubic graphs. This supports the idea that subcubic graphs are most likely to give α^{2EC} in general.

5   New Lower Bounds for the Integrality Gap for 2EC

In this section we discuss a family of subcubic graphs which asymptotically give a ratio of $\frac{8}{7}$ for $OPT(G)$ and $OPT_{LP}(G)$, thus improving on the previous best known lower bound for α^{2EC} of $\frac{9}{8}$ [4]. In our computational study, we

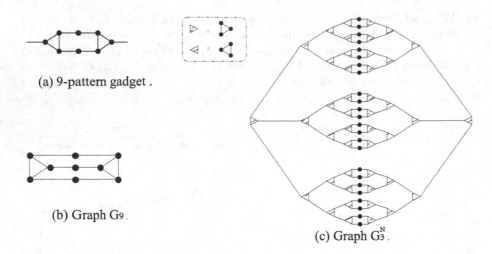

(a) 9-pattern gadget .

(b) Graph G_9 .

(c) Graph G_3^N .

Fig. 3. Illustrations for the family G_t^N

were able to find a pattern which gives relatively high ratio between $OPT(G)$ and $OPT_{LP}(G)$. Inspired by this finding, we designed the following family of graphs by continuously replacing all degree 2 nodes with the 9-pattern gadget shown in Figure 3(a), starting from G_9 (shown in Figure 3(b)). The family of graphs generated from the above operation is referred to as $\mathcal{F}^N$. Executing the replacement for t times gives us the graph $G_t^N \in \mathcal{F}^N (t \geq 0)$. Figure 3(c) shows the graph $G_3^N \in \mathcal{F}^N$, which was obtained by repeating the replacement three times. The following theorem is proved through a series of calculations (see [13]).

Theorem 3. *For the family of graphs $\mathcal{F}^N$, the following hold :*

$$\lim_{t \to \infty} \frac{OPT(G_t^N)}{|V(G_t^N)|} = \frac{8}{7}, \quad \lim_{t \to \infty} \frac{OPT(G_t^N)}{OPT_{LP}(G_t^N)} = \frac{8}{7}.$$

From the above discussion, Corollary 3 naturally follows from Theorem 3.

Corollary 3. *The integrality gap α^{2EC} for 2EC is at least $\frac{8}{7}$, even when restricted to subcubic bridgeless graphs.*

Acknowledgement. The authors would like to thank the anonomous reviewers for their valuable suggestions and comments.

References

1. Csaba, B., Karpinski, M., Krysta, P.: Approximability of dense and sparse instances of minimum 2-connectivity, tsp and path problems. In: Eppstein, D. (ed.) SODA, ACM/SIAM, pp. 74–83 (2002)

2. Alexander, A., Boyd, S., Elliott-Magwood, P.: On the integrality gap of the 2-edge connected subgraph problem. Technical Report TR-2006-04, SITE, University of Ottawa, Ottawa, Canada (2006)
3. Mömke, T., Svensson, O.: Approximating graphic tsp by matchings. In: Ostrovsky, R. (ed.) IEEE FOCS, pp. 560–569 (2011)
4. Sebő, A., Vygen, J.: Shorter tours by nicer ears. CoRR abs/1201.1870 (2012)
5. Khuller, S., Vishkin, U.: Biconnectivity approximations and graph carvings. J. ACM 41(2), 214–235 (1994)
6. Cheriyan, J., Sebő, A., Szigeti, Z.: Improving on the 1.5 approximation of a smallest 2-edge connected spanning subgraph. SIAM J. Discrete Math. 14, 170–180 (2001)
7. Vempala, S., Vetta, A.: Factor 4/3 approximations for minimum 2-connected subgraphs. In: Jansen, K., Khuller, S. (eds.) APPROX 2000. LNCS, vol. 1913, pp. 262–273. Springer, Heidelberg (2000)
8. Krysta, P., Kumar, V.S.A.: Approximation algorithms for minimum size 2-connectivity problems. In: Ferreira, A., Reichel, H. (eds.) STACS 2001. LNCS, vol. 2010, pp. 431–442. Springer, Heidelberg (2001)
9. Huh, W.T.: Finding 2-edge connected spanning subgraphs. Oper. Res. Lett. 32(3), 212–216 (2004)
10. Boyd, S., Iwata, S., Takazawa, K.: Finding 2-factors closer to tsp tours in cubic graphs. SIAM J. Discrete Math. 27(2), 918–939 (2013)
11. Hoffman, A.J.: Some recent applications of the theory of linear inequalities to extremal combinatorial analysis. In: Combinatorial Analysis, pp. 113–127 (1960)
12. Schrijver, A.: Chapters 11-12. In: Combinatorial Optimization. Springer (2003)
13. Sun, Y.: Theoretical and experimental studies on the minimum size 2-edge-connected spanning subgraph problem. Master's thesis, University of Ottawa, Ottawa, Canada (2013)
14. McKay, B.D.: Practical graph isomorphism. Congressus Numerantium 30, 45–87 (1981)

Box-Constrained Mixed-Integer Polynomial Optimization Using Separable Underestimators

Christoph Buchheim[1] and Claudia D'Ambrosio[2]

[1] Fakultät für Mathematik, Technische Universität Dortmund
Vogelpothsweg 87, 44227 Dortmund, Germany
`christoph.buchheim@tu-dortmund.de`
[2] LIX CNRS (UMR7161), École Polytechnique
91128 Palaiseau Cedex, France
`dambrosio@lix.polytechnique.fr`

Abstract. We propose a novel approach to computing lower bounds for box-constrained mixed-integer polynomial minimization problems. Instead of considering convex relaxations, as in most common approaches, we determine a separable underestimator of the polynomial objective function, which can then be minimized easily over the feasible set even without relaxing integrality. The main feature of our approach is the fast computation of a good separable underestimator; this is achieved by computing tight underestimators monomialwise after an appropriate shifting of the entire polynomial. If the total degree of the polynomial objective function is bounded, it suffices to consider finitely many monomials, the optimal underestimators can then be computed offline and hardcoded. For the quartic case, we determine all optimal monomial underestimators analytically.

In the case of pure integer problems, we perform an extensive experimental evaluation of our approach. It turns out that the proposed branch-and-bound algorithm clearly outperforms all standard software for mixed-integer optimization when variable domains contain more than two values, while still being competitive in the binary case. Compared to approaches based on linearization, our algorithm suffers significantly less from large numbers of monomials. It could minimize complete random polynomials on ten variables with domain $\{-10, ..., 10\}$ in less than a minute on average, while no other approach was able to solve any such instance within one hour of running time.

1 Introduction

In the last decade, mixed-integer non-linear programming (MINLP) has increasingly moved into the focus of the mathematical optimization community. Such problems are usually very hard both from a theoretical and practical point of view. Often this is true even in the linear case due to the integrality constraints, but non-linearities render such problems even harder, in particular when these non-linearities are non-convex. The standard way of dealing with non-convex problems is to consider convex relaxations, often combined with linearization.

J. Lee and J. Vygen (Eds.): IPCO 2014, LNCS 8494, pp. 198–209, 2014.

The feasible set is thus replaced by a larger convex set and the objective function is underestimated by a convex function. For an overview of the standard methods for MINLP, see [1, 2, 3, 4].

In this paper, we restrict ourselves to box-constrained problems and follow a different approach: we consider separable non-convex underestimators of the objective function instead of convex ones. In [5], this idea has been applied to quadratic combinatorial optimization problems. Here, we consider a very different class of problems, namely the minimization of polynomials subject to box and integrality constraints. More formally, let $l, u \in \mathbb{R}^n$ with $l \leq u$ and consider the box $[l, u] := \prod_{i=1}^{n} [l_i, u_i]$. Moreover, let $f = \sum_{\alpha \in A} c_\alpha x^\alpha$ be an arbitrary polynomial of total degree $d \in \mathbb{N}$ and $I \subseteq \{1, \ldots, n\}$. We aim at solving the problem

$$\begin{aligned} \min \quad & f(x) \\ \text{s.t.} \quad & x \in [l, u] \\ & x_i \in \mathbb{Z} \quad \forall i \in I, \end{aligned} \tag{1}$$

where we may assume $l_i, u_i \in \mathbb{Z}$ for all $i \in I$.

This problem is already NP-hard in the quadratic case, both when all variables are binary and when all variables are continuous. Various types of convex relaxations and reformulations have been proposed for the quadratic case in the literature. The resulting convex problems may be linear [6], semidefinite [7, 8] or other types of tractable problems [9, 10]. However, for the case of general polynomials, few specific approaches exist. In the binary case, the problem is usually linearized by adding new variables for all appearing monomials; it is also possible to reduce the polynomial problem to a quadratic one [11, 12]. For the general integer case, a binary expansion may be applied in order to reduce the problem to a binary one, but in practice this approach suffers from a large number of variables and disadvantageous numerical properties.

In the general integer (or mixed-integer) context, the most common approach found in optimization software is a combination of convex underestimators and bound tightening techniques. The convex underestimators are derived from the building blocks of the objective function. In our approach, we use a similar idea, computing underestimators monomialwise. However, as we are aiming at separable underestimators, our approach differs significantly from the convexification approach: on the one hand, using convex functions one can generally expect to obtain tighter underestimators than using separable functions. On the other hand, a separable underestimator can be minimized very quickly, even taking the integrality constraints into account. Our experimental results show that our approach significantly outperforms standard software for mixed-integer non-linear optimization such as BARON [13], Couenne [14], and SCIP [15].

Our method also compares favorably to GloptiPoly [16], which represents an important class of approaches to continuous polynomial optimization based on hierarchies of semidefinite programming relaxations; see, e.g., [17, 18, 19]. These approaches do not allow explicit integrality constraints, but a discrete variable domain can be modeled by a polynomial equation. The main drawback is the long running time needed to solve the resulting semidefinite programs. However, in our approach we can use tools such as GloptiPoly to compute optimal separable underestimators for each monomial offline, assuming that the maximal degree d is fixed. The separable underestimator of f is then determined as the sum of these monomialwise underestimators.

The basic ideas of our approach are explained in more detail in Section 2. In Section 3, we discuss some features of our branch-and-bound algorithm. Finally, we report results of an extensive experimental evaluation in Section 4.

2   Separable Underestimators for Polynomials

Our objective is to underestimate a general multivariate polynomial $f \colon \mathbb{R}^n \to \mathbb{R}$ of degree $d := \deg f$ by a separable polynomial $g(x) = \sum_{i=1}^n g_i(x_i)$, where all functions $g_i \colon \mathbb{R} \to \mathbb{R}$ are univariate polynomials of degree at most $\bar{d}$. The degree $\bar{d}$ depends on the degree of the original polynomial f. In order to ensure that a feasible separable underestimator exists, we need that $\bar{d} \geq d$ if d is even and $\bar{d} \geq d + 1$ if d is odd. On the other hand, as we will see, a small degree $\bar{d}$ is preferable in our approach. In our algorithm, we thus set $\bar{d} := 2 \left\lceil \frac{1}{2} d \right\rceil$.

Since we aim at using separable underestimators in order to derive lower bounds for Problem (1), we need to answer the following two questions: how can a good separable underestimator be computed, and how can it be minimized? We start with a discussion of the latter question.

2.1   Minimizing a Separable Polynomial

The first observation needed for the minimization of the underestimator g is

Lemma 1

$$
\begin{aligned}
\min \quad & g(x) \\
\text{s.t.} \quad & x \in [l, u] \\
& x_i \in \mathbb{Z} \quad \forall i \in I
\end{aligned}
\quad = \quad
\sum_{i=1}^n
\begin{cases}
\min \quad & g_i(x) \\
\text{s.t.} \quad & x_i \in [l_i, u_i] \\
& x_i \in \mathbb{Z} \quad \text{if } i \in I.
\end{cases}
$$

This result is easy to verify. It means that we can minimize g by solving n univariate minimization problems involving the g_i, where the problem is a discrete problem if $i \in I$ and a continuous problem otherwise. We would like to point out here that we never relax any integrality constraint in our approach, instead the two cases $i \in I$ and $i \notin I$ are solved in a slightly different way.

In the case of a continuous variable ($i \notin I$), the minimizer of g_i over $[l_i, u_i]$ must either be on the boundary or it must satisfy the first order condition $g_i'(x) = 0$. As g_i' is a polynomial of degree at most $\bar{d} - 1$, it has at most $\bar{d} - 1$ real zeroes, which can be computed by closed formulae if $\bar{d} \leq 5$ and numerically otherwise. In summary, we obtain a set of at most $\bar{d} + 1$ candidates for the desired minimizer. We can thus find a minimizer of g_i by enumerating these candidates. In the case $i \in I$, it is easy to see that an integer minimizer of g_i over $[l_i, u_i]$ must be a continuous minimizer of g_i rounded up or down. We can thus apply the same approach as in the continuous case, round up and down all continuous candidates, and end up with at most $2\bar{d}$ many candidates. For an illustration, see Figure 1. In all cases, we thus obtain a minimizer x_i^* for g_i, such that x^* is a mixed-integer minimizer of g by Lemma 1.

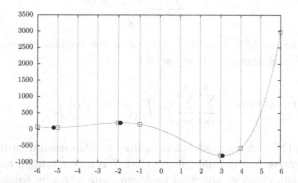

Fig. 1. A univariate polynomial of degree five, its stationary points (circles), and the candidates for being integer optimizers (squares)

2.2 Computing Tight Separable Underestimators

The best possible separable underestimator would be one that yields the tightest lower bound on f, i.e., a solution of

$$\max \quad \min_{\substack{x \in [l,u] \\ x_i \in \mathbb{Z} \text{ for } i \in I}} \quad g(x)$$

$$\text{s.t.} \quad g(x) \le f(x) \ \ \forall x \in [l, u]$$
$$g \text{ separable polynomial of degree at most } \bar{d}.$$

However, aiming for an optimal underestimator in this sense is too ambitious, as even deciding whether $g = 0$ is a feasible solution would require deciding nonnegativity of f, which is well-known to be an intractable problem for general f.

Our strategy to deal with this difficulty is to split up the polynomial into its monomials and to replace each monomial by an optimal separable underestimator. The result will not be an optimal underestimator of the entire polynomial, but hopefully still yield tight lower bounds. However, since we add up different underestimators, it is not appropriate anymore to maximize the minimum of the underestimator on the feasible region. Instead, we now search for an underestimator that is as close to f as possible on average, on the feasible region $[l, u]$. This leads to the following problem formulation:

$$\max \quad \int_{[l,u]} g(x) \, dx$$

$$\text{s.t.} \quad g(x) \le f(x) \ \ \forall x \in [l, u] \tag{P}$$
$$g \text{ separable polynomial of degree at most } \bar{d}.$$

Note that the variables in this optimization problem are the coefficients of the univariate polynomials g_i. Collecting the constant terms of all g_i in a separate variable c_0, we can write

$$g(x) = \sum_{i=1}^{n} \sum_{j=1}^{\bar{d}} c_{ij} x_i^j + c_0$$

so that (P) contains $n\bar{d} + 1$ variables, namely the coefficients c_{ij} and c_0. The objective function is linear in these variables, as

$$\int_{[l,u]} g(x)\, dx = \sum_{i=1}^{n} \sum_{j=1}^{\bar{d}} c_{ij} \int_{[l,u]} x_i^j\, dx + c_0 \int_{[l,u]} 1\, dx \, .$$

Hence Problem (P) can be modeled as a linear optimization problem over the cone of non-negative polynomials, using that f is a constant in this context. In particular, one can consider the dual problem, which can be calculated as

$$\begin{aligned}
\min \quad & \int f\, d\mu \\
\text{s.t.} \quad & \int x_i^j\, d\mu = \int_{[l,u]} x_i^j\, dx \text{ for } i = 1, \ldots, n, \; j = 1, \ldots, \bar{d} \\
& \int 1\, d\mu = \int_{[l,u]} 1\, dx \\
& \mu \text{ Borel-measure with support on } [l, u].
\end{aligned} \qquad \text{(D)}$$

First observe that Problem (P) is always feasible: as f is continuous, its minimum m over $[l, u]$ exists and we can choose the constant function $g(x) := m$ as separable underestimator of f. In order to find a better solution to Problem (P), we first translate the problem such that the origin becomes the center of the feasible region. More formally, we consider the shifted polynomial $\bar{f}$ defined by $\bar{f}(x) := f(x+t)$, where $t_i := \frac{1}{2}(l_i + u_i)$ for all $i = 1, \ldots, n$. The new polynomial $\bar{f}$ has the same degree as f, but not the same monomials in general, as one may obtain all submonomials of the given monomials of f. In the next step, we solve Problem (P) separately for each monomial, as mentioned above, over the feasible set $[l - t, u - t]$. The corresponding monomial underestimators are summed up to obtain a separable underestimator $\bar{g}$ of $\bar{f}$ over $[l - t, u - t]$, which is then turned into a separable underestimator g of f over $[l, u]$ by defining $g(x) := \bar{g}(x - t)$.

As $l - t = -(u - t)$ by definition of t, we may thus assume that the feasible set in Problem (P) is symmetric with respect to the origin, i.e., that $l = -u$. If f is a monomial, we can further normalize by scaling the feasible box:

Lemma 2. *Let $u > 0$ and let $\sum_{i=1}^{n} \sum_{j=1}^{\bar{d}} c_{ij} x_i^j + c_0$ be an optimal solution to (P) over the box $[-1, 1]^n$ for a monomial f. Then $f(u) \cdot \left(\sum_{i=1}^{n} \sum_{j=1}^{\bar{d}} \frac{c_{ij}}{u_i^j} x_i^j + c_0 \right)$ is an optimal solution to (P) over $[-u, u]$.*

Clearly, if $u_i = 0$ for some i, the corresponding variable x_i can be eliminated, hence the above lemma can still be applied in this case.

This approach shows why we need to integrate over the entire box $[l, u]$ in the objective function of Problem (P) instead of restricting ourselves to the set of all mixed-integer feasible solutions: otherwise, the scaling would produce a problem that still depends on the given bounds l and u.

In summary, we need to solve Problem (P) only for monomials over $[-1, 1]^{\bar{d}}$ in our approach. This implies that, as soon as we fix any bound on the degree of f and hence on $\bar{d}$, we end up with a finite set of problems of type (P) that have to be solved: for each possible monomial x^α with $|\alpha| \leq \bar{d}$, we need to solve (P) for both $f(x) := x^\alpha$ and $f(x) := -x^\alpha$ over $[-1, 1]^{\bar{d}}$.

In particular, we do not solve Problem (P) in each node of the enumeration tree, but we solve it offline for the finitely many cases under consideration and hardcode the results. At runtime, it suffices to replace each monomial in $\bar{f}$ by its optimal underestimator. No optimization algorithm is required here, the algorithm only needs to shift and scale. From a theoretical point of view, the computation of the given underestimators thus takes constant time if $\bar{d}$ is bounded. From a practical point of view and for large $\bar{d}$, it might however be preferable to replace the constraint $g(x) \leq f(x)$ by requiring that $f - g$ be an sos-polynomial. Problem (P) then becomes tractable and the resulting underestimators will still be feasible and hopefully tight.

Of course, one may also try to solve Problem (P) analytically. For showing optimality, one may use the dual problem (D) and the following lemma:

Lemma 3 (Complementary Slackness). *Let g be feasible for (P) and let μ be feasible for (D). If $\int (f - g) \, d\mu = 0$, then g is optimal for (P).*

This approach can be used to show the results stated in the following section.

2.3   Optimal Quartic Underestimators

In the following, we determine all optimal underestimators according to (P) for monomials of degree at most four; for the proofs, we refer to the full version of this paper [20]. All proofs make use of Lemma 3, where a corresponding feasible solution for the dual problem (D) is constructed explicitly.

Up to symmetry, 12 monomials have to be considered, where the cases of positive and negative coefficients have to be distinguished. Out of the resulting 24 cases, namely the monomials $\pm x_1^k$ (for $k = 0, \dots, 4$), $\pm x_1 x_2$, $\pm x_1^2 x_2$, $\pm x_1 x_2 x_3$, $\pm x_1^3 x_2$, $\pm x_1^2 x_2^2$, $\pm x_1^2 x_2 x_3$, and $\pm x_1 x_2 x_3 x_4$, the first ten are already separable and hence trivial. A further nine cases are covered by the following result.

Theorem 1. *Let $\alpha \in \mathbb{N}_0^n$ such that $d := \sum_{i=1}^n \alpha_i$ is even and $d \leq 4$. Assume that at least one α_i is odd, or that $c \leq 0$. Then an optimal solution of (P) for $f(x) := c x^\alpha$ over $[-1, 1]^n$ is*

$$g(x) := -\frac{|c|}{d} \sum_{i=1}^n \alpha_i x_i^d .$$

From our proof given in [20], it follows that these optimal underestimators are independent of the domain $[-u, u]$, as $f - g$ turns out to be globally non-negative.

Moreover, these underestimators remain optimal for any $\bar{d} \geq d$. Next, we consider the only remaining even degree case for $\bar{d} \leq 4$.

Theorem 2. *Let $c \geq 0$. Then an optimal solution of* (P) *for $f(x) := cx_1^2 x_2^2$ over $[-1, 1]^n$ for $\bar{d} = 4$ is*

$$g(x) = -\tfrac{1}{2}x_1^4 + \tfrac{2}{3}x_1^2 - \tfrac{1}{2}x_2^4 + \tfrac{2}{3}x_2^2 - \tfrac{2}{9} \, .$$

See Fig. 2 for an illustration. In this case, the proof again shows optimality independently of the domain, but not of the degree $\bar{d}$. In fact, when allowing degree $\bar{d} = 6$, a better underestimator can be computed.

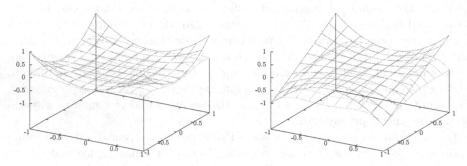

Fig. 2. Optimal separable underestimators of degree $\bar{d} \leq 4$ for $f(x) = x_1^2 x_2^2$ (left) and $f(x) = x_1^2 x_2$ (right)

The following results complete the quartic case.

Theorem 3. *An optimal solution of* (P) *for $f(x) := cx_1 x_2 x_3$ over $[-1, 1]^n$ is*

$$g(x) = -\tfrac{|c|}{6}\left(\sqrt{\tfrac{5}{3}}(x_1^4 + x_2^4 + x_3^4) + \sqrt{\tfrac{3}{5}}(x_1^2 + x_2^2 + x_3^2) \right) .$$

Theorem 4. *An optimal solution of* (P) *for $f(x) := cx_1^2 x_2$ over $[-1, 1]^n$ has the form*

$$g(x) = |c|c_{14}x_1^4 + |c|c_{12}x_1^2 + |c|c_{24}x_2^4 + cc_{23}x_2^3 + |c|c_{22}x_2^2 + cc_{21}x_2 + |c|c_0$$

with

$$c_{14} \approx -0.820950196623141 \qquad c_{24} \approx +0.052169786129554$$
$$c_{12} \approx +0.472574632153429 \qquad c_{23} \approx +0.057375067121109$$
$$\qquad\qquad\qquad\qquad\qquad\qquad c_{22} \approx -0.311590671084123$$
$$c_0 \approx -0.070759806839502 \qquad c_{21} \approx +0.272799782607049 \, .$$

Unlike in all other quartic cases, the optimal local underestimator in Theorem 4 is not valid globally, but requires $x_2 \in [-1, 1]$. See Fig. 2 again.

3   Branch-and-Bound Algorithm

The computation of lower bounds based on separable underestimators as devised in Section 2 can be embedded into a branch-and-bound framework in a straightforward way. Branching can be realized by splitting up the domain of a variable. This leads to improved lower bounds in two ways: on a smaller feasible box, one may obtain a tighter separable underestimator, moreover, even the same underestimator can lead to tighter bounds on smaller feasible sets.

In our implementation, we use the following simple branching rule: we always choose a variable x_i with maximal $u_i - l_i$, i.e., with largest domain. Then we produce two subproblems by splitting up the domain in the middle. Experiments with other branching strategies did not lead to any improvements so far; this is left as future work. The same is true for the enumeration scheme. Currently, we use a straightforward depth first strategy.

As we do not relax integrality when minimizing the separable underestimator, the resulting minimizer is always a feasible solution for our original problem (1). In particular, we obtain an upper bound on (1) in every node of the enumeration tree. We do not apply any further primal heuristics.

4   Experimental Results

In this section, we evaluate the performance of the proposed branch-and-bound algorithm on pure integer instances with polynomials f of degree $d = 4$. We generated different sets of random instances by using the following parameters:

- The number of variables is $n \in \{10, 15, 20, 25, 30, 35, 40\}$.
- For the bounds on the variables we consider
  - $[l_i, u_i] = [-10, 10]$ for all $i = 1, \ldots, n$.
  - $[l_i, u_i] = [-1, 1]$ for all $i = 1, \ldots, n$ (ternary instances).
  - $[l_i, u_i] = [0, 1]$ for all $i = 1, \ldots, n$ (binary instances).
- The polynomial f either consists of m monomials with m being a multiple of n, or of all possible monomials of degree ≤ 4. We will call the latter instances *complete*.

The coefficients of each monomial were chosen uniformly at random from $[-1, 1]$. For generating a single monomial, we randomly selected four integer values from $\{0, 1, \ldots, n\}$, where repetitions were allowed, so as to obtain the variables that are present in the monomial. The value 0 represents that no variable was chosen, thus allowing the monomial to have a degree strictly smaller than 4. In the complete case, a random coefficient is generated for each possible monomial of degree ≤ 4. For each possible combination of parameters, we generated 10 different instances.

We implemented the proposed branch-and-bound scheme, which we will call PolyOpt, in C++. All experiments were run on a cluster of 64-bit Intel(R) Xeon(R) CPU E5-2670 0 processors running at 2.60GHz with 20480 KB cache. We compared PolyOpt with three solvers for general non-convex mixed integer programming problems, namely BARON [13], Couenne [14, 21], and SCIP [15], with a simple complete enumeration method, and with GloptiPoly [16], a solver for continuous polynomial optimization. For the latter, we modeled the integrality requirement on the variables as a polynomial constraint, as GloptiPoly developers suggested. We used modeling languages to formulate our problems and solve them through the MINLP solvers. More precisely, we used the GAMS interface [22] for BARON and the AMPL interface [23] for Couenne and SCIP. A time limit of one CPU hour was set for all the methods.

The results are presented in Tables 1–3. Each table is organized as follows. The first and the second column represent the number of variables n and of monomials m, respectively. The next columns correspond to the different methods that are compared, i.e., enumeration, SCIP, Couenne, BARON, GloptiPoly, and PolyOpt. Each entry states the average CPU time needed to solve a single instance to global optimality. In parenthesis we report the number of solved instances, out of 10 candidates. Note that the time for unsolved instances is not taken into account when computing the average. The entry *** means that none of the instances was solved within the time limit. The first block in each table contains the results for complete instances.

Table 1 shows the results for instances with variables in $[-10, 10]$. PolyOpt performs best in terms of number of solved instances. As expected, the complete enumeration cannot solve any of the instances. Also GloptiPoly cannot solve any instance; the integrality requirement has to be modeled by a high-degree polynomial, making the approach inefficient from a practical point of view. SCIP, Couenne, and BARON can solve 38, 23, and 32 instances out of 210 instances, respectively. PolyOpt can solve 90 instances and, in particular, it performs much better that the other methods on denser instances. Already instances with 20 variables and 60 monomials or 25 variables and 50 monomials turn out to be very challenging for all the other approaches. Focusing on the complete instances, PolyOpt is the only method that can solve instances on 10 variables. For instance classes not shown in the table, all tested methods failed to solve a single instance ($n = 20$ with $m \geq 180$, $n = 25$ with $m \geq 100$, and complete instances for $n \geq 15$).

Table 1. Results for bounds $[-10, 10]$

n	m	enum	scip	couenne	baron	gloptipoly	polyopt
10	1001	***	***	***	***	***	33.7 (10)
20	20	***	0.7 (10)	0.4 (10)	1.4 (10)	***	9.6 (10)
20	40	***	305.6 (10)	730.2 (3)	396.0 (9)	***	97.3 (10)
20	60	***	3585.7 (1)	***	***	***	362.0 (10)
20	80	***	***	***	***	***	900.8 (10)
20	100	***	***	***	***	***	2066.4 (9)
20	120	***	***	***	***	***	2126.8 (6)
20	140	***	***	***	***	***	2718.8 (2)
20	160	***	***	***	***	***	2827.3 (1)
25	25	***	3.0 (10)	1.0 (10)	7.5 (10)	***	27.2 (10)
25	50	***	1552.8 (7)	880.9 (1)	1662.2 (3)	***	813.4 (10)
25	75	***	***	***	***	***	2292.7 (2)

In Table 2 we show the results for instances with variables in $[-1, 1]$. This is the simplest integer case beyond binary variables; our objective was to understand whether the good performance of PolyOpt shown in Table 1 depends on the large range of feasible integer values. This is not the case: PolyOpt shows the best performance also for ternary instances. It solves all complete instances with $n \leq 15$ and all other instances with $n \leq 20$. For $n = 25$ it is still able to solve 83 out of 100 instances. Both the complete enumeration and GloptiPoly solve all complete instances for $n \leq 15$ and some sparse instances with $n = 20$. The MINLP solvers cannot solve as many instances as PolyOpt and, even for the instances that they could solve, in general the CPU time needed is larger than the one needed by PolyOpt, the only exception being the very sparse instances. The table, thus, confirms the superiority of PolyOpt also for general integer instances with small variable domains.

Table 2. Results for bounds $[-1, 1]$

n	m	enum	scip	couenne	baron	gloptipoly	polyopt
10	1001	3.2 (10)	***	125.2 (8)	612.8 (10)	2.6 (10)	1.1 (10)
15	3876	3000.9 (10)	***	***	***	139.58 (10)	357.04 (10)
20	20	3220.6 (8)	0.6 (10)	0.3 (10)	0.6 (10)	2412.7 (7)	0.6 (10)
20	40	***	12.9 (10)	4.9 (10)	28.5 (10)	2814.1 (10)	4.4 (10)
20	60	***	104.9 (10)	34.6 (10)	207.1 (10)	2915.8 (7)	13.8 (10)
20	80	***	411.8 (10)	136.5 (10)	628.4 (10)	2915.7 (5)	28.7 (10)
20	100	***	1080.7 (10)	127.8 (10)	1568.8 (10)	3110.7 (4)	42.0 (10)
20	120	***	2048.9 (8)	259.6 (10)	2645.7 (7)	3141.2 (6)	51.3 (10)
20	140	***	2580.0 (1)	331.0 (10)	1486.4 (3)	2849.9 (7)	67.9 (10)
20	160	***	***	366.7 (10)	2499.3 (1)	2952.5 (9)	79.8 (10)
20	180	***	***	973.2 (10)	***	2998.7 (7)	116.3 (10)
20	200	***	***	1216.0 (10)	***	3041.2 (6)	120.6 (10)
25	25	***	1.1 (10)	0.6 (10)	1.7 (10)	***	4.2 (10)
25	50	***	67.8 (10)	37.3 (10)	290.2 (10)	***	51.1 (10)
25	75	***	853.5 (10)	525.7 (10)	2256.7 (8)	***	274.0 (10)
25	100	***	2154.9 (4)	913.8 (10)	3009.9 (1)	***	605.9 (10)
25	125	***	***	1016.1 (9)	***	***	948.3 (10)
25	150	***	***	1957.9 (6)	***	***	1211.7 (10)
25	175	***	***	2639.3 (1)	***	***	2132.8 (10)
25	200	***	***	***	***	***	2792.5 (9)
25	225	***	***	***	***	***	2857.6 (7)
25	250	***	***	***	***	***	3224.3 (1)

Note that the optimal solutions of almost 75% of the instances solved by PolyOpt as reported in Tables 1–2 are attained at a vertex of the feasible box, i.e., none of the variables assumes a value different from both the lower and the upper bound. We are currently studying alternative ways to generate random instances with optimal solutions lying in the interior of the feasible box, e.g., by adding a quadratic regularization term to the objective function. However, first experiments show that this does not change the general picture obtained from the above results.

Finally we show the results for binary instances in Table 3. PolyOpt still outperforms the other methods on the complete instances, though for $n \geq 30$ none of the methods can solve any of the instances. Concerning the sparse instances, we report only the case $n = 40$, as for smaller instances PolyOpt, SCIP, and Couenne solve all the problems and show comparable CPU times (while BARON, the complete enumeration, and GloptiPoly results are worse). For $n = 40$, it turns out that PolyOpt cannot solve all instances, while SCIP and Couenne can. The latter approaches also need less CPU time for all classes of sparse instances. This can be explained by the fact that SCIP and Couenne have several features tailored for the binary case, such as specific heuristics and cuts. On the contrary, in PolyOpt we implemented just one special feature: we exploit the fact that $x_i^e = x_i$ for all $e \in \mathbb{N}$ in the binary case. Thus, we substitute each x_i^e with x_i, potentially obtaining monomials of lower degree.

Table 3. Results for bounds $[0, 1]$

n	m	enum	scip	couenne	baron	gloptipoly	polyopt
10	1001	0.0 (10)	8.8 (10)	16.9 (10)	26.9 (10)	2.4 (10)	0.1 (10)
15	3876	5.0 (10)	264.5 (10)	448.2 (10)	***	123.0 (10)	1.4 (10)
20	10626	480.4 (10)	***	***	***	2551.2 (10)	99.5 (10)
25	23751	***	***	***	***	***	1488.3 (7)
40	40	***	0.1 (10)	0.0 (10)	0.6 (10)	***	65.3 (9)
40	80	***	0.8 (10)	1.1 (10)	11.2 (10)	***	94.3 (10)
40	120	***	4.0 (10)	6.3 (10)	289.9 (10)	***	637.9 (10)
40	160	***	7.1 (10)	11.4 (10)	1073.6 (10)	***	569.8 (9)
40	200	***	12.7 (10)	23.0 (10)	1832.6 (7)	***	798.3 (10)
40	240	***	21.3 (10)	40.8 (10)	3019.7 (2)	***	790.3 (8)
40	280	***	46.3 (10)	91.9 (10)	***	***	1185.6 (7)
40	320	***	88.1 (10)	175.5 (10)	***	***	1212.9 (5)
40	360	***	117.6 (10)	212.6 (10)	***	***	1692.9 (9)
40	400	***	234.0 (10)	381.7 (10)	***	***	2140.3 (7)

In conclusion, PolyOpt shows a very satisfactory performance on general integer instances, especially in the dense case, while being competitive on binary instances. It is left as future work to evaluate its performance for mixed-integer instances and for polynomials of higher degree.

An important open question is how to extend our approach to problems having non-trivial constraints. An obvious idea would be to apply Lagrangian relaxation. In the case of linear constraints, the additional term in the objective function would be separable and hence not deteriorate the quality of the underestimators, but the additional task of optimizing the Lagrangian multipliers has to be addressed in an appropriate way. This is also left as future work.

References

[1] D'Ambrosio, C., Lodi, A.: Mixed integer nonlinear programming tools: a practical overview. 4OR 9(4), 329–349 (2011)

[2] Burer, S., Letchford, A.N.: Non-convex mixed-integer nonlinear programming: a survey. Surveys in Operations Research and Management Science 17, 97–106 (2012)

[3] Belotti, P., Kirches, C., Leyffer, S., Linderoth, J., Luedtke, J., Mahajan, A.: Mixed-integer nonlinear optimization. Acta Numerica 22, 1–131 (2013)

[4] D'Ambrosio, C., Lodi, A.: Mixed integer nonlinear programming tools: an updated practical overview. Annals OR 204(1), 301–320 (2013)

[5] Buchheim, C., Traversi, E.: Separable non-convex underestimators for binary quadratic programming. In: Bonifaci, V., Demetrescu, C., Marchetti-Spaccamela, A. (eds.) SEA 2013. LNCS, vol. 7933, pp. 236–247. Springer, Heidelberg (2013)

[6] Billionnet, A., Elloumi, S., Lambert, A.: Extending the QCR method to general mixed integer programs. Mathematical Programming 131(1), 381–401 (2012)

[7] Vandenbussche, D., Nemhauser, G.L.: A branch-and-cut algorithm for nonconvex quadratic programs with box constraints. Mathematical Programming 102(3), 559–575 (2005)

[8] Buchheim, C., Wiegele, A.: Semidefinite relaxations for non-convex quadratic mixed-integer programming. Mathematical Programming 141(1-2), 435–452 (2013)

[9] Burer, S.: Optimizing a polyhedral-semidefinite relaxation of completely positive programs. Mathematical Programming Computation 2(1), 1–19 (2010)

[10] Buchheim, C., De Santis, M., Palagi, L., Piacentini, M.: An exact algorithm for nonconvex quadratic integer minimization using ellipsoidal relaxations. SIAM Journal on Optimization 23(3), 1867–1889 (2013)

[11] Rosenberg, I.G.: Reduction of bivalent maximization to the quadratic case. Cahiers du Centre d'Etudes de Recherche Opérationelle 17, 71–74 (1975)

[12] Buchheim, C., Rinaldi, G.: Efficient reduction of polynomial zero-one optimization to the quadratic case. SIAM Journal on Optimization 18(4), 1398–1413 (2007)

[13] Tawarmalani, M., Sahinidis, N.V.: A polyhedral branch-and-cut approach to global optimization. Mathematical Programming 103, 225–249 (2005)

[14] Belotti, P., Lee, J., Liberti, L., Margot, F., Wächter, A.: Branching and bounds tightening techniques for non-convex MINLP. Optimization Methods and Software 24, 597–634 (2009)

[15] SCIP, http://scip.zib.de/scip.shtml

[16] Henrion, D., Lasserre, J.B., Loefberg, J.: Gloptipoly 3: moments, optimization and semidefinite programming. Optimization Methods and Software 24, 761–779 (2009)

[17] Shor, N.Z.: Class of global minimum bounds of polynomial functions. Cybernetics 23, 731–734 (1987)

[18] Lasserre, J.B.: Convergent SDP-relaxations in polynomial optimization with sparsity. SIAM Journal on Optimization 17, 822–843 (2006)

[19] Parrilo, P.A., Sturmfels, B.: Minimizing polynomial functions. Algorithmic and quantitative real algebraic geometry, DIMACS Series in Discrete Mathematics and Theoretical Computer Science 60, 83–99 (2001)

[20] Buchheim, C., D'Ambrosio, C.: Box-constrained mixed-integer polynomial optimization using separable underestimators (full version, in preparation)

[21] COUENNE (v. 0.1), http://projects.coin-or.org/Couenne

[22] GAMS, http://www.gams.com/

[23] AMPL, http://www.ampl.com/

Submodular Maximization Meets Streaming: Matchings, Matroids, and More*

Amit Chakrabarti and Sagar Kale

Department of Computer Science,
Dartmouth College, Hanover, NH 03755, USA
{ac,sag}@cs.dartmouth.edu

Abstract. We study the problem of finding a maximum matching in a graph given by an input stream listing its edges in some arbitrary order, where the quantity to be maximized is given by a monotone submodular function on subsets of edges. This problem, which we call maximum submodular-function matching (MSM), is a natural generalization of maximum weight matching (MWM). We give two incomparable algorithms for this problem with space usage falling in the semi-streaming range—they store only $O(n)$ edges, using $O(n \log n)$ working memory—that achieve approximation ratios of 7.75 in a single pass and $(3 + \varepsilon)$ in $O(\varepsilon^{-3})$ passes respectively. The operations of these algorithms mimic those of known MWM algorithms. We identify a general framework that allows this kind of adaptation to a broader setting of constrained submodular maximization.

Note. A full version of this extended abstract [1] can be found online at the following URL: http://arxiv.org/abs/1309.2038.

1 Introduction

The explosion of data—in particular graph data—over the past decade has motivated a number of researchers to revisit several algorithmic problems on graphs with a view towards designing *space efficient* algorithms that process their inputs in *streaming fashion*, i.e., via sequential access alone, though perhaps in multiple passes. In particular, a series of recent works [2–7] have studied the maximum cardinality matching (MCM) problem and its natural generalization, the maximum weight matching (MWM) problem, on graph streams.

We study a further generalization of MWM: the *maximum submodular-function matching* (MSM) problem, which does not seem to have been studied in previous work. However, there has been plenty of work on general (non-streaming) algorithms for submodular maximization under constraints more general than matchings; see Section 1.2. Our techniques too lead to results for a wider class of constraints, including hypermatchings and intersection of matroids.

A *submodular function* on a ground set $\mathcal{X}$ is defined to be a function $f : 2^{\mathcal{X}} \to \mathbb{R}$ that satisfies $f(X \cup \{x\}) - f(X) \leqslant f(Y \cup \{x\}) - f(Y)$, for all $Y \subseteq X \subset \mathcal{X}$

---

* Supported in part by NSF grant CCF-1217375.

J. Lee and J. Vygen (Eds.): IPCO 2014, LNCS 8494, pp. 210–221, 2014.

and $x \in \mathcal{X} \setminus X$; f is *monotone* if $f(Y) \leqslant f(X)$ whenever $Y \subseteq X \subseteq \mathcal{X}$ and *proper* if $f(\emptyset) = 0$. An instance of MSM consists of a graph $G = (V, E)$ on vertex set $V = [n] := \{1, 2, \ldots, n\}$ and a non-negative monotone proper submodular function f whose ground set is E, i.e., $f : 2^E \rightarrow \mathbb{R}_+$. The goal is to output a matching $M^* \subseteq E$ that maximizes $f(M^*)$; we shall refer to such a matching as an f-MSM of G. For a real number $\alpha \geqslant 1$, an α-approximate f-MSM of G is a matching $M \subseteq E$ such that $f(M) \geqslant \alpha^{-1} f(M^*)$.

Our concern is with graph *streams*: the input graph is described by a stream of edges $\{u, v\}$, with $u, v \in [n]$. We assume that the number of vertices, n, is known in advance and that each edge in E appears exactly once in the stream. The order of edge arrivals is arbitrary, possibly adversarial. We seek algorithms for MSM that use only quasi-linear working memory—i.e., $O(n(\log n)^{O(1)})$ bits of storage, with $O(n \log n)$ being the holy grail—and process each edge very quickly, ideally in $O(1)$ time. Algorithms with such guarantees have come to be known as *semi-streaming* algorithms [2]. Notice that $\Omega(n \log n)$ bits are necessary simply to store a matching that saturates $\Omega(n)$ vertices. To handle general f, we make the standard assumption that f is presented by a *value oracle*.

1.1 Our Results

We give two incomparable semi-streaming approximation algorithms for the MSM problem on graph streams, formally stated in the two theorems below. For brevity, "submodular f," means a non-negative monotone proper submodular function f, presented by a value oracle.

Theorem 1. *For every submodular f, there is a one-pass semi-streaming algorithm that outputs a 7.75-approximate f-MSM of an n-vertex input graph, storing at most $O(n)$ edges at all times.*

Theorem 2. *For every submodular f, and every constant $\varepsilon > 0$, there is a multi-pass semi-streaming algorithm that makes $O(\varepsilon^{-3})$ passes over an n-vertex graph stream and outputs a $(3 + \varepsilon)$-approximate f-MSM of the graph. This algorithm stores only a matching in the input graph at all times; in particular it stores only $O(n)$ edges.*

Perhaps more important than these specific approximation ratios is the technique behind these results. We identify a general framework for matching algorithms in graph streams. We show that whenever an MWM algorithm fits this framework, it can be adapted to the broader setting of MSM. The two theorems above then follow by revisiting two recent MWM algorithms—those of Zelke [4] and McGregor [3] respectively—and showing that they fit our framework.

To adapt an MWM algorithm, we need to assign a weight $w(e)$ to each edge e we encounter: using $w(e) = f(\{e\})$ is too naïve to be useful, but we show that $w(e) := f(I \cup \{e\}) - f(I)$ works, where I is our "current solution." For multi-pass algorithms, this choice causes weights to change from one pass to another; with appropriate analysis we can prove a good approximation ratio.

Our framework is not deeply wedded to matchings: it is general enough to capture set maximization problems constrained to abstract "independent sets," for a very general notion of independence. We obtain results for maximization over intersection of p matroids and MWM and f-MSM algorithms for matchings in hypergraphs, given a bound p on the size of hyperedges (in both cases, p is a constant). Here, we have a maximum-submodular problem (MSIS, say—the "IS" stands for "independent set"), and a maximum-weight problem (MWIS, say) where the submodular function is modular. The results are summarized below. We comment on the proofs briefly in Section 5.

Theorem 3. *For every submodular f, the MWIS and f-MSIS problems, with independent sets being given either by a hypermatching constraint in p-hypergraphs or by the intersection of p matroids, there are near-linear-space streaming algorithms giving the following approximation ratios.*

Problem type	MWIS	MSIS
One pass: p-hypergraphs; p matroids	$2(p + \sqrt{p(p-1)}) - 1$	$4p$
$O(\varepsilon^{-3} \log p)$ passes: p-hypergraphs; p partition matroids	$p + \varepsilon$	$p + 1 + \varepsilon$

The one-pass MWIS result for matroids was already known from the work of Badanidiyuru Varadaraja [8]; the remaining results in Theorem 3 are novel.

We can strengthen some of our results by considering the *curvature* $\mathrm{curv}(f)$ of the submodular function f, defined as $\min\{c : \forall A \subseteq E \,\forall e \in E \backslash A$, we have $f(A \cup \{e\}) - f(A) \geqslant (1 - c)f(\{e\})\}$. This measures how far f is from being modular: $\mathrm{curv}(f) \in [0, 1]$ and $\mathrm{curv}(f) = 0$ iff f is modular. Our next result, whose proof we omit in this extended abstract, gives approximation ratios for f-MSM and f-MSIS that gradually improve to those for Zelke's MWM algorithm and Badanidiyuru Varadaraja's MWIS algorithm as $\mathrm{curv}(f) \rightarrow 0$.

Theorem 4. *For every submodular f, the approximation ratios for f-MSM in Theorem 1 and the one-pass approximation ratio for f-MSIS in Theorem 3 can be improved to $\min\{7.75, 5.585/(1 - \mathrm{curv}(f))\}$ and $\min\{4p, (2(p + \sqrt{p(p-1)}) - 1)/(1 - \mathrm{curv}(f))\}$ respectively.*

1.2   Context and Related Work

We place Theorems 1 and 2 in context. As noted before, our work is the first to consider the MSM problem. For MWM, the one-pass semi-streaming algorithms of Feigenbaum *et al.* [2], McGregor [3], Zelke [4], and Epstein *et al.* [5] achieved approximation ratios of 6, 5.828, 5.585, and 4.911 respectively; all of these algorithms except the last fit our aforementioned framework. McGregor's algorithm extends to give a ratio of $(2 + \varepsilon)$ in $O(\varepsilon^{-3})$ passes. Very recently, Ahn and Guha [9] gave a $(1 + \varepsilon)$ approximation with multiple passes, using a very different algorithm that falls outside our framework.

For a more detailed summary of streaming MCM and MWM algorithms, please see Ahn and Guha [9] or the full version of this paper [1].

Besides streaming algorithms, another useful viewpoint is to consider f-MSM as the problem of maximizing the submodular function $f(S)$ subject to S being a matching. This makes MSM an instance of constrained submodular maximization, which is a heavily-studied topic in optimization. Matroids form an important class of constraints. One can consider more general "independence systems," such as the intersection of p different matroids on the same ground set E (called a p-intersection system) or, even more generally, a p-*system*, wherein

$$\forall A \subseteq E : \frac{\max\{|S| : S \subseteq A,\ S \text{ maximally independent}\}}{\min\{|S| : S \subseteq A,\ S \text{ maximally independent}\}} \leqslant p.$$

This last generalization captures the constraint of S being a matching: matchings form a 2-system. All of these classes of problems were studied in the seminal work of Fisher, Nemhauser and Wolsey [10,11]. More recently, Calinescu *et al.* [12] gave a polynomial time $(e/(e-1))$-approximation algorithm for maximizing f subject to a matroid constraint. Lee *et al.* [13,14] considered maximization over p-intersection systems and gave a *local search* algorithm with ratio $p+\varepsilon$, improving upon $p+1$ [11], known for the more general p-systems. Recently, Feldman *et al.* [15] gave a local-search-based $(p+\varepsilon)$-approximation for "p-exchange systems," which captures matchings (with $p=2$). Therefore they improved—after a span of over 30 years—the best known approximation ratio for f-MSM from 3 [11] to $2+\varepsilon$. We invite the reader to compare with our Theorem 2.

Very recently, and concurrent with our work, Badanidiyuru and Vondrák [16] gave a $(p+1+\varepsilon)$-approximation algorithm for submodular maximization over a p-system that can be thought of as using $O(\varepsilon^{-2}\log^2(m/\varepsilon))$ passes, where $m = |E|$. Our result in Theorem 3 uses fewer passes (for constant ε), but handles a smaller class of constraints than p-systems.

Feldman *et al.* [15] and Badanidiyuru and Vondrák [16] give detailed summaries of results on constrained submodular maximization.

1.3 Motivation and Significance of Our Results

Our study of MSM is inspired in part by its applicability to the Word Alignment Problem (WAP) from computational linguistics, as studied in Lin and Bilmes [17]. We also submit that MSM considered as a data streaming problem is a novel and pleasing marriage of several important theoretical ideas: submodularity, data streaming, matching theory, and matroids.

In applications such as big data analytics, a good solution obtained *quickly* may be preferred over a theoretically stronger guarantee given by a slower algorithm. Our algorithms in this work should be seen in this light: they are significant because they are *faster* algorithms with slightly worse approximation ratios than best known offline algorithms. Moreover, they are able to handle input presented in streaming fashion, a clear advantage when handling big data. The classic greedy or local-search strategies are unsuitable for streamed input.

2 Preliminaries

We start by making our model of computation precise. The input is an n-vertex graph stream, defined as a sequence $\sigma = \langle e_1, e_2, \ldots, e_m \rangle$ of distinct edges, where each $e_i = (u_i, v_i) \in [n] \times [n]$ and $u_i < v_i$. We put $V = [n], E = \{e_1, \ldots, e_m\}$, and $G = (V, E)$. The submodular function $f : 2^E \to \mathbb{R}_+$, which is part of the problem specification, is given by an entity external to the stream, called the *value oracle* for f, or the f-oracle. A data stream algorithm, after reading each edge from the input stream, is allowed to make an arbitrary number of *calls* to the f-oracle, sending a subset $S \subseteq E$ and getting $f(S)$ in return. (In fact the algorithms we design here make only $O(1)$ such calls on average.) The algorithm is charged for the space required to describe S. It is deemed to *fail* or *abort* if it ever tries to obtain $f(S)$ with $S \not\subseteq E$; this prevents it from "cheating" and learning about E indirectly from oracle calls.

We cannot solve MSM exactly: the very special case MCM cannot be approximated any better than $e/(e-1)$ in the semi-streaming setting [7]. As shown in the full paper [1], even the non–streaming version does not admit a polynomial-time approximation better than $e/(e-1)$.

2.1 A Framework for Streaming MSM and MSIS Algorithms

We proceed to describe a generic streaming algorithm for f-MSM, which defines the framework alluded to in Section 1.1. In fact, as noted towards the end of Section 1.1, our framework applies to the much more general problem of f-MSIS (Maximum Submodular Independent Set), an instance of which is given by a submodular $f : 2^E \to \mathbb{R}_+$ and a collection $\mathcal{I} \subseteq 2^E$ of *independent sets* such that $\emptyset \in \mathcal{I}$. We put $m := |E|$ and $n := \max_{I \in \mathcal{I}} |I|$. We assume that independence (i.e., membership in $\mathcal{I}$) can be tested easily; we require no other structural property of $\mathcal{I}$. For the special case of MSM, $\mathcal{I}$ is the collection of matchings in a graph with edge set E.

The generic algorithm for f-MSIS starts with a given independent set P (possibly empty) and then proceeds to make *one* pass over the input stream σ, attempting to end up with an improved independent set I by the end of the pass. The algorithm processes the elements in E in a *pretend stream order* that consists of an arbitrary permutation of the elements in P, followed by the elements in $E \setminus P$ in the same order as σ. Throughout, the algorithm maintains a "current solution" $I \in \mathcal{I}$, a set $S \subseteq E$ of "shadow elements" (this term is borrowed from Zelke [4]), and a weight $w(e)$ for each element e it has processed. The intuition behind shadow elements is to have more scope to improve the current solution. The algorithm bases its decisions on a parameter $\gamma > 0$, which we eventually tune to optimize our approximation ratio. For a set $A \subseteq E$, we denote $w(A) := \sum_{e \in A} w(e)$. An *augmenting pair* for a set $I \in \mathcal{I}$ is a pair of sets (A, J) such that $J \subseteq I$ and $(I \setminus J) \cup A \in \mathcal{I}$. For $e \in E$, define $A + e$ to be $A \cup \{e\}$.

Notice that PROCESS-ELEMENT maintains the invariant that $w(e)$ is defined for all $e \in I \cup S$. Therefore, Line 8 does not access an element weight before defining it. The algorithm need only remember the weights of elements in $I \cup$

Algorithm 1. Generic One-Pass Independent Set Improvement Alg. for f-MSIS

1: **function** IMPROVE-SOLUTION(σ, P, γ)
2:     $I \leftarrow \emptyset, S \leftarrow \emptyset$
3:     **foreach** $e \in P$ in some arbitrary order **do** $w(e) \leftarrow f(I + e) - f(I), I \leftarrow I + e$
4:     **foreach** $e \in \sigma \setminus P$ in the order given by σ **do** PROCESS-ELEMENT(e, I, S)
5:     **return** I

6: **procedure** PROCESS-ELEMENT(e, I, S) ▷ Note: Assigns $w(e)$, modifies I and S.
7:     $w(e) \leftarrow f(I \cup S + e) - f(I \cup S)$
8:     $(A, J) \leftarrow$ a well-chosen augmenting pair for I with $A \subseteq I \cup S + e$ and $w(A) \geqslant$
    $(1 + \gamma)w(J)$
9:     $S \leftarrow$ a well-chosen subset of $(S \setminus A) \cup J$
10:     $I \leftarrow (I \setminus J) \cup A$ ▷ Augment independent set I using A.

S. Therefore, the space usage of the algorithm is bounded by $O((|P| + |I| + |S|) \log m) = O((n + |S|) \log m)$, since P and I are independent sets.

To instantiate this generic algorithm, one must specify the precise logic used in Lines 8 and 9. If the algorithm is for MWIS rather than MSIS, then $w(e)$ values are already given and assignments to those values (see Lines 3 and 7) should be ignored.

Definition 5. *We say that an MWIS algorithm is* compliant *if each pass instantiates Algorithm 1 in the above sense, i.e., it starts with some solution $P \in \mathcal{I}$ computed in the previous pass and calls* IMPROVE-SOLUTION(σ, P, γ). *The parameter γ need not be the same for all passes.*

Definition 6. *For a submodular f, we define an f-extension of a compliant MWIS algorithm $\mathcal{A}$ to be Algorithm 1, with the logic used in Lines 8 to 9 being borrowed from $\mathcal{A}$, and with values of the parameter γ possibly differing from those used by $\mathcal{A}$.*

Lemma 7 (Modular to submodular). *Let $\mathcal{A}$ be a one-pass compliant MWIS algorithm that computes a C_γ-approximate MWIS when run with parameter γ. Then, for every non-negative monotone proper submodular f, its f-extension with parameter γ computes a $(C_\gamma + 1 + 1/\gamma)$-approximate f-MSIS.*

3   A One-Pass Solution via Compliant Algorithms

Consider the f-extension with parameter γ of a particular one-pass compliant algorithm. Let I denote its output and I^* be an f-MSIS. Let I_e, S_e denote the contents of the variables I, S in Algorithm 1 just before element e is processed. Let $K = (\bigcup_{e \in E} I_e) \setminus I$ denote the set of elements that were added to the current solution at some point but were *killed* and did not make it to the final output. Then $\bigcup_{e \in E} S_e \subseteq I \cup K$, because an element can become a shadow element only when it was removed from the current solution at some point (see Line 9 of Algorithm 1). Hence

$$\bigcup_{e \in E}(I_e \cup S_e) \subseteq I \cup K. \tag{1}$$

Lemma 8. *For an f-extension of a compliant algorithm, $w(K) \leqslant w(I)/\gamma$.*

Proof. Let A_e, J_e be the sets A, J chosen at Line 8 when processing e. Each augmentation by A_e (Line 10) increases the weight of the current solution by $w(A_e) - w(J_e) \geqslant \gamma w(J_e)$. Hence, $w(I)/\gamma \geqslant \sum_{e \in E} w(J_e)$.

The set $\bigcup_{e \in E} J_e$ consists of elements that were removed from the current solution at some point. Thus, it includes K (the inclusion may be proper: K does not contain elements that were removed from the current solution, reinserted, and eventually ended up in I). Therefore, $w(K) \leqslant w(\bigcup_{e \in E} J_e) \leqslant \sum_{e \in E} w(J_e) \leqslant w(I)/\gamma$. ∎

Lemma 9. *For an f-extension of a compliant algorithm, we have $w(I) \leqslant f(I)$.*

Proof. Let $e_1^I, e_2^I, \ldots, e_s^I$ be an enumeration of I in order of processing, where $s = |I|$. The logic of Algorithm 1 ensures that an element once removed from the shadow set can never return to the current solution (though elements can move between the two arbitrarily). Thus, $I \cap (I_{e_i^I} \cup S_{e_i^I}) = \{e_1^I, e_2^I, \ldots, e_{i-1}^I\}$. Since $I \cap (I_{e_i^I} \cup S_{e_i^I}) \subseteq (I_{e_i^I} \cup S_{e_i^I})$ and f is submodular, we have

$$f(\{e_1^I, e_2^I, \ldots, e_i^I\}) - f(\{e_1^I, e_2^I, \ldots, e_{i-1}^I\}) \geqslant f(I_{e_i^I} \cup S_{e_i^I} + e_i^I) - f(I_{e_i^I} \cup S_{e_i^I}) = w(e_i^I).$$

Summing this over $i \in [s]$ gives $f(I) = f(I) - f(\emptyset) \geqslant \sum_{i=1}^s w(e_i^I) = w(I)$. ∎

Lemma 10. *For an f-extension of a compliant algorithm, we have $f(I^*) \leqslant (1/\gamma + 1)f(I) + w(I^*)$.*

Proof. Let $e_1^B, \ldots, e_b^B$ be an enumeration of $B := I \cup K$ in order of processing. The set $B_i := \{e_1^B, \ldots, e_{i-1}^B\}$ consists of elements inserted into the current solution before e_i^B was processed. Meanwhile $I_{e_i^B} \cup S_{e_i^B}$ is the subset of these elements that were not removed from S before e_i^B was processed. Thus, $B_i \supseteq I_{e_i^B} \cup S_{e_i^B}$ for all $i \in [b]$. By submodularity of f, $f(\{e_1^B, e_2^B, \ldots, e_i^B\}) - f(\{e_1^B, e_2^B, \ldots, e_{i-1}^B\}) \leqslant f(I_{e_i^B} \cup S_{e_i^B} + e_i^B) - f(I_{e_i^B} \cup S_{e_i^B}) = w(e_i^B)$. Summing this over $i \in [b]$ gives $f(B) = f(B) - f(\emptyset) \leqslant w(B)$. Thus, we have

$$f(I \cup K) \leqslant w(I \cup K) = w(I) + w(K) \leqslant f(I) + w(I)/\gamma = (1/\gamma + 1)f(I), \quad (2)$$

where the last two inequalities use Lemma 9 and Lemma 8 respectively.

Now we bound $f(I^*)$. Let $I^* \setminus (I \cup K) = \{e_1^{I^*}, e_2^{I^*}, \ldots, e_t^{I^*}\}$; this enumeration is in arbitrary order. Put $D_0 = I \cup K$, $D_i = I \cup K \cup \{e_1^{I^*}, \ldots, e_i^{I^*}\}$ for $i \in [t]$. By Equation (1), $D_{i-1} \supseteq I \cup K \supseteq I_{e_i^{I^*}} \cup S_{e_i^{I^*}}$. Appealing to submodularity again,

$$f(D_i) - f(D_{i-1}) \leqslant f(I_{e_i^{I^*}} \cup S_{e_i^{I^*}} + e_i^{I^*}) - f(I_{e_i^{I^*}} \cup S_{e_i^{I^*}}) = w(e_i^{I^*}).$$

Summing this over $i \in [t]$ gives $f(D_t) - f(D_0) \leqslant w(I^* \setminus (I \cup K)) \leqslant w(I^*)$. In other words, $f(I \cup K \cup I^*) - f(I \cup K) \leqslant w(I^*)$. By monotonicity of f and Equation (2), we have

$$f(I^*) \leqslant f(I \cup K \cup I^*) \leqslant f(I \cup K) + w(I^*) \leqslant (1/\gamma + 1)f(I) + w(I^*). \quad (3)$$

Proof (Lemma 7). Since the compliant algorithm $\mathcal{A}$ outputs a C_γ-approximate MWIS, it satisfies $w(I^*) \leqslant C_\gamma w(I)$ for any weight assignment; in particular, the weights assigned by its f-extension. Using Theorem 10 and Theorem 9, we conclude that $f(I^*) \leqslant (C_\gamma + 1 + 1/\gamma)f(I)$.

Proof (Theorem 1). Recall that an "independent set" is just a matching in the setting of MWM and MSM. Zelke's algorithm [4] chooses the augmenting pair (A, J) as follows: A is chosen from an $O(1)$-sized "neighborhood" of the edge e being processed, and J is set to be $M \upharpoonright A$: the set of edges in M that share a vertex with some edge in A. At most two shadow edges are stored per edge in the current matching, thus $|S| = O(n)$ and a space bound of $O(n \log n)$ bits.

Zelke's algorithm is compliant with $C_\gamma = 2(1 + \gamma) + (1/\gamma + 1) - \gamma/(1 + \gamma)^2$ [4, Theorem 3]. By Lemma 7, its f-extension yields an approximation ratio of $2(1 + \gamma)^2/\gamma - \gamma/(1 + \gamma)^2$, which attains a minimum value of 7.75 at $\gamma = 1$. This proves the theorem.

4   A Multi-pass MSM Algorithm

In this section we prove Theorem 2. For this we first review McGregor's multi-pass MWM algorithm [3], which is compliant. Our algorithm is simply its f-extension, as explained in Section 2.1.

To describe McGregor's algorithm with respect to our framework (Algorithm 1), we need only explain the two choices we make inside PROCESS-EDGE. We always set $S = \emptyset$ in Line 9. In Line 8, we choose the augmenting pair (A, J) so that $A = \{e\}$ if possible, and $A = \emptyset$ otherwise, and $J = M \upharpoonright A$. Recall that $M \upharpoonright A$ denotes the set of edges in matching M that share a vertex with some edge in set A. This describes a single pass. The overall algorithm starts with an empty matching and repeatedly invokes IMPROVE-MATCHING with $\gamma = 1/\sqrt{2}$ for the first pass and $\gamma = 2\varepsilon/3$ for the remaining passes. It stops when the multiplicative improvement made in a pass drops below a certain well-chosen rational function of γ. McGregor analyzes this algorithm to show that it makes at most $O(\varepsilon^{-3})$ passes and terminates with a $(2 + 2\varepsilon)$-approximate MWM.

In our f-extension, we use $\gamma = 1$ for the first pass and $\gamma = \varepsilon/3$ for the remaining passes. We lay out the logic of the resulting f-MSM algorithm explicitly in Algorithm 2. The function IMPROVE-MATCHING is exactly as in Algorithm 1 except that it calls PROCESS-EDGE(e, M), since S is never used.

Let M^i denote the matching M computed by Algorithm 2 at the end of its ith pass over σ. When an edge e is added to M in Line 12, we say that e is *born* and that it *kills* the (at most two) edges in $M \upharpoonright \{e\}$. Notice that during pass $i > 1$, thanks to the *pretend stream order* in which edges are processed (cf. the discussion at the start of Section 2.1), initially all edges in M^{i-1} are born without killing anybody.[1] For the rest of the pass these edges are never considered for addition to M.

---

[1] This subtlety appears to have been missed in McGregor's analysis [3] and it creates a gap in his argument. Using a pretend stream order as we do fixes that gap.

Algorithm 2. Multi-Pass Algorithm for f-MSM

1: **function** MULTI-PASS-MSM(σ)
2:     $M \leftarrow$ IMPROVE-MATCHING$(\sigma, \emptyset, 1)$ ▷ Obtains 8-approximate f-MSM.
3:     $\gamma \leftarrow \varepsilon/3, \ \kappa \leftarrow \gamma^3/(2 + 3\gamma + \gamma^2 - \gamma^3)$
4:     **repeat**
5:         $w_{\text{prev}} \leftarrow f(M)$
6:             $M \leftarrow$ IMPROVE-MATCHING(σ, M, γ)
7:     **until** $w(M)/w_{\text{prev}} \leqslant 1 + \kappa$
8:     **return** M

9: **procedure** PROCESS-EDGE(e, M) ▷ Compare with Algorithm 1.
10:     $w(e) \leftarrow f(M + e) - f(M)$
11:     **if** $w(e) \geqslant (1 + \gamma)w(M \upharpoonright \{e\})$ **then**
12:         $M \leftarrow M \setminus (M \upharpoonright \{e\}) + e$

Let K^i denote the set of edges killed during pass i (some of them may be born during a subsequent pass). Then $M^i \cup K^i$ is exactly the set of edges born in pass i. These edges can be made the nodes of a collection of disjoint rooted *killing trees*[2] where the parent of a killed edge e is the edge e' that killed it. The set of roots of these killing trees is precisely M^i. Let $T^i(e)$ denote the set of strict descendants of $e \in M^i$ in its killing tree. Then $K^i = \bigcup_{e \in M^i} T^i(e)$.

Let $B^i = M^i \cap M^{i-1}$ denote the set of edges that pass i retains in the matching from the previous pass. By the preceding discussion, it follows that $T^i(e) = \emptyset$ for all $e \in B^i$.

4.1   Analysis

We now analyze Algorithm 2. As before, let M^* denote an optimal solution to the f-MSM instance. The first pass of the algorithm is McGregor's [3] compliant algorithm with $C_\gamma = 1/\gamma + 3 + 2\gamma$. Applying Lemma 7 and using $\gamma = 1$, we get:

Lemma 11. *We have $f(M^1) \geqslant f(M^*)/8$.*

Define τ to be the number of passes made by Algorithm 2. Let $w_i(e)$ denote the weight assigned to edge e in Line 10 during the ith pass. For the rest of this section, γ denotes the parameter value used by passes 2 through τ, and κ denotes the corresponding value assigned at Line 3. To analyze the result of those passes, we first borrow three results—stated in the next three lemmas—from McGregor's analysis [3, Lemma 3 and Theorem 3], which in turn borrows from the Feigenbaum *et al.* analysis [2, Theorem 2].

Lemma 12. *For all $i \in [2, \tau]$ and all $e \in M^i$, we have $w_i(T^i(e)) \leqslant w_i(e)/\gamma$.*

Proof. Directly analogous to Lemma 8.

Lemma 13. *We have $w_\tau(B^\tau)/w_\tau(M^\tau) \geqslant (\gamma - \kappa)/(\gamma + \gamma\kappa)$.*

---

[2] Feigenbaum *et al.* [2] and McGregor [3] used the evocative term "trail of the dead."

Proof (sketch). The logic in Lines 11 to 12 ensures that, for all $i \in [2, \tau]$, we have $w_i(M^i \setminus B^i) \geqslant (1 + \gamma)w_i(M^{i-1} \setminus B^i)$. In particular, this holds at $i = \tau$.

During the initial phase of pass $i \geqslant 2$, the set M is *monotonically* built up from $\emptyset$ to M^{i-1} according to a pretend stream order and weights are assigned to edges in M^{i-1} according to Line 10. Because of this monotonicity, summing the weights of these edges causes the f terms to telescope, giving $w_i(M^{i-1}) = f(M^{i-1})$. So the stopping criterion in Line 7 ensures that $w_\tau(M^\tau)/w_\tau(M^{\tau-1}) \leqslant 1 + \kappa$. Combining this with the inequality in the last paragraph (at $i = \tau$) yields the lemma after some straightforward algebra.

Lemma 14. *We have $w_i(M^*) \leqslant (1+\gamma) \sum_{e \in M^i} (w_i(T^i(e)) + 2w_i(e)), \forall i \in [2, \tau]$.*

Proof (remarks only). This lemma has a rather creative proof, wherein the weights of edges in M^* are *charged* to edges in $M^i \cup K^i$ using a careful *charge transfer scheme*. This scheme is described in McGregor's analysis [3] and also in the full version of our paper.

We may now analyze Algorithm 2, thereby proving Theorem 2.

Proof (Theorem 2, sketch). Due to lack of space we prove only the approximation ratio. We note that bounding the number of passes does require some care. As noted earlier, $T^i(e) = \emptyset$ for all $e \in B^i$ and $K^i = \bigcup_{e \in M^i} T^i(e)$. Therefore, $K^\tau = \bigcup_{e \in M^\tau \setminus B^\tau} T^\tau(e)$, which gives

$$w_\tau(K^\tau) = \sum_{e \in M^\tau \setminus B^\tau} w_\tau(T^\tau(e)) \leqslant \sum_{e \in M^\tau \setminus B^\tau} \frac{w_\tau(e)}{\gamma} = \frac{w_\tau(M^\tau) - w_\tau(B^\tau)}{\gamma}, \quad (4)$$

where the inequality follows from Lemma 12. Using Equation (4) and relating the first and third terms in Equation (2), we get

$$f(M^\tau \cup K^\tau) \leqslant w_\tau(M^\tau) + \frac{w_\tau(M^\tau) - w_\tau(B^\tau)}{\gamma} = \left(1 + \frac{1}{\gamma}\right) w_\tau(M^\tau) - \frac{1}{\gamma} w_\tau(B^\tau).$$
$$(5)$$

Using Lemma 14, we now get

$$\begin{aligned}
w_\tau(M^*) &\leqslant (1 + \gamma) \sum_{e \in M^\tau} (w_\tau(T^\tau(e)) + 2w_\tau(e)) \\
&= (1 + \gamma) \left[\left(\sum_{e \in M^\tau \setminus B^\tau} w_\tau(T^\tau(e)) \right) + 2w_\tau(M^\tau) \right] \quad \because \bigcup_{e \in B^\tau} T^\tau(e) = \emptyset \\
&\leqslant (1 + \gamma) \left[\left(\sum_{e \in M^\tau \setminus B^\tau} \frac{w_\tau(e)}{\gamma} \right) + 2w_\tau(M^\tau) \right] \quad \text{by Lemma 12} \\
&= (1/\gamma + 1)(w_\tau(M^\tau) - w_\tau(B^\tau)) + (2 + 2\gamma)w_\tau(M^\tau) \\
&= (1/\gamma + 3 + 2\gamma) w_\tau(M^\tau) - (1 + 1/\gamma) w_\tau(B^\tau).
\end{aligned}$$

By using Equation (3), we have $f(M^*) \leqslant f(M^\tau \cup K^\tau) + w_\tau(M^*)$. So adding Equation (5) and above inequality, we get

$$f(M^*) \leqslant w_\tau(M^\tau)\left[1 + 1/\gamma + 1/\gamma + 3 + 2\gamma\right] - w_\tau(B^\tau)\left[1 + 2/\gamma\right]$$

$$\leqslant \left[4 + \frac{2}{\gamma} + 2\gamma - \left(1 + \frac{2}{\gamma}\right)\frac{\gamma - \kappa}{\gamma + \gamma\kappa}\right] w_\tau(M^\tau) \qquad \text{by Lemma 13}$$

$$= (3 + 3\gamma)w_\tau(M^\tau) \qquad\qquad\qquad\qquad \text{substituting } \kappa$$

$$\leqslant (3 + \varepsilon)f(M^\tau) \qquad\qquad\qquad\qquad\qquad \text{by Lemma 9}$$

and this shows that Algorithm 2 computes a $(3 + \varepsilon)$-approximate f-MSM.

5   Generalizations: Multiple Matroids, Hypergraphs

In the previous two sections, we have given most of the details of the proofs of our results for MSM (Theorems 1 and 2). We close with some brief remarks on Theorem 3 which generalizes our results to maximization over (one or more) matroids and to MSM in p-hypergraphs. Detailed algorithms and proofs appear in the full version of the paper.

Given p matroids $M_1 = (E, \mathcal{I}_1), \ldots, M_p = (E, \mathcal{I}_p)$ over a common ground set E, and a submodular $f : 2^E \to \mathbb{R}_+$, we consider the problem of finding $\mathrm{argmax}_{I \in \mathcal{I}_1 \cap \cdots \cap \mathcal{I}_p} f(I)$, in the streaming model, where elements of E arrive in the stream and membership in each independence system $\mathcal{I}_i$ is given by a corresponding oracle. Badanidiyuru Varadaraja [8] gave a one-pass $(2(p + \sqrt{p(p-1)}) - 1)$-approximation algorithm for this problem when f is modular. His algorithm is compliant. Hence, by Lemma 7, we get an approximation ratio of $4p$ when f is submodular (by setting $\gamma = 1$). Our algorithm uses $O(n(\log m)^{O(1)})$ memory, where $m := |E|$ and $n := \max_{I \in \mathcal{I}_1 \cap \cdots \cap \mathcal{I}_p} |I|$.

In the multipass setting, we are unable to extend Badanidiyuru Varadaraja's charging scheme (cf. our remarks following Lemma 14) to argue that his algorithm can be used in Line 6 of Algorithm 2. But we can give a simpler charging scheme for partition matroids. We get the approximation ratios $p + \varepsilon$ and $p + 1 + \varepsilon$ for the modular and submodular cases, respectively, by setting $\gamma = \varepsilon/(p+1)$ for both cases, and $\kappa = \gamma^3/((p-1)(1+\gamma)^2 - \gamma^3)$ for the modular case and $\kappa = \gamma^3/(p + (2p-1)\gamma + (p-1)\gamma^2 - \gamma^3)$ for the submodular case, in Algorithm 2.

A hypergraph where every hyperedge contains at most p vertices is called a p-hypergraph. A matching is a pairwise disjoint subset of hyperedges. Most of the work required to derive the results claimed in Theorem 3 for hypergraphs goes towards appropriately generalizing Lemma 14 by designing a generalized charging scheme. Again, details appear in the full version of the paper.

References

1. Chakrabarti, A., Kale, S.: Submodular maximization meets streaming: Matchings, matroids, and more. arXiv preprint arXiv:1309.2038 (2013)
2. Feigenbaum, J., Kannan, S., McGregor, A., Suri, S., Zhang, J.: On graph problems in a semi-streaming model. Theor. Comput. Sci. 348(2), 207–216 (2005)

3. McGregor, A.: Finding graph matchings in data streams. In: Chekuri, C., Jansen, K., Rolim, J.D.P., Trevisan, L. (eds.) APPROX 2005 and RANDOM 2005. LNCS, vol. 3624, pp. 170–181. Springer, Heidelberg (2005)
4. Zelke, M.: Weighted matching in the semi-streaming model. In: Proc. 25th International Symposium on Theoretical Aspects of Computer Science, STACS 2008, pp. 669–680 (2008)
5. Epstein, L., Levin, A., Mestre, J., Segev, D.: Improved approximation guarantees for weighted matching in the semi-streaming model. SIAM Journal on Discrete Mathematics 25(3), 1251–1265 (2011)
6. Goel, A., Kapralov, M., Khanna, S.: On the communication and streaming complexity of maximum bipartite matching. In: Proc. 23rd Annual ACM-SIAM Symposium on Discrete Algorithms, SODA 2012, pp. 468–485. SIAM (2012)
7. Kapralov, M.: Better bounds for matchings in the streaming model. In: Proc. 24th Annual ACM-SIAM Symposium on Discrete Algorithms, SODA 2013. SIAM (2013)
8. Badanidiyuru Varadaraja, A.: Buyback problem: approximate matroid intersection with cancellation costs. In: Aceto, L., Henzinger, M., Sgall, J. (eds.) ICALP 2011, Part I. LNCS, vol. 6755, pp. 379–390. Springer, Heidelberg (2011)
9. Ahn, K.J., Guha, S.: Linear programming in the semi-streaming model with application to the maximum matching problem. In: Aceto, L., Henzinger, M., Sgall, J. (eds.) ICALP 2011, Part II. LNCS, vol. 6756, pp. 526–538. Springer, Heidelberg (2011)
10. Nemhauser, G., Wolsey, L., Fisher, M.: An analysis of approximations for maximizing submodular set functions—I. Mathematical Programming 14(1), 265–294 (1978)
11. Fisher, M., Nemhauser, G., Wolsey, L.: An analysis of approximations for maximizing submodular set functions—II. In: Balinski, M., Hoffman, A. (eds.) Polyhedral Combinatorics. Mathematical Programming Studies, vol. 8, pp. 73–87. Springer, Heidelberg (1978)
12. Calinescu, G., Chekuri, C., Pál, M., Vondrák, J.: Maximizing a monotone submodular function subject to a matroid constraint. SIAM J. Comput. 40(6), 1740–1766 (2011)
13. Lee, J., Mirrokni, V.S., Nagarajan, V., Sviridenko, M.: Non-monotone submodular maximization under matroid and knapsack constraints. In: Proc. 41st Annual ACM Symposium on the Theory of Computing, STOC 2009, pp. 323–332. ACM, Bethesda (2009)
14. Lee, J., Sviridenko, M., Vondrák, J.: Submodular maximization over multiple matroids via generalized exchange properties. Mathematics of Operations Research 35(4), 795–806 (2010)
15. Feldman, M., Naor, J(S.), Schwartz, R., Ward, J.: Improved approximations for k-exchange systems. In: Demetrescu, C., Halldórsson, M.M. (eds.) ESA 2011. LNCS, vol. 6942, pp. 784–798. Springer, Heidelberg (2011)
16. Badanidiyuru, A., Vondrák, J.: Fast algorithms for maximizing submodular functions. In: Proc. 25th Annual ACM-SIAM Symposium on Discrete Algorithms, SODA 2014. SIAM (2014)
17. Lin, H., Bilmes, J.: Word alignment via submodular maximization over matroids. In: Proceedings of the 49th Annual Meeting of the Association for Computational Linguistics: Human Language Technologies: short papers, HLT 2011, vol. 2, pp. 170–175. Association for Computational Linguistics, Stroudsburg (2011)

The All-or-Nothing Flow Problem in Directed Graphs with Symmetric Demand Pairs

Chandra Chekuri[1,*] and Alina Ene[2,3,**]

[1] Dept. of Computer Science, University of Illinois at Urbana-Champaign
chekuri@illinois.edu
[2] Center of Computational Intractability, Princeton University
[3] Dept. of Computer Science and DIMAP, University of Warwick
aene@cs.princeton.edu

Abstract. We study the approximability of the All-or-Nothing multicommodity flow problem in directed graphs with symmetric demand pairs (SymANF). The input consists of a directed graph $G = (V, E)$ and a collection of (unordered) pairs of nodes $\mathcal{M} = \{s_1t_1, s_2t_2, \ldots, s_kt_k\}$. A subset $\mathcal{M}'$ of the pairs is *routable* if there is a feasible multicommodity flow in G such that, for each pair $s_it_i \in \mathcal{M}'$, the amount of flow from s_i to t_i is at least one *and* the amount of flow from t_i to s_i is at least one. The goal is to find a maximum cardinality subset of the given pairs that can be routed. Our main result is a poly-logarithmic approximation with constant congestion for SymANF. We obtain this result by extending the well-linked decomposition framework of [6] to the directed graph setting with symmetric demand pairs. We point out the importance of studying routing problems in this setting and the relevance of our result to future work.

1 Introduction

We consider some fundamental maximum throughput routing problems in *directed* graphs. In this setting, we are given a capacitated directed graph $G = (V, E)$ with n nodes and m edges. We are also given source-destination pairs of nodes $(s_1, t_1), (s_2, t_2), \ldots, (s_k, t_k)$. The goal is to select a largest subset of the pairs that are simultaneously *routable* subject to the capacities; a set of pairs is routable if there is a multicommodity flow for the pairs satisfying certain constraints that vary from problem to problem (e.g., integrality, unsplittability, edge or node capacities). Two well-studied optimization problems in this context are the Maximum Edge Disjoint Paths (MEDP) and the All-or-Nothing Flow (ANF) problem. In MEDP, a set of pairs is routable if the pairs can be connected using edge-disjoint paths. In ANF, a set of pairs is routable if there is a feasible multicommodity flow that fractionally routes one unit of flow from s_i to t_i for each

* Supported in part by NSF grants CCF-1016684 and CCF-1319376. Part of this work done while the author was on sabbatical at TTI Chicago in Fall 2012.
** Part of this work was done while the author was an intern at TTI Chicago in Fall 2012. Supported in part by NSF grants CCF-1016684 and CCF-0844872.

J. Lee and J. Vygen (Eds.): IPCO 2014, LNCS 8494, pp. 222–233, 2014.
© Springer International Publishing Switzerland 2014

routed pair (s_i, t_i). ANF, introduced in [9,5], can be seen as a relaxed version of MEDP where the flow for the routed pairs is not required to be integral.

MEDP and ANF are both **NP**-hard and their approximability has attracted substantial attention. Over the last decade, several non-trivial results on both upper bounds and lower bounds have led to a much better understanding of these problems. At a high level, one can summarize this progress as follows. MEDP and ANF admit poly-logarithmic approximation in *undirected* graphs if one allows constant congestion[1]; in fact, a congestion of 2 is sufficient for MEDP [11] and for ANF no extra congestion is needed [5]. Moreover, both problems are hard to to approximate to within a factor of $\Omega(\log^{\frac{1-\varepsilon}{c+1}} n)$ for any constant congestion $c \geq 1$ [1]; the hardness is under the assumption that **NP** $\not\subseteq$ **ZPTIME**$(n^{\text{polylog}(n)})$. In sharp contrast, in *directed* graphs both problems are hard to approximate to within a *polynomial* factor for any constant congestion $c \geq 1$; the hardness factor is $n^{\Omega(1/c)}$ [10]. The upper bounds and lower bounds on the approximability are closely related to corresponding integrality gap bounds on a multicommodity flow relaxation for these problems.

In this paper, we initiate the study of maximum throughput routing problems in directed graphs in the setting where the demand pairs are *symmetric*. Informally, in a symmetric demand pair instance, the input pairs are *unordered* and a pair $s_i t_i$ is routed only if both the ordered pairs (s_i, t_i) and (t_i, s_i) are routed. In particular, we focus our attention on the SymANF problem. The input consists of a directed graph $G = (V, E)$ and a collection of (unordered) pairs of nodes $\mathcal{M} = \{s_1 t_1, s_2 t_2, \ldots, s_k t_k\}$. A subset $\mathcal{M}'$ of the pairs is *routable* if there is a feasible multicommodity flow in G such that, for each pair $s_i t_i \in \mathcal{M}'$, the amount of flow from s_i to t_i is one unit *and* the amount of flow from t_i to s_i is one unit[2]. We will assume without loss of generality that G has only node-capacities; this allows us to relate to, and use ideas from, (directed) treewidth. The goal is to find a maximum cardinality subset of the given pairs that can be routed. Our main result is the following theorem that gives a poly-logarithmic approximation with constant congestion for SymANF.

Theorem 1. *There is a polynomial time algorithm that, given any instance of the* SymANF *problem in directed graphs, it routes* $\Omega(\text{OPT}/\log^2 k)$ *pairs with constant node congestion, where* OPT *is the value of an optimal fractional solution for the instance.*

---

[1] A routing has congestion c if it violates the capacities by a factor of at most c.

[2] There are alternative ways to define routability that captures symmetry. One option is to require a flow of $1/2$ unit in each direction which is compatible with a total of one unit of flow entering and leaving each terminal. Another option is to require that for any orientation of the demand pairs, there is a feasible multicommodity for the pairs with one unit for each pair in the direction given by the orientation. For simplicity we require one unit of flow in each direction which results in a factor of 2 loss in the congestion when compared to other models.

The congestion that we guarantee is 64. We believe that the congestion can be improved, but we have not attempted to optimize the constant. Our algorithm uses a natural LP relaxation for the problem as a starting point and we also show a poly-logarithmic upper bound on the integrality gap of the relaxation. Some simple and natural extensions such as handling capacitated graphs and pairs with demand values can be handled via known techniques and we do not address them in this version.

We observe that, via existing results on the hardness of ANF in undirected graphs with congestion [1], one can conclude that SymANF with congestion c is hard to approximate to within a factor of $\log^{\Omega(1/c)} n$ for any fixed c unless $\mathbf{NP} \subseteq \mathbf{ZPTIME}\left(n^{\mathrm{polylog}\, n}\right)$.

A strong motivation for this work, in addition to understanding the complexity of routing problems in directed graphs, is the connection between structural graph theory and routing problems. Recent results on routing have led to progress in graph theory via the connection to treewidth [3,4]. Routing in directed graphs with symmetric demands is, in a similar vein, related to the notion of *directed treewidth* [12,13]. We defer a detailed discussion of this relationship, other connections, and related work to a longer version of this paper.

1.1   Overview of the Algorithm and Technical Contributions

Let $(G, \mathcal{M})$ be an instance of SymANF. Let $\mathcal{T}$ be the set of all nodes that participate in the pairs of $\mathcal{M}$; we refer to the nodes in $\mathcal{T}$ as the *terminals*. Our algorithm for SymANF in directed graphs follows the framework of Chekuri, Khanna, and Shepherd [5,6] for the ANF problem in undirected graphs. In a nutshell, the framework decomposes an arbitrary instance of ANF into several instances that are *flow-well-linked*. The set of terminals $\mathcal{T} = \{s_1, t_1, \ldots, s_k, t_k\}$ is *flow-well-linked* if *any* matching on the terminals is routable. This is essentially equivalent (modulo a factor of 2 in congestion) to saying that the G admits a symmetric product multicommodity flow where the weight on each terminal is 1 and is 0 on every non-terminal. If the terminals are flow-well-linked, we can route all the input pairs. Thus the heart of the matter is showing that an arbitrary instance can be decomposed into well-linked instances without losing too much flow.

The decomposition has two main components. The first step is a weaker decomposition in which we take a fractional solution to a natural multicommodity flow based LP (described in Section 2.1) and use it to decompose the instance into instances that are only *fractionally* flow-well-linked. More precisely, there is a weight function $\pi : \mathcal{T} \rightarrow [0, 1]$ and the terminals are flow-well-linked with respect to these weights; if all terminals have weight 1 then they are flow-well-linked. The second step is a clustering step in which we take a fractionally flow-well-linked instance and we identify a large subset of the pairs such that their endpoints are flow-well-linked. In this paper, we show how to implement these two steps for the SymANF problem in directed graphs. In the first step, we extend the approach of [6] to our setting; we defer this extension to a longer version of this paper. We note that the approximation factor that we lose in the decomposition is proportional to the flow-cut gap; for symmetric instances, the flow-cut gap is only polylog(k).

The second step poses several technical difficulties in directed graphs and it is our main technical contribution. We briefly highlight some of the difficulties involved in the clustering step, and we refer the reader to Section 4 for an outline of our approach. Chekuri, Khanna, and Shepherd [6] gave a simple clustering technique for *edge-capacitated* undirected graphs. Roughly speaking, the approach is to take a spanning tree and to partition it into edge-disjoint subtrees where each subtree gathers roughly a unit weight from π. These subtrees are then used to find the desired flow-well-linked subset of pairs/terminals; one terminal is picked from each subtree. The clustering step is more involved in *node-capacitated* undirected graphs. The spanning tree approach, combined with some preprocessing to reduce the degree, gives a clustering for node-capacitated graphs with slightly weaker parameters [6]. In [7], the authors gave a stronger clustering for the node-capacitated setting; this approach is more involved than the spanning tree clustering and it exploits a connection between well-linked sets and treewidth; recent work [4] obtains a stronger result but requires more involved ideas. In directed graphs, there is no simple clustering process akin to using a spanning tree (or even an arborescence). Instead, our approach exploits the connection between well-linked sets and directed treewidth. However, the main challenge is to make this algorithmic. We also mention that, in addition to finding a large flow-well-linked set Y from a fractionally flow-well-linked set X, we also need to ensure that Y contains a large enough matching from the original set of pairs. For this purpose, we rely on a flow augmentation tool developed in [8]. These difficulties are also the reason why we are only able to obtain a constant congestion for SymANF while ANF admits a poly-logarithmic ratio with congestion 1 in edge-capacitated graphs [5] and with congestion $(1+\varepsilon)$ in node-capacitated graphs [6].

Organization: Section 2 introduces the main definitions and technical tools that we use, and it describes the approximation algorithm for SymANF. Section 3 and Section 4 outline the well-linked decomposition and clustering technique for directed graphs with symmetric demand pairs that underlie the algorithm.

2 Approximation Algorithm for SymANF

2.1 Preliminaries and Setup

In the following, we work with an instance $(G, \mathcal{M})$ of the SymANF problem, where $G = (V, E)$ is a directed graph and $\mathcal{M} = \{s_1 t_1, \dots, s_k t_k\}$ is a collection of node pairs. We refer to the nodes participating in the pairs of $\mathcal{M}$ as terminals, and we use $\mathcal{T}$ to denote the set of all terminals. We assume that the pairs $\mathcal{M}$ form a perfect matching on $\mathcal{T}$ and each terminal is a leaf in G, i.e., each terminal is connected to a single neighbor using an edge in each direction. One can reduce an arbitrary instance to an instance that satisfies these assumptions as follows. If a node v participates in several pairs, we make a copy of v for each of the pairs it participates in, and attach the copy v' to v using an edge in each direction; finally we replace v by v' in the pair. Similarly, if a terminal is not a leaf, we make a copy of the terminal, we attach the copy to the original node as a leaf, and we replace the terminal by its copy in the pairs that contain it. Note that, if a set of pairs was routable in the original instance, then it is routable in the new instance with congestion at most 2.

LP Relaxation: We consider a natural multicommodity flow relaxation for the SymANF problem. For each ordered pair (u, v) of nodes of G, let $\mathcal{P}(u, v)$ be the set of all paths in G from u to v. Since $\mathcal{M}$ forms a matching on $\mathcal{T}$, for all $i \neq j$, the sets $\mathcal{P}(s_i, t_i)$, $\mathcal{P}(t_i, s_i)$, $\mathcal{P}(s_j, t_j)$, and $\mathcal{P}(t_j, s_j)$ are pairwise disjoint. Let $\mathcal{P} = \bigcup_{i=1}^{k} (\mathcal{P}(s_i, t_i) \cup \mathcal{P}(t_i, s_i))$. For each path $p \in \mathcal{P}$, we have a variable $f(p)$ that is equal to the amount of flow on p. For each unordered pair $s_i t_i \in \mathcal{M}$ we have a variable x_i to indicate whether to route the pair or not. The LP relaxation ensures the symmetry constraint: there is a flow from s_i to t_i of value x_i *and* a flow from t_i to s_i of value x_i. Recall that we will be working with the node-capacitated problem and each node has unit capacity.

$$
\begin{array}{lll}
& \text{(symANF-LP)} & \\
\max & \displaystyle\sum_{i=1}^{k} x_i & \\
\text{s.t.} & \displaystyle\sum_{p \in \mathcal{P}(s_i, t_i)} f(p) \geq x_i & 1 \leq i \leq k \\
& \displaystyle\sum_{p \in \mathcal{P}(t_i, s_i)} f(p) \geq x_i & 1 \leq i \leq k \\
& \displaystyle\sum_{p:\, v \in p} f(p) \leq 1 & v \in V(G) \\
& x_i \leq 1 & 1 \leq i \leq k \\
& f(p) \geq 0 & p \in \mathcal{P}
\end{array}
$$

The dual of the symANF-LP relaxation has polynomially many variables and exponentially many constraints. The separation oracle for the dual is the shortest path problem. Thus we can solve the relaxation in polynomial time. Alternatively, we can write an equivalent LP relaxation that is polynomial sized.

Multicommodity Flows and Node Separators: Let $G = (V, E, \mathrm{cap})$ be a directed node-capacitated graph with node capacities given by cap. A *multicommodity flow* instance in G is a demand vector $\mathbf{d}$ that assigns a demand value $d(u, v) \in \mathbb{R}_+$ to each ordered pair (u, v) of nodes of G. The instance is *symmetric* if $d(u, v) = d(v, u)$ for all pairs (u, v). The instance is a *product* multicommodity flow instance if $d(u, v) = w(u)w(v)$, where $w : V \to \mathbb{R}_+$ is a weight function on the nodes of G. Note that a product multicommodity flow instance is symmetric. In the following, we only consider symmetric multicommodity flow instances. The instance is *routable* if there is a feasible flow that routes $d(u, v)$ units of flow from u to v for each pair (u, v). The *maximum concurrent flow* for **d** is the maximum value $\lambda \geq 0$ such that $\lambda\mathbf{d}$ is routable. The demand separated by a set $C \subseteq V$ of nodes, denoted by $\mathrm{dem}_\mathbf{d}(C)$, is the total demand of all of the unordered pairs uv such that u and v are *not* in the same strongly connected component of $G - C$. A *sparsest node separator* is a set C for which the ratio $\mathrm{cap}(C)/\mathrm{dem}_\mathbf{d}(C)$ is minimized. The *flow-cut gap* in G is the maximum value, over all symmetric instances **d** in G, of the ratio between the minimum sparsity of a node separator and the maximum concurrent flow. A *node separation* in G is a partition (A, B, C) of the nodes of G such that there is no edge of G from A to B (note that there can be an edge of G from B to A).

Well-Linked Sets: There are two notions of well-linkedness that have been used for routing problems in undirected graphs [6]; one is based on a flow requirement and the other is based on a cut requirement. In the following, we define directed versions of these two notions and we show some basic properties of these notions.

Flow-well-linked sets: Let G be a directed graph with unit capacities on the nodes. We define a fractional version of flow-well-linkedness as follows. Let $\pi : X \rightarrow [0, 1]$ be a weight function on X. Let **d** be the following demand vector: $d(u, v) = \pi(u)\pi(v)/\pi(X)$ for each ordered pair (u, v) of nodes in X. The set X is π-**flow-well-linked** in G iff **d** is routable in G. For a scalar $c \in [0, 1]$ we say that X is c-flow-well-linked if X is π-flow-well-linked, where $\pi(v) = c$ for each vertex $v \in X$.

Cut-well-linked sets: A set $X \subseteq V$ is **cut-well-linked** in G iff, for any two disjoint subsets Y and Z of X of equal size, there are $|Y|$ node-disjoint paths from Y to Z in G. Recall that a node is a leaf in G if it is connected to a single neighbor using an edge in each direction. If the nodes of X are leaves in G, an equivalent definition is the following. The set X is cut-well-linked iff, for any node separation (A, B, C) satisfying $X \cap C = \emptyset$, we have $|C| \geq \min\{|X \cap A|, |X \cap B|\}$. We define a fractional version of cut-well-linkedness as follows. Let X be a set of nodes of G and let $\pi : X \rightarrow [0, 1]$ be a weight function on X. Suppose that all the nodes in X are leaves of G. The set X is π-**cut-well-linked** in G if, for any node separation (A, B, C), we have $|C| \geq \min\{\pi(A), \pi(B)\}$. Note that, since the nodes in X are leaves, it suffices to check this condition for separations (A, B, C) for which $\pi(C) = 0$. Now consider a set X that contains nodes that are not leaves. For each node $x \in X$, we add a new node x' and connect x' to x using two edges, one in each direction. Let X' be the set of new nodes, let G' be the resulting graph, and let $\pi' : X' \rightarrow [0, 1]$ be the weight function $\pi'(x') = \pi(x)$ for each node $x \in X$. The set X is π-cut-well-linked in G iff X' is π'-cut-well-linked in G'.

Well-Linked Decomposition: The following theorem is an extension to directed graphs of the well-linked decomposition technique introduced by [6] for routing problems in undirected graphs. The proof follows the outline of the approach in [6] and it is deferred to a longer version of this paper. The decomposition algorithm is given in Section 3.

Theorem 2. *Let* OPT *be the value of a solution to the* symANF-LP *relaxation for a given instance* $(G, \mathcal{M})$ *of* SymANF. *Let* $\alpha = \alpha(G) \geq 1$ *be an upper bound on the worst case flow-cut gap for product multicommodity flows in* G. *There is a partition of* G *into node-disjoint induced subgraphs* $G_1, G_2, \ldots, G_\ell$ *and weight functions* $\pi_i : V(G_i) \to \mathbb{R}_+$ *with the following properties. Let* $\mathcal{M}_i$ *be the induced pairs of* $\mathcal{M}$ *in* G_i *and let* X_i *be the endpoints of the pairs in* $\mathcal{M}_i$. *We have*

(a) $\pi_i(u) = \pi_i(v)$ *for each pair* $uv \in \mathcal{M}_i$.
(b) X_i *is* π_i-*flow-well-linked in* G_i.
(c) $\sum_{i=1}^{\ell} \pi_i(X_i) = \Omega(\text{OPT}/(\alpha \log \text{OPT})) = \Omega(\text{OPT}/\log^2 k)$.

Moreover, such a partition is computable in polynomial time if there is a polynomial time algorithm for computing a node separator with sparsity at most $\alpha(G)$ *times the maximum concurrent flow.*

From Fractional Well-Linked Sets to Well-Linked Sets: We give an outline of the proof of the following theorem in Section 4.

Theorem 3. *Let* X *be a* π-*flow-well-linked set in* G *and let* $\mathcal{M}$ *be a perfect matching on* X *such that* $\pi(u) = \pi(v)$ *for each pair* $uv \in \mathcal{M}$. *There is a matching* $\mathcal{M}' \subseteq \mathcal{M}$ *on a set* $X' \subseteq X$ *such that* X' *is* $1/32$-*flow-well-linked in* G *and* $|\mathcal{M}'| = 2|X'| = \Omega(\pi(X))$. *Moreover, given* X *and* $\mathcal{M}$, *we can construct* X' *and* $\mathcal{M}'$ *in polynomial time.*

Routing a Flow-Well-Linked Instance: Finally, we observe that, if an instance of SymANF is c-flow-well-linked for some $c \leq 1$, then we can route all of the pairs with congestion at most $2/c$.

Proposition 1. *Let* $(G, \mathcal{M})$ *be an instance of* SymANF *and let* X *be the set of all vertices that participate in the pairs of* $\mathcal{M}$. *If* X *is* c-*flow-well-linked for some* $c \leq 1$, *then we can route all of the pairs of* $\mathcal{M}$ *with congestion at most* $2/c$.

2.2   The Approximation Algorithm for SymANF

In this section, we describe our algorithm for SymANF. Let $(G, \mathcal{M})$ be an instance of SymANF. The algorithm is the following.

(1) Solve the relaxation symANF-LP to get an optimal fractional solution (x, f) for the instance $(G, \mathcal{M})$.
(2) Use the well-linked decomposition (Theorem 2) to get a collection $(G_1, \mathcal{M}_1, \pi_1)$, $\ldots, (G_\ell, \mathcal{M}_\ell, \pi_\ell)$ of disjoint instances and weight functions.
(3) For each instance $(G_i, \mathcal{M}_i, \pi_i)$ in the decomposition, use the clustering technique (Theorem 3) to get an instance $(G_i, \mathcal{M}'_i)$.
(4) For each instance $(G_i, \mathcal{M}'_i)$, route all of the pairs of $\mathcal{M}'_i$ in G_i (Proposition 1). Output the union of these routings.

The number of pairs routed by the algorithm is $\sum_{i=1}^{\ell} |\mathcal{M}'_i| = \sum_{i=1}^{\ell} \Omega(\pi(V(\mathcal{M}_i))) = \Omega(\text{OPT}/\log^2 k)$. Since each instance $(G_i, \mathcal{M}'_i)$ is $1/32$-flow-well-linked, the routing in G_i has congestion at most 64. Since the instances are node disjoint, the congestion of the final routing is at most 64. This completes the proof of Theorem 1.

Decomposition Algorithm

Input: Strongly connected subgraph H.
Output: Node-disjoint subgraphs $H_1, H_2, \ldots, H_\ell$ with associated weight functions $\pi_1, \pi_2, \ldots, \pi_\ell$, where each H_i is a node-induced subgraph of H.

(1) Suppose that $0 < \mathbf{w}(H) \leq \alpha \log \mathrm{OPT}$. Let $\pi(u) = w(u; H)/(8\alpha \log \mathrm{OPT})$ for each node $u \in V(H)$. Stop and output H and π.

(2) Suppose that $\mathbf{w}(H) > \alpha \log \mathrm{OPT}$. Let $\mathbf{d}$ be the following demand vector: $d(u, v) = w(u; H)w(v; H)/\mathbf{w}(H)$ for each ordered pair (u, v) of nodes in H. Let λ be the maximum concurrent flow for $\mathbf{d}$.

    (a) If $\lambda \geq 1/(8\alpha \log \mathrm{OPT})$, stop the recursive procedure. Let $\pi(u) = w(u; H)/(8\alpha \log \mathrm{OPT})$ for each node $u \in V(H)$. Output H and π.

    (b) Otherwise find a node separation (A, B, C) such that $|C| \leq \min\left\{\sum_{a \in A} w(a; H), \sum_{b \in B} w(b; H)\right\} / (4 \log \mathrm{OPT})$. Recursively decompose each strongly connected component of $H - C$. Output the decompositions of the strongly connected components.

Fig. 1. Well-linked decomposition algorithm

3 Well-Linked Decomposition

In this section, we give the decomposition algorithm guaranteed by Theorem 2. We follow the notation and the approach introduced in [6] for edge and node-capacitated multicommodity flow problems in undirected graphs.

Let (x, f) be a solution to the symANF-LP with value $\mathrm{OPT} = \sum_{i=1}^{k} x_i$. The flow f is a symmetric multicommodity flow; as before, we view f as a path-based flow. Let H be a node-induced subgraph of G. For each ordered pair (u, v) of nodes in H, let $\gamma(u, v; H)$ be the total amount of f-flow on paths p from u to v that are completely contained in H. For each unordered pair uv of nodes in H, let $\gamma'(u, v; H) = \gamma'(v, u; H) = \min\{\gamma(u, v; H), \gamma(v, u; H)\}$. For each node u in H, let $w(u; H) = \sum_{v \in V(H)} \gamma'(u, v; H)$. Let $\mathbf{w}(H) = \sum_{u \in V(H)} w(u; H)$.

The algorithm is given in Figure 1. We apply the algorithm to each strongly connected component of G in order to get a decomposition of G into node-induced disjoint subgraphs $G_1, G_2, \ldots, G_\ell$ with associated weight functions π_1, $\pi_2, \ldots, \pi_\ell$. The resulting decomposition satisfies the conditions of Theorem 2, and we defer the proof to a longer version of this paper.

4 From Fractional Well-Linked Sets to Well-Linked Sets

In this section, we prove Theorem 3. We prove the theorem in two steps. In the first step, we show that there exists a set Y of cardinality $\Omega(\pi(X))$ such that Y is $\Omega(1)$-flow-well-linked. Additionally, the set Y can send flow to X and receive flow from X. In the second step, we use Y to select a matching $\mathcal{M}' \subseteq \mathcal{M}$ of size $\Omega(|Y|)$. Before we give details of this procedure we first give an intuitive (and non-constructive) argument that motivates the approach. This is partly inspired by the work in [7] and differs from the low-degree spanning tree based clustering that has been the main approach in the undirected case. The reader can skip the following paragraph and go straight to the technical proof.

Intuitive Argument: Suppose G has a set X that is π-well-linked. One can show that the directed treewidth[3] of G is within a constant factor of the largest well-linked set in G. One can adapt a similar argument to show that the directed treewidth of G is $\Omega(\pi(X))$ where X is π-cut-well-linked. Applying approximate duality again implies that there is a well-linked set Z in G such that $|Z| = \Omega(\pi(X))$. Since Z's existence was shown, essentially, via the π-cut-well-linkedness of X, it is intuitive that there is such a Z that is reachable from X in the following sense. There is a flow from X to Z where each node in Z receives one unit of flow and each node v in X sends $\pi(v)$ units of flow. Similarly there is a flow from Z to X. From this property and the fact that X is π-flow-linked, one can argue that Z is $\Omega(1)$-flow-well-linked. We then have to identify a subset $X' \subset X$ that is flow-well-linked. Moreover, for the SymANF problem we need to ensure that for the initial matching M on X there is a sufficiently large sub-matching of M induced on X'. These latter arguments require an incremental flow-augmentation technique from [8]. The main technical challenge is to *efficiently* find a Z reachable from X as described above. Surprisingly, we are able to show that a simply greedy iterative approach based on the intuition of the existence argument, with a careful argument, works to give the desired set Z modulo constant congestion. We believe that this is a useful technical building block for further work in this area. Now we give an outline of the formal argument.

First Step: Finding a Large Well-Linked Set. In the first step, we find a set Y with the following properties:

Theorem 4. *Let G be a directed graph. Let X be a set of nodes of G and let $\pi : X \to (0,1]$ be a weight function on X. Suppose that X is π-flow-well-linked in G. There is a polynomial time algorithm that constructs a set $Y \subseteq V(G)$ with the following properties. We have*

(P_1) $|Y| = \lfloor \pi(X)/8 \rfloor$.
(P_2) Y *is $1/4$-flow-well-linked in G.*

Additionally, for any subset $X' \subseteq X$ such that $\pi(X') \leq \pi(X)/15$, we have

(Q_1) *There is a single commodity flow in G from X' to Y such that each node $x \in X'$ sends $\pi(x)/64$ units of flow and each node in Y receives at most one unit of flow.*
(Q_2) *There is a single commodity flow in G from Y to X' such that each node $x \in X'$ receives $\pi(x)/64$ units of flow and each node in Y sends at most one unit of flow.*

---

[3] We refer the reader to [12] for the definition of directed treewidth.

The main ingredient in the proof of Theorem 4 is the following lemma. The lemma shows that, if we have a set X that is π-flow-well-linked, then there exists a set Y of size $\Omega(\pi(X))$ such that Y is $\Omega(1)$-flow-well-linked. The main idea behind the lemma is the following. If X is π-cut-well-linked and Z is a node separator of size less than $\pi(X)/4$, there is a unique strongly connected component $\beta(Z)$ of $G - Z$ whose π-weight is more than half the weight of X. The main insight is that, if we consider the set Y of size $\lfloor \pi(X)/4 \rfloor$ for which $|Y \cup \beta(Y)|$ is minimum, this gives us the desired set. This gives us a non-constructive proof of the existence of such a set Y, and we show that a simple iterative procedure will construct such a set in polynomial time.

Lemma 1. *Let G be a directed graph. Let X be a set of nodes of G and let $\pi : X \to (0, 1]$ be a weight function on X. Suppose that X is π-cut-well-linked in G. There is a polynomial time algorithm that constructs a set $Y \subseteq V(G)$ with the following properties. We have*

(R_1) $|Y| = \lfloor \pi(X)/4 \rfloor$.

(R_2) *There is a single commodity flow in G from X to Y such that each node $x \in X$ sends at most $\pi(x)$ units of flow and each node in Y receives one unit of flow.*

(R_3) *There is a single commodity flow in G from Y to X such that each node in Y sends one unit of flow and each node $x \in X$ receives at most $\pi(x)$ units of flow.*

Now we are ready to sketch the proof of Theorem 4.

Proof of Theorem 4: Since X is π-flow-well-linked in G, the set X is $(\pi/2)$-cut-well-linked in G. By Lemma 1, there is a set Y with the following properties.

- $|Y| = \lfloor \pi(X)/8 \rfloor$.
- There is a single commodity flow f_1 in G from X to Y such that each node $x \in X$ sends at most $\pi(x)/2$ units of flow and each node in Y receives one unit of flow.
- There is a single commodity flow f_2 in G from Y to X such that each node in Y sends one unit of flow and each node $x \in X$ receives at most $\pi(x)/2$ units of flow.

One can verify that the properties above imply that Y is $1/4$-flow-well-linked, and thus Y satisfies the conditions (P_1) and (P_2) in the theorem statement.

Consider a set $X' \subseteq X$ such that $\pi(X') \leq \pi(X)/15$. We can verify that X' and Y satisfy conditions (Q_1) and (Q_2) as follows. Let $X_1 \subseteq X$ be the set of all nodes $x \in X$ such that x sends at least $\pi(x)/32$ units of flow in f_1. Let $X_2 \subseteq X$ be the set of all nodes $x \in X$ such that x receives at least $\pi(x)/32$ units of flow in f_2. It is straightforward to show that $\pi(X_1) \geq \pi(X)/15$ and $\pi(X_2) \geq \pi(X)/15$. Therefore $\pi(X') \leq \pi(X_1)$ and $\pi(X') \leq \pi(X_2)$. Consider the following multicommodity flow instance $\mathbf{d}$: $d(x', x) = \pi(x')\pi(x)/(32\pi(X_1))$ for each pair $(x', x) \in X' \times X_1$, $d(x, x') = \pi(x)\pi(x')/(32\pi(X_2))$ for each pair $(x, x') \in X_2 \times X'$, and $d(\cdot)$ is zero for all other pairs. Since $d(a, b) \leq \pi(a)\pi(b)/\pi(X)$ for all pairs of nodes (a, b), there is a feasible flow g that routes $\mathbf{d}$. The flow g satisfies the following properties:

- Each node $x \in X_1$ receives $\pi(x)\pi(X')/(32\pi(X_1)) \leq \pi(x)/32$ units of flow.
- Each node $x \in X_2$ sends $\pi(x)\pi(X')/(32\pi(X_2)) \leq \pi(x)/32$ units of flow.
- Each node $x' \in X'$ sends and receives $\pi(x')/32$ units of flow.

By combining the flows f_1 and g, we get a congestion two flow from X' to Y in which each node in Y receives at most one unit of flow and each node $x' \in X'$ sends $\pi(x')/32$ units of flow. Similarly, by combining the flows f_2 and g, we get a congestion two flow from Y to X' in which each node in Y sends at most one unit of flow and each node $x' \in X'$ receives $\pi(x')/32$ units of flow. We scale down these flows by a factor of two to get feasible flows.   □

Second Step: Finding a Matching. Let Y be the set guaranteed by Theorem 4. Using Y, we select a matching $\mathcal{M}' \subseteq \mathcal{M}$ as follows. Using a flow augmentation technique from [8], we can identify a large matching whose terminals can send one unit of flow to Y and receive one unit of flow from Y. More precisely, we have the following lemma.

Lemma 2. There is a matching $\mathcal{M}' \subseteq \mathcal{M}$ with the following properties. Let X_1' be a set of nodes containing exactly one node from each pair in $\mathcal{M}'$, and let $X_2' = V(\mathcal{M}') - X_1'$ be the partners of the nodes in X_1'. We have

(C_1) $|\mathcal{M}'| = \Omega(|Y|)$.
(C_2) There is a feasible single-commodity flow in G in which each node in X_1' sends one unit of flow to Y.
(C_3) There is a feasible single-commodity flow in G in which each node in X_1' receives one unit of flow from Y.
(C_4) There is a feasible single-commodity flow in G in which each node in X_2' sends one unit of flow to Y.
(C_5) There is a feasible single-commodity flow in G in which each node in X_2' receives one unit of flow from Y.

Let $\mathcal{M}'$ be the set of pairs guaranteed by Lemma 2 and let X' be the set of terminals participating in the pairs of $\mathcal{M}'$. In order to complete the proof of Theorem 3, it suffices to verify that X' is 1/32-flow-well-linked. Note that the properties $(C_2) - (C_5)$ gives us the following flows: a congestion two flow from X' to Y in which each node in X' sends one unit of flow and each node in Y receives at most two units of flow, and a congestion two flow from Y to X' in which each node in Y sends at most two units of flow and each node in X' receives one unit of flow. We scale these flows by a factor of 8 to ensure that each node in Y sends and receives at most 1/4 units of flow. Using these flows and the fact that Y is 1/4-flow-well-linked, one can verify that X' is 1/32-flow-well-linked.

References

1. Andrews, M., Chuzhoy, J., Guruswami, V., Khanna, S., Talwar, K., Zhang, L.: Inapproximability of edge-disjoint paths and low congestion routing on undirected graphs. Combinatorica 30(5), 485–520 (2010)

2. Andrews, M., Chuzhoy, J., Khanna, S., Zhang, L.: Hardness of the undirected edge-disjoint paths problem with congestion. In: Proc. of IEEE FOCS, pp. 226–241 (2005)
3. Chekuri, C., Chuzhoy, J.: Large-treewidth graph decompositions and applications. In: Proc. of ACM STOC (2013)
4. Chekuri, C., Chuzhoy, J.: Polynomial bounds for the grid-minor theorem. In: Proc. of ACM STOC (2014)
5. Chekuri, C., Khanna, S., Shepherd, F.B.: The all-or-nothing multicommodity flow problem. SIAM Journal on Computing 42(4), 1467–1493 (2013)
6. Chekuri, C., Khanna, S., Shepherd, F.B.: Multicommodity flow, well-linked terminals, and routing problems. In: Proc. of ACM STOC, pp. 183–192 (2005)
7. Chekuri, C., Khanna, S., Shepherd, F.B.: Well-linked terminals for node-capacitated routing problems (2005) (Manuscript)
8. Chekuri, C., Khanna, S., Shepherd, F.B.: An $O(\sqrt{n})$ approximation and integrality gap for disjoint paths and unsplittable flow. Theory of Computing 2(7), 137–146 (2006)
9. Chekuri, C., Mydlarz, M., Shepherd, F.B.: Multicommodity demand flow in a tree and packing integer programs. ACM Transactions on Algorithms 3(3), 27 (2007)
10. Chuzhoy, J., Guruswami, V., Khanna, S., Talwar, K.: Hardness of routing with congestion in directed graphs. In: Proc. of ACM STOC, pp. 165–178 (2007)
11. Chuzhoy, J., Li, S.: A polylogarithimic approximation algorithm for edge-disjoint paths with congestion 2. In: Proc. of IEEE FOCS (2012)
12. Johnson, T., Robertson, N., Seymour, P.D., Thomas, R.: Directed tree-width. Journal of Combinatorial Theory, Series B 82(1), 138–154 (2001)
13. Reed, B.: Introducing directed tree width. Electronic Notes in Discrete Mathematics 3, 222–229 (1999)

Reverse Split Rank*

Michele Conforti[1], Alberto Del Pia[2], Marco Di Summa[1], and Yuri Faenza[3]

[1] Dipartimento di Matematica, Università degli Studi di Padova, Italy
[2] Business Analytics and Mathematical Sciences Department,
IBM Watson Research Center, Yorktown Heights, NY, USA
[3] DISOPT, Institut de mathématiques d'analyse et applications, EPFL, Switzerland

Abstract. The *reverse split rank* of an integral polytope P is defined as the supremum of the split ranks of all rational polyhedra whose integer hull is P. Already in $\mathbb{R}^3$ there exist polytopes with infinite reverse split rank. We give a geometric characterization of the integral polytopes in $\mathbb{R}^n$ with infinite reverse split rank.

1   Introduction

The problem of finding or approximating the integer hull of a rational polyhedron is crucial in Integer Programming (see, e.g., [16,21]). In this paper we consider one of the most well-known procedures used for this purpose: the split inequalities.

Given an integral polyhedron $P \subseteq \mathbb{R}^n$, a *relaxation* of P is a rational polyhedron $Q \subseteq \mathbb{R}^n$ such that $P \cap \mathbb{Z}^n = Q \cap \mathbb{Z}^n$, i.e., $\mathrm{conv}(Q \cap \mathbb{Z}^n) = P$ (where "conv" denotes the convex hull operator). A *split* $S \subseteq \mathbb{R}^n$ is a set of the form $S = \{x \in \mathbb{R}^n : \beta \le ax \le \beta+1\}$ for some primitive vector $a \in \mathbb{Z}^n$ (i.e., an integer vector whose entries have greatest common divisor equal to 1) and some integer number β. Note that a split does not contain any integer point in its interior $\mathrm{int}\, S$. Therefore, if Q is a rational polyhedron and S is a split, then the set $\mathrm{conv}(Q \setminus \mathrm{int}\, S)$ contains the same integer points as Q. The *split closure* $SC(Q)$ of Q is defined as

$$SC(Q) = \bigcap_{S \text{ split}} \mathrm{conv}(Q \setminus \mathrm{int}\, S).$$

A shown in [9], if Q is a rational polyhedron, its split closure $SC(Q)$ is a rational polyhedron containing the same integer points as Q. For $k \in \mathbb{N}$, the *k-th split closure* of Q is $SC^k(Q) = SC(SC^{k-1}(Q))$, with $SC^0(Q) = Q$. If Q is a rational polyhedron, then there is an integer k such that $SC^k(Q) = \mathrm{conv}(Q \cap \mathbb{Z}^n)$ (see [9]); the minimum k for which this happens is called the *split rank* of Q, and we denote it by $s(Q)$.

* This work was supported by the *Progetto di Eccellenza 2008–2009* of *Fondazione Cassa di Risparmio di Padova e Rovigo.* Yuri Faenza's research was supported by the German Research Foundation (DFG) within the Priority Programme 1307 Algorithm Engineering.

J. Lee and J. Vygen (Eds.): IPCO 2014, LNCS 8494, pp. 234–248, 2014.
© Springer International Publishing Switzerland 2014

While the split rank of all rational polytopes in $\mathbb{R}^2$ is bounded by a constant, there is no bound for the split rank of all rational polytopes in $\mathbb{R}^3$. Furthermore, even if the set of integer points in Q is fixed, there might be no constant bounding the split rank of Q. For instance, let $P \subseteq \mathbb{R}^3$ be the convex hull of the points $(0,0,0)$, $(2,0,0)$ and $(0,2,0)$. For every $t \geq 0$, the polyhedron $Q_t = \text{conv}(P,(1/2,1/2,t))$ is a relaxation of P. As shown in [7], $s(Q_t) \to +\infty$ as $t \to +\infty$.

In this paper we aim at understanding which polytopes admit relaxations with arbitrarily high split rank. For this purpose, given an integral polytope P, we define the *reverse split rank* of P, denoted $s^*(P)$, as the supremum of the split ranks of all relaxations of P:

$$s^*(P) = \sup\{s(Q) : Q \text{ is a relaxation of } P\}.$$

Note that the polytope P given in the above example satisfies $s^*(P) = +\infty$.

Our main result is now stated. Given a subset $K \subseteq \mathbb{R}^n$, we denote by $\text{int } K$ its interior and by $\text{relint } K$ its relative interior. We say that K is (relatively) lattice-free if there are no integer points in its (relative) interior. (See, e.g., [18].)

Theorem 1. *Let $P \subseteq \mathbb{R}^n$ be an integral polytope. Then $s^*(P) = +\infty$ if and only if there exist a nonempty face F of P and a nonzero rational linear subspace $L \subseteq \mathbb{R}^n$ such that*

(i) $P + L$ *is relatively lattice-free,*
(ii) $F + L$ *is relatively lattice-free,*
(iii) $\text{relint}(F + L)$ *is not contained in the interior of any split.*

Note that for the polytope P given in the example above, conditions (i)–(iii) are satisfied by taking $F = P$ and L equal to the line generated by the vector $(0,0,1)$.

The analogous concept of *reverse Chvátal–Gomory (CG) rank* of an integral polyhedron P was introduced in [6]. We recall that an inequality $cx \leq \lfloor \delta \rfloor$ is a *CG inequality* for a polyhedron $Q \subseteq \mathbb{R}^n$ if c is an integer vector and $cx \leq \delta$ is valid for Q. Alternatively, a CG inequality is a split inequality in which the split $S = \{x \in \mathbb{R}^n : \beta \leq ax \leq \beta + 1\}$ is such that one of the half-spaces $\{x \in \mathbb{R}^n : ax \leq \beta\}$ and $\{x \in \mathbb{R}^n : ax \geq \beta + 1\}$ does not intersect Q. The *CG closure*, the *CG rank* $r(Q)$, and the *reverse CG rank* $r^*(Q)$ of Q are defined as for the split inequalities. (The CG rank of a rational polyhedron was proved to be finite in [20].) In [6] the following characterization was proved (here we state the result for polytopes).

Theorem 2 ([6]). *Let $P \subseteq \mathbb{R}^n$ be a nonempty integral polytope. Then $r^*(P) = +\infty$ if and only if there exists a one-dimensional rational linear subspace L such that $P + L$ is relatively lattice-free.*

Since every CG inequality is a split inequality, we have $s(Q) \leq r(Q)$ for every rational polyhedron Q, and $s^*(P) \leq r^*(P)$ for every integral polyhedron P. This explains why the conditions of Theorem 1 are a strengthening of those of

Theorem 2. Indeed, there are examples of integral polytopes with finite reverse split rank but infinite reverse CG rank: e.g., the polytope defined as the convex hull of points $(0,0)$ and $(0,1)$ in $\mathbb{R}^2$ (see [6]).

The comparison between Theorem 1 and Theorem 2 suggests that there is some "gap" between the CG rank and the split rank. This is not surprising, as the literature already offers results in this direction. For instance, if we consider a rational polytope contained in the cube $[0,1]^n$, it is known that its split rank is at most n [2], while its CG rank can be as high as n^2 (see [19]; weaker results were previously given in [12,17]). Some more details about the differences between the statements of Theorem 1 and Theorem 2 will be given at the end of the paper.

We remark that, despite the similarity between the statements of Theorem 1 and Theorem 2, the proof of the former result (which we give here) needs more sophisticated tools and is more involved.

The rest of the paper is organized as follows. In Sect. 2 we recall some known facts, and also present a result which, beside being used in the proof of Theorem 1, seems to be of its own interest (Lemma 6). The sufficiency of conditions (i)–(iii) of Theorem 1 is proved in Sect. 3, while the necessity of the conditions is shown in Sect. 4. We conclude with some observations in Sect. 5.

2   Basic Facts

In this section we introduce some notation and present some basic facts that will be used in the proof of Theorem 1. We refer the reader to a textbook, e.g. [21], for standard preliminaries that do not appear here.

Given a point $x \in \mathbb{R}^n$ and a number $r > 0$, we denote by $B(x,r)$ the closed ball of radius r centered at x. We write aff P to indicate the affine hull of a polyhedron $P \subseteq \mathbb{R}^n$ and lin P to denote its lineality space. The angle between two vectors $v, w \in \mathbb{R}^n$ is denoted by $\phi(v,w)$. Given linear subspaces $L_1, \ldots, L_k$ of $\mathbb{R}^n$, we indicate with $\langle L_1, \ldots, L_k \rangle$ the linear subspace of $\mathbb{R}^n$ generated by the union of $L_1, \ldots, L_k$. (With little abuse, if L is a subspace of $\mathbb{R}^n$ and $v \in \mathbb{R}^n$, we write $\langle L, v \rangle$ instead of $\langle L, \langle v \rangle \rangle$.) Finally, $L^\perp$ is the orthogonal complement of a linear subspace $L \subseteq \mathbb{R}^n$.

2.1   Unimodular Transformations

A *unimodular transformation* $u : \mathbb{R}^n \to \mathbb{R}^n$ maps a point $x \in \mathbb{R}^n$ to $u(x) = Ux + v$, where U is an $n \times n$ unimodular matrix (i.e., a square integer matrix with $|\det(U)| = 1$) and $v \in \mathbb{Z}^n$. It is well-known (see e.g. [21]) that U is a unimodular matrix if and only if so is U^{-1}. Furthermore, a unimodular transformation is a bijection of both $\mathbb{R}^n$ and $\mathbb{Z}^n$. It follows that if $Q \subseteq \mathbb{R}^n$ is a rational polyhedron and $u : \mathbb{R}^n \to \mathbb{R}^n$ is a unimodular transformation, then the split rank of Q coincides with the split rank of $u(Q)$.

The following basic fact will prove useful: if $L \subseteq \mathbb{R}^n$ is a rational linear subspace of dimension d, then there exists a unimodular transformation that maps L to the hyperplane $\{x \in \mathbb{R}^n : x_{d+1} = \cdots = x_n = 0\}$; in other words, L is equivalent to $\mathbb{R}^d$ up to a unimodular transformation.

2.2   Some Properties of Chvátal and Split Rank

We will use the following result (see [1, Lemma 10]) and its easy corollary.

Lemma 3. *For every $n \in \mathbb{N}$ there exists a number $\theta(n)$ such that the following holds: for every rational polyhedron $Q \subseteq \mathbb{R}^n$ and every $c \in \mathbb{Z}^n$ and $\delta, \delta' \in \mathbb{R}$ with $\delta' \geq \delta$ such that $cx \leq \delta$ is valid for $\mathrm{conv}(Q \cap \mathbb{Z}^n)$ and $cx \leq \delta'$ is valid for Q, the inequality $cx \leq \delta$ is valid for the p-th CG closure of Q, where $p = (\lfloor \delta' \rfloor - \lfloor \delta \rfloor)\theta(n) + 1$.*

Corollary 4. *Given an integral polytope $P \subseteq \mathbb{R}^n$ and a bounded set B containing P, there exists an integer N such that $r(Q) \leq N$ for all relaxations Q of P contained in B.*

We also need the following lemma.

Lemma 5. *Let $Q \subseteq \mathbb{R}^n$ be a rational polyhedron contained in a split S, where $S = \{x \in \mathbb{R}^n : \beta \leq ax \leq \beta + 1\}$. Let Q^0 (resp., Q^1) be the face of Q induced by the inequality $ax \geq \beta$ (resp., $ax \leq \beta + 1$). Then $s(Q) \leq \max\{s(Q^0), s(Q^1)\} + 1$.*

Proof. For $i = 0, 1$, since Q^i is a face of Q, we have $SC(Q^i) = SC(Q) \cap Q^i$ (see [9]). Then, for $k = \max\{s(Q^0), s(Q^1)\}$, both $SC^k(Q) \cap Q^0$ and $SC^k(Q) \cap Q^1$ are integral polyhedra. It follows that after another application of the split closure (actually, the split S is sufficient) we obtain an integral polyhedron.   □

2.3   Compactness

The proof of Theorem 1 exploits the notion of compactness and sequential compactness, which we recall here. A subset K of a topological space is *compact* if every collection of open sets covering K contains a finite subcollection which still covers K. It is well-known that a subset of $\mathbb{R}^n$ is compact (with respect to the usual topology of $\mathbb{R}^n$) if and only if it is closed and bounded. For a normed space (such as $\mathbb{R}^n$) the notion of compactness coincides with that of sequential compactness: a set K is *sequentially compact* if every sequence $(x_i)_{i \in \mathbb{N}}$ of elements of K admits a subsequence that converges to an element of K.

2.4   On Integer Points around Linear Subspaces

A result given in [4], based on Dirichlet's lemma (see, e.g., [21]), shows that for each line passing through the origin there are integer points arbitrarily close to the line and arbitrarily far from the origin. We give here a strengthening of that result, showing that for every line passing through the origin the integer points that are "very close" to the line are not too far from each other. Furthermore, this result is presented in a more general version, valid for every linear subspace. This result will be used in the proof of Theorem 1, but we find it interesting in its own right. The proof can be found in the journal version of the paper.

Lemma 6. *Let $L \subseteq \mathbb{R}^n$ be a linear subspace and fix $\delta > 0$. Then there exists $R > 0$ such that, for every $x \in L$, there is an integer point y satisfying $\|y - x\| \leq R$ and $\mathrm{dist}(y, L) \leq \delta$.*

3 Proof of Sufficiency

In this section we prove that if F and L satisfying conditions (i)–(iii) of Theorem 1 exist, then P has infinite reverse split rank.

By hypothesis, F and P are nonempty. Since L is a rational subspace, it admits a basis $v^1, \ldots, v^k \in \mathbb{Z}^n$. Fix $\bar{x} \in \operatorname{relint} F$, and for $\lambda \geq 0$ define

$$Q_F^\lambda = \operatorname{conv}(F, \bar{x} \pm \lambda v^1, \ldots, \bar{x} \pm \lambda v^k),$$

$$Q_P^\lambda = \operatorname{conv}(P, \bar{x} \pm \lambda v^1, \ldots, \bar{x} \pm \lambda v^k).$$

Clearly $Q_F^\lambda \subseteq Q_P^\lambda$ for every $\lambda \geq 0$. As $\bar{x} \in \operatorname{relint} F$ and $F + L$ is relatively lattice-free, it follows that Q_F^λ is a relaxation of F for every $\lambda \geq 0$. It can be checked that, since also $P + L$ is relatively lattice-free, Q_P^λ is a relaxation of P for every $\lambda \geq 0$.

Let $r > 0$ be the radius of the largest ball in $\operatorname{aff} F$ centered at $\bar{x}$ and contained in F, and let $R > 0$ be the length of the longest segment passing through $\bar{x}$ and contained in F. We will show below that, for each $\lambda \geq 0$, $SC(Q_F^\lambda)$ contains the two points $\bar{x} \pm \min\{(\lambda - 1), \frac{r}{2R}\lambda\}v^i$ for every $i = 1, \ldots, k$. As $Q_F^\lambda \subseteq Q_P^\lambda$, we have $SC(Q_F^\lambda) \subseteq SC(Q_P^\lambda)$. As λ was chosen arbitrarily, this implies that $s(Q_P^\lambda) \to +\infty$ as $\lambda \to +\infty$, hence $s^*(P) = +\infty$.

It remains to prove that $SC(Q_F^\lambda)$ contains the two points $\bar{x} \pm \min\{(\lambda - 1), \frac{r}{2R}\lambda\}v^i$ for every $i = 1, \ldots, k$. To do so, we prove that for every split S, the set $\operatorname{conv}(Q_F^\lambda \setminus \operatorname{int} S)$ contains the two points $\bar{x} \pm (\lambda - 1)v^i$ or the two points $\bar{x} \pm \frac{r}{2R}\lambda v^i$, for every $i = 1, \ldots, k$. To simplify notation, for fixed S and λ we define $T = \operatorname{conv}(Q_F^\lambda \setminus \operatorname{int} S)$, omitting the dependence on S and λ.

Case 1. Let S be a split such that there exists a vector $\bar{v} \in \{v^1, \ldots, v^k\}$ *not in* $\operatorname{lin} S$. In this case we show that T contains the point $\bar{x} + (\lambda - 1)v^i$ for every $i = 1, \ldots, k$. Symmetrically, T will also contain the point $\bar{x} - (\lambda - 1)v^i$ for every $i = 1, \ldots, k$.

Let $i \in \{1, \ldots, k\}$ be such that $v^i \notin \operatorname{lin} S$. As $v^i \in \mathbb{Z}^n$, it is easy to check that $\operatorname{int} S$ can contain at most one of the points $\bar{x} + \lambda v^i$ and $\bar{x} + (\lambda - 1)v^i$. Thus T contains the point $\bar{x} + \lambda v^i$ or the point $\bar{x} + (\lambda - 1)v^i$. As $\bar{x} \in F$, it follows that T must contain $\bar{x} + (\lambda - 1)v^i$.

Now let $j \in \{1, \ldots, k\}$ be such that $v^j \in \operatorname{lin} S$. If $\bar{x} + (\lambda - 1)v^j \notin \operatorname{int} S$ we are done, thus assume that $\bar{x} + (\lambda - 1)v^j \in \operatorname{int} S$. Since the three points $\bar{x} + \lambda v^j$, $\bar{x} \pm \lambda \bar{v}$ are in Q_F^λ, also are their convex combinations $\bar{x} + (\lambda - 1)v^j \pm \bar{v}$. As $\bar{v} \in \mathbb{Z}^n$, $\bar{v} \notin \operatorname{lin} S$, and $\bar{x} + (\lambda - 1)v^j \in \operatorname{int} S$, it is easy to check that both points $\bar{x} + (\lambda - 1)v^j \pm \bar{v}$ are not in $\operatorname{int} S$, and therefore are in T. Therefore also their convex combination $\bar{x} + (\lambda - 1)v^j$ is in T.

Case 2. Let S be a split such that $v^i \in \operatorname{lin} S$ for every $i = 1, \ldots, k$. In this case we show that T contains the point $\bar{x} + \frac{r}{2R}\lambda v^i$ for every $i = 1, \ldots, k$. Symmetrically, T will also contain the point $\bar{x} - \frac{r}{2R}\lambda v^i$ for every $i = 1, \ldots, k$.

Let $\tilde{v} \in \{v^1, \ldots, v^k\}$. If $\bar{x} + \lambda \tilde{v} \notin \operatorname{int} S$, then the statement follows trivially, as $\bar{x} \in F$. Thus we now assume that $\bar{x} + \lambda \tilde{v} \in \operatorname{int} S$, which implies that also $\bar{x} \in \operatorname{int} S$.

Since, by (iii), relint$(F+L)$ is not contained in int S and $F \cap$ int $S \neq \emptyset$, $F+L$ is not contained in S. As $v^i \in \lin S$ for every $i = 1, \ldots, k$, this implies that F is not contained in S. Therefore wlog $ax \geq \beta$ is not valid for F, where $a \in \mathbb{Z}^n$, $\beta \in \mathbb{Z}$ are such that $S = \{x \in \mathbb{R}^n : \beta \leq ax \leq \beta \mid 1\}$. Since F is integral, there exists an integer point w of F that satisfies $ax < \beta$, thus $aw \leq \beta - 1$. Let w' be the point defined as the intersection of the boundary of S with the segment with endpoints w and $\bar{x}$. (See Fig. 1.)

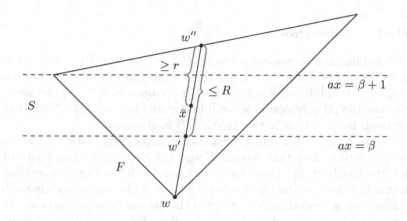

Fig. 1. Illustration of Case 2

We show that Q_F^λ contains the point $w' + \frac{\lambda}{2}\tilde{v}$. Since $aw \leq \beta - 1$, we have $aw' = \beta$. As $\beta < a\bar{x} < \beta + 1$, it follows that the convex hull of points w and $\bar{x} + \lambda\tilde{v}$ contains a point $w' + \lambda'\tilde{v}$, with $\lambda' \geq \frac{\lambda}{2}$. Thus the point $w' + \frac{\lambda}{2}\tilde{v}$ is in Q_F^λ.

Note that $w' + \frac{\lambda}{2}\tilde{v} \in T$. We finally show that T contains the point $\bar{x} + \frac{r}{2R}\lambda\tilde{v}$. Let w'' be the point different from w that belongs to the intersection of the boundary of F with the line passing through w and $\bar{x}$. The segment $[\bar{x}, w'']$ is contained in F, thus the distance between $\bar{x}$ and w'' is at least r. The segment $[w', w'']$ is contained in F and contains $\bar{x}$, thus the distance between w' and w'' is at most R. Therefore the convex hull of points w'' and $w' + \frac{\lambda}{2}\tilde{v}$ contains a point $\bar{x} + \lambda''\tilde{v}$, with $\lambda'' \geq \frac{r}{2R}\lambda$. Thus T contains the point $\bar{x} + \frac{r}{2R}\lambda\tilde{v}$.

4   Proof of Necessity

In this section we prove that if an integral polytope P has infinite reverse split rank, then F and L satisfying conditions (i)–(iii) of Theorem 1 exist. We remark that if $P = \emptyset$ then its reverse split rank is finite, as this is the case even for the reverse CG rank (see [8,6]). Therefore in this section we assume that $P \neq \emptyset$.

In order to prove the necessity of conditions (i)–(iii), we need to extend the notion of relaxation and reverse split rank to *rational* polyhedra. Indeed, when dealing with a non-full-dimensional integral polytope P in Sect. 4.7, we will approximate P with a non-integral full-dimensional polytope containing the same integer points as P.

Given a rational polyhedron $P \subseteq \mathbb{R}^n$, we call relaxation of P a rational polyhedron $Q \subseteq \mathbb{R}^n$ such that $P \subseteq Q$ and $P \cap \mathbb{Z}^n = Q \cap \mathbb{Z}^n$. The reverse split rank of a rational polyhedron P is defined as follows:

$$s^*(P) = \sup\{s(Q) : Q \text{ is a relaxation of } P\}.$$

In the following we prove that if a nonempty rational polyhedron has infinite reverse split rank, then F and L satisfying conditions (i)–(iii) of Theorem 1 exist.

4.1   Outline of the Proof

Given a full-dimensional rational polytope $P \subseteq \mathbb{R}^n$ with $s^*(P) = +\infty$, we prove conditions (i)–(iii) of Theorem 1 under the assumption that the result holds for all (possibly non-full-dimensional) rational polytopes in $\mathbb{R}^{n-1}$. (The case of a non-full-dimensional polytope in $\mathbb{R}^n$ will be treated in Sect. 4.7.) Note that the theorem holds for $n = 1$, as in this case $s^*(P)$ is always finite.

So let $P \subseteq \mathbb{R}^n$ be a full-dimensional rational polytope with $s^*(P) = +\infty$. We now give a procedure that returns F and L satisfying the conditions of the theorem. We justify it and prove its correctness in the rest of this section. We remark that at this stage the linear subspace returned by the procedure might be non-rational, but we will show at the end of this section how to choose a rational subspace. Also, we point out that the procedure below is not an "executable algorithm", but only a theoretical proof of the existence of F and L as required.

1. Fix a point $\bar{x} \in \operatorname{int} P$; choose a sequence $(Q_i)_{i \in \mathbb{N}}$ of relaxations of P such that $\sup_i s(Q_i) = +\infty$; initialize $k = 1$, $L_0 = \{0\}$, and $S = P$.
2. Choose a sequence of points $(x_i)_{i \in \mathbb{N}}$ such that $x_i \in Q_i$ for every $i \in \mathbb{N}$ and $\sup_i \operatorname{dist}(x_i, S) = +\infty$; let w_i be the projection of $x_i - \bar{x}$ onto $L_{k-1}^\perp$; define $\bar{v}$ as the limit of some subsequence of the sequence $\left(\frac{w_i}{\|w_i\|}\right)_{i \in \mathbb{N}}$ (and assume wlog that this subsequence coincides with the original sequence); define $L_k = \langle L_{k-1}, \bar{v} \rangle$.
3. If $P + L_k$ is not contained in any split, then return $F = P$ and $L = L_k$, and stop; otherwise, let S be a split such that $P + L_k \subseteq S$, where $S = \{x \in \mathbb{R}^n : \beta \le ax \le \beta + 1\}$.
4. If there exists $M \in \mathbb{R}$ such that $Q_i \subseteq \{x \in \mathbb{R}^n : \beta - M \le ax \le \beta + M\}$ for every $i \in \mathbb{N}$, then choose $j \in \{0, 1\}$ such that $P^j := P \cap \{x \in \mathbb{R}^n : ax = \beta + j\}$ has infinite reverse split rank (when viewed as a polytope in the affine space $H = \{x \in \mathbb{R}^n : ax = \beta + j\}$); since H is a rational subspace and we assumed that the result holds in dimension $n - 1$, there exist F and L satisfying conditions (i)–(iii) of the theorem with respect to the space H; return F and L, and stop. Otherwise, if no M as above exists, set $k \leftarrow k + 1$ and go to 2.

In order to prove the correctness of the above procedure, we will show the following:

(a) in step 2, a sequence $(x_i)_{i \in \mathbb{N}}$ and a vector $\bar{v}$ as required can be found;
(b) the procedure terminates (either in step 3 or step 4);
(c) if the procedure terminates in step 4, then there exists $j \in \{0, 1\}$ such that P^j has infinite reverse split rank, and the output is correct;
(d) if the procedure terminates in step 3, then the output is correct.

4.2   Proof of (a)

We prove that at every execution of step 2 a sequence $(x_i)_{i \in \mathbb{N}}$ and a vector $\bar{v}$ as required can be found.

Consider first the iteration $k = 1$; in this case, $S = P$. Since $\sup_i s(Q_i) = +\infty$, we also have $\sup_i r(Q_i) = +\infty$. By Corollary 4, there is no bounded set containing every Q_i for $i \in \mathbb{N}$. Then there is a sequence of points $(x_i)_{i \in \mathbb{N}}$ such that $x_i \in Q_i$ for every $i \in \mathbb{N}$ and $\sup_i \text{dist}(x_i, P) = +\infty$. Define $v_i = \frac{x_i - \bar{x}}{\|x_i - \bar{x}\|}$ for $i \in \mathbb{N}$. Since every v_i belongs to the unit sphere, which is a compact set, the sequence $(v_i)_{i \in \mathbb{N}}$ has a subsequence converging to some unit-norm vector $\bar{v}$. Wlog, we assume that this subsequence coincides with the original sequence.

Assume now that we are at the k-th iteration ($k \geq 2$). Then the algorithm has determined a split $S \subseteq \mathbb{R}^n$ such that $P + L_{k-1} \subseteq S = \{x \in \mathbb{R}^n : \beta \leq ax \leq \beta + 1\}$. Furthermore, we know that there is no $M \in \mathbb{R}$ such that $Q_i \subseteq \{x \in \mathbb{R}^n : \beta - M \leq ax \leq \beta + M\}$ for every $i \in \mathbb{N}$. This implies that there is a sequence of points $(x_i)_{i \in \mathbb{N}}$ such that $x_i \in Q_i$ for $i \in \mathbb{N}$ and $\sup_i \text{dist}(x_i, S) = +\infty$. For $i \in \mathbb{N}$, let w_i be the projection of the vector $x_i - \bar{x}$ onto the space $L_{k-1}^{\perp}$. Note that $\sup_i \|w_i\| = +\infty$. Since the elements of the sequence $(w_i)_{i \in \mathbb{N}}$ belong to the intersection of $L_{k-1}^{\perp}$ with the unit sphere, and this intersection gives a compact set, there is a subsequence converging to some unit-norm vector belonging to $L_{k-1}^{\perp}$, which we call $\bar{v}$. We assume wlog that this subsequence coincides with the original sequence.

4.3   Proof of (b)

In order to show that the procedure terminates after a finite number of iterations, it is sufficient to observe that at every iteration in step 2 we select a vector $\bar{v} \in L_{k-1}^{\perp}$, thus the dimension of $L_k = \langle L_{k-1}, \bar{v} \rangle$ is k. In particular, the procedure terminates after at most n iterations, as for $k = n$ no split S can be found in step 3.

4.4   Proof of (c)

We now prove that if the procedure terminates in step 4, then there exists $j \in \{0, 1\}$ such that P^j has infinite reverse split rank (when viewed as a polytope in the affine space $\{x \in \mathbb{R}^n : ax = \beta + j\}$), and the output is correct.

Since $Q_i \subseteq \{x \in \mathbb{R}^n : \beta - M \leq ax \leq \beta + M\}$ for every $i \in \mathbb{N}$, by Lemma 3 there exists a number N such that, for each $i \in \mathbb{N}$, N iterations of the CG closure operator (hence, also of the split closure operator) applied to Q_i are sufficient to obtain a relaxation of P contained in S. For $i \in \mathbb{N}$, let $\tilde{Q}_i$ be the relaxation of P obtained this way. Then we have $\sup_i s(\tilde{Q}_i) = +\infty$.

Recall that P^0 and P^1 are the faces of P induced by equations $ax = \beta$ and $ax = \beta + 1$, respectively. Similarly, for $i \in \mathbb{N}$, let $\tilde{Q}_i^0$ and $\tilde{Q}_i^1$ be the faces of $\tilde{Q}_i$ induced by equations $ax = \beta$ and $ax = \beta + 1$, respectively. Since $\tilde{Q}_i \subseteq S$, by Lemma 5 we have $s(\tilde{Q}_i) \leq \max\{s(\tilde{Q}_i^0), s(\tilde{Q}_i^1)\} + 1$. Then there exists $j \in \{0,1\}$ such that $\sup_i s(\tilde{Q}_i^j) = +\infty$. Since every relaxation $\tilde{Q}_i^j$ is contained in the affine space $H = \{x \in \mathbb{R}^n : ax = \beta + j\}$, we have $s^*(P^j) = +\infty$ with respect to the ambient space H (which is equivalent to $\mathbb{R}^{n-1}$ under some unimodular transformation). Let H^* be the translation of H passing through the origin. Since H is a rational space of dimension $n - 1$, by induction there exist a nonempty face F of P^j and a nonzero linear subspace $L \subseteq H^*$ satisfying conditions (i)–(iii) for P^j: specifically, $P^j + L$ is relatively lattice-free, $F + L$ is relatively lattice-free, and $\mathrm{relint}(F + L)$ is not contained in the interior of any $(n-1)$–dimensional split in the affine space H.

We show that F and L satisfy conditions (i)–(iii) for P, too. First, note that F is a nonempty face of P and L is a nonzero linear subspace of $\mathbb{R}^n$. To show (i), observe that $L \subseteq H^* = \mathrm{lin}\, S$; since $P \subseteq S$, then $P + L \subseteq S$; thus $P + L$ is lattice-free. Condition (ii) is clearly satisfied. To prove (iii), assume by contradiction that there is an n-dimensional split T such that $\mathrm{relint}(F + L) \subseteq \mathrm{int}\, T$. Then $\mathrm{lin}\, T \neq \mathrm{lin}\, S$. This implies that $T \cap H$ is contained in some $(n - 1)$–dimensional split U living in H. But then, with respect to the ambient space H, we would have $\mathrm{relint}(F + L) \subseteq \mathrm{int}\, U$, a contradiction.

4.5   Proof of (d)

We now prove that if the procedure terminates in step 3, then the output is correct. Note that it is sufficient to prove that $P + L_k$ is lattice-free at every iteration of the algorithm.

The subspace L_1 is constructed following the same procedure as in the proof of Theorem 2 given in [6, Section 3.2]. Therefore, with the same arguments as in [6], one proves that $P + L_1$ is lattice-free.

We now assume $k \in \{2, \ldots, n\}$; see Figs. 2 and 3 to follow the proof. Recall that $L_k = \langle L_{k-1}, \bar{v} \rangle$ and $\bar{v} \in L_{k-1}^\perp$. Suppose by contradiction that there is an integer point $\bar{z} \in \mathrm{int}(P + L_k) = \mathrm{int}\, P + L_k$. Then there exists a vector $u \in L_{k-1}$ such that $z_0 := \bar{z} + u \in \mathrm{int}\, P + \langle \bar{v} \rangle$. Let $x_0 \in \mathrm{int}\, P$ be such that $x_0 = z_0 - d\bar{v}$, where wlog we can assume $d > 0$. Note that since $\|\bar{v}\| = 1$, $d = \|z_0 - x_0\|$. Let $r > 0$ be such that $B(x_0, r) \subseteq P$. Furthermore, denote by $\pi : \mathbb{R}^n \to L_{k-1}^\perp$ the orthogonal projection onto $L_{k-1}^\perp$.

Recall that there is a split S such that $P + L_{k-1} \subseteq S$ (step 3 of the previous iteration). Define $H = \mathrm{lin}\, S$ and $H_0 = z_0 + H$. We can assume wlog that H_0 does not intersect P: if this is not the case, we can choose a different integer point $\bar{z} \in \mathrm{int}(P + L_k)$ so that this condition is satisfied. Let w denote the unit-norm vector which is orthogonal to H and forms an acute angle with $\bar{v}$ (recall that $\bar{v} \notin H$, thus $\bar{v}$ and $\bar{w}$ cannot be orthogonal). We define $\alpha = \phi(\bar{v}, \pi(w))$; note that $0 \leq \alpha < \pi/2$.

We need the following two claims, whose proof we defer to the journal version of the paper.

Claim 7. *For every $\delta > 0$ there exists $M > 0$ such that the following holds: for every $x \in L_{k-1}$ and for every $y_1, \ldots, y_{k-1}$ satisfying $y_t \in x + L_t$ and $\mathrm{dist}(y_t, x + L_{t-1}) \geq M$ for $t = 1, \ldots, k-1$, one has $\mathrm{dist}(\mathrm{conv}(x, y_1, \ldots, y_{k-1}), H \cap \mathbb{Z}^n) \leq \delta$.*

Claim 8. *For every $x \in \mathrm{int}\, P$, $M' > 0$ and $\varepsilon > 0$, there exist an index $i \in \mathbb{N}$ and points $y_1, \ldots, y_k$ satisfying $y_t \in Q_i \cap (x_0 + L_t)$, $\mathrm{dist}(y_t, x + L_{t-1}) \geq M'$ for $t = 1, \ldots, k$, and $\phi(\pi(y_k - x_0), \bar{v}) \leq \varepsilon$.*

By Claim 7 with $\delta = r/8$, we obtain $M > 0$ such that the condition of the claim is satisfied. Define $M' = \max\{2M, 2d\}$. By Claim 8, there exists $i \in \mathbb{N}$ and points $y_1, \ldots, y_k$ satisfying $y_t \in Q_i \cap (x_0 + L_t)$ and $\mathrm{dist}(y_t, x_0 + L_{t-1}) \geq M'$ for $t = 1, \ldots, k$. Furthermore, again because of Claim 8, we can enforce the condition

$$\beta := \phi(\pi(y_k - x_0), \bar{v}) \leq \arctan\left(\tan\alpha + \frac{r}{8d}\right) - \alpha \tag{1}$$

(see Fig. 3). Note that the value on the right-hand-side of (1) is nonnegative, as $0 \leq \alpha < \pi/2$.

For $\rho > 0$, define $B'(\rho) = B(0, \rho) \cap L_{k-1}^{\perp} \cap H$. For $t = 1, \ldots, k-1$, let $\tilde{y}_t$ be the midpoint of the segment $[x_0, y_t]$. Note that

$$Q_i \supseteq \mathrm{conv}\left(B(x_0, r) \cup \{y_1, \ldots, y_{k-1}\}\right)$$
$$\supseteq C := \mathrm{conv}\left(x_0 + B'(r/2), \tilde{y}_1 + B'(r/2), \ldots, \tilde{y}_{k-1} + B'(r/2)\right).$$

Let x' be the unique point in $[x_0, y_k] \cap H_0$, and, for $i = 1, \ldots, k-1$, let y_t' be the unique point in $[\tilde{y}_t, y_k] \cap H_0$. Since $\mathrm{dist}(y_k, x_0 + L_{k-1}) \geq M' \geq 2d \geq 2\,\mathrm{dist}(x_0 + L_{k-1}, x' + L_{k-1})$,

$$\mathrm{conv}(C, y_k) \cap H_0 \supseteq C' := \mathrm{conv}(x' + B'(r/4), y_1' + B'(r/4), \ldots, y_k' + B'(r/4)). \tag{2}$$

Moreover, as $B(x_0, r) \subseteq P \subseteq Q_i$ and $B(\tilde{y}_t, r/2) \subseteq Q_i$ for $t = 1, \ldots, k-1$, we have

$$B(x', r/2) \subseteq Q_i \quad \text{and} \quad B(y_t', r/4) \subseteq Q_i \quad \text{for } t = 1, \ldots, k-1. \tag{3}$$

Let x'' be the projection of x' onto the space $z_0 + L_{k-1}$. We claim that

$$\|x'' - x'\| = \|\pi(x'' - x')\| \leq d\tan(\alpha + \beta) - d\tan\alpha \leq r/8$$

(see again Fig. 3). The equality holds because $x'' - x' \in L_{k-1}^{\perp}$ by construction; the first inequality describes the worst case (which is the one depicted in the figure), i.e., when $\|\pi(x'' - x')\|$ is as large as possible; the last bound follows from (1).

Now define $y_1'', \ldots, y_{k-1}''$ as the orthogonal projections of $y_1', \ldots, y_{k-1}'$ onto $z_0 + L_{k-1} = \bar{z} + L_{k-1}$. Note that $y_1'', \ldots, y_{k-1}''$ are obtained by translating $y_1', \ldots, y_{k-1}'$ by vector $x'' - x'$. By the definition of C' given in (2), $y_1'', \ldots, y_{k-1}''$ still belong to C'. One verifies that $y_t'' \in x'' + L_t$ and $\mathrm{dist}(y_t'', x'' + L_{t-1}) \geq M'/2 \geq M$ for $t = 1, \ldots, k-1$. Since $\bar{z} + L_{k-1}$ is a translation of L_{k-1} by an integer vector, by the choice of M given by Claim 7 there is an integer point $p \in \bar{z} + H = H_0$ at distance at most $\delta = r/8$ from the set $\mathrm{conv}(x'', y_1'', \ldots, y_{k-1}'')$.

We claim that $p \in Q_i$. To see this, first observe that

$$\mathrm{dist}(p, \mathrm{conv}(x', y_1', \ldots, y_{k-1}')) \leq \mathrm{dist}(p, \mathrm{conv}(x'', y_1'', \ldots, y_{k-1}'')) + \|x'' - x'\|$$
$$\leq \frac{r}{8} + \frac{r}{8} = \frac{r}{4}. \tag{4}$$

Now from (3) we obtain that $\mathrm{conv}(x', y_1', \ldots, y_{k-1}') + B(0, r/4) \subseteq Q_i$ and thus, by (4), $p \in Q_i$. This is a contradiction, as p is an integer point in $Q_i \setminus P$ (p does not belong to P because $p \in H_0$ and $H_0 \cap P = \varnothing$ by assumption).

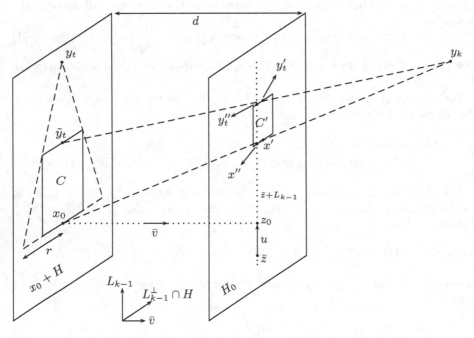

Fig. 2. Illustration of the proof of (d)

4.6   Rationality of L

As mentioned above, our procedure might return a non-rational linear subspace L. Note that this cannot be the case if the procedure terminates in step 4, as in this case the rationality of L follows from the fact that we assumed that the theorem holds in $\mathbb{R}^{n-1}$. Therefore we now assume that the procedure terminates in step 3, and show that we can replace L with a suitable nonzero rational linear subspace $\tilde{L}$ and still have conditions (i)–(iii) fulfilled.

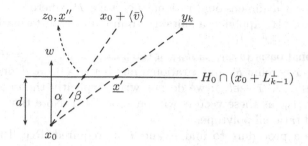

Fig. 3. Illustration of the proof of (d). The space $x_0 + L_{k-1}^\perp$ is represented. Underlined symbols indicate points that do not necessarily belong to $x_0 + L_{k-1}^\perp$; in other words, their orthogonal projection onto $x_0 + L_{k-1}^\perp$ is represented.

We use a result in [4, Theorem 2] (see also [15]), which states that a maximal lattice-free convex set in $\mathbb{R}^n$ is either an irrational hyperplane or a polyhedron $\tilde{P} + \tilde{L}$, where $\tilde{P}$ is a polytope and $\tilde{L}$ is a rational linear subspace. Since $P + L$ is lattice-free, it is contained in a maximal lattice-free convex set. Since it is full-dimensional, it is not contained in an irrational hyperplane. Then $P \subseteq \tilde{P} + \tilde{L}$, with $\tilde{P}, \tilde{L}$ as above. Moreover, $\tilde{L} \neq \{0\}$, as it contains L. Since we are assuming that the procedure terminates in step 3 (thus $F = P$), conditions (i)–(iii) are satisfied if L is replaced with $\tilde{L}$.

4.7 The Non-full-Dimensional Case

The proof of the necessity of Theorem 1 given above covers the case of a full-dimensional rational polytope $P \subseteq \mathbb{R}^n$, assuming the result true both for full-dimensional and non-full-dimensional rational polytopes in $\mathbb{R}^{n-1}$. We now deal with the case of a non-full-dimensional polytope in $\mathbb{R}^n$. For this purpose, we will take a non-full-dimensional polytope P and make it full-dimensional by "growing" it along directions orthogonal to its affine hull. This will be done in such a way that no integer point is added to P. The idea is then to use the proof of the full-dimensional case given above. (We remark that even if we start from an integral polytope P, the new polytope that we construct will not be integral. This is why at the beginning of Sect. 4 we extended the notion of reverse split rank to rational polyhedra.)

Note that if P_I is the convex hull of integer points in a rational polytope P, it is not true (in general) that $s^*(P_I) = +\infty$ implies $s^*(P) = +\infty$. However, the key fact underlying our approach is the following:

Given a non-full-dimensional rational polytope P with $s^*(P) = +\infty$, it is possible to "enlarge" P and obtain a full-dimensional polytope P' containing the same integer points as P, in such a way that $s^*(P') = +\infty$.

Now, let P be a d–dimensional rational polytope P, where $d < n$. Assume that $s^*(P) = +\infty$. By applying a suitable unimodular transformation, we can assume that aff $P = \mathbb{R}^d \times \{0\}^{n-d}$.

Given a rational basis $\{b_{d+1}, \ldots, b_n\}$ of the subspace $\text{aff}(P)^\perp = \{0\}^d \times \mathbb{R}^{n-d}$, a rational point $\bar{x} \in \text{relint}\, P$, and a rational number $\varepsilon > 0$, we define $P(\bar{x}, \varepsilon) = \text{conv}(P, \bar{x} + \varepsilon b_{d+1}, \ldots, \bar{x} + \varepsilon b_n)$; we do not write explicitly the dependence on vectors $b_{d+1}, \ldots, b_n$, as these vectors will be soon fixed. Note that $P(\bar{x}, \varepsilon)$ is a full-dimensional rational polytope.

We now give a procedure to find F and L as required. Recall that we are assuming by induction that the theorem is true for both full-dimensional and non-full-dimensional rational polytopes in $\mathbb{R}^{n-1}$.

1. Fix a rational point $\bar{x} \in \text{relint}\, P$; choose a rational basis $\{b_{d+1}, \ldots, b_n\}$ of $\text{aff}(P)^\perp$, a rational number $\varepsilon > 0$, and a sequence of rational polyhedra $(Q_i)_{i \in \mathbb{N}}$ such that:
    (i) $P(\bar{x}, \varepsilon)$ has the same integer points as P,
    (ii) Q_i is a relaxation of $P(\bar{x}, \varepsilon)$ (and thus of P) for every $i \in \mathbb{N}$,
    (iii) $\sup_i s(Q_i) = +\infty$;
   initialize $k = 1$, $L_0 = \{0\}$, and $S = P(\bar{x}, \varepsilon)$.

2. Construct a sequence of points $(x_i)_{i \in \mathbb{N}}$ such that $x_i \in Q_i$ for every $i \in \mathbb{N}$ and $\sup_i \text{dist}(x_i, S) = +\infty$; let w_i be the projection of $x_i - \bar{x}$ onto $L_{k-1}^\perp$; define $\bar{v}$ as the limit of some subsequence of the sequence $\left(\frac{w_i}{\|w_i\|}\right)_{i \in \mathbb{N}}$ (and assume that this subsequence coincides with the original sequence); define $L_k = \langle L_{k-1}, \bar{v} \rangle$.

3. If, for every strictly positive rational number $\varepsilon' \le \varepsilon$, $P(\bar{x}, \varepsilon') + L_k$ is not contained in any split, then choose a rational subspace $L \supseteq L_k$ such that $P(\bar{x}, \varepsilon) + L$ is lattice-free, return $F = P$ and L, and stop; otherwise, let $S = \{x \in \mathbb{R}^n : \beta \le ax \le \beta + 1\}$ be a split such that $P(\bar{x}, \varepsilon') + L_k \subseteq S$ for some strictly positive rational number $\varepsilon' \le \varepsilon$, and update $\varepsilon \leftarrow \varepsilon'$.

4. If there exists $M \in \mathbb{R}$ such that $Q_i \subseteq \{x \in \mathbb{R}^n : \beta - M \le ax \le \beta + M\}$ for every $i \in \mathbb{N}$, then choose $j \in \{0, 1\}$ such that $P^j := P \cap \{x \in \mathbb{R}^n : ax = \beta + j\}$ has infinite reverse split rank (when viewed as a polytope in the affine space $\{x \in \mathbb{R}^n : ax = \beta + j\}$), then F and L exist by induction; return F and L, and stop. Otherwise, set $k \leftarrow k + 1$, and go to 2.

The following facts, which we prove in the journal version of the paper, imply the correctness of the procedure:

  − in step 1, a basis $\{b_{d+1}, \ldots, b_n\}$, a number ε, and a sequence $(Q_i)_{i \in \mathbb{N}}$ satisfying (i)–(iii) do exist;
  − if we stop in step 3, then F and L are correctly determined;
  − if the condition of step 4 is true, then there exists $j \in \{0, 1\}$ such that P^j has infinite reverse split rank (when viewed as a polytope in the affine space $\{x \in \mathbb{R}^n : ax = \beta + j\}$).

5   Concluding Remarks

As illustrated in the introduction, Theorem 1 has strong similarities with Theorem 2, which characterizes the integral polyhedra with infinite reverse CG rank. One of the differences between the two statements is that in Theorem 2 L is a one-dimensional subspace. It can be shown that L cannot be assumed to have dimension one in Theorem 1 (see the journal version of the paper).

Moreover, one notices that in order to determine whether a polytope has infinite reverse split rank, all faces need to be considered, while this is not the case for the reverse CG rank ($F = P$ is the only interesting face in that case). We now show that this "complication" is necessary. Let $P \subseteq \mathbb{R}^4$ be defined as the convex hull of points $(0,0,0,0)$, $(1,0,0,0)$, $(1,2,0,0)$, and $(1,0,2,0)$. If F is the face of P induced by equation $x_1 = 1$ and $L = \langle (0,0,0,1) \rangle$, then the conditions of the theorem are satisfied, and thus $s^*(P) = +\infty$. However, the conditions are not satisfied if we choose $F = P$ and the same L, as $\operatorname{relint}(P+L)$ is contained in the interior of the split $\{x \in \mathbb{R}^4 : 0 \le x_1 \le 1\}$. Indeed one can verify that there is no nonzero subspace L' such that the conditions are satisfied with $F = P$.

In the context of mixed-integer programming, a result of Del Pia [10] implies necessary and sufficient conditions for an inequality to have finite split rank for a given polyhedron (see also [5,11]). His result seems to have some similarities with the conditions of Theorem 1. This is why we believe that the notion of finite reverse split rank in the pure integer case and that of finite split rank in the mixed-integer case are related to each other. This is the object of current research.

References

1. Averkov, G., Conforti, M., Del Pia, A., Di Summa, M., Faenza, Y.: On the convergence of the affine hull of the Chvátal-Gomory closures. SIAM Journal on Discrete Mathematics 27, 1492–1502 (2013)
2. Balas, E.: Disjunctive programming: Properties of the convex hull of feasible points. Discrete Applied Mathematics 89, 3–44 (1998)
3. Barvinok, A.: A Course in Convexity, Grad. Stud. Math. 54, AMS, Providence, RI (2002)
4. Basu, A., Conforti, M., Cornuéjols, G., Zambelli, G.: Maximal lattice-free convex sets in linear subspaces. Mathematics of Operations Research 35, 704–720 (2010)
5. Basu, A., Cornuéjols, G., Margot, F.: Intersection cuts with infinite split rank. Mathematics of Operations Research 37, 21–40 (2012)
6. Conforti, M., Del Pia, A., Di Summa, M., Faenza, Y., Grappe, R.: Reverse Chvátal–Gomory rank. In: Goemans, M., Correa, J. (eds.) IPCO 2013. LNCS, vol. 7801, pp. 133–144. Springer, Heidelberg (2013)
7. Conforti, M., Del Pia, A., Di Summa, M., Faenza, Y., Grappe, R.: Reverse Chvátal–Gomory rank (submitted, 2014)
8. Cook, W., Coullard, C.R., Túran, G.: On the complexity of cutting-plane proofs. Discrete Applied Mathematics 18, 25–38 (1987)
9. Cook, W., Kannan, R., Schrijver, A.: Chvátal closures for mixed integer programming problems. Mathematical Programming 174, 155–174 (1990)

10. Del Pia, A.: On the rank of disjunctive cuts. Mathematics of Operations Research 37, 372–378 (2012)
11. Dey, S.S., Louveaux, Q.: Split rank of triangle and quadrilateral inequalities. Mathematics of Operations Research 36, 432–461 (2011)
12. Eisenbrand, F., Schulz, A.S.: Bounds on the Chvátal rank of polytopes in the 0/1 cube. Combinatorica 23, 245–261 (2003)
13. Kannan, R., Lovasz, L.: Covering minima and lattice-point-free convex bodies. Ann. Math., Second Series 128, 577–602 (1988)
14. Khintchine, A.: A quantitative formulation of Kronecker's theory of approximation. Izv. Acad. Nauk SSSR, Ser. Mat. 12, 113–122 (1948) (in Russian)
15. Lovász, L.: Geometry of Numbers and Integer Programming. In: Iri, M., Tanabe, K. (eds.) Mathematical Programming: Recent Developements and Applications, pp. 177–210. Kluwer (1989)
16. Nemhauser, G.L., Wolsey, L.A.: Integer and Combinatorial Optimization. Wiley-Interscience, New York (1988)
17. Pokutta, S., Stauffer, G.: Lower bounds for the Chvátal–Gomory rank in the 0/1 cube. Operations Research Letters 39, 200–203 (2011)
18. Rockafellar, R.T.: Convex Analysis. Princeton University Press, Princeton (1970)
19. Rothvoß, T., Sanitá, L.: 0/1 polytopes with quadratic chvátal rank. In: Goemans, M., Correa, J. (eds.) IPCO 2013. LNCS, vol. 7801, pp. 349–361. Springer, Heidelberg (2013)
20. Schrijver, A.: On cutting planes. Annals of Discrete Mathematics 9, 291–296 (1980)
21. Schrijver, A.: Theory of Linear and Integer Programming. John Wiley (1986)

Strong LP Formulations for Scheduling Splittable Jobs on Unrelated Machines[*]

José R. Correa[1], Alberto Marchetti-Spaccamela[2], Jannik Matuschke[1],
Leen Stougie[3], Ola Svensson[4], Víctor Verdugo[1], and José Verschae[1]

[1] Departamento de Ingeniería Industrial, Universidad de Chile
[2] Department of Computer and System Sciences, Sapienza University of Rome
[3] Department of Econometrics and Operations Research, VU Amsterdam & CWI
[4] School of Computer and Communication Sciences, EPFL

Abstract. We study a natural generalization of the problem of minimizing makespan on unrelated machines in which jobs may be split into parts. The different parts of a job can be (simultaneously) processed on different machines, but each part requires a setup time before it can be processed. First we show that a natural adaptation of the seminal approximation algorithm for unrelated machine scheduling [11] yields a 3-approximation algorithm, equal to the integrality gap of the corresponding LP relaxation. Through a stronger LP relaxation, obtained by applying a lift-and-project procedure, we are able to improve both the integrality gap and the implied approximation factor to $1 + \phi$, where $\phi \approx 1.618$ is the golden ratio. This ratio decreases to 2 in the restricted assignment setting, matching the result for the classic version. Interestingly, we show that our problem cannot be approximated within a factor better than $\frac{e}{e-1} \approx 1.582$ (unless $\mathcal{P} = \mathcal{NP}$). This provides some evidence that it is harder than the classic version, which is only known to be inapproximable within a factor $1.5 - \varepsilon$. Since our $1 + \phi$ bound remains tight when considering the seemingly stronger machine configuration LP, we propose a new *job based* configuration LP that has an infinite number of variables, one for each possible way a job may be split and processed on the machines. Using convex duality we show that this infinite LP has a finite representation and can be solved in polynomial time to any accuracy, rendering it a promising relaxation for obtaining better algorithms.

1 Introduction

The unrelated machine scheduling problem, $R||C_{\max}$ in the three-field notation of [8], has attracted significant attention within the scientific community. The problem is to find a schedule of jobs with machine-dependent processing times

---

[*] This work was partially supported by Nucleo Milenio Información y Coordinación en Redes ICM/FIC P10-024F, by EU-IRSES grant EUSACOU, by the DFG Priority Programme "Algorithm Engineering" (SPP 1307), by the Berlin Mathematical School, by ERC Starting Grant 335288-OptApprox, and by FONDECYT project 3130407.

J. Lee and J. Vygen (Eds.): IPCO 2014, LNCS 8494, pp. 249–260, 2014.

that minimizes the makespan, i.e., the maximum machine load. In [11] a polynomial time linear programming based rounding algorithm was shown to give an approximation guarantee of 2, and a lower bound of 3/2 on the approximation ratio of any polynomial time algorithm was shown, assuming $\mathcal{P} \neq \mathcal{NP}$.

A natural generalization of this problem is to allow jobs to be *split* and processed on multiple machines simultaneously, where in addition a *setup* has to be performed on every machine processing the job. This generalized scheduling problem finds applications in production planning, e.g., in textile and semiconductor industries [18,10], and disaster relief operations [21]. Formally, we are given a set of m machines M and a set of n jobs J with *processing times* $p_{ij} \in \mathbb{Z}_+$ and *setup times* $s_{ij} \in \mathbb{Z}_+$ for every $i \in M$ and $j \in J$. A *schedule* corresponds to a vector $x \in [0,1]^{M \times J}$, where x_{ij} denotes the fraction of job j that is assigned to machine i, satisfying $\sum_{i \in M} x_{ij} = 1$ for all $j \in J$. If job j is processed (partially) on machine i then a setup of length s_{ij} has to be performed on the machine. During the setup of a job, the machine is occupied and thus no other job can be processed nor be set up. This results in the definition of *load* of machine $i \in M$ as $\sum_{j:x_{ij}>0}(x_{ij}p_{ij} + s_{ij})$. The objective is to minimize the *makespan*, the maximum load of the schedule. We denote this problem by $R|\text{split,setup}|C_{\max}$. Note that by setting $p_{ij} = 0$ and interpreting the setup times s_{ij} as processing requirements we obtain $R||C_{\max}$.

Related Work. Reducing the approximability gap for $R||C_{\max}$ is a prominent open question [23]. Since the seminal work by Lenstra et al. [11] there has been a considerable amount of effort leading to partial solutions to this question. In the *restricted assignment* problem, the processing times are of the form $p_{ij} \in \{p_j, \infty\}$ for all $i, j \in J$. A special case of this setting, in which each job can only be assigned to two machines, was considered by Ebenlendr et al. [6]. They note that while the lower bound of 3/2 still holds, a 7/4-approximation can be obtained. Svensson [19] shows that the general restricted assignment problem is approximable within a factor of $33/17 + \varepsilon \approx 1.9412 + \varepsilon$, breaking the barrier of 2. This algorithm is based on a *machine configuration* linear programming relaxation where each variable indicates the subset of jobs assigned to a given machine. On the other hand, this relaxation has an integrality gap of 2 for general unrelated machines [22]. Configuration LPs have also been studied extensively for the max-min version of the problem [22,3,7,9,2,14].

Most work concerned with scheduling splittable jobs focuses on heuristics. Theoretical results on the subject are not only scarce, but also restricted to the special case of identical machines. In particular, Xing and Zhang [24] describe a $(1.75 - 1/m)$-approximation for makespan minimization, that was later improved to 5/3 by Chen et al. [4]. The objective of minimizing the sum of completion times is studied by Schalekamp et al. [16], who gave a polynomial time algorithm in the case of 2 machines, and a 2.781-approximation algorithm for arbitrary m. This was later improved to $2 + \varepsilon$ in [5], even in the presence of weights.

Another setting that comes close to job splitting is preemptive scheduling with setup times [17,12,15], which does not allow simultaneous processing of parts of the same job. We also refer to the survey [1] and references therein for results on other scheduling problems with setup costs.

Our Contribution. Due to the novelty of the considered problem, our aim is to advance the understanding of its approximability, in particular in comparison to $R||C_{\max}$. We first study the integrality gap of a natural generalization of the LP relaxation by Lenstra et al. [11] to our setting and notice that their rounding technique does not work in our case. This is because it might assign a job with very large processing time to a single machine, while the optimal solution splits this job. On the other hand, assigning jobs by only following the fractional solution given by the LP might incur a large number of setups (belonging to different jobs) to a single machine. We get around these two extreme cases by adapting the technique from [11] so as to only round variables exceeding a certain threshold while guaranteeing that only one additional setup time is required per machine. This yields a 3-approximation algorithm presented in § 2. Additionally, we show that the integrality gap of this LP is exactly 3, and therefore our algorithm is best possible for this LP.

In § 3 we improve the approximation ratio by tightening our LP relaxation with a lift-and-project approach. We refine our previous analysis by balancing the rounding threshold, resulting in a $(1 + \phi)$-approximation, where $\phi \approx 1.618$ is the golden ratio. Surprisingly, we can show that this number is best possible for this LP; even for the seemingly stronger machine configuration LP mentioned above. This suggests that considerably different techniques are necessary to match the 2-approximation algorithm for $R||C_{\max}$. Indeed, we also show in § 5 that it is $\mathcal{NP}$-hard to approximate within a factor $\frac{e}{e-1} \approx 1.582$, a larger lower bound than the 3/2 hardness result known for $R||C_{\max}$. For the restricted assignment case, where $s_{ij} \in \{s_j, \infty\}$ and $p_{ij} \in \{p_j, \infty\}$, we obtain a 2-approximation algorithm, matching the 2-approximation of [11] in § 4. We remark that the solutions produced by all our algorithms have the property that at most one split job is processed on each machine. This property may be desirable in practice since in manufacturing systems setups require labor causing additional expenses.

As the integrality gaps of all mentioned relaxations are no better than $1 + \phi$, we propose a novel *job based* configuration LP relaxation in § 6 that has the potential to lead to better guarantees. Instead of considering machine configurations that assign jobs to machines, we introduce *job configurations*, describing the assignment of a particular job to the machines. The resulting LP cuts away worst-case solutions of the other LPs considered in this paper, rendering it a promising candidate for obtaining better approximation ratios. While the job configuration LP has an infinite set of variables, we show that we can restrict a priori to a finite subset. Applying discretization techniques we can approximately solve the LP within a factor of $(1 + \varepsilon)$ by separation over the dual constraints. Finally, we study the projection of this polytope to the *assignment space* and derive an explicit set of inequalities that defines this polytope. An interesting open problem is to determine the integrality gap of the job configuration LP.

2 A 3-Approximation Algorithm

Our 3-approximation algorithm is based on a generalization of the LP by Lenstra, Shmoys, and Tardos [11]. Let C^* be a guess on the optimal makespan. Consider the following feasibility LP, whose variable x_{ij} denotes the fraction of job j assigned to machine i. Notice that the LP is a relaxation, since it allows the setups to be performed fractionally.

$$[\text{LST}]: \qquad \sum_{i \in M} x_{ij} = 1 \qquad \qquad \text{for all } j \in J, \qquad (1)$$

$$\sum_{j \in J} x_{ij}(p_{ij} + s_{ij}) \leq C^* \qquad \qquad \text{for all } i \in M, \qquad (2)$$

$$x_{ij} = 0 \qquad \text{for all } i \in M, j \in J : s_{ij} > C^*, \qquad (3)$$

$$x_{ij} \geq 0 \qquad \qquad \text{for all } i \in M, j \in J.$$

Let x be a feasible extreme solution. We define the bipartite graph $G(x) = (J \cup M, E(x))$, where $E(x) = \{ij : 0 < x_{ij}\}$. Using the same arguments as in [11], not repeated here, we can show the following property.

Lemma 1. *For every extreme solution x of* [LST], *each connected component of $G(x)$ is a pseudotree; a tree plus at most one edge that creates a single cycle.*

We show how to round an extreme solution x of [LST]. Let $E_+ = \{ij \in E(x) : x_{ij} > 1/2\}$ and $J_+ = \{j \in J : \text{there exists } i \in M \text{ with } ij \in E_+\}$, i.e., those jobs that the fractional solution x assigns to some machine by a factor of more than $1/2$. In our rounding procedure each job $j \in J_+$ is completely assigned to the machine $i \in M$ if $x_{ij} > 1/2$. We now show how to assign the rest of the jobs.

Let us call $G'(x)$ the subgraph of $G(x)$ induced by $(J \cup M) \setminus J_+$. Notice that every edge ij in $G'(x)$ satisfies that $0 < x_{ij} \leq 1/2$. Also, since $G'(x)$ is a subgraph of $G(x)$ every connected component of $G'(x)$ is a pseudotree.

Definition 1. *Given $A \subseteq E(G'(x))$, we say that a machine $i \in M$ is A-balanced, if there exists at most one job $j \in J \setminus J_+$ such that $ij \in A$. We say that a job $j \in J \setminus J_+$ is A-processed if there is at most one machine $i \in M$ such that $ij \notin A$ and $x_{ij} > 0$.*

In what follows we seek to find a subset $A \subseteq E(G'(x))$ such that each job $j \in J \setminus J_+$ is A-processed and each machine is A-balanced. We will show that this is enough for a 3-approximation, by assigning each job $j \in J \setminus J_+$ to machine i by a fraction of at most $2x_{ij}$ for each $ij \in A$, and not assigning it anywhere else. Since every job $j \in J \setminus J_+$ is A-processed and $x_{ij} \leq 1/2$ for all $i \in M$, job j will be completely assigned. Also, since each machine is A-balanced, the load of each machine i will be affected by at most the setup-time of one job j. This setup time s_{ij} is at most C^* by restriction (3). This and the fact that the processing time of a job on each machine is at most doubled are the basic ingredients to show the approximation factor of 3.

Construction of the Set A. In the following, we denote by (T, r) a rooted tree T with root r. Consider a connected component T of $G'(x)$. Since $G'(x)$ is a subgraph of $G(x)$, Lemma 1 implies that T is a pseudotree. We denote by $C = j_1 i_1 j_2 i_2 \cdots j_\ell i_\ell j_1$ the only cycle of T (if it exists), which must be of even length. (If such a cycle does not exist we choose any path in T from j_1 to some i_ℓ.) Here the jobs are $J(C) = \{j_1, \ldots, j_\ell\}$ and the machines are $M(C) = \{i_1, \ldots, i_\ell\}$. In the cycle, we define the matching $K_C = \{(j_k, i_k) : k \in \{1, \ldots, \ell\}\}$. In the forest $T \setminus K_C$, we denote by (T_u, u) the tree rooted in u, for every $u \in M(C)$. Notice that by deleting the matching, no two vertices in $M(C)$ will be in the same component of $T \setminus K_C$.

For every $u \in M(C)$, directing the edges of (T_u, u) away from the root, we obtain the directed tree of which each level consists either entirely of machine-nodes or entirely of job-nodes. We delete all edges going out of machine nodes, i.e. all edges entering job-nodes. The remaining edges we denote by A_u. We define $A := K_C \cup \bigcup_{u \in M(C)} A_u$. We obtain the following to lemmas.

Lemma 2. *Every job $j \in J \setminus J_+$ is A-processed.*

Lemma 3. *Every machine $i \in M$ is A-balanced.*

Given set A, we apply the following rounding algorithm that constructs a new assignment $\tilde{x}$. The algorithm also outputs a binary vector $\tilde{y}_{ij} \in \{0, 1\}$ which indicates whether job j is (partially) assigned to machine i or not.

Algorithm 1. Rounding(x)

1: Construct the graphs $G(x)$, $G'(x)$, and the set A as above.
2: For all $ij \in E_+$, $\tilde{x}_{ij} \leftarrow 1$ and $\tilde{y}_{ij} \leftarrow 1$;
3: For all $ij \in A$, $\tilde{x}_{ij} \leftarrow \dfrac{x_{ij}}{\sum_{k:kj \in A} x_{kj}}$ and $\tilde{y}_{ij} \leftarrow 1$;
4: For all $ij \in E \setminus (E_+ \cup A)$, $\tilde{x}_{ij} \leftarrow 0$ and $\tilde{y}_{ij} \leftarrow 0$.

Theorem 1. *There exists a 3-approximation algorithm for $R|split, setup|C_{\max}$.*

Proof. Our algorithm first finds the smallest value C^* for which [LST] is feasible. This can be easily done with a binary search procedure. For that value C^*, let x be an extreme point of [LST], and consider the output $\tilde{x}, \tilde{y}$ of algorithm Rounding(x). Clearly $\tilde{x}, \tilde{y}$ can be computed in polynomial time. We show that the schedule that assigns a fraction $\tilde{x}_{ij}$ of job j to machine i has a makespan of at most $3C^*$. This implies the theorem since $C^* \leq \text{OPT}$.

First we show that $\tilde{x} \geq 0$ defines a valid assignment, i.e., $\sum_{i \in M} \tilde{x}_{ij} = 1$ for all j. Indeed, this directly follows by the algorithm Rounding(x): If $j \in J_+$, then there exists a unique machine $i \in M$ with $ij \in E_+$ and therefore j is completely assigned to machine i. If $j \notin J_+$, then $\sum_{i \in M} \tilde{x}_{ij} = 1$ by construction.

Now we show that the makespan of the solution is at most $3C^*$. First notice that for every $ij \in E_+$ we have that $1 = \tilde{x}_{ij} = \tilde{y}_{ij} \leq 2x_{ij}$, because $ij \in E_+$

implies that $x_{ij} > 1/2$. On the other hand, for every $j \in J \setminus J_+$ we have that $\sum_{k:kj\in A} x_{kj} \geq 1/2$, because at most one machine that processes j fractionally is not considered in A. We conclude that $\tilde{x} \leq 2x$. Then for every $i \in M$ it holds that

$$
\sum_{j\in J}(\tilde{x}_{ij}p_{ij} + \tilde{y}_{ij}s_{ij}) = \sum_{j:ij\in E_+} (\tilde{x}_{ij}p_{ij} + \tilde{y}_{ij}s_{ij}) + \sum_{j:ij\in A} (\tilde{x}_{ij}p_{ij} + \tilde{y}_{ij}s_{ij})
$$

$$
\leq \sum_{j:ij\in E_+} 2x_{ij}(p_{ij}+s_{ij}) + \sum_{j:ij\in A} (2x_{ij}p_{ij}+s_{ij})
$$

$$
\leq 2C^* + \sum_{j:ij\in A} s_{ij}.
$$

Recall that machine i is A-balanced, and therefore there is at most one job j with $ij \in A$. Also, $ij \in A$ implies that $ij \in E(x) = \{ij : x_{ij} > 0\}$, and hence, by (3), $s_{ij} \leq C^*$. We conclude that $\sum_{j:ij\in A} s_{ij} \leq C^*$, proving the theorem. □

We finish this section by noting that our analysis is tight. Specifically, it can be shown that the gap between the LP solution and the optimum can be arbitrarily close to 3.

Theorem 2. *For any $\varepsilon > 0$, there exists an instance such that $(3-\varepsilon)C^* \leq OPT$, where C^* is the smallest number such that [LST] is feasible.*

3   A $(1 + \phi)$-Approximation Algorithm

In this section we refine the previous algorithm and improve the approximation ratio. Since [LST] has a gap of 3, we strengthen it in order to obtain a stronger LP. To this end notice that inequalities (2) in [LST] are the LP relaxation of the following restrictions of the mixed integer linear program with binary variables y_{ij} for machine i and job j:

$$
\sum_{j\in J}(x_{ij}p_{ij} + y_{ij}s_{ij}) \leq C^* \qquad\qquad \text{for all } i \in M, \qquad (4)
$$

$$
x_{ij} \leq y_{ij} \qquad\qquad \text{for all } i \in M \text{ and } j \in J. \qquad (5)
$$

A stronger relaxation is obtained by applying a lift and project step [13] to the first inequality. For some fixed choice ij multiplying both sides of the i-th inequality (4) by the corresponding variable y_{ij} implies (by leaving out terms)

$$
y_{ij}x_{ij}p_{ij} + y_{ij}^2 s_{ij} \leq y_{ij}C^*.
$$

In case $C^* - s_{ij} > 0$, this inequality implies the valid linear inequality

$$
x_{ij}\frac{p_{ij}}{C^* - s_{ij}} \leq y_{ij}, \qquad (6)
$$

since every feasible integer solution has $y_{ij}x_{ij} = x_{ij}$ and $y_{ij}^2 = y_{ij}$. Note that, in optimal solutions of the LP relaxation, y_{ij} attains the smallest value that satisfies (5) and (6). We define $\alpha_{ij} = \max\left\{1, \frac{p_{ij}}{C^* - s_{ij}}\right\}$ if $C^* > s_{ij}$, and $\alpha_{ij} = 1$ otherwise, and substitute y_{ij} by $\alpha_{ij}x_{ij}$ to obtain the strengthened LP relaxation

$$[\text{LST}_{\text{strong}}]: \qquad \sum_{i \in M} x_{ij} = 1 \qquad\qquad \text{for all } j \in J, \quad (7)$$

$$\sum_{j \in J} x_{ij}(p_{ij} + \alpha_{ij}s_{ij}) \leq C^* \qquad\qquad \text{for all } i \in M, \quad (8)$$

$$x_{ij} = 0 \qquad \text{for all } i \in M, j \in J : s_{ij} > C^*, \quad (9)$$

$$x_{ij} \geq 0 \qquad\qquad\qquad \text{for all } i \in M, j \in J.$$

Notice that this LP is at least as strong as [LST] since $\alpha_{ij} \geq 1$ and, therefore, the C^* values used in [LST] and [LST$_{\text{strong}}$] might differ. Again binary search allows us to find the minimum C^* for which [LST$_{\text{strong}}$] is feasible.

Let x be an extreme point solution of this LP. We use a rounding approach similar to the one in the previous section. Proofs that are the same as in that section will be skipped. Consider the graph $G(x)$. As before, each connected component of $G(x)$ is a pseudotree, using the same arguments that justified Lemma 1. Also, we define again a set of jobs J_+ that the LP assigns to one machine by a sufficiently large fraction. In the previous section this fraction was $1/2$. Now we parameterize it by $\beta \in (1/2, 1)$, to be chosen later. We define $E_+ = \{j \in E(x) : x_{ij} > \beta\}$ and $J_+ = \{j \in J : \text{there exists } i \in M \text{ with } ij \in E_+\}$.

Consider the subgraph $G'(x)$ of $G(x)$ induced by the set of nodes $(J \cup M) \setminus J_+$. Let A be a set constructed as in the previous section. Then every machine is A-balanced and every job is A-processed. Now we apply the algorithm Rounding(x) of the last section to obtain a new assignment $\tilde{x}, \tilde{y}$. We show that for $\beta = \phi - 1$ this is a solution with makespan $(1 + \phi)C^*$, where $\phi = (1 + \sqrt{5})/2 \approx 1.618$ is the golden ratio. We need the following technical lemma.

Lemma 4. *Let β be a real number such that $1/2 < \beta < 1$. Then*

$$\max_{0 \leq \mu \leq 1}\left\{\mu + \max\left\{\frac{1}{\beta}, \frac{1-\mu}{1-\beta}\right\}\right\} = \max\left\{\frac{1}{1-\beta}, 1 + \frac{1}{\beta}\right\}.$$

Theorem 3. *There exists a $(1 + \phi)$-approximation algorithm for the problem $R|\text{split,setup}|C_{\max}$.*

Proof. Let x be an extreme point solution of [LST$_{\text{strong}}$], and let $\tilde{x}, \tilde{y}$ be the output of algorithm Rounding(x) described in § 2. The fact that $\tilde{x}, \tilde{y}$ correspond to a feasible assignment follows from the same argument as in the proof of Theorem 1. We now show that the makespan of this solution is at most $(1+\phi)C^*$, which implies the approximation factor.

For any edge $ij \in E_+$, we have $x_{ij} > \beta$ and hence $1 = \tilde{x}_{ij} = \tilde{y}_{ij} \leq 1/\beta \cdot x_{ij}$. Additionally, for every $j \in J \setminus J_+$, we have again, by the choice of A, that it is A-processed. Hence, $\sum_{k:kj \notin A} x_{kj} \leq \beta$, because at most one machine that processes

j fractionally is not considered in A. Thus, $\sum_{k:kj\in A} x_{kj} \geq 1-\beta$, which implies that $\tilde{x}_{ij} \leq x_{ij}/(1-\beta)$. Hence, for machine i,

$$\sum_{j\in J}(\tilde{x}_{ij}p_{ij} + \tilde{y}_{ij}s_{ij}) = \sum_{j:ij\in E_+}(\tilde{x}_{ij}p_{ij} + \tilde{y}_{ij}s_{ij}) + \sum_{j:ij\in A}(\tilde{x}_{ij}p_{ij} + \tilde{y}_{ij}s_{ij})$$

$$\leq \frac{1}{\beta}\sum_{j:ij\in E_+} x_{ij}(p_{ij} + s_{ij}) + \frac{1}{1-\beta}\sum_{j:ij\in A} x_{ij}p_{ij} + \sum_{j:ij\in A} s_{ij}.$$

Since machine i is A-balanced, there exists at most one job j with $ij \in A$ (if there is no such job then i has load at most C^*/β). Let $j(i)$ be that job, and define $\mu_i = s_{ij(i)}/C^*$. Then notice that

$$x_{ij(i)}(p_{ij(i)} + \alpha_{ij(i)}s_{ij(i)}) \geq x_{ij(i)}p_{ij(i)}\left(1 + \frac{s_{ij(i)}}{C^* - s_{ij(i)}}\right)$$

$$= x_{ij(i)}p_{ij(i)}\left(1 + \frac{\mu_i}{1-\mu_i}\right) = x_{ij(i)}p_{ij(i)}\frac{1}{1-\mu_i}.$$

Combining the last two inequalities we obtain that

$$\sum_{j\in J}(\tilde{x}_{ij}p_{ij} + \tilde{y}_{ij}s_{ij}) \leq \frac{1}{\beta}\sum_{j:ij\in E_+} x_{ij}(p_{ij} + s_{ij}) + \frac{1}{1-\beta}x_{ij(i)}p_{ij(i)} + s_{ij(i)}$$

$$\leq \frac{1}{\beta}\sum_{j:ij\in E_+} x_{ij}(p_{ij} + s_{ij}) + \frac{1-\mu_i}{1-\beta}x_{ij(i)}(p_{ij(i)} + \alpha_{ij(i)}s_{ij(i)}) + \mu_i C^*$$

$$\leq \max\left\{\frac{1}{\beta}, \frac{1-\mu_i}{1-\beta}\right\}\sum_{j\in J} x_{ij}(p_{ij} + \alpha_{ij}s_{ij}) + \mu_i C^*$$

$$\leq C^*\left(\mu_i + \max\left\{\frac{1}{\beta}, \frac{1-\mu_i}{1-\beta}\right\}\right).$$

Therefore, by the previous lemma we have that the load of each machine is at most $C^* \cdot \max\{1/(1-\beta), 1+1/\beta\}$. The approximation factor is minimized when $1/(1-\beta) = 1 + 1/\beta$, hence $\beta = (-1+\sqrt{5})/2 = (1+\sqrt{5})/2 - 1 = \phi - 1$. Thus, the approximation ratio is $1 + 1/(\phi - 1) = 1 + \phi$. $\qquad\square$

We close this section by showing that $1 + \phi$ is the best approximation ratio achievable by [LST$_{\text{strong}}$].

Theorem 4. *For any $\varepsilon > 0$, there exists an instance such that $C^*(1+\phi-\varepsilon) \leq OPT$, where C^* is the smallest number such that [LST$_{\text{strong}}$] is feasible.*

Proof. Consider the instance depicted in Fig. 1. It consists of two disjoint sets of jobs J and J'. Each job $j_\ell \in J$ forms a pair with its corresponding job $j'_\ell \in J'$. Each such pair is associated with a *parent machine* i_p^ℓ such that both j_ℓ and j'_ℓ can be processed on this machine with setup time $s_{i_p^\ell j_\ell} = s_{i_p^\ell j'_\ell} = \phi/2$ and $p_{i_p^\ell j_\ell} = p_{i_p^\ell j'_\ell} = 0$. Each job j of each pair is furthermore associated with a *child machine* $i_c(j)$ such that $s_{i_c(j)j} = 0$ and $p_{i_c(j)j} = \phi + 1$. $\qquad\square$

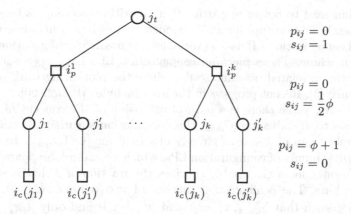

$$p_{ij} = 0$$
$$s_{ij} = 1$$

$$p_{ij} = 0$$
$$s_{ij} = \frac{1}{2}\phi$$

$$p_{ij} = \phi + 1$$
$$s_{ij} = 0$$

Fig. 1. Example showing that [LST$_{\text{strong}}$] has a gap of $1 + \phi$

4 A 2-Approximation Algorithm for Restricted Assignment

We also consider the restricted assignment case, where for every $j \in J$ there are values p_j and s_j such that $p_{ij} \in \{p_j, \infty\}$ and $s_{ij} \in \{s_j, \infty\}$ for all $i \in M$. For this setting we obtain an improved approximation ratio of 2, also based on rounding the [LST$_{\text{strong}}$] relaxation. After constructing the same graph $G(x)$, we distribute the processing requirement of each job to the machine corresponding to its child nodes. Although this might increase the processing requirement of a job on the child machines by more than a factor 2, we show that increasing the load of these machines by C^* suffices to completely process the job and its setup.

Theorem 5. *There exists a 2-approximation algorithm for scheduling splittable jobs on unrelated machines under restricted assignment.*

5 Hardness of Approximation

By reducing from MAX k-COVER, we derive an inapproximability bound of $e/(e-1) \approx 1.582$ for $R|$split,setup$|C_{\max}$, indicating that the problem might indeed be harder from an approximation point of view compared to the classic $R||C_{\max}$, for which $3/2$ is the best known lower bound.

Theorem 6. *For any $\varepsilon > 0$, there is no $\left(\frac{e}{e-1} - \varepsilon\right)$-approximation algorithm for $R|$split,setup$|C_{\max}$ unless $\mathcal{P} = \mathcal{NP}$.*

6 A Job Configuration LP

A basic tool of combinatorial optimization is to design stronger linear programs based on certain configurations. These LPs often provide improved integrality gaps and thus lead to better approximation algorithms as long as they can be solved efficiently and be appropriately rounded. In machine scheduling the most widely used configuration LP uses as variables the possible configurations of jobs in a given machine. These *machine configuration* LPs have been successfully studied for the unrelated machine setting since the pioneering work of Bansal and Sviridenko [3]. Recent progress in the area includes [19,6,22,20].

Unfortunately, while there is a natural extension of the concept of machine configurations to $R|\text{split,setup}|C_{\max}$, this formulation surprisingly exhibits the same integrality gap of $1 + \phi$ as already observed for $[\text{LST}_{\text{strong}}]$. Instead, we introduce a new family of configuration LPs, which we call *job configuration* LPs. A configuration f for a given job j specifies the fraction of j that is scheduled on each machine. The configuration consists of two vectors $x^f \in [0,1]^M$ and $y^f \in \{0,1\}^M$ such that $\sum_{i \in M} x_i^f = 1$ and $y_i^f = 1$ if and only if $x_i^f > 0$. On machine $i \in M$ configuration f requires time $t_i^f := p_{ij}x_i^f + s_{ij}y_i^f$. Let $\mathcal{F}_j$ be the set of configurations for job j with $t_i^f \leq C$ for all $i \in M$. Then every feasible solution to $R|\text{split,setup}|C_{\max}$ with makespan C corresponds to an integer solution of

$$[\text{CLP}]: \qquad \sum_{f \in \mathcal{F}_j} \lambda_f = 1 \qquad \qquad \text{for all } j \in J,$$

$$\sum_{j \in J} \sum_{f \in \mathcal{F}_j} \lambda_f t_i^f \leq C \qquad \qquad \text{for all } i \in M,$$

$$\lambda_f \geq 0 \qquad \qquad \text{for all } f \in \bigcup_{j \in J} \mathcal{F}_j.$$

Note that this formulation has infinitely many variables. However, by investigating the separation problem of the convex dual of [CLP], we can show that we can restrict [CLP] without loss of generality to the finite subset of so-called maximal configurations. A configuration $f \in \mathcal{F}_j$ is *maximal*, if there is at most one machine $i \in M$ with $0 < x_i^f < x_{ij}^{\max}$, where $x_{ij}^{\max} := (C - s_{ij})/p_{ij}$.

Theorem 7. [CLP] *is feasible if and only if the restriction of* [CLP] *to maximal configurations is feasible.*

It can further be shown that after discretizing the configurations, the dual separation problem can be solved in polynomial time, implying that [CLP] can be solved efficiently up to a factor $(1 + \varepsilon)$. Henceforth, we will restrict $\mathcal{F}_j$ to the set of maximal configurations for each job $j \in J$.

Projection of the Job Configuration LP. Observe that any convex combination of job configurations λ can be translated into a pair of vectors $x^\lambda, y^\lambda \in [0,1]^{M \times J}$ in the assignment space by setting

$$x_{ij}^\lambda := \sum_{f \in \mathcal{F}_j} \lambda_f x_i^f \quad \text{and} \quad y_{ij}^\lambda := \sum_{f \in \mathcal{F}_j} \lambda_f y_i^f.$$

We show that applying this projection to [CLP] leads to assignment vectors described by the following set of inequalities:

$$[\text{CLP}_{\text{proj}}]: \quad \sum_{j \in J}(p_{ij}x_{ij} + s_{ij}y_{ij}) \leq C \qquad \text{for all } i \in M, \tag{10}$$

$$\sum_{i \in M}(\beta_i x_{ij} + \gamma_i y_{ij}) \geq M(j, \beta, \gamma) \quad \text{for all } j \in J, \ \beta, \gamma \in \mathbb{R}^M, \tag{11}$$

with $M(j, \beta, \gamma) := \min\left\{\sum_{i \in M}(\beta_i x_i^f + \gamma_i y_i^f) : f \in \mathcal{F}_j\right\}$.

Theorem 8. *If $\lambda \in$ [CLP] then $(x^\lambda, y^\lambda) \in$ [CLP$_{\text{proj}}$]. Conversely, if $(x, y) \in$ [CLP$_{\text{proj}}$] then there exists $\lambda \in$ [CLP] such that $x = x^\lambda$ and $y = y^\lambda$.*

We conclude by showing that already a very special class of [CLP$_{\text{proj}}$]-inequalities is sufficient to eliminate the gap in the worst-case instances of [LST$_{\text{strong}}$]. For a set of machines $S \subseteq M$ let $L(j, S) := \sum_{i \in M \setminus S} \max\left\{\frac{C - s_{ij}}{p_{ij}}, 0\right\}$ be the maximum fraction of job j that can be processed within time C by the machines in $M \setminus S$. The following inequalities are satisfied by the vector x, y induced by any feasible solution to $R|\text{split,setup}|C_{\max}$ with makespan at most C.

$$\frac{\sum_{i \in S'} x_{ij}}{1 - L(j, S \cup S')} + \sum_{i \in S} y_{ij} \geq 1 \quad \text{for all } j \in J \text{ and } S, S' \subseteq M \text{ with } L(j, S \cup S') < 1.$$

Interestingly, these inequalities can be seen as a special case of inequalities (11) by setting $\beta_i = \frac{1}{1 - L(j, S \cup S')}$ for $i \in S'$ and $\gamma_i = 1$ for $i \in S$. Furthermore, consider the example instance given in the proof of Theorem 4 (cf. Fig. 1). If $C < 1 + \phi$, then $L(j, \{i_p(j)\}) = C/p_{i_c(j)j} < 1$ and therefore $y_{i_p(j)j} = 1$ for all $j \in J \cup J'$ in any feasible solution to [CLP$_{\text{proj}}$]. This immediately implies infeasibility of [CLP$_{\text{proj}}$] for $C < 1 + \phi$. We also note that the exact same argument applies to the worst-case instance of the machine configuration LP.

References

1. Allahverdi, A., Ng, C., Cheng, T., Kovalyov, M.: A survey of scheduling problems with setup times or costs. Eur. J. Oper. Res. 187, 985–1032 (2008)
2. Asadpour, A., Feige, U., Saberi, A.: Santa claus meets hypergraph matchings. ACM Trans. Algorithms 24, 24:1–24:9 (2012)
3. Bansal, N., Sviridenko, M.: The Santa Claus problem. In: STOC, pp. 31–40 (2006)
4. Chen, B., Ye, Y., Zhang, J.: Lot-sizing scheduling with batch setup times. J. Sched. 9, 299–310 (2006)
5. Correa, J.R., Verdugo, V., Verschae, J.: Approximation algorithms for scheduling splitting jobs with setup times. In: Talk in MAPSP (2013)
6. Ebenlendr, T., Krčál, M., Sgall, J.: Graph balancing: A special case of scheduling unrelated parallel machines. Algorithmica (2012), doi:10.1007/s00453-012-9668-9
7. Feige, U.: On allocations that maximize fairness. In: SODA, pp. 287–293 (2008)
8. Graham, R., Lawler, E., Lenstra, J., Kan, A.: Optimization and approximation in deterministic sequencing and scheduling: a survey. Ann. Discrete Math. 5, 287–326 (1979)

9. Haeupler, B., Saha, B., Srinivasan, A.: New constructive aspects of the Lovász Local Lemma. J. ACM 58, 28:1–28 (2011)
10. Kim, D.-W., Na, D.-G., Frank Chen, F.: Unrelated parallel machine scheduling with setup times and a total weighted tardiness objective. Robot. Com. -Int. Manuf. 19, 173–181 (2003)
11. Lenstra, J.K., Shmoys, D.B., Tardos, E.: Approximation algorithms for scheduling unrelated parallel machines. Math. Program. 46, 259–271 (1990)
12. Liu, Z., Cheng, T.C.E.: Minimizing total completion time subject to job release dates and preemption penalties. J. Sched. 7, 313–327 (2004)
13. Lovász, L., Schrijver, A.: Cones of matrices and set-functions and 0-1 optimization. SIAM J. Optimiz. 1, 166–190 (1991)
14. Polacek, L., Svensson, O.: Quasi-polynomial local search for restricted max-min fair allocation. In: Czumaj, A., Mehlhorn, K., Pitts, A., Wattenhofer, R. (eds.) ICALP 2012, Part I. LNCS, vol. 7391, pp. 726–737. Springer, Heidelberg (2012)
15. Potts, C.N., Wassenhove, L.N.V.: Integrating scheduling with batching and lot-sizing: A review of algorithms and complexity. J. Oper. Res. Soc. 43, 395–406 (1992)
16. Schalekamp, F., Sitters, R., van der Ster, S., Stougie, L., Verdugo, V., van Zuylen, A.: Split scheduling with uniform setup times. Arxiv (2012)
17. Schuurman, P., Woeginger, G.J.: Preemptive scheduling with job-dependent setup times. In: SODA, pp. 759–767 (1999)
18. Serafini, P.: Scheduling jobs on several machines with the job splitting property. Oper. Res. 44, 617–628 (1996)
19. Svensson, O.: Santa claus schedules jobs on unrelated machines. SIAM J. Comput. 41, 1318–1341 (2012)
20. Sviridenko, M., Wiese, A.: Approximating the configuration-lp for minimizing weighted sum of completion times on unrelated machines. In: Goemans, M., Correa, J. (eds.) IPCO 2013. LNCS, vol. 7801, pp. 387–398. Springer, Heidelberg (2013)
21. van der Ster, S.: The allocation of scarce resources in disaster relief. MSc-Thesis in Operations Research at VU University Amsterdam (2010)
22. Verschae, J., Wiese, A.: On the configuration-LP for scheduling on unrelated machines. In: Demetrescu, C., Halldórsson, M.M. (eds.) ESA 2011. LNCS, vol. 6942, pp. 530–542. Springer, Heidelberg (2011)
23. Williamson, D.P., Shmoys, D.B.: The Design of Approximation Algorithms. Cambridge University Press (2011)
24. Xing, W., Zhang, J.: Parallel machine scheduling with splitting jobs. Discrete Appl. Math. 103, 259–269 (2000)

How Good Are Sparse Cutting-Planes?[*]

Santanu S. Dey, Marco Molinaro, and Qianyi Wang

School of Industrial and Systems Engineering, Georgia Institute of Technology

Abstract. Sparse cutting-planes are often the ones used in mixed-integer programing (MIP) solvers, since they help in solving the linear programs encountered during branch-&-bound more efficiently. However, how well can we approximate the integer hull by just using sparse cutting-planes? In order to understand this question better, given a polyope $\mathbf{P}$ (e.g. the integer hull of a MIP), let $\mathbf{P}^k$ be its best approximation using cuts with at most k non-zero coefficients. We consider $d(\mathbf{P}, \mathbf{P}^k) = \max_{x \in \mathbf{P}^k} (\min_{y \in \mathbf{P}} \|x - y\|)$ as a measure of the quality of sparse cuts.

In our first result, we present general upper bounds on $d(\mathbf{P}, \mathbf{P}^k)$ which depend on the number of vertices in the polytope and exhibits three phases as k increases. Our bounds imply that if $\mathbf{P}$ has polynomially many vertices, using half sparsity already approximates it very well. Second, we present a lower bound on $d(\mathbf{P}, \mathbf{P}^k)$ for random polytopes that show that the upper bounds are quite tight. Third, we show that for a class of hard packing IPs, sparse cutting-planes do not approximate the integer hull well. Finally, we show that using sparse cutting-planes in extended formulations is at least as good as using them in the original polyhedron, and give an example where the former is actually much better.

1 Introduction

Most successful mixed integer linear programming (MILP) solvers are based on branch-&-bound and cutting-plane (cut) algorithms. Since MILPs belong to the class of NP-hard problems, one does not expect the size of branch-&-bound tree to be small (polynomial is size) for every instance. In the case where the branch-&-bound tree is not small, a large number of linear programs must be solved. It is well-known that dense cutting-planes are difficult for linear programming solvers to handle. Therefore, most commercial MILPs solvers consider sparsity of cuts as an important criterion for cutting-plane selection and use [4,1,7].

Surprisingly, very few studies have been conducted on the topic of sparse cutting-planes. Apart from cutting-plane techniques that are based on generation of cuts from single rows (which implicitly lead to sparse cuts if the underlying row is sparse), to the best of our knowledge only the paper [2] explicitly discusses methods to generate sparse cutting-planes.

The use of sparse cutting-planes may be viewed as a compromise between two competing objectives. As discussed above, on the one hand, the use of sparse

---

[*] Santanu S. Dey and Qianyi Wang were partially supported by NSF grant CMMI-1149400.

J. Lee and J. Vygen (Eds.): IPCO 2014, LNCS 8494, pp. 261–272, 2014.

cutting-planes aids in solving the linear programs encountered in the branch-&-bound tree faster. On the other hand, it is possible that 'important' facet-defining or valid inequalities for the convex hull of the feasible solutions are dense and thus without adding these cuts, one may not be able to attain significant integrality gap closure. This may lead to a larger branch-&-bound tree and thus result in the solution time to increase.

It is challenging to simultaneously study both the competing objectives in relation to cutting-plane sparsity. Therefore, a first approach to understanding usage of sparse cutting-planes is the following: *If we are able to separate and use valid inequalities with a given level of sparsity (as against completely dense cuts), how much does this cost in terms of loss in closure of integrality gap?*

Considered more abstractly, the problem reduces to a purely geometric question: Given a polytope $\mathbf{P}$ (which represents the convex hull of feasible solutions of a MILP), how well is $\mathbf{P}$ approximated by the use of sparse valid inequalities. In this paper we will study polytopes contained in the $[0, 1]^n$ hypercube. This is without loss of generality since one can always translate and scale a polytope to be contained in the $[0, 1]^n$ hypercube.

1.1 Preliminaries

A cut $ax \leq b$ is called *k-sparse* if the vector a has at most k nonzero components. Given a set $\mathbf{P} \subseteq \mathbb{R}^n$, define $\mathbf{P}^k$ as the best outer-approximation obtained from k-sparse cuts, that is, it is the intersection of all k-sparse cuts valid for $\mathbf{P}$.

For integers k and n, let $[n] := \{1, \ldots, n\}$ and let $\binom{[n]}{k}$ be the set of all subsets of $[n]$ of cardinality k. Given a k-subset of indices $I \subseteq [n]$, define $\mathbb{R}^I = \{x \in \mathbb{R}^n : x_i = 0 \text{ for all } i \in I\}$. An equivalent and handy definition of $\mathbf{P}^k$ is the following: $\mathbf{P}^k = \bigcap_{I \in \binom{[n]}{k}} \left(\mathbf{P} + \mathbb{R}^I\right)$. Thus, if $\mathbf{P}$ is a polytope, $\mathbf{P}^k$ is also a polytope.

1.2 Measure of Approximation

There are several natural measures to compare the quality of approximation provided by $\mathbf{P}^k$ in relation to $\mathbf{P}$. For example, one may consider objective value ratio: maximum over all costs c of expression $\frac{z^{c,k}}{z^c}$, where $z^{c,k}$ is the value of maximizing c over $\mathbf{P}^k$, and z^c is the same for $\mathbf{P}$. We discard this ratio, since this ratio can become infinity and not provide any useful information[1]. Similarly, we may compare the volumes of $\mathbf{P}$ and $\mathbf{P}^k$. However, this ratio is not useful if $\mathbf{P}$ is not full-dimensional and $\mathbf{P}^k$ is.

In order to have a useful measure that is well-defined for all polytopes contained in $[0, 1]^n$, we consider the following *distance measure*:

$$\mathrm{d}(\mathbf{P}, \mathbf{P}^k) := \max_{x \in \mathbf{P}^k} \left(\min_{y \in \mathbf{P}} \|x - y\|\right),$$

where $\|\cdot\|$ is the ℓ_2 norm. It is easily verified that there is a vertex of $\mathbf{P}^k$ attaining the maximum above. Thus, alternatively the distance measure can be interpreted as the Euclidean distance between $\mathbf{P}$ and the farthest vertex of $\mathbf{P}^k$ from $\mathbf{P}$.

---

[1] Take $\mathbf{P} = \mathrm{conv}\{(0,0), (0,1), (1,1)\}$ and compare with $\mathbf{P}^1$ wrt $c = (1, -1)$.

Observation 1 (d(P, P^k) is an upper bound on depth of cut). *Suppose $\alpha x \leq \beta$ is a valid inequality for* **P** *where* $\|\alpha\| = 1$. *Let the depth of this cut be the smallest $\gamma \geq 0$ such that $\alpha x \leq \beta + \gamma$ is valid for* **P**^{k}. *It is straightforward to verify that $\gamma \leq d(\mathbf{P}, \mathbf{P}^k)$. Therefore, the distance measure gives an upper bound on additive error when optimizing a (normalized) linear function over* **P** *and* **P**^{k}.

Observation 2 (Comparing d(P, P^k) to $\sqrt{n}$). *Notice that the largest distance between any two points in the $[0, 1]^n$ hypercube is at most $\sqrt{n}$. Therefore in the rest of the paper we will compare the value of $d(\mathbf{P}, \mathbf{P}^k)$ to $\sqrt{n}$.*

1.3 Some Examples

In order to build some intuition we begin with some examples in this section. Let $\mathbf{P} := \{x \in [0, 1]^n : ax \leq b\}$ where a is a non-negative vector. It is straightforward to verify that in this case, $\mathbf{P}^k := \{x \in [0, 1]^n : a^I x \leq b \ \forall I \in \binom{[n]}{k}\}$, where $a_j^I := a_j$ if $j \in I$ and $a_j^I = 0$ otherwise.

Example 1: Consider the simplex $\mathbf{P} = \{x \in [0,1]^n : \sum_{i=1}^n x_i \leq 1\}$. Using the above observation, we have that $\mathbf{P}^k = \text{conv}\{e^1, e^2, \ldots, e^n, \frac{1}{k}e\}$, where e^j is the unit vector in the direction of the jth coordinate and e is the all ones vector. Therefore the distance measure between **P** and **P**^{k} is $\sqrt{n}(\frac{1}{k} - \frac{1}{n}) \approx \frac{\sqrt{n}}{k}$, attained by the points $\frac{1}{n}e \in \mathbf{P}$ and $\frac{1}{k}e \in \mathbf{P}^k$. This is quite nice because with $k \approx \sqrt{n}$ (which is pretty reasonably sparse) we get a constant distance. Observe also that the *rate of change of the distance measure* follows a 'single pattern' - we call this a *single phase example*. See Figure 1(a) for $d(\mathbf{P}, \mathbf{P}^k)$ plotted against k (in blue) and $k \cdot d(\mathbf{P}, \mathbf{P}^k)$ plotted against k (in green).

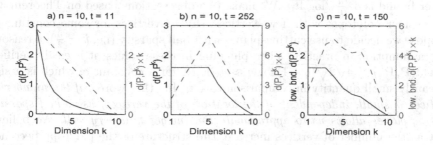

Fig. 1. (a) Sparsity is good. (b) Sparsity is not so good. (c) Example with three phases.

Example 2: Consider the set $\mathbf{P} = \{x \in [0,1]^n : \sum_i x_i \leq \frac{n}{2}\}$. We have that $\mathbf{P}^k := \{x \in [0, 1]^n : \sum_{i \in I} x_i \leq \frac{n}{2}, \ \forall I \in \binom{[n]}{k}\}$. Therefore, for all $k \in \{1, \ldots, n/2\}$ we have $\mathbf{P}^k = [0, 1]^n$ and hence $d(\mathbf{P}, \mathbf{P}^k) = \sqrt{n}/2$. Thus, we stay with distance $\Omega(\sqrt{n})$ (the worst possible for polytopes in $[0, 1]^n$) even with $\Theta(n)$ sparsity. Also observe that for $k > \frac{n}{2}$, we have $d(\mathbf{P}, \mathbf{P}^k) = \frac{n\sqrt{n}}{2k} - \frac{\sqrt{n}}{2}$. Clearly the rate of change of the distance measure has *two phases*, first phase of k between 1 and $\frac{n}{2}$ and the second phase of k between $\frac{n}{2}$ and n. See Figure 1(b) for the plot of $d(\mathbf{P}, \mathbf{P}^k)$ against k (in blue) and of $k \cdot d(\mathbf{P}, \mathbf{P}^k)$ against k (in green).

Example 3: We present an experimental example in dimension $n = 10$. The polytope $\mathbf{P}$ is now set as the convex hull of 150 binary points randomly selected from the hyperplane $\{x \in \mathbb{R}^{10} : \sum_{i=1}^{10} x_i = 5\}$. We experimentally computed lower bounds on $d(\mathbf{P}, \mathbf{P}^k)$ which are plotted in Figure 1(c) as the blue line (details appear in the full version of the paper). Notice that there are now three phases, which are more discernible in the plot between the lower bound on $k \cdot d(\mathbf{P}, \mathbf{P}^k)$ and k (in green).

The above examples serve to illustrate the fact that different polytopes, behave very differently when we try and approximate them using sparse inequalities. We note here that in all our additional experiments, albeit in small dimensions, we have usually found at most three phases as in the previous examples.

2 Main Results

2.1 Upper Bounds

Surprisingly, it appears that the complicated behavior of $d(\mathbf{P}, \mathbf{P}^k)$ as k changes can be described to some extent in closed form. Our first result is nontrivial upper bounds on $d(\mathbf{P}, \mathbf{P}^k)$ for general polytopes. This is proven in Section 3.

Theorem 3 (Upper Bound on $d(\mathbf{P}, \mathbf{P}^k)$). *Let $n \geq 2$. Let $\mathbf{P} \subseteq [0,1]^n$ be the convex hull of points $\{p^1, \ldots, p^t\}$. Then*

1. $d(\mathbf{P}, \mathbf{P}^k) \leq 4 \max \left\{ \frac{n^{1/4}}{\sqrt{k}} \sqrt{8 \max_{i \in [t]} \|p^i\|} \sqrt{\log 4tn}, \frac{8\sqrt{n}}{3k} \log 4tn \right\}$
2. $d(\mathbf{P}, \mathbf{P}^k) \leq 2\sqrt{n} \left(\frac{n}{k} - 1 \right)$.

Since $\max_{i \in \{1, \ldots, t\}} \|p^i\| \leq \sqrt{n}$ and the first upper bound yields nontrivial values only when $k \geq 8 \log 4tn$, a simpler (although weaker) expression for the first upper bound is $4 \frac{\sqrt{n}}{\sqrt{k}} \sqrt{\log 4tn}$. We make two observations based on Theorem 3.

Consider polytopes with 'few' vertices, say n^q vertices for some constant q. Suppose we decide to use cutting-planes with half sparsity (i.e. $k = \frac{n}{2}$), a reasonable assumption in practice. Then plugging in these values, it is easily verified that $d(\mathbf{P}, \mathbf{P}^k) \leq 4\sqrt{2} \sqrt{(q+1) \log n} \approx c\sqrt{\log n}$ for a constant c, which is a significantly small quantity in comparison to $\sqrt{n}$. In other words, *if the number of vertices is small, independent of the location of the vertices, using half sparsity cutting-planes allows us to approximate the integer hull very well.* We believe that as the number of vertices increase, the structure of the polytope becomes more important in determining $d(\mathbf{P}, \mathbf{P}^k)$ and Theorem 3 only captures the worst-case scenario. Overall, Theorem 3 presents a theoretical justification for the use of sparse cutting-planes in many cases.

Theorem 3 supports the existence of three phases in the behavior of $d(\mathbf{P}, \mathbf{P}^k)$ as k varies: **(Small k)** When $k \leq 16 \log 4tn$ the (simplified) upper bounds are larger than $\sqrt{n}$, indicating that 'no progress' is made in approximating the shape of $\mathbf{P}$ (this is seen Examples 2 and 3). **(Medium k)** When $16 \log 4tn \leq k \lesssim n - \sqrt{n \log 4tn}$ the first upper bound in Theorem 3 dominates. **(Large k)** When $k \gtrsim n - \sqrt{n \log 4tn}$ the upper bound $2\sqrt{n} \left(\frac{n}{k} - 1 \right)$ dominates. In particular, in this phase, $k \cdot d(\mathbf{P}, \mathbf{P}^k) \leq 2n^{3/2} - 2\sqrt{n}k$, i.e., the upper bound times k is a linear function of k. All the examples in Section 1 illustrate this behaviour.

2.2 Lower Bounds

How good is the quality of the upper bound presented in Theorem 3? Let us first consider the second upper bound in Theorem 3. Then observe that for the second example in Section 1, this upper bound is tight up to a constant factor for k between the values of $\frac{n}{2}$ and n.

We study lower bounds on $d(\mathbf{P}, \mathbf{P}^k)$ for random polytopes in Section 4 that show that the first upper bound in Theorem 3 is also quite tight.

Theorem 4. Let $X^1, X^2, \ldots, X^t$ be independent uniformly random points in $\{0,1\}^n$, and let $\mathbf{P} = conv(X^1, X^2, \ldots, X^t)$. Then for t and k satisfying $(2k^2 \log n + 2)^2 \leq t \leq e^n$ we have with probability at least $1/4$

$$d(\mathbf{P}, \mathbf{P}^k) \geq \min\left\{ \frac{\sqrt{n}}{\sqrt{k}} \frac{\sqrt{\log(t/2)}}{78\sqrt{\log n}}, \frac{\sqrt{n}}{8} \right\} \left(\frac{1}{2} - \frac{1}{k^{3/2}} \right) - 3\sqrt{\log t}.$$

Let us compare this lower bound with the simpler expression $4\frac{\sqrt{n}}{\sqrt{k}}\sqrt{\log tn}$ for the first part of the upper bound of Theorem 3. We focus on the case where the minimum in the lower bound is acheived by the first term. Then comparing the leading term $\sqrt{\frac{n}{k}} \frac{\sqrt{\log t}}{2 \cdot 78\sqrt{\log n}}$ in the lower bound with the upper bound, we see that these quantities match up to a factor of $624\frac{\sqrt{\log(tn)}\sqrt{\log n}}{\sqrt{\log t}}$, showing that for many 0/1 polytopes the first upper bound of Theorem 3 is quite tight. We also remark that the in order to simplify the exposition we did not try to optimize constants and lower order terms in our bounds.

The main technical tool for proving this lower bound is a new anticoncentration result for linear combinations aX, where the X_i's are independent Bernoulli random variables. The main difference from standard anticoncentration results is that the latter focus on variation around the standard deviation; in this case, standard tools such as the Berry-Esseen Theorem or the Paley-Zygmund Inequality can be used to obtain constant-probability anticoncentration. However, we need to control the behavior of aX much further from its standard deviation, where we cannot hope to get constant-probability anticoncentration.

Lemma 1. Let $X_1, X_2, \ldots, X_n$ be independent random variables with X_i taking value 0 with probability 1/2 and value 1 with probability 1/2. Then for every $a \in [-1,1]^n$ and $\alpha \in [0, \frac{\sqrt{n}}{8}]$,

$$\Pr\left(aX \geq \mathbb{E}[aX] + \frac{\alpha}{2\sqrt{n}} \left(1 - \frac{1}{n^2}\right) \|a\|_1 - \frac{1}{2n^2} \right) \geq \left(e^{-50\alpha^2} - e^{-100\alpha^2} \right)^{60 \log n}.$$

2.3 Hard Packing Integer Programs

We also study well-known, randomly generated, hard packing integer program instances (see for instance [5]). Given parameters $n, m, M \in \mathbb{N}$, the convex hull of the packing IP is given by $\mathbf{P} = conv(\{x \in \{0,1\}^n : A^j x \leq \frac{\sum_i A_i^j}{2}, \forall j \in [m]\})$, where the A_i^j's are chosen independently and uniformly in the set $\{0, 1, \ldots, M\}$. Let (n, m, M)-PIP denote the distribution over the generated $\mathbf{P}$'s.

The following result shows the limitation of sparse cuts for these instances.

Theorem 5. *Consider* $n, m, M \in \mathbb{N}$ *such that* $n \geq 50$ *and* $8 \log 8n \leq m \leq n$. *Let* $\mathbf{P}$ *be sampled from the distribution* (n, m, M)-PIP. *Then with probability at least* $1/2$, $d(\mathbf{P}, \mathbf{P}^k) \geq \frac{\sqrt{n}}{2} \left(\frac{2}{\max\{\alpha, 1\}}(1 - \epsilon)^2 - (1 + \epsilon') \right)$, *where* $c = k/n$ *and*

$$\frac{1}{\alpha} = \frac{M}{2(M+1)} \left[\frac{n - 2\sqrt{n \log 8m}}{c((2 - c)n + 1) + 2\sqrt{10cnm}} \right], \quad \epsilon = \frac{24\sqrt{\log 4n^2 m}}{\sqrt{n}}, \quad \epsilon' = \frac{3\sqrt{\log 8n}}{\sqrt{m} - 2\sqrt{\log 8n}}.$$

Notice that when m is sufficiently large, and n reasonably larger than m, we have ϵ and ϵ' approximately 0, and the above bound reduces to approximately $\frac{\sqrt{n}}{2} \left(\left(\frac{M}{M+1} \right) \left(\frac{n}{k(2-n/k)} \right) - 1 \right) \approx \frac{\sqrt{n}}{2} \left(\frac{n}{k(2-n/k)} - 1 \right)$, which is within a constant factor of the upper bound from Theorem 3. The poor behavior of sparse cuts gives an indication for the hardness of these instances and suggests that denser cuts should be explored in this case.

One interesting feature of this result is that it works directly with the IP formulation, not relying on an explicit linear description of the convex hull.

2.4 Sparse Cutting-Planes and Extended Formulations

Let $\text{proj}_x : \mathbb{R}^n \times \mathbb{R}^m \to \mathbb{R}^n$ denote the projection operator onto the first n coordinates. We say that a set $\mathbf{Q} \subseteq \mathbb{R}^n \times \mathbb{R}^m$ is an *extended formulation* of $\mathbf{P} \subseteq \mathbb{R}^n$ if $\mathbf{P} = \text{proj}_x(\mathbf{Q})$.

As our final result we remark that using sparse cutting-planes in extended formulations is at least as good as using them in the original polyhedron, and sometime much better; proofs are provided in the full version of the paper.

Lemma 2. *Consider a polyhedron* $\mathbf{P} \subseteq \mathbb{R}^n$ *and an extended formulation* $\mathbf{Q} \subseteq \mathbb{R}^n \times \mathbb{R}^m$ *for it. Then* $\text{proj}_x(\mathbf{Q}^k) \subseteq (\text{proj}_x(\mathbf{Q}))^k = \mathbf{P}^k$.

Lemma 3. *Consider* $n \in \mathbb{N}$ *and assume it is a power of 2. Then there is a polytope* $\mathbf{P} \subseteq \mathbb{R}^n$ *such that: 1)* $d(\mathbf{P}, \mathbf{P}^k) = \sqrt{n/2}$ *for all* $k \leq n/2$; *2) there is an extended formulation* $\mathbf{Q} \subseteq \mathbb{R}^n \times \mathbb{R}^{2n-1}$ *of* $\mathbf{P}$ *such that* $\text{proj}_x(\mathbf{Q}^3) = \mathbf{P}$.

3   Upper Bound

In this section we prove Theorem 3. In fact we prove the same bound for polytopes in $[-1, 1]^n$, which is a slightly stronger result. The following well-known property is crucial for the constructions used in both parts of the theorem.

Observation 6 (Section 2.5.1 of [3]). *Consider a compact convex set* $S \subseteq \mathbb{R}^n$. *Let* $\bar{x}$ *be a point outside* S *and let* $\bar{y}$ *be the closest point to* $\bar{x}$ *in* S. *Then setting* $a = \bar{x} - \bar{y}$, *the inequality* $ax \leq a\bar{y}$ *is valid for* S *and cuts* $\bar{x}$ *off.*

3.1 Proof of First Part of Theorem 3

Consider a polytope $\mathbf{P} = \text{conv}\{p^1, p^2, \ldots, p^t\}$ in $[-1, 1]^n$. Define

$$\lambda^* = \max\left\{\frac{n^{1/4}}{\sqrt{k}}\sqrt{8\max_i \|p^i\|}\sqrt{\log 4tn}, \frac{8\sqrt{n}}{3k}\log 4tn\right\}.$$

In order to show that $d(\mathbf{P}, \mathbf{P}^k)$ is at most $4\lambda^*$ we show that every point at distance more than $4\lambda^*$ from $\mathbf{P}$ is cut off by a valid inequality for $\mathbf{P}^k$. Assume until the end of this section that $4\lambda^*$ is at most $\sqrt{n}$, otherwise the result is trivial; in particular, this implies that the second term in the definition of λ^* is at most $\sqrt{n}/4$ and hence $k \geq 8\log 4tn$.

So let $u \in \mathbb{R}^n$ be a point at distance more than $4\lambda^*$ from $\mathbf{P}$. Let $v \in \mathbf{P}$ be the closest point in $\mathbf{P}$ to $\mathbf{P}^k$. We can write $u = v + \lambda d$ for some vector d with $\|d\|_2 = 1$ and $\lambda > 4\lambda^*$. From Observation 6, inequality $dx \leq dv$ is valid for $\mathbf{P}$, so in particular $dp^i \leq dv$ for all $i \in [t]$; in addition, it that this inequality cuts off u: $du = dv + \lambda > dv$. The idea is to use this extra slack factor λ in the previous equation to show we can 'sparsify' the inequality $dx \leq dv$ while maintaining separation of $\mathbf{P}$ and u. It then suffices to prove the following lemma.

Lemma 4. *There is a k-sparse vector $\tilde{d} \in \mathbb{R}^n$ such that $\tilde{d}p^i \leq \tilde{d}v + \frac{\lambda}{2}$ for all $i \in [t]$, and $\tilde{d}u > \tilde{d}v + \frac{\lambda}{2}$.*

To prove the lemma we construct a random vector $\tilde{D} \in \mathbb{R}^n$ which, with non-zero probability, is k-sparse and satisfies the two other requirements of the lemma. Let $\alpha = \frac{k}{2\sqrt{n}}$. Define $\tilde{D}$ as the random vector with independent coordinates, where $\tilde{D}_i$ is defined as follows: if $\alpha|d_i| \geq 1$, then $\tilde{D}_i = d_i$ with probability 1; if $\alpha|d_i| < 1$, then $\tilde{D}_i$ takes value $\text{sign}(d_i)/\alpha$ with probability $\alpha|d_i|$ and takes value 0 with probability $1 - \alpha|d_i|$. (For convenience we define $\text{sign}(0) = 1$.)

The next observation follows directly from the definition of $\tilde{D}$.

Observation 7. *For every vector $a \in \mathbb{R}^n$ the following hold:*
1. $\mathbb{E}[\tilde{D}a] = da$
2. $\text{Var}(\tilde{D}a) \leq \frac{1}{\alpha}\sum_{i \in [n]} a_i^2|d_i|$
3. $|\tilde{D}_i a_i - \mathbb{E}[\tilde{D}_i a_i]| \leq \frac{|a_i|}{\alpha}$.

Claim. With probability at least $1 - 1/4n$, $\tilde{D}$ is k-sparse.

Proof. Construct the vector $a \in \mathbb{R}^n$ as follows: if $\alpha|d_i| \geq 1$ then $a_i = 1/d_i$, and if $\alpha|d_i| < 1$ then $a_i = \alpha/\text{sign}(d_i)$. Notice that $\tilde{D}a$ equals the number of non-zero coordinates of $\tilde{D}$ and $\mathbb{E}[\tilde{D}a] \leq \alpha\|d\|_1 \leq k/2$. Also, from Observation 7 we have

$$\text{Var}(\tilde{D}a) \leq \frac{1}{\alpha}\sum_{i \in [n]} a_i^2|d_i| \leq \alpha\|d\|_1 \leq \frac{k}{2}.$$

Then using Bernstein's inequality ([6], Appendix A.2) we obtain

$$\Pr(\tilde{D}a \geq k) \leq \exp\left(-\min\left\{\frac{k^2}{8k}, \frac{3k}{8}\right\}\right) \leq \frac{1}{4n},$$

where the last inequality uses our assumption that $k \geq 8\log 4tn$. $\qquad\square$

We now show that property 1 required by Lemma 4 holds for $\tilde{D}$ with high probability.

Claim. $\Pr(\max_{i \in [t]}[\tilde{D}(p^i - v) - d(p^i - v)] > 2\lambda^*) \le 1/4n$.

Proof. Define the centered random variable $Z = \tilde{D} - d$. To make the analysis cleaner, notice that $\max_{i \in [t]} Z(p^i - v) \le 2\max_{i \in [t]} |Zp^i|$; this is because $\max_{i \in [t]} Z(p^i - v) \le \max_{i \in [t]} |Zp^i| + |Zv|$, and because for all $a \in \mathbb{R}^n$ we have $|av| \le \max_{p \in \mathbf{P}} |ap| = \max_{i \in [t]} |ap^i|$ (since $v \in \mathbf{P}$).

Therefore our goal is to upper bound the probability that the process $\max_{i \in [t]} |Zp^i|$ is larger then λ^*. Fix $i \in [t]$. By Bernstein's inequality,

$$\Pr(|Zp^i| \ge \lambda^*) \le \exp\left(-\min\left\{\frac{(\lambda^*)^2}{4\mathrm{Var}(|Zp^i|)}, \frac{3\lambda^*}{4M}\right\}\right), \tag{1}$$

where M is an upper bound on $\max_j |Z_j p_j^i|$.

To bound the terms in the right-hand side, from Obersvation 7 we have

$$\mathrm{Var}(Zp^i) = \mathrm{Var}(\tilde{D}p^i) \le \frac{1}{\alpha}\sum_j (p_j^i)^2 |d_j| \le \frac{1}{\alpha}\sum_j p_j^i |d_j| \le \frac{1}{\alpha}\|p^i\|\|d\| = \frac{1}{\alpha}\|p^i\|,$$

where the second inequality follows from the fact $p^i \in [0,1]^n$, and the third inequality follows from the Cauchy-Schwarz inequality. Moreover, it is not difficulty to see that for every random variable W, $\mathrm{Var}(|W|) \le \mathrm{Var}(W)$. Using the first term in the definition of λ^*, we then have

$$\frac{(\lambda^*)^2}{\mathrm{Var}(|Zp^i|)} \ge 4\log 4tn.$$

In addition, for every coordinate j we have $|Z_j p_j^i| = |\tilde{D}_j p_j^i - \mathbb{E}[\tilde{D}_j p_j^i]| \le 1/\alpha$, where the inequality follows from Observation 7. Then we can set $M = 1/\alpha$ and using the second term in the definition of λ^* we get $\frac{\lambda^*}{M} \ge \frac{4}{3}\log 4tn$. Therefore, replacing these bounds in inequality (1) gives $\Pr(|Zp^i| \ge \lambda^*) \le \frac{1}{4tn}$.

Taking a union bound over all $i \in [t]$ gives that $\Pr(\max_{i \in [t]} |Zp^i| \ge \lambda^*) \le 1/4n$. This concludes the proof of the claim. $\square$

Claim. $\Pr(\tilde{D}(u - v) \le \lambda/2) \le 1 - 1/(2n - 1)$.

Proof. Recall $u - v = \lambda d$, hence it is equivalent to bound $\Pr(\tilde{D}d \le 1/2)$. First, $\mathbb{E}[\tilde{D}d] = dd = 1$. Also, from Observation 7 we have $\tilde{D}d \le |\tilde{D}d - dd| + |dd| \le \frac{1}{\alpha}\sum_i |d_i| + 1 \le \frac{2n}{k} + 1 \le n$, where the last inequality uses the assumption $k \ge 8\log 4tn$. Then employing Markov's inequality to the non-negative random variable $n - \tilde{D}d$, we get $\Pr(\tilde{D}d \le 1/2) \le 1 - \frac{1}{2n-1}$. This concludes the proof. $\square$

Proof of Lemma 4. Employ the previous three claims and union bound to find a realization of $\tilde{D}$ that is k-sparse and satisfies requirements 1 and 2 of the lemma.

This concludes the proof of the first part of Theorem 3.

Observation 8. *Notice that in the above proof λ^* is set by Claim 3.1, and need to be essentially $\mathbb{E}[\max_{i \in [t]}(\tilde{D} - d)p^i]$. There is a vast literature on bounds on the supremum of stochastic processes [6], and improved bounds for structured $\mathbf{P}$'s are possible (for instance, via the* generic chaining *method).*

3.2 Proof of Second Part of Theorem 3

The main tool for proving this upper bound is the following lemma, which shows that when $\mathbf{P}$ is 'simple', and we have a stronger control over the distance of a point $\bar{x}$ to $\mathbf{P}$, then there is a k-sparse inequality that cuts $\bar{x}$ off.

Lemma 5. *Consider a hyperplane $H = \{x \in \mathbb{R}^n : ax \le b\}$ and let $\mathbf{P} = H \cap [-1,1]^n$. Let $\bar{x} \in [-1,1]^n$ be such that $d(\bar{x}, H) > 2\sqrt{n}(\frac{n}{k} - 1)$. Then $\bar{x} \notin \mathbf{P}^k$.*

Proof. Assume without loss of generality that $\|a\|_2 = 1$. Let $\bar{y}$ be the point in H closes to $\bar{x}$, and notice that $\bar{x} = \bar{y} + \lambda a$ where $\lambda > \sqrt{n}(\frac{n}{k} - 1)$.

For any set $I \in \binom{[n]}{k}$, the inequality $\sum_{i \in I} a_i x_i \le b + \sum_{i \notin I : a_i \ge 0} a_i - \sum_{i \notin I : a_i < 0} a_i$ is valid for $\mathbf{P}$; since it is k-sparse, it is also valid for $\mathbf{P}^k$. Averaging out this inequality over all $I \in \binom{[n]}{k}$, we get that the following is valid for $\mathbf{P}^k$:

$$\frac{k}{n} ax \le b + \left(1 - \frac{k}{n}\right) \left(\sum_{i:a_i \ge 0} a_i - \sum_{i:a_i < 0} a_i\right) \equiv ax \le b + \left(\frac{n}{k} - 1\right)(b + \|a\|_1).$$

We claim that $\bar{x}$ violates this inequality. First notice that $a\bar{x} = a\bar{y} + \lambda = b + \lambda > b + 2\sqrt{n}\left(\frac{n}{k} - 1\right)$, hence it suffices to show $b + \|a\|_1 \le 2\sqrt{n}$. Our assumption on $\bar{x}$ implies that $\mathbf{P} \ne [-1,1]^n$, and hence $b < \max_{x \in [-1,1]} ax = \|a\|_1$; this gives $b + \|a\|_1 \le 2\|a\|_1 \le 2\sqrt{n}\|a\|_2 = 2\sqrt{n}$, thus concluding the proof. $\square$

To prove the second part of Theorem 3 consider a point $\bar{x}$ of distance greater than $2\sqrt{n}(\frac{n}{k} - 1)$ from $\mathbf{P}$; we show $\bar{x} \notin \mathbf{P}^k$. Let $\bar{y}$ be the closest point to $\bar{x}$ in $\mathbf{P}$. Let $a = \bar{x} - \bar{y}$. From Observation 6 we have that $ax \le a\bar{y}$ is valid for $\mathbf{P}$. Define $H' = \{x \in \mathbb{R}^n : ax \le a\bar{y}\}$ and $\mathbf{P}' = H' \cap [-1,1]^n$. Notice that $d(\bar{x}, H') = d(\bar{x}, \bar{y}) > 2\sqrt{n}(\frac{n}{k} - 1)$. Then Lemma 5 guarantees that $\bar{x}$ does not belong to $\mathbf{P}'^k$. But $\mathbf{P} \subseteq \mathbf{P}'$, so by monotonicity of the k-sparse closure we have $\mathbf{P}^k \subseteq \mathbf{P}'^k$; this shows that $\bar{x} \notin \mathbf{P}^k$, thus concluding the proof.

4 Lower Bound

In this section we prove Theorem 4. The proof is based on the 'bad' polytope of Example 2 and proceeds in two steps. First, for a random 0/1 polytope $\mathbf{P}$ we show that with good probability the facets $dx \le d_0$ for $\mathbf{P}^k$ have d_0 being large, namely $d_0 \gtrsim \left(\frac{1}{2} + \frac{\sqrt{\log t}}{\sqrt{k}}\right) \sum_i d_i$; therefore, with good probability the point $\bar{p} \approx (\frac{1}{2} + \frac{\sqrt{\log t}}{\sqrt{k}})e$ belongs to $\mathbf{P}^k$. In the second step, we show that with good probability the distance from $\bar{p}$ to $\mathbf{P}$ is at least $\approx \sqrt{\frac{n}{k}}\sqrt{\log t}$, by showing that the inequality $\sum_i x_i \lesssim \frac{n}{2} + \sqrt{n}$ is valid for $\mathbf{P}$.

We now proceed with the proof. Consider the random set $\mathbf{X} = \{X^1, X^2, \dots, X^t\}$ where the X^i's are independent uniform random points in $\{0, 1\}^n$, and define the random 0/1 polytope $\mathbf{P} = \text{conv}(\mathbf{X})$.

We say that a 0/1 polytope $\mathbf{P}$ is α-*tough* if for every facet $dx \leq d_0$ of $\mathbf{P}^k$ we have $d_0 \geq \frac{\sum_i d_i}{2} + \frac{\alpha}{k^2}(1 - \frac{1}{k^2})\|d\|_1 - \|d\|_\infty/2k^2$. To get a handle on α-toughness of random 0/1 polytopes, define $\mathcal{D}$ as the set of all integral vectors $\ell \in \mathbb{R}^n$ that are k-sparse and satisfy $\|\ell\|_\infty \leq (k+1)^{(k+1)/2}$. The following claim, shows that all the facets of $\mathbf{P}^k$ come from the set $\mathcal{D}$; it follows directly from applying Corollary 26 in [8] to each term $\mathbf{P} + \mathbb{R}^{\bar{I}}$ in the definition of $\mathbf{P}^k$ from Section 1.1.

Lemma 6. *Let $\mathbf{Q} \subseteq \mathbb{R}^n$ be a 0/1 polytope. Then there is a subset $\mathcal{D}' \subseteq \mathcal{D}$ such that $\mathbf{Q}^k = \{x : dx \leq \max_{y \in \mathbf{P}^k} dy, \ d \in \mathcal{D}'\}$.*

Now we can analyze the probability that $\mathbf{P}$ is α-tough.

Lemma 7. *If $1 \leq \alpha^2 \leq \min\left\{\frac{\log(t/2)}{6000 \log n}, \frac{k}{64}\right\}$ and $k \leq n - 1$, then $\mathbf{P}$ is α-tough with probability at least $1/2$.*

Proof. Let E be the event that for all $d \in \mathcal{D}$ we have $\max_{i \in [t]} dX^i \geq \frac{1}{2}\sum_j d_j + \frac{\alpha}{2\sqrt{k}}(1 - \frac{1}{k^2})\|d\|_1 - \|d\|_\infty/2k^2$. Because of Lemma 6, whenever E holds we have that $\mathbf{P}$ is α-tough and thus it suffices to show $\Pr(E) \geq 1/2$.

Fix $d \in \mathcal{D}$. Since d is k-sparse, we can apply Lemma 1 to $d/\|d\|_\infty$ restricted to the coordinates in its support to obtain that

$$\Pr\left(dX^i \geq \frac{\sum_i d_i}{2} + \frac{\alpha}{2\sqrt{k}}\left(1 - \frac{1}{k^2}\right)\|d\|_1 - \frac{\|d\|_\infty}{2k^2}\right) \geq \left(e^{-50\alpha^2} - e^{-100\alpha^2}\right)^{60 \log n}$$

$$\geq e^{-100\alpha^2 \cdot 60 \log n} \geq \frac{1}{t^{1/2}},$$

where the second inequality follows from our lower bound on α and the last inequality follows from our upper bound on α. By independence of the X^i's,

$$\Pr\left(\max_{i \in [t]} dX^i < \frac{\sum_i d_i}{2} + \frac{\alpha}{2\sqrt{k}}\left(1 - \frac{1}{k^2}\right)\|d\|_1 - \frac{\|d\|_\infty}{2k^2}\right) \leq \left(1 - \frac{1}{t^{1/2}}\right)^t \leq e^{-t^{1/2}},$$

where the second inequality follows from the fact that $(1 - x) \leq e^{-x}$ for all x.

Finally notice that $|\mathcal{D}| = \binom{n}{k}(k+1)^{(k+1)^2/2}$ and that, by our assumption on the size of t and $k \leq n - 1$, $e^{-t^{1/2}} \leq (1/2)|\mathcal{D}|$. Therefore, taking a union bound over all $d \in \mathcal{D}$ of the previous displayed inequality gives $\Pr(E) \geq 1/2$, concluding the proof of the lemma. $\qquad\square$

The next lemma takes care of the second step of the argument; its simple proof is based on Bernstein's and is deferred to the full version of the paper.

Lemma 8. *With probability at least $3/4$, the inequality $\sum_j x_j \leq \frac{n}{2} + 3\sqrt{n \log t}$ is valid for $\mathbf{P}$.*

Lemma 9. *Suppose that the polytope* $\mathbf{Q}$ *is* α-*tough for* $\alpha \geq 1$ *and that the inequality* $\sum_i x_i \leq \frac{n}{2} + 3\sqrt{n \log t}$ *is valid for* $\mathbf{Q}$. *Then we have* $d(\mathbf{Q}, \mathbf{Q}^k) \geq \sqrt{n} \left(\frac{\alpha}{2\sqrt{k}} - \frac{\alpha}{k^2} - \frac{3\sqrt{\log t}}{\sqrt{n}} \right)$.

Proof. We first show that the point $\bar{q} = (\frac{1}{2} + \frac{\alpha}{2\sqrt{k}} - \frac{\alpha}{k^2})e$ belongs to $\mathbf{P}$. Let $dx \leq d_0$ be facet for $\mathbf{P}$. Then we have

$$d\bar{q} = \frac{\sum_i d_i}{2} + \alpha \left(\frac{1}{2\sqrt{k}} - \frac{1}{k^2} \right) \sum_i d_i \leq \frac{\sum_i d_i}{2} + \alpha \left(\frac{1}{2\sqrt{k}} - \frac{1}{k^2} \right) \|d\|_1$$

$$\leq \frac{\sum_i d_i}{2} + \alpha \left(\frac{1}{2\sqrt{k}} - \frac{1}{2k^2} \right) \|d\|_1 - \frac{\|d\|_\infty}{2k^2},$$

where the first inequality uses the fact that $\frac{1}{2\sqrt{k}} - \frac{1}{k^2} \geq 0$ and the second inequality uses $\alpha \geq 1$ and $\|d\|_1 \geq \|d\|_\infty$. Since $\mathbf{Q}$ is α-tough it follows that $\bar{q}$ satisfies $dx \leq d_0$; since this holds for all facets of $\mathbf{Q}$, we have $\bar{q} \in \mathbf{Q}$.

Now define the halfspace $H = \{x : \sum_i x_i \leq \frac{n}{2} + 3\sqrt{n \log t}\}$. By assumption $\mathbf{Q} \subseteq H$, and hence $d(\mathbf{Q}, \mathbf{Q}^k) \geq d(H, \mathbf{Q}^k)$. But it is easy to see that the point in H closest to $\bar{q}$ is the point $\tilde{q} = (\frac{1}{2} + \frac{3\sqrt{\log t}}{\sqrt{n}})e$. This gives that $d(\mathbf{Q}, \mathbf{Q}^k) \geq d(H, \mathbf{Q}^k) \geq d(\bar{q}, \tilde{q}) \geq \sqrt{n} \left(\frac{\alpha}{2\sqrt{k}} - \frac{\alpha}{k^2} - \frac{3\sqrt{\log t}}{\sqrt{n}} \right)$. This concludes the proof. $\square$

We now conclude the proof of Theorem 4. Set $\bar{\alpha}^2 = \min \left\{ \frac{\log(t/2)}{6000 \log n}, \frac{k}{64} \right\}$. Taking union bound over Lemmas 7 and 8, with probability at least $1/4$ $\mathbf{P}$ is $\bar{\alpha}$-tough and the inequality inequality $\sum_i x_i \leq \frac{n}{2} + 3\sqrt{n \log t}$ is valid for it. Then from Lemma 9 we get that with probability at least $1/4$, $d(\mathbf{P}, \mathbf{P}^k) \geq \sqrt{n} \left(\frac{\bar{\alpha}}{2\sqrt{k}} - \frac{\bar{\alpha}}{k^2} - \frac{3\sqrt{\log t}}{\sqrt{n}} \right)$, and the result follows by plugging in the value of $\bar{\alpha}$.

5 Hard Packing Integer Programs

In this section we prove Theorem 5; missing proof are presented in the full version of the paper. With overload in notation, we use $\binom{[n]}{k}$ to denote the set of vectors in $\{0,1\}^n$ with exactly k 1's.

Let $\mathbf{P}$ be a random polytope sampled from the distribution (n, m, M)-PIP and consider the corresponding random vectors A^j's. The idea of the proof is to show that with constant probability $\mathbf{P}$ behaves like Example 2, by showing that the cut $\sum_i x_i \lesssim \frac{n}{2}$ is valid for it and that it approximately contains 0/1 points with many 1's.

We start with a couple of lemmas that are proved via Bernstein's inequality.

Lemma 10. *With probability at least* $1 - \frac{1}{8}$ *we have* $|\sum_{i=1}^n A_i^j - \frac{nM}{2}| \leq M\sqrt{n \log 8m}$ *for all* $j \in [m]$.

Lemma 11. *With probability at least* $1 - \frac{1}{4}$ *the cut* $(1 - \frac{2\sqrt{\log 8n}}{\sqrt{m}}) \sum_i x_i \leq \frac{n}{2} + \frac{\sqrt{n \log 8}}{\sqrt{m}}$ *is valid for* $\mathbf{P}$.

The next lemma shows that with constant probability $\mathbf{P}$ almost contains *all* 0/1 points with many 1's.

Lemma 12. *With probability at least* $1 - \frac{1}{8}$ *we have*

$$A^j \bar{x} \le \frac{(M+1)c(2n - cn + 1)}{2} + (M+1)\sqrt{10cnm}, \qquad \forall j \in [m], \forall \bar{x} \in \binom{[n]}{cn}.$$

Lemma 13. *Consider a 0/1 polytope* $\mathbf{Q} = conv(\{x \in \{0,1\}^n : a^j x \le b_j,\ j = 1, 2, \ldots, m\})$ *where* $n \ge 20$, $m \le n$, $a_i^j \in [0, M]$ *for all* i, j, *and* $b_j \ge \frac{nM}{12}$ *for all* i. *Consider* $1 < \alpha \le 2\sqrt{n}$ *and let* $\bar{x} \in \{0,1\}^n$ *be such that for all* j, $a^j \bar{x} \le \alpha b_j$. *Then the point* $\frac{1}{\alpha}(1 - \epsilon)^2 \bar{x}$ *belongs to* $\mathbf{Q}$ *as long as* $\frac{12\sqrt{\log 4n^2 m}}{\sqrt{n}} \le \epsilon \le \frac{1}{2}$.

Proof of Theorem 5. Recall the definitions of $\alpha, \epsilon, \epsilon'$, and $c = k/n$ from the statement of the theorem. Let E be the event that Lemmas 10, 11 and 12 hold; notice that $\Pr(E) \ge 1/2$. For the rest of the proof we fix a $\mathbf{P}$ (and the associated A^j's) where E holds and prove a lower bound on $\mathrm{d}(\mathbf{P}, \mathbf{P}^k)$.

Consider a set $I \in \binom{[n]}{cn}$ and let $\bar{x}$ be the incidence vector of I (i.e. $\bar{x}_i = 1$ if $i \in I$ and $\bar{x}_i = 0$ if $i \notin I$). Since the bounds from Lemmas 10 and 12 hold for our $\mathbf{P}$, straightforward calculations show that $A^j \bar{x} \le \alpha \frac{1}{2} \sum_i A_i^j$ for all $j \in [m]$. Therefore, from Lemma 13 we have that the point $\frac{1}{\max\{\alpha, 1\}}(1 - \epsilon)^2 \bar{x}$ belongs to $\mathbf{P}$. This means that the point $\tilde{x} = \frac{1}{\max\{\alpha, 1\}}(1 - \epsilon)^2 e$ belongs to $\mathbf{P} + \mathbb{R}^{\bar{I}}$ (see Section 1.1). Since this holds for every $I \in \binom{[n]}{cn}$, we have $\tilde{x} \in \mathbf{P}^k$.

Let $\tilde{y}$ be the point in $\mathbf{P}$ closest to $\tilde{x}$. Let $a = (1 - \frac{2\sqrt{\log 8n}}{\sqrt{m}})$ and $b = \frac{n}{2} + \sqrt{n}\log 8m$, so that the cut in Lemma 11 is given by $aex \le b$. From Cauchy–Schwarz we have that $\mathrm{d}(\tilde{x}, \tilde{y}) \ge \frac{ae\tilde{x} - ae\tilde{y}}{\|ae\|} = \frac{e\tilde{x}}{\sqrt{n}} - \frac{ae\tilde{y}}{a\sqrt{n}}$.

By definition of $\tilde{x}$ we have $e\tilde{x} = \frac{1}{\max\{\alpha, 1\}}(1 - \epsilon)^2 n$. From the fact the cut $aex \le b$ is valid for $\mathbf{P}$ and $\tilde{y} \in \mathbf{P}$, we have $ae\tilde{y} \le b$. Simple calculations show that $\frac{b}{a\sqrt{n}} \le \frac{n}{2}(1 + \epsilon')$. Plugging these values in we get that $\mathrm{d}(\mathbf{P}, \mathbf{P}^k) = \mathrm{d}(\tilde{x}, \tilde{y}) \ge \frac{\sqrt{n}}{2}\left(\frac{2(1 - \epsilon)^2}{\max\{\alpha, 1\}} - (1 + \epsilon')\right)$. Theorem 5 follows from the definition of α, ϵ and ϵ'.

References

1. Achterberg, T.: Personal communication
2. Andersen, K., Weismantel, R.: Zero-coefficient cuts. In: Eisenbrand, F., Shepherd, F.B. (eds.) IPCO 2010. LNCS, vol. 6080, pp. 57–70. Springer, Heidelberg (2010)
3. Boyd, S., Vandenberghe, L.: Convex Optimization. Cambridge University Press (2004)
4. Gu, Z.: Personal communication
5. Kaparis, K., Letchford, A.N.: Separation algorithms for 0-1 knapsack polytopes. Mathematical Programming 124(1-2), 69–91 (2010)
6. Koltchinskii, V.: Oracle Inequalities in Empirical Risk Minimization and Sparse Recovery Problems. Springer (2011)
7. Narisetty, A.: Personal communication
8. Ziegler, G.M.: Lectures on Polytopes. Springer (1995)

Short Tours through Large Linear Forests

Uriel Feige[1,*], R. Ravi[2,**], and Mohit Singh[3]

[1] Department of Computer Science, The Weizmann Institute, Rehovot, Israel
`uriel.feige@weizmann.ac.il`
[2] Tepper School of Business, Carnegie Mellon University
`ravi@cmu.edu`
[3] Microsoft Research, Redmond, USA
`mohits@microsoft.com`

Abstract. A tour in a graph is a connected walk that visits every vertex at least once, and returns to the starting vertex. Vishnoi (2012) proved that every connected d-regular graph with n vertices has a tour of length at most $(1 + o(1))n$, where the $o(1)$ term (slowly) tends to 0 as d grows. His proof is based on van-der-Warden's conjecture (proved independently by Egorychev (1981) and by Falikman (1981)) regarding the permanent of doubly stochastic matrices. We provide an exponential improvement in the rate of decrease of the $o(1)$ term (thus increasing the range of d for which the upper bound on the tour length is nontrivial). Our proof does not use the van-der-Warden conjecture, and instead is related to the linear arboricity conjecture of Akiyama, Exoo and Harary (1981), or alternatively, to a conjecture of Magnant and Martin (2009) regarding the path cover number of regular graphs. More generally, for arbitrary connected graphs, our techniques provide an upper bound on the minimum tour length, expressed as a function of their maximum, average, and minimum degrees. Our bound is best possible up to a term that tends to 0 as the minimum degree grows.

1 Introduction

A *tour* in a graph is a connected walk that starts at a vertex, visits every vertex of the graph at least once, and returns to the starting vertex. The *length* of the tour is the number of steps of the corresponding walk. Vishnoi [19] proved the following theorem.

Theorem 1. *[19] Every n-vertex d-regular connected graph has a tour of length at most $\left(1 + \sqrt{\frac{64}{\log d}}\right) n$. Moreover, there is a randomized polynomial time algorithm that with high probability finds such a tour.*

The existential part of the proof of Theorem 1 is based on van-der-Warden's conjecture (proved independently by Egorychev [8] and by Falikman [9]) regarding the permanent of doubly stochastic matrices. (See also Section 3 of [2] for related results.) The algorithmic part is based on randomized algorithms for approximating the permanent [11]. We provide the following strengthening of Theorem 1.

---

* Work supported in part by the Israel Science Foundation (grant No. 621/12) and by the I-CORE Program of the Planning and Budgeting Committee and The Israel Science Foundation (grant No. 4/11).
** Work supported in part by NSF grant CCF-1218382.

J. Lee and J. Vygen (Eds.): IPCO 2014, LNCS 8494, pp. 273–284, 2014.
© Springer International Publishing Switzerland 2014

Theorem 2. *Every n-vertex d-regular connected graph has a tour of length at most* $\left(1 + O\left(\frac{1}{\sqrt{d}}\right)\right) n$. *Moreover, there is a randomized polynomial time algorithm that finds such a tour.*

Our proof does not use the van-der-Warden conjecture, and instead works by constructing a large *linear forest*. A linear forest is an acyclic subgraph where the degree of any vertex is at most two. Equivalently, a linear forest is a vertex disjoint union of paths. We prove the following theorem about existence of large linear forests in d-regular graphs.

Theorem 3. *Every n-vertex d-regular graph has a linear forest of size* $\left(1 - O\left(\frac{1}{\sqrt{d}}\right)\right) n$, *and moreover, such a linear forest can be found in randomized polynomial time.*

A linear forest with a large size given by Theorem 3 is useful for constructing a spanning tree with few odd-degree nodes: indeed, extending the forest to any spanning tree introduces odd degree nodes of the same order as the number of components of the forest. For an even cardinality set T of vertices, a T-join is a collection of edges which has odd-degree exactly on vertices in T. Following Christofides' algorithm [6], it then suffices to construct a T-join on these few odd degree nodes. We show that, in a graph whose minimum degree is large, there is always a small size T-join when $|T|$ is small in the following theorem.

Theorem 4. *Let $G(V, E)$ be an arbitrary connected graph with n vertices and minimum degree δ, and let $T \subset V$ be an arbitrary set of vertices of even cardinality. Then there is a T-join with fewer than $2|T| + \frac{3n}{\delta+1}$ edges.*

The above theorem then, along with Theorem 3, directly gives Theorem 2. We also observe that Theorem 4 can be thought of as a generalization of the classical result [15], up to additive constant terms, that every graph with minimum degree δ has diameter at most $\frac{3n}{\delta+1}$. This follows since when $T = \{u, v\}$ then the smallest T-join is exactly the shortest path between u and v.

We observe that certain unproved conjectures (specifically, the linear arboricity conjecture of [1], or alternatively, a conjecture of [13] regarding the path cover number of regular graphs) would imply that every d-regular graph has a linear forest of size $(1 - O(\frac{1}{d}))n$ and hence, a tour of length at most $\left(1 + O\left(\frac{1}{d}\right)\right) n$. This bound would be best possible up to the hidden constant in the O notation, as there are d-regular graphs in which every tour is of length $(1 + \Omega(\frac{1}{d}))n$. For more details, see Section 4.

In contrast to the result of Vishnoi [19], our results extend naturally to nearly regular graphs and we prove the following general theorem.

Theorem 5. *Let G be a connected n-vertex graph with maximum degree Δ, average degree d, and minimum degree δ. Then G has a tour of length at most*

$$\left(1 + \frac{\Delta - d}{\Delta} + O\left(\frac{1}{\sqrt{\Delta}}\right) + O\left(\frac{1}{\delta}\right)\right) n.$$

Moreover, there is a randomized polynomial time algorithm that finds such a tour.

Theorem 5 provides tours not much larger than n when Δ is close to d, and δ is fairly large. The term $\frac{\Delta - d}{\Delta}$ is best possible, but the error terms $O\left(\frac{1}{\sqrt{\Delta}}\right) + O\left(\frac{1}{\delta}\right)$ can possibly be improved. See more details in Section 3.1.

1.1   Related Work

There has been extensive recent work on approximation algorithms for the graph-TSP problem, which is the same as that of finding a minimum length tour of a given undirected graph. While Christofides' algorithm [6] gives a $\frac{3}{2}$ approximation even for graph-TSP, a small but constant improvement was presented by Oveis-Gharan et al. [17]. Mömke and Svensson [14] improved this significantly while further improvements by Sebö and Vygen [18] have brought the current best approximation factor for graph-TSP down to $\frac{7}{5}$. The methods of Mömke and Svensson [14] also give a $\frac{4}{3}$ approximation algorithm for minimum length tours in subcubic 2-connected graphs.

Another line of work has focused on graph theoretic methods to obtain improved approximation factors: Boyd et al. [5] showed a $\frac{4}{3}$ approximation for 2-connected cubic graphs; Correa et al [7] gave an algorithm that finds a tour of length at most $(\frac{4}{3} - \frac{1}{61236})n$ in n-node 2-connected cubic graphs, while Karp and Ravi [12] gave an algorithm that finds a tour of length at most $\frac{9n}{7}$ in cubic bipartite graphs. For general, d-regular connected graphs, Vishnoi [19] gave an algorithm for finding tours of length at most $(1 + \frac{8}{\sqrt{\log d}})n$.

2   Small T-joins in Regular Graphs

In this section we prove Theorem 4 which follows directly from the following strengthening.

Theorem 6. *Let $G(V, E)$ be an arbitrary connected graph with n vertices and minimum degree δ, and let $T \subseteq V$ be an arbitrary set of vertices of even cardinality. Then there is a T-join with fewer than $2|T| + \frac{3n}{\delta+1} - 2\nu$ edges, where ν is the number of connected components in the T-join. Moreover, such a T-join can be found in polynomial time.*

Proof. Given $u, v \in V$, let $d(u, v)$ denote the number of edges along the shortest path between u and v in G. Consider the following iterative procedure for constructing a set $S \subset V$ together with a set P of *virtual edges*, each of length 3. Initially, place an arbitrary vertex v in S. Thereafter, in every iteration, consider an arbitrary vertex (say, u) whose distance from S is exactly 3. If there is no such vertex the procedure ends. Given such a vertex u, let w be an arbitrary vertex in S with $d(w, u) = 3$. Add u to S and the virtual edge (w, u) to P. This completes the description of the iteration.

Observe that necessarily $|S| \leq \frac{n}{\delta+1}$, because every vertex of S excludes all its neighbors from being in S, and the neighborhoods of vertices in S are disjoint. Observe also that the graph $G'(S, P)$ induced on S and the virtual edges is a tree.

Associate with every vertex $v \in T \setminus S$ the vertex $u \in S$ that is closest to v (breaking ties arbitrarily), and observe that $d(u, v) \leq 2$ (due to the maximality of S). Add an auxiliary edge (u, v) to P, with length $d(u, v)$. Consider now the tree T' whose vertices are $T \cup S$, and whose edge set is P. The total number of virtual edges in T' is exactly $|S \cup T| - 1$, exactly $|S| - 1$ of these virtual edges have length 3, and the remaining edges have length at most 2. Within T', find the unique T-join (where a tree edge is in the T-join iff each of the two connected components that are formed by removing it has an odd number of vertices from T). Let ν' denote the number of connected components (with respect to T') in this T-join. Then the number of virtual edges in the T-join is exactly $|S \cup T| - \nu'$, and their total length is at most $3(|S|-1) + 2|T \setminus S| - 2(\nu' - 1) < 3|S| + 2|T| - 2\nu'$.

Now replace the virtual edges of the T-join by edges along the corresponding shortest paths in G. The total number of edges needed is less than $3|S| + 2|T| - 2\nu'$. In the process of replacing virtual edges by paths, the same edge of G might be introduced multiple times. If so, any double occurrence of an edge is removed (as this does not change the parity of degrees), so as to make the resulting T-join a simple subgraph of G with no parallel edges. The removal of a set of edges parallel to each other might add 1 to the number of connected components, but decreases the number of edges in the T-join by at least 2. Hence if the final number of connected components in the T-join is ν, then the total number of edges in the T-join is less than $3|S| + 2|T| - 2\nu \leq 2|T| + \frac{3n}{\delta+1} - 2\nu$, as desired.

We now prove the following corollary.

Corollary 1. *Let G be a connected graph with n vertices and minimum degree δ, and let F be a linear forest in G. Then given F, one can find in polynomial time a tour of G of length smaller than $2n - |F| + \frac{5n}{\delta+1}$.*

Proof. Without loss of generality, assume that F is a maximal linear forest. This implies that isolated vertices cannot be neighbors of each other or neighbors of endpoints of paths, and endpoints of different paths cannot be neighbors of each other. The forest F induces in G exactly $n - |F|$ connected components, where a connected component is either a path or an isolated vertex.

We first describe a process for adding edges from G to the forest so that it becomes connected. In the process we may add the same edge more than once, and hence we shall obtain a connected spanning multigraph. The governing consideration in deciding which edges to add is that of keeping the number of odd degree vertices as small as possible (in particular, all odd degree vertices will be of degree one). The rules for adding edges are as follows:

1. If a component is an isolated vertex v, add an arbitrary edge incident to v (hence v joins some other connected component), and double this edge. Hence the number of connected components drops by one, the number of edges grows by two, and the number of odd degree vertices does not change.
2. If there are two vertices u and v of degree one in different connected components C_u and C_v with $d(u,v) = 2$ then connect them by a shortest path. Hence the number of connected components drops by one, the number of edges grows by two, and the number of odd degree vertices drops by two. Observe that the path between u and v might go through another component C', in which case the number of connected components should have dropped by two. However, for uniformity of the analysis (and without affecting its correctness) we shall count C' as a component distinct from the new component formed by C_u and C_v.

When none of the above two rules applies, let q denote the number of remaining connected components (each of which has two vertices of degree one).

The number of edges (including parallel edges) added by the above procedure is exactly $2(n - |F| - q)$.

Observe that if we take one vertex of degree one from each remaining component, no two such vertices share a neighbor (otherwise rule 2 above would apply). Hence $q \leq \frac{n}{\delta+1}$.

Let T denote the set of odd degree vertices that still remain, and note that $|T| = 2q \leq \frac{2n}{\delta+1}$. Now find a T-join in G using Theorem 6. This T-join has less than $2|T| + \frac{3n}{\delta+1} - 2\nu$ edges, where ν denotes the number of connected components of the T-join.

The union of the q components and the T-join is an Eulerian subgraph of G with at most ν connected components. It can be made connected (and kept Eulerian) by adding $\nu - 1$ pairs of parallel edges. Thereafter, an Eulerian cycle can serve as a tour of G. The total number of edges (counting multiplicities) in this union is less than

$$|F| + 2(n - |F| - q) + 4q + \frac{3n}{\delta + 1} - 2\nu + 2(\nu - 1) < 2n - |F| + \frac{5n}{\delta + 1}$$

proving the theorem.

3   Large Linear Forests and Fractional Arboricity

The *fractional linear arboricity* of a graph G is the minimum number of linear forests needed to cover every edge where we are allowed to pick a linear forest fractionally. Given Corollary 1, the proof of Theorem 2 would follow from a lower bound on the size of the maximum linear forest in regular graphs. Indeed, we prove a stronger result and show that fractional linear arboricity of any d-regular graph is at most $\frac{d - O(\sqrt{d})}{2}$.

Theorem 7. *There exists a randomized algorithm that given a d-regular graph $G = (V, E)$ returns a linear forest F such that for each edge $e \in E$, the probability $e \in F$ is at least $\frac{2}{d + O(\sqrt{d})}$. Thus the fractional arboricity of any d-regular graph is at most $\frac{d - O(\sqrt{d})}{2}$.*

Before we prove Theorem 7, we prove Theorem 3.

Proof (Proof of Theorem 3). Sample a random linear forest F as given by Theorem 7. The expected size of the forest F is at least

$$\sum_{e \in E} Pr[e \in F] \geq |E| \cdot \frac{2}{d + O(\sqrt{d})} \geq \frac{nd}{2} \cdot \frac{2}{d + O(\sqrt{d})} \geq \left(1 - O\left(\frac{1}{\sqrt{d}}\right)\right) n$$

as required.

We can now prove Theorem 2.

Proof (Proof of Theorem 2). Theorem 3 implies that every n-vertex d-regular graph has a linear forest of size $(1 - O(\sqrt{\frac{1}{d}}))n$, and moreover, such a linear forest can be found in polynomial time. Plugging this value of $|F|$ in Corollary 1 proves the theorem.

Thus it remains to prove Theorem 7.

Proof (Proof of Theorem 7). We shall now describe an iterative algorithm for constructing a random linear forest F. We shall assume for simplicity that n is a power of 2. This assumption has negligible effect on our bounds.

In the beginning of iteration i for $i = 1, 2, \ldots$ we have a directed graph $G_i = (V_i, E_i)$ where the maximum out/in-degree of every vertex is d_i where $d_i \simeq \frac{d}{2^i}$ and there are no parallel arcs (but there can be two anti-parallel arcs). The vertex set of G_i is obtained by identifying vertices of G; thus each vertex of V_i corresponds to subset of vertices in V and these subsets form a partition of V. We also maintain that edges included in F up to iteration $i - 1$ have both their endpoints contracted to the same vertex in V_i. Indeed, the edges of included so far in F within any subset corresponding to a vertex in V_i form a path. For $i = 1$, we initialize G_1 as follows. If d is even, we pick an Eulerian tour traversing each edge of G exactly once and orient the edges by picking an orientation of the tour and we set $d_1 = \frac{d}{2}$. If d is odd, we first add a perfect matching of auxiliary edges to G. Observe that the multiplicity of any edge is at most two after adding the matching. Now we pick an Eulerian orientation which traverses any two parallel edges right after each other. Now consider the orientation of the edges as given by the Eulerian tour. Clearly, there are no parallel arcs as any edge of multiplicity two is oriented as a pair of anti-parallel arcs. In this case, we set $d_1 = \frac{d+1}{2}$.

In each iteration i, we do the following steps.

1. Pair the vertices of G_i in an arbitrary manner. Then form a directed bipartite graph D_i with bipartition $L_i \cup R_i = V_i$, where from each pair of vertices one vertex is included in L_i and the other in R_i, uniformly at random and independently for each pair. Remove all arcs with both endpoints in L_i or both endpoints in R_i to obtain a directed bipartite graph.

2. Next, scan all vertices one by one, and if a vertex has current in- or out-degree more than $d_{i+1} = \lceil \frac{d_i}{2} \rceil$, delete a uniformly random set of in- or out-edges until the degree is exactly d_{i+1}. After this pass, all vertices have in- and out-degree at most d_{i+1} but some may have a strictly smaller degree.

3. Consider the bipartite graph formed by edges directed from L_i to R_i. This bipartite graph has maximum degree d_{i+1}. Add *auxiliary* edges between vertices of L_i and R_i of degree less than d_{i+1} in an arbitrary manner (allowing also parallel edges), until a regular bipartite multi-graph of degree d_{i+1} is obtained.

4. Legally color the edges of this regular bipartite multi-graph with d_{i+1} colors, thus obtaining d_{i+1} perfect matchings. Select uniformly at random one of the color classes as the perfect matching N_i.

5. Let N_i' denote the set of edges which go from R_i to L_i and are anti-parallel to an edge in N_i. Now do one of the following steps.
   (a) Select N_i' as matching M_i with probability $\frac{2}{d_{i+1}}$ and end the algorithm.
   (b) Otherwise, with probability $1 - \frac{2}{d_{i+1}}$, let $M_i = N_i$, and remove all arcs of N_i'. Remove all arcs that go from L_i to R_i and unify the endpoints of M_i. Thus in the contracted graph, we only retain edges that go from R_i to L_i, and the out/in-degree of each vertex is at most d_{i+1}. Observe that the contracted graph is a simple graph, with no parallel edges and no self loops. This contracted graph serves as G_{i+1} for the next iteration $i + 1$.

If step 5(a) is never invoked, the algorithm ends after $\log n$ iterations, as by then the whole graph is contracted to a single vertex. The final output F of the algorithm is the union of all M_i, excluding all the auxiliary edges.

Proposition 1. *The output of the algorithm is a linear forest.*

Proof. Add to the output of the algorithm also all the auxiliary edges that were used in order to extend the M_i matchings into perfect matchings. Considering the graph induced on the edges of all M_i and the auxiliary edges, it can be verified by induction that every vertex in iteration i corresponds to a directed path with exactly 2^i vertices, where the operation performed in iteration i matches such paths in pairs, and concatenates the two members of a pair with a directed (original or auxiliary) edge. Hence with the auxiliary edges the final output of the algorithm is a collection of vertex disjoint paths. Removing the auxiliary edges leaves a vertex disjoint set of paths (some of which might be isolated vertices), which by definition is a linear forest.

Lemma 1. *For any i, an edge e is deleted in Step 2 with probability at most $\frac{8}{\sqrt{d_{i+1}}}$.*

Proof. Let $e = (u, v)$ be any arc $e \in G_i$. The probability that $e \in E(R_i, L_i)$ is at least $\frac{1}{4}$ (exactly $\frac{1}{2}$ if u and v are paired and exactly $\frac{1}{4}$ otherwise), and likewise for $e \in E(L_i, R_i)$. Let us first condition on the event that either $e \in E(R_i, L_i)$ or $e \in E(L_i, R_i)$ and calculate the probability that e is deleted in Step (2) of iteration i. This can happen if the out-degree of u or in-degree of v is more than d_{i+1}. Let us concentrate on the event that e is deleted due to high out-degree at u. For each pair $p = \{w, w'\}$ of vertices in G_i, let X_p denote the indicator random variable for the event that either (u, w) or (u, w') is in $E(R_i, L_i)$. If u has out arcs to both w and w', then $X_p = 1$ with probability one. If u has an out-arc to exactly one of w or w', then $X_p = 1$ with probability $\frac{1}{2}$. Moreover, these random variables are independent since we make decisions for each pair independently. Thus the out-degree of u is $X = \sum_p X_p$ where $E[X] \leq \sum_p E[X_p] \leq d_{i+1}$. Since X is a sum of $\{0, 1\}$-valued independent Bernoulli random variables, standard Chernoff bounds imply that

$$Pr[deg^{out}(u) \geq d_{i+1} + c\sqrt{d_{i+1}}] \leq e^{-\frac{c^2}{3}}.$$

Since we delete a random set of required number of edges at u, we obtain that

$Pr[e$ is deleted due to high out-degree at $u]$

$$\leq \sum_{c=0}^{\infty} \frac{(c+1)\sqrt{d_{i+1}}}{d_i + c\sqrt{d_{i+1}}} Pr[d_{i+1} + c\sqrt{d_{i+1}} \leq deg^{out}(u) \leq d_{i+1} + (c+1)\sqrt{d_{i+1}}]$$

$$\leq \sum_{c=0}^{\infty} \frac{(c+1)\sqrt{d_{i+1}}}{d_i + c\sqrt{d_{i+1}}} \cdot e^{-\frac{c^2}{3}} \leq \frac{1}{\sqrt{d_{i+1}}} \sum_{c=0}^{\infty} (c+1) \cdot e^{-\frac{c^2}{3}} \leq \frac{4}{\sqrt{d_{i+1}}}$$

An identical bound holds for the event that e is deleted due to high in-degree at v. Thus, from the union bound, it follows that

$$Pr[e \text{ is deleted in Step 2}] \leq \frac{8}{\sqrt{d_{i+1}}}.$$

We now lower bound the probability that any e is included in F. We prove the following lemma for the first $T = \frac{\log d}{2}$ iterations. For this lemma, we assume that d is also a power of two for ease of analysis. This assumption has a negligible effect on the bounds.

Lemma 2. *For any* $1 \leq i \leq T$,

$$Pr[e \in F | e \in G_i] \geq \frac{1}{d_i + \left(2^{\frac{i}{2}+1} + 30(\sqrt{2}+1)\right)\sqrt{d_i}}$$

Proof. The proof is by reverse induction on i. For $i = T$, we have $2^{T/2} = d^{1/4}$ and $d_T = \frac{d}{2^T} = \sqrt{d}$. Thus we have that

$$\frac{1}{d_T + \left(2^{\frac{T}{2}+1} + 30(\sqrt{2}+1)\right)\sqrt{d_T}} = \frac{1}{d_T + (2\sqrt{d_T} + 30(\sqrt{2}+1))\sqrt{d_T}} \leq \frac{1}{3d_T}.$$

We will show that

$$Pr[e \in F | e \in G_T] \geq \frac{1}{3d_T} \geq \frac{1}{d_T + \left(2^{\frac{T}{2}+1} + 30(\sqrt{2}+1)\right)\sqrt{d_T}}$$

which will prove the base case.

The chance that $e \in E(L_i, R_i)$ is at least $\frac{1}{4}$. From Lemma 1, e is removed in Step 2 with probability at most $\frac{8}{\sqrt{d_{T+1}}}$. Since, each color class is chosen with probability $\frac{1}{d_{T+1}}$, we obtain that

$$Pr[e \in N_T | e \in G_T] \geq \left(\frac{1}{4} - \frac{8}{\sqrt{d_{T+1}}}\right) \cdot \frac{1}{d_{T+1}}$$

But then it is included in M_T with probability $1 - \frac{2}{d_{T+1}}$ independent of the earlier events. Thus

$$Pr[e \in M_T | e \in G_T] \geq \left(\frac{1}{4} - \frac{8}{\sqrt{d_{T+1}}}\right)\frac{1}{d_{T+1}} \cdot \left(1 - \frac{2}{d_{T+1}}\right) \geq \frac{1}{6d_{T+1}} = \frac{1}{3d_T}$$

for sufficiently large values of d. This proves the base case of the induction.

Now consider any i and let $e \in G_i$. Then e can be included in F in the following three events. Firstly, if it is in $E(L_i, R_i)$, then it can included in N_i and then chosen in M_i. Secondly, if it is in $E(R_i, L_i)$ and if one of the anti-parallel edges to it is chosen in N_i then it is included in N_i' and can be chosen in M_i. Lastly, if it is $E(R_i, L_i)$ but M_i is chosen as N_i and it is not deleted in Step 5, it can be chosen in G_{i+1}. We will calculate the probabilities of the first two events and apply induction to the last event to prove the inductive hypothesis. We have the following inequalities. All events are conditioned on the event that $e \in G_i$.

$$\begin{aligned}Pr[e \in F] =&Pr[e \in E(L_i, R_i)] \cdot Pr[e \in F | e \in E(L_i, R_i)] \\ &+ Pr[e \in E(R_i, L_i)] \cdot Pr[e \in M_i | e \in E(R_i, L_i)] \\ &+ Pr[e \in E(R_i, L_i)] \cdot Pr[e \in G_{i+1} | e \in E(R_i, L_i)] \cdot Pr[e \in F | e \in G_{i+1}]\end{aligned}$$

Now, we calculate each of terms.

$$Pr[e \in E(L_i, R_i)] \geq \frac{1}{4}$$

$$Pr[e \in F | e \in E(L_i, R_i)] \geq \left(1 - \frac{8}{\sqrt{d_{i+1}}}\right) \cdot \frac{1}{d_{i+1}} \cdot \left(1 - \frac{2}{d_{i+1}}\right)$$

Now $Pr[e \in E(R_i, L_i)] \geq \frac{1}{4}$. Conditioned on the event that $e \in E(R_i, L_i)$, let r be the multiplicity of the anti-parallel arc to e in $E(L_i, R_i)$ after the addition of dummy edges in Step 3. Then, one of these arcs in selected in N_i with probability $\frac{r}{d_{i+1}}$.

$$Pr[e \in M_i | e \in E(R_i, L_i)] \geq \left(1 - \frac{8}{\sqrt{d_{i+1}}}\right) \cdot \frac{r}{d_{i+1}} \cdot \frac{2}{d_{i+1}}$$

In the last case, e is included in G_{i+1} if one of the anti-parallel edges is not chosen in N_i but we chose $M_i = N_i$.

$$Pr[e \in G_{i+1} | e \in E(R_i, L_i)] \geq \left(1 - \frac{8}{\sqrt{d_{i+1}}}\right) \cdot \left(1 - \frac{r}{d_{i+1}}\right) \cdot \left(1 - \frac{2}{d_{i+1}}\right)$$

By induction, we have

$$Pr[e \in F | e \in G_{i+1}] \geq \frac{1}{d_{i+1} + f(i+1)\sqrt{d_{i+1}}}$$

where we let $f(j) = 2^{j/2+1} + 30(\sqrt{2} + 1)$ for any j for ease of notation.

Combining all the above inequalities, we obtain that

$$Pr[e \in F | e \in G_i] \geq \frac{1}{4} \cdot \left(1 - \frac{8}{\sqrt{d_{i+1}}}\right) \cdot \frac{1}{d_{i+1}} \cdot \left(1 - \frac{2}{d_{i+1}}\right) + \frac{1}{4} \cdot \left(1 - \frac{8}{\sqrt{d_{i+1}}}\right) \cdot \frac{r}{d_{i+1}} \cdot \frac{2}{d_{i+1}}$$

$$+ \frac{1}{4} \cdot \left(1 - \frac{8}{\sqrt{d_{i+1}}}\right) \cdot \left(1 - \frac{r}{d_{i+1}}\right) \cdot \left(1 - \frac{2}{d_{i+1}}\right) \cdot \frac{1}{d_{i+1} + f(i+1)\sqrt{d_{i+1}}}$$

We first notice that the coefficient at r is always positive and hence the expression is minimized when $r = 0$. Simplifying we get

$$Pr[e \in F | e \in G_i] \geq \frac{1}{4} \cdot \left(1 - \frac{8}{\sqrt{d_{i+1}}}\right) \left(1 - \frac{2}{d_{i+1}}\right) \left(\frac{1}{d_{i+1}} + \frac{1}{d_{i+1} + f(i+1)\sqrt{d_{i+1}}}\right)$$

$$\geq \frac{1}{4} \cdot \left(1 - \frac{10}{\sqrt{d_{i+1}}}\right) \left(\frac{1}{d_{i+1}} + \frac{1}{d_{i+1} + f(i+1)\sqrt{d_{i+1}}}\right)$$

$$\geq \frac{1}{4} \left(\frac{1}{d_{i+1} + 30\sqrt{d_{i+1}}} + \frac{1}{d_{i+1} + f(i+1)\sqrt{d_{i+1}}}\right)$$

where the last inequality is true for large enough d. Using the fact that $d_{i+1} = \frac{d_i}{2}$ and simplifying the above expression further, a simple check shows that

$$Pr[e \in F | e \in G_i] \geq \frac{1}{2d_i + 60\sqrt{2}\sqrt{d_i}} + \frac{1}{2d_i + 2\sqrt{2}f(i+1)\sqrt{d_i}}$$

$$\geq \frac{1}{d_i + \left(2^{\frac{i}{2}+1} + 30(\sqrt{2}+1)\right)\sqrt{d_i}}$$

This completes the proof of lemma by induction.

Using the fact $d_1 \geq \frac{d}{2}$, we obtain that

$$Pr[e \in F] = Pr[e \in F | e \in G_1] \geq \frac{1}{d_1 + \left(2^{\frac{3}{2}} + 30(\sqrt{2}+1)\right)\sqrt{d_1}} \geq \frac{2}{d + 120\sqrt{d}}$$

This completes the proof of Theorem 7.

3.1   Extensions to Nearly Regular Graphs

Here we prove Theorem 5.

Proof (Proof of Theorem 5). Let G be a connected graph of maximum degree Δ, average degree d, and minimum degree δ. Simply replacing d by Δ in the proof of Theorem 3 establishes that the fractional linear arboricity of G is $\frac{\Delta}{2} + O(\sqrt{\Delta})$. As G has $dn/2$ edges, this implies that it has at least one linear forest F with $n(\frac{d}{\Delta} - O(\frac{1}{\sqrt{\Delta}}))$ edges. Moreover, such a linear forest can be found in polynomial time by sampling linear forests from the distribution generated by the algorithm appearing in the proof of Theorem 3. Given such a linear forest, Corollary 1 finds a tour of length $2n - |F| + O(\frac{n}{\delta}) = \left(1 + \frac{\Delta - d}{\Delta} + O\left(\sqrt{\frac{1}{\Delta}}\right) + O(\frac{1}{\delta})\right)n$, as specified in Theorem 5.

The term $\frac{\Delta - d}{\Delta}$ in Theorem 5 is best possible, as shown by the following example. Let G be a bipartite graph in which the larger side has $\frac{\Delta n}{\Delta + \delta}$ vertices of degree δ, whereas the smaller side has $\frac{\delta n}{\Delta + \delta}$ vertices of degree Δ. As a tour must visit every vertex in the large side, the length of the shortest tour is at least $\frac{2\Delta n}{\Delta + \delta}$. The average degree is $d = \frac{2\Delta \delta}{\Delta + \delta}$, and a simple manipulation show that expressing the minimum tour length as a function of d gives $\left(1 + \frac{\Delta - d}{\Delta}\right)n$, as desired.

The term $O\left(\sqrt{\frac{1}{\Delta}}\right)$ in Theorem 5 is carried over from Theorem 2, and possibly can be replaced by $O(\frac{1}{\Delta})$. See the discussion in Section 4.

An interesting question is whether the term $O(\frac{1}{\delta})$ can be replaced by $O(\frac{1}{d})$. If so, the bound in Theorem 5 would become independent of δ. Our proof of Theorem 5 uses Corollary 1, and there the term $O(\frac{1}{\delta})$ cannot be replaced by $O(\frac{1}{d})$. Consider for example a path of length (roughly) $n/2$ in which one endpoint is connected to a triangle and the other to a clique of size $n/2$. This graph has a Hamiltonian path (and hence a linear forest of size $n - 1$), but the shortest tour is of length roughly $3n/2$, despite the fact that its average degree is very high, roughly $n/4$. However, the above graph does not show that the term $O(\frac{1}{\delta})$ cannot be replaced by $O(\frac{1}{d})$ in Theorem 5, because for this graph $\frac{\Delta - d}{\Delta} \simeq \frac{1}{2}$.

4   Some Conjectures

Linear arboricity of a graph G is a covering of all its edges by linear forests. The linear arboricity conjecture of [1] states that every d-regular graph has a linear arboricity with $\lceil \frac{d+1}{2} \rceil$ linear forests. If true, then one of these forests must be of size at least $(1 - \frac{2}{d+2})n$, and Theorem 4 would then imply that every d-regular graph has a tour of length $(1 + O(\frac{1}{d}))n$. The linear arboricity conjecture has been proved for small values of d, and is known to be true up to low order terms for large values of d (see [3] or [4]). The known upper bounds on the linear arboricity number translate to a $\left(1 - O\left(\left(\frac{\log d}{d}\right)^{1/3}\right)\right)n$ lower bound on the sizes of linear forests, which is weaker than the bound that we prove in Theorem 3.

The path cover number of a graph G is the minimum number of vertex disjoint paths required to cover the vertices of G. Magnant and Martin [13] conjecture that the path cover number of d-regular graphs is at most $\frac{n}{d+1}$ (and even smaller if the graph is required to be connected). They prove the conjecture for all $d \leq 5$. Observe that every path cover is a linear forest, and that the size of the forest plus the respective cover number is exactly n. Hence the path cover number conjecture, if true, could be combined with our Corollary 1 to show that every d-regular graph has a tour of length $(1 + O(\frac{1}{d}))n$.

A *minimum Hamiltonian completion* of a graph is the minimum size set of edges that, when added to the graph, makes it Hamiltonian [10]. The size of such a set is exactly one more than the size of a minimum path cover of the graph.

An upper bound of $(1 + O(\frac{1}{d}))n$ on the shortest tour length is the best that one can hope for, and likewise, a lower bound of $(1 - \Omega(\frac{1}{d}))n$ on the largest linear forest would be best possible. This can be demonstrated by taking a d-regular tree of depth $\ell \simeq \log_d n$ (the root node has d children whereas internal nodes have $d - 1$ children), and converting it to a d-regular graph as follows (assume for simplicity that d is odd). Add a single child to each leaf, connect this child to every sibling of its parent leaf (by now original leaves have degree d), and add a matching on the set of newly added children in which two such vertices can be matched if they are children of sibling leaves (so now all vertices have degree d). In this d-regular graph, a path can contain at most two vertices from the penultimate level of the tree (the parents of the leaves). It follows that a path cover contains at least $\Omega(n/d)$ paths, implying the desired lower bound on the length of a tour and upper bound on size of the linear forest.

Acknowledgements. Part of this work was done while the first two authors were visiting Microsoft Research in Redmond, Washington. We thank Noga Alon and Jeff Kahn for directing us to relevant literature.

References

1. Akiyama, J., Exoo, G., Harary, F.: Covering and packing in graphs IV: linear arboricity. Networks 11, 69–72 (1981)
2. Alon, N.: Problems and results in extremal combinatorics, Part I. Discrete Math. 273, 31–53 (2003)

3. Alon, N., Spencer, J.: The Probabilistic Method. John Wiley and Sons (2004)
4. Alon, N., Teague, V.J., Wormald, N.C.: Linear arboricity and linear k-arboricity of regular graphs. Graphs and Combinatorics 17, 11–16 (2001)
5. Boyd, S., Sitters, R., van der Ster, S., Stougie, L.: TSP on cubic and subcubic graphs. In: Günlük, O., Woeginger, G.J. (eds.) IPCO 2011. LNCS, vol. 6655, pp. 65–77. Springer, Heidelberg (2011)
6. Christofides, N.: Worst-case analysis of a new heuristic for the travelling salesman problem. Technical Report 388, Graduate School of Industrial Administration, Carnegie Mellon University (1976)
7. Correa, J.R., Larré, O., Soto, J.A.: TSP tours in cubic graphs: beyond 4/3. In: Epstein, L., Ferragina, P. (eds.) ESA 2012. LNCS, vol. 7501, pp. 790–801. Springer, Heidelberg (2012)
8. Egorychev, G.P.: The solution of Van der Waerden's problem for permanents. Advances in Mathematics 42(3), 299–305 (1981)
9. Falikman, D.I.: Proof of the Van der Waerden conjecture regarding the permanent of a doubly stochastic matrix. Mathematical Notes 29(6), 475–479 (1981)
10. Franzblau, D.S., Raychaudhuri, A.: Optimal Hamiltonian completions and path covers for trees, and a reduction to maximum flow. ANZIAM Journal 44(2), 193–204 (2002)
11. Jerrum, M., Sinclair, A., Vigoda, E.: A polynomial-time approximation algorithm for the permanent of a matrix with nonnegative entries. J. ACM 51(4), 671–697 (2004)
12. Karp, J., Ravi, R.: A 9/7-Approximation Algorithm for Graphic TSP in Cubic Bipartite Graphs. arXiv:1311.3640 (November 2013)
13. Magnant, C., Martin, D.M.: A note on the path cover number of regular graphs. Australasian Journal of Combinatorics 43, 211–217 (2009)
14. Mömke, T., Svensson, O.: Approximating graphic TSP by matchings. In: Proceedings of Foundations of Computer Science (FOCS), pp. 560–569 (2011)
15. Moon, J.W.: On the diameter of a graph. The Michigan Mathematical Journal 12(3), 349–351 (1965)
16. Moser, R.A., Tardos, G.: A constructive proof of the general Lovász local lemma. J. ACM 57(2) (2010)
17. Saberi, A., Gharan, S.O., Singh, M.: A randomized rounding approach to the traveling salesman problem. In: Proceedings of Foundations of Computer Science (FOCS), pp. 550–559 (2011)
18. Sebö, A., Vygen, J.: Shorter Tours by Nicer Ears. CoRR, abs/1201.1870 (2012)
19. Vishnoi, N.K.: A Permanent Approach to the Traveling Salesman Problem. In: Proceedings of Foundations of Computer Science (FOCS), pp. 76–80 (2012)

Linear Programming Hierarchies Suffice for Directed Steiner Tree

Zachary Friggstad[1], Jochen Könemann[2,*], Young Kun-Ko[3], Anand Louis[4,**],
Mohammad Shadravan[2,*], and Madhur Tulsiani[5,***]

[1] Department of Computing Science, University of Alberta
[2] Department of Combinatorics and Optimization, University of Waterloo
[3] Department of Computer Science, Princeton University
[4] College of Computing, Georgia Tech.
[5] Toyota Technical Institute at Chicago

Abstract. We demonstrate that ℓ rounds of the Sherali-Adams hierarchy and 2ℓ rounds of the Lovász-Schrijver hierarchy suffice to reduce the integrality gap of a natural LP relaxation for Directed Steiner Tree in ℓ-layered graphs from $\Omega(\sqrt{k})$ to $O(\ell \cdot \log k)$ where k is the number of terminals. This is an improvement over Rothvoss' result that 2ℓ rounds of the considerably stronger Lasserre SDP hierarchy reduce the integrality gap of a similar formulation to $O(\ell \cdot \log k)$.

We also observe that Directed Steiner Tree instances with 3 layers of edges have only an $O(\log k)$ integrality gap in the standard LP relaxation, complementing the known fact that the gap can be as large as $\Omega(\sqrt{k})$ in graphs with 4 layers.

1 Introduction

In the Directed Steiner Tree (DST) problem, we are given a directed graph $G = (V, E)$ with edge costs $c_e \geq 0, e \in E$. Furthermore, we are given a root node $r \in V$ and a collection of terminals $X \subseteq V$ and the goal is to find the cheapest collection of edges $F \subseteq E$ such that there is an $r - t$ path using only edges in F for every terminal $t \in X$. The nodes in $V - (X \cup \{r\})$ are called *Steiner nodes*. Throughout (except in Section 2.1), we will let $n = |V|$, $m = |E|$, and $k = |X|$ and we let OPT_G denote the optimum solution cost to the DST instance in graph G.

If $X \cup \{r\} = V$, then the problem is simply the minimum-cost arborescence problem which can be solved efficiently. However, the general case is well-known to be NP-hard. In fact, the problem can be seen to generalize the Group Steiner Tree problem, which cannot be approximated within $O(\log^{2-\epsilon}(n))$ for any constant $\epsilon > 0$ unless NP $\subseteq$ DTIME($n^{\mathrm{polylog}(n)}$) [7].

* Supported by NSERC grant no. 288340.
** Supported by NSF award CCF-1217793. Part of work done while the author was a summer intern at TTI-Chicago.
*** Research supported by NSF Career Award CCF-1254044.

J. Lee and J. Vygen (Eds.): IPCO 2014, LNCS 8494, pp. 285–296, 2014.
© Springer International Publishing Switzerland 2014

Definition 1. *Say that an instance $G = (V, E)$ of DST with terminals X is ℓ-layered if V can be partitioned as $V_0, V_1, \ldots, V_\ell$ where $V_0 = \{r\}, V_\ell = X$ and every edge $uv \in E$ has $u \in V_i$ and $v \in V_{i+1}$ for some $0 \le i < \ell$.*

For any DST instance G and any integer $\ell \ge 1$, Zelikovsky showed that there is an ℓ-layered DST instance H with at most $\ell \cdot n$ nodes such that $OPT_G \le OPT_H \le \ell \cdot k^{1/\ell} \cdot OPT_G$ and that a DST solution in H naturally corresponds to a DST solution in G with the same cost [11,1]. Charikar et al. [2] exploit this fact and present an $O(\ell^2 k^{1/\ell} \log k)$-approximation[1] with running time $\mathrm{poly}(n, k^\ell)$ for any integer $\ell \ge 1$. In particular, this can be used to obtain an $O(\log^3 k)$-approximation in quasi-polynomial time and for any constant $\epsilon > 0$ a polynomial-time $O(k^\epsilon)$-approximation. Finding a polynomial-time polylogarithmic approximation is an important open problem.

A natural linear programming (LP) relaxation for Directed Steiner Tree is given by LP (P0).

$$\min \quad \sum_{e \in E} c_e x_e \qquad\qquad (\text{P0})$$

$$\text{s.t.} \quad x(\delta^{\mathrm{in}}(S)) \ge 1 \quad \forall\, S \subseteq V - r, S \cap X \ne \emptyset \qquad (1)$$

$$x_e \in [0, 1] \quad \forall e \in E$$

Zosin and Khuller [12] demonstrated that the integrality gap of this relaxation can, unfortunately, be as bad as $\Omega(\sqrt{k})$, even in instances where G is a 4-layered graph. Recently, Rothvoss [9] showed that 2ℓ rounds of the Lasserre semidefinite programming (SDP) hierarchy suffice to reduce the integrality gap of a similar LP relaxation to only $O(\ell \cdot \log k)$ in ℓ-layered graphs. The LP he considers is an extended formulation of (P0) with polynomially many constraints plus additional constraints of the form $x(\delta^{in}(v)) \le 1$ for each non-root node v.

A related problem that will appear frequently throughout this paper is the Group Steiner Tree (GST) problem mentioned above. In this, we are given an undirected graph with edge costs, a root node r, and a collection of subsets $X_1, X_2, \ldots, X_k$ of nodes called *terminal groups*. The goal is to find the cheapest subset of edges F such that for every group X_i, there is a path from r to some node in X_i using only edges in F. Unlike DST, the integrality gap of the natural LP relaxation (GST-LP) (introduced in Section 3) is polylogarithmically bounded.

Theorem 1 (Garg, Konjevod, and Ravi [4]). *The integrality gap of LP (GST-LP) is $O(\min\{\ell, \log n\} \cdot \log k)$ in GST instances that have n nodes, k terminal groups, and are trees with height ℓ when rooted at r.*

---

[1] The algorithm in [2] is presented as an $O(\ell k^{1/\ell} \log k)$-approximation and relied on an incorrect claim in [11]. A correction to this claim was made in [1] which gives the stated DST approximation bound.

Only the bound of $O(\log n \log k)$ is explicitly shown in [4] but the bound $O(\ell \cdot \log k)$ easily follows from their techniques[2].

Hierarchies of convex programming relaxations, a.k.a. "lift-and-project" methods, have recently been used successfully in the design of approximation algorithms. For the sake of space, we omit a general introduction to lift-and-project procedure and will only include the specifics of the Sherali-Adams hierarchy and Lovász-Schrijver LP hierarchy as needed to describe our result. For more information, we direct the reader to an introduction and survey by Chlamtáč and Tulsiani [3] and note that a more recent application of the Sherali-Adams hieararchy by Gupta, Talwar and Witmer obtains a 2-approximation for the non-uniform Sparsest Cut problem in graphs with bounded treewidth [5] .

1.1 Our Results and Techniques

Using the ellipsoid method, it is possible to design a separation oracle for the ℓ-th level lift of (P0) in the Sheral-Adams and Lovász-Schrijver hierarchies with running time being polynomial in n and k^ℓ. However, we will start with a much simpler LP relaxation with only polynomially many constraints.

$$\min \qquad \sum_{e \in E} c_e x_e \qquad\qquad\qquad\qquad\qquad\qquad (P1)$$

$$\text{s.t.} \quad x(\delta^{in}(t)) \geq 1 \qquad t \in X \qquad\qquad\qquad\qquad (2)$$

$$x(\delta^{in}(v)) \leq 1 \qquad \forall\, v \in V - r \qquad\qquad\qquad (3)$$

$$x(\delta^{in}(v)) \geq x_e \quad \forall\, v \in V - (X \cup \{r\}), e \in \delta^{out}(v) \qquad (4)$$

$$x_e \in [0,1] \quad \forall e \in E$$

This is a relaxation in the sense that integer solutions corresponding to *minimal* DST solutions are feasible. That is, any minimal DST solution F is a branching, so every node has indegree 1 in F which justifies the inclusion of Constraints (3). Similarly, if some Steiner node v has no incoming edges in F then, by minimality of F, v also has outdegree 0 which justifies Constraints (4).

Our main result is the following. The notation $\mathrm{SA}_t(\mathcal{P})$ and $\mathrm{LS}_t(\mathcal{P})$ (defined properly in Section 2.1 and Section 2.2 respectively) refers to the t-th level lift of polytope $\mathcal{P} \subseteq [0,1]^m$ in the Sherali-Adams hierarchy and Lovász-Schrijver hierarchy respectively, which can be optimized over in time that is polynomial in the size of LP (P1) and m^ℓ. Thus, we consider the following LPs.

$$\min \left\{ \sum_{e \in E} c_e \cdot y_{\{e\}} : y \in \mathrm{SA}_\ell(\mathcal{P}) \right\} \qquad\qquad \text{(SA-LP)}$$

$$\min \left\{ \sum_{e \in E} c_e \cdot y_e : y \in \mathrm{LS}_{2\ell}(\mathcal{P}) \right\} \qquad\qquad \text{(LS-LP)}$$

---

[2] [4] groups nodes together in their analysis so that the tree has height $h = O(\log n)$. They then prove the gap is $O(h \cdot \log k)$. One could skip the grouping argument to directly prove the $O(\ell \cdot \log k)$ bound.

where $\mathcal{P}$ is the polytope given by the constraints of the LP relaxation (P1).

Theorem 2. *Then the integrality gap of LP (SA-LP) is $O(\ell \cdot \log k)$ in ℓ-layered instances of DST.*

Theorem 3. *Then the integrality gap of LP (LS-LP) is $O(\ell \cdot \log k)$ in ℓ-layered instances of DST.*

Note that Theorems 2 and 3 are incomparable; fewer rounds are used in the stronger Sherali-Adams hierarchy. For the sake of space, we will only present the proof of Theorem 2 in this extended abstract. The proof of Theorem 3 is deferred to the full version.

We can also find feasible DST solutions witnessing these integrality gap upper bounds.

Theorem 4. *Given oracle access to some fixed $y \in \mathrm{LS}_{2\ell}(\mathcal{P})$ or $y \in \mathrm{SA}_{\ell}(\mathcal{P})$, with high probability we can find a Directed Steiner Tree solution in time $O(\mathrm{poly}(n))$ of cost at most $O(\ell \cdot \log k)$ times the cost of y^*.*

Rothvoss proved an analogous result for the Lasserre SDP hieararchy [9], but his arguments relied on a particular decomposition theorem proven by Karlin, Mathieu, and Nguyen [8]. This decomposition theorem does not hold in weaker LP hierarchies.

The algorithm for rounding a point in $\mathrm{SA}_{\ell}(\mathcal{P})$ lifted LP is quite different from the algorithm for rounding a point in $\mathrm{LS}_{2\ell}(\mathcal{P})$. At a high level, we prove Theorem 2 by mapping a point y^* in the Sherali-Adams lifted polytope into an LP solution with the same cost as y^* for a related Group Steiner Tree instance. Using Theorem 1, we find a GST solution with cost $O(\ell \cdot \log k)$ times the cost of y^* and this will naturally correspond to a DST solution in G. This construction does not need to made explicit; one can emulate the GST rounding algorithm in [4] in an expected $O(\mathrm{poly}(n))$ steps given oracle access to $y \in \mathrm{SA}_{\ell}(\mathcal{P})$.

However, these techniques do not seem to help in proving Theorem 3. We prove Theorem 3 by employing a different algorithm to round LP (LS-LP). Roughly speaking, we start from the terminals, then iteratively extend the paths by adding edges in a bottom-up fashion guided by probabilities given by the LP.

As a warmup, we also obtain the following interesting bound that shows lift-and-project techniques are not necessary for graphs having 3 layers.

Theorem 5. *The integrality gap of LP (P0) is $O(\log k)$ in 3-layered graphs.*

As with Theorem 2, this is obtained by mapping a point in LP (P0) to an LP solution for the corresponding GST instance. However, the restriction to only 3 layers allows us to accomplish this without the use of hierarchies. In contrast, the integrality gap of LP (P0) is $\Omega(\sqrt{k})$ in some graphs with 4 layers [12].

The paper is organizes as follows. Section 2 describes the hierarchies and introduces some additional notation. The proof of Theorem 5 is presented in Section 3. The proof of Theorem 2 is outlined in Section 4. Finally, the rounding algorithms for both hierarchies are described in Section 5.

2 Preliminaries

2.1 The Sherali-Adams Hierarchy

Consider a polytope $\mathcal{P} \subseteq \mathbb{R}^n$ specified by m linear constraints $\sum_{i=1}^n A_{j,i} \cdot x_i \geq b_j, 1 \leq j \leq m$. Suppose the "box constraints" $0 \leq x_i$ and $x_i \leq 1$ (equivalently, $-x_i \geq -1$) appear among these constraints for each $1 \leq i \leq n$.

For $t \geq 0$, let $\mathscr{P}_t([n]) = \{S \subseteq \{1, \ldots, n\} : |S| \leq t\}$ denote the collection of subsets of $\{1, \ldots, n\}$ of size at most t. We also let $\mathbb{R}^{\mathscr{P}_t([n])}$ denote $\mathbb{R}^\alpha$ where $\alpha = |\mathscr{P}_t([n])| = n^{O(t)}$. We index a vector $y \in \mathbb{R}^{\mathscr{P}_t([n])}$ by sets in $\mathscr{P}_t([n])$. The Sherali-Adams hieararchy (introduced in [10]) is described as follows.

Definition 2. $SA_t(\mathcal{P})$ *is the set of vectors* $y \in \mathbb{R}^{\mathscr{P}_{t+1}([n])}$ *satisfying* $y_\emptyset = 1$ *and*

$$\sum_{H \subseteq J} (-1)^{|H|} \cdot \left(\sum_{i=1}^n A_{j,i} \cdot y_{I \cup H \cup \{i\}} - b_j \cdot y_{I \cup H} \right) \geq 0 \tag{5}$$

for each $j = 1, \ldots, m$ *and each pair of subsets of indices* $I, J \subseteq \{1, \ldots, n\}$ *having* $|I| + |J| \leq t$.

If $\mathcal{P}$ is described by m linear constraints over n variables, then $SA_t(\mathcal{P})$ has $n^{O(t)}$ variables and $n^{O(t)}m$ constraints. So, we can solve the LP

$$\min \left\{ \sum_{i=1}^n c_i \cdot y_{\{i\}} : y \in SA_\ell(\mathcal{P}) \right\}$$

with only poly(n^ℓ) overhead over the running time of solving $\min\{c^T x : x \in \mathcal{P}\}$.

We only use some of the many well-known properties of the Sherali-Adams hierarchy.

Lemma 1. *Suppose* $y \in SA_t(\mathcal{P})$ *for some* $t \geq 0$. *Then the following hold.*

- *For any* $A \in \mathscr{P}_t([n])$ *such that* $y_A > 0$, *let* $y' \in \mathbb{R}^{\mathscr{P}_{t+1-|A|}([n])}$ *be defined by* $y'_I = \frac{y_{I \cup A}}{y_A}$. *Then* $y' \in SA_{t-|A|}(\mathcal{P})$.
- *For any* $A \subseteq B \subseteq [n]$ *with* $|B| \leq t+1$, *we have* $y_B \leq y_A$.

Furthermore, $\mathcal{P}$ *and the projection of* $SA_t(\mathcal{P})$ *to the singletons have the same integral solutions.*

In particular, the last statement implies that if $\mathcal{P}$ is an LP relaxation of a $\{0, 1\}$ integer program, then $SA_t(\mathcal{P})$ is also a relaxation for the same integer program for any $t \geq 0$.

2.2 The Lovász-Schrijver Hierarchy

Given a convex set $\mathcal{P} \subseteq [0, 1]^n$, we convert it to a cone in $\mathbb{R}^{n+1}$ as follows.

$$cone(\mathcal{P}) = \{\mathbf{y} = (\lambda, \lambda x_1, \ldots, \lambda x_n) \mid \lambda \geq 0, (x_1, \ldots, x_n) \in \mathcal{P}\}$$

With a linear program given by constraints $\sum_{i=1}^n A_{j,i} \cdot x_i \geq b_j, 1 \leq j \leq m$, this is is accomplished by *homogenizing* the constraints with a new variable x_0, yielding the cone $\{x \in \mathbb{R}^{d+1} : \sum_{i=1}^n A_{j,i} \cdot x_i \geq b_j \cdot x_0, 1 \leq j \leq m$ and $x_0 \geq 0\}$.

Definition 3. *For a cone $K \subseteq \mathbb{R}^{d+1}$ we define the set $N(K)$ (also a cone in $\mathbb{R}^{d+1}$) as follows: a vector $\mathbf{y} = (y_0, \ldots, y_d) \in \mathbb{R}^{d+1}$ is in $N(K)$ if and only if there is a matrix $Y \in \mathbb{R}^{(d+1) \times (d+1)}$ such that*

1. *Y is symmetric*
2. *For every $i \in \{0, 1, \ldots, d\}$, $Y_{0,i} = Y_{i,i} = y_i$*
3. *Each row Y_i is an element of K*
4. *Each vector $Y_0 - Y_i$ is an element of K*

The matrix Y is said to be a protection matrix *for $y \in N(K)$. For $t \geq 0$ we recursively define the cone $N^t(K)$ as $N^0(K) = K$ and $N^t(K) = N(N^{t-1}(K))$.*

Then we project the cone back to the desired space.

Definition 4. $LS_t(\mathcal{P})$ *is the set of vectors $\mathbf{y} \in N^t(cone(\mathcal{P}))$ with $y_0 = 1$.*

Lovász and Schrijver [13] showed that if we start from a LP relaxation of a 0-1 integer program with n variables, then $LS_n(\mathcal{P})$ is a tight relaxation in the sense that the only feasible solutions are convex combinations of integral solutions. In addition, if we start with a LP relaxation with $poly(n)$ inequalities, we can obtain an optimal solution over the set of solutions given by t levels of LS in $n^{O(t)}$ time.

One key fact that is derived easily from Definition 3 is the following.

Lemma 2. *If $\mathbf{y} = (1, \mathbf{x}) \in LS_t(\mathcal{P})$ with protection matrix Y, for any $i \in \{1, \ldots, n\}$ such that $x_i > 0$ consider the vector $\mathbf{y}' = \frac{1}{x_i} Y_i$. Then $\mathbf{y}' \in LS_{t-1}(\mathcal{P})$ with $y_i' = 1$.*

2.3 Notation

Suppose G is an ℓ-layered instance of Directed Steiner Tree with root r, terminals X, and layers $\{r\} = V_0, V_1, \ldots, V_\ell = X$. We will assume every $v \in V$ can be reached by r. In particular, for every $v \in V_1$ we have $rv \in E$.

Say a path in G is *rooted* if it begins at r. The notation $\langle v_j, v_{j+1}, v_{j+2}, \ldots, v_i \rangle$ refers to a path in G that follows edges $v_j v_{j+1}, v_{j+1} v_{j+2}, \ldots, v_{i-1} v_i \in E$ in succession. The subscript of a vertex in this notation will always indicate which layer the node lies in. The notation $\langle e_j, e_{j+1}, e_{j+2}, \ldots, e_i \rangle$ refers to a path in G that follows edges $e_j, e_{j+1}, \ldots, e_i \in E$ in succession. The subscript of an edge in this notation will always indicate which layer the (directed) edge starts from.

For any node $v \in V(G)$ we let

$$Q(v) = \{\langle r, v_1, v_2, \ldots, v_i \rangle : v_i = v\}$$

and for any $e \in E(G)$ we let

$$Q(e) = \{\langle r, v_1, v_2, \ldots, v_i \rangle : v_{i-1} v_i = e\}$$

denote all rooted paths ending at node v or ending with edge e, respectively. More generally, for a vertex v and another vertex u or an edge e, we let $Q(v, u)$

and $Q(v, e)$ denote all paths starting at v and ending at u or ending with edge e, respectively. We let $Q(e, v)$ denote all paths starting with edge e and ending at v. It will also sometimes be convenient to think of a path as a set of edges $\{v_j v_{j+1}, \dots, v_{i-1} v_i\}$.

Definition 5. *Suppose $G = (V, E)$ is an ℓ-layered instance of DST with root r and k terminals X. Then we consider the Group Steiner Tree instance on a tree $\mathcal{T}(G)$ with terminal groups $X_t, t \in X$ defined as follows.*

- *The vertex set of $\mathcal{T}(G)$ consists of all rooted paths $\cup_{v \in V} Q(v)$ in G.*
- *For any rooted path $P \neq \langle r \rangle$, we connect P to its maximal proper rooted subpath and give this edge cost c_e, where $P \in Q(e)$. Denote this edge in $\mathcal{T}(G)$ by $m(P)$.*
- *For each terminal $t \in X$, we let $X_t = Q(t)$: the set of all $r - t$ paths in G.*

This construction is illustrated in Figure 1. We will not explicitly construct $\mathcal{T}(G)$ in our rounding algorithm described in Section 5. It is simply a tool for analysis.

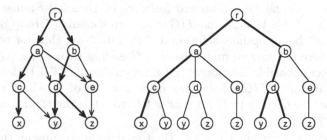

Fig. 1. A 3-layered DST instance with terminals $X = \{x, y, z\}$ (left) and the corresponding GST instance $\mathcal{T}(G)$ (right). Each node in $\mathcal{T}(G)$ corresponds to a path P in G and is labelled in the figure with the endpoint of P in G. A terminal group in $\mathcal{T}(G)$ in the figure consists of all leaf nodes with a common label. A DST solution and its corresponding GST solution are drawn with bold edges.

The following is immediate from the construction of $\mathcal{T}(G)$.

Lemma 3. *Let $|V| = n$. The graph $\mathcal{T}(G)$ constructed from an ℓ-layered Directed Steiner Tree instance G is a tree with height ℓ when rooted at $\langle r \rangle$. For every GST solution in $\mathcal{T}(T)$ there is a DST solution in G of no greater cost, and vice-versa.*

3 Rounding for 3-Layered Graphs

We first demonstrate that the natural LP relaxation (P0) for Directed Steiner Tree has an integrality gap of $O(\log k)$ in 3-layered graphs without using any lift-and-project machinery. As mentioned earlier, this complements the observation of Zosin and Khuller [12] that the integrality gap is $\Omega(\sqrt{k})$ in some 4-layered instances.

We show this by directly embedding a solution to the Directed Steiner Tree LP relaxation (P0) for some 3-layered instance G into a feasible LP solution to the Group Steiner Tree LP (GST-LP) on instance $\mathcal{T}(G)$. The reason we can do this with 3-layered instances is essentially due to the fact that for any edge $e = uv$ that either $v \in X$ or $|Q(e)| = 1$ (Figure 1 also helps illustrate this). This property does not hold in general for instances with at least 4 layers.

Consider a Group Steiner Tree instance $H = (V, E)$ with root r, terminal groups $X_1, X_2, \ldots, X_k \subseteq V$, and edge costs $c_e, e \in E$. The LP relaxation we consider for Group Steiner Tree is the following.

$$\min \quad \sum_{e \in E} c_e z_e \qquad\qquad\qquad\qquad \text{(GST-LP)}$$

$$\text{s.t.} \quad z(\delta(S)) \geq 1 \qquad \forall S \subseteq V - r, X_i \subseteq S \text{ for some group } X_i \qquad (6)$$

$$z \geq 0$$

Now we can prove Theorem 5.

Proof. Let $G = (V, E)$ be a 3-layered instance of Directed Steiner Tree with layers $\{r\} = V_0, V_1, V_2, V_3 = X$ and $\mathcal{T}(G)$ the corresponding Group Steiner Tree instance. Let x^* be an optimal solution to LP (P0). Note that for edge $uv \in E$ with $v \notin X$ there is a unique rooted path in G ending with e (i.e. $|Q(e)| = 1$).

We construct a feasible solution z^* to LP relaxation (GST-LP) for the Group Steiner Tree instance $\mathcal{T}(G)$. For every edge $e = uv$ of G where $v \notin X$, set $z^*_{m(P)} := x^*_e$ where $Q(e) = \{P\}$. All that is left to set is the the z^*-value for the leaf edges of $\mathcal{T}(G)$.

To do this, fix a terminal $t \in X$. By the max-flow/min-cut theorem and Constraints (1), there is a flow f^t sending 1 unit of flow from r to t satisfying $f^t_e \leq x^*_e$ for every edge e. Furthermore, for each $e \in \delta^{in}(t)$ we may assume that $x^*_e = f^t_e$, otherwise we could reduce x^*_e while maintaining feasibility. Consider any path decomposition of f^t and say that this decomposition places weight w^t_P on a path $P \in Q(t)$. That is, $f^t_e = \sum_{P \in Q(t):e \in P} w^t_P$ for every edge $e \in G$. Then we set $z^*_{m(P)} := w^t_P$ for each $P \in Q(t)$.

We claim that z^* is a feasible solution for LP (GST-LP) with cost equal to $\sum_{e \in E} c_e x^*_e$. If so, then by Theorem 1, there is a Group Steiner Tree solution of cost at most $O(\log k)$ times the cost of x^*. We conclude by using Lemma 3 to note that there is then a Directed Steiner Tree solution of cost at most $O(\log k)$.

To see why z^* is feasible, we prove for every group t that there is a flow g^t of value 1 from $\langle r \rangle$ to the nodes in X_t with $g^t_{m(P)} \leq z^*_{m(P)}$ for every edge $m(P), P$ of H. By the max-flow/min-cut theorem, this means every constraint of (GST-LP) is satisfied by z^*. That such a flow exists essentially follows from the path decomposition of the flow f^t. Recall that a path decomposition of f^t placed weight w^t_P on $P \in Q(t)$. So, for each group X_t we define a flow g^t in $\mathcal{T}(G)$ by $g^t_{m(P)} = \sum_{P^* \in Q(t):P \subseteq P^*} w^t_{P^*}$.

Verifying that g^t is one unit of $r - X_t$ flow satisfying $g^t \leq z^*$ is straightforward; the details are left to the full version. It is also easy to see that the total z^*-value for paths ending with a copy of an edge e in G is equal to x_e^*, so the x^* and z^* have the same cost.

4   Sherali-Adams Gap for ℓ-Layered Graphs

Our basic approach for proving Theorem 2 is similar to our approach for Theorem 5. Let $\mathcal{P}$ denote the polytope defined by the constraints of LP (P1). We show how to embed a point y^* in the Sherali-Adams lift of LP (P1), namely $\mathrm{SA}_\ell(\mathcal{P})$, for an ℓ-layered instance G to a feasible solution to LP (GST-LP) for the corresponding Group Steiner Tree instance $\mathcal{T}(G)$.

Describing the embedding is straightforward. For every edge $m(P)$ in $\mathcal{T}(G)$, simply set $z_{m(P)}^* := y_P^*$. The rest of our analysis shows that z^* is feasible for LP (GST-LP) for instance $\mathcal{T}(G)$ and the cost of z^* in (GST-LP) is equal to $\sum_{e \in E} c_e \cdot y_{\{e\}}^*$.

Before delving into the proofs of these statements, we note a technical result about the structure of Sherali-Adams solutions which will be very helpful.

Lemma 4. *Suppose $0 \leq i < j \leq \ell$. For any node $v \in V_i$, any edge $e = uw$ with $w \in V_j$, and any $y \in \mathrm{SA}_{j-i}(\mathcal{P})$ we have $\sum_{P \in Q(v,e)} y_P \leq y_{\{e\}}$. Furthermore, if $v = r$ then this bound holds with equality.*

The proof is deferred to the full version of this paper. Note that $|P| = j - i$ for any $P \in Q(v, e)$ so it is valid to index $y \in \mathrm{SA}_{j-i}(\mathcal{P})$ with P in the sum.

4.1   Cost Analysis

The cost bound is an easy consequence of Lemma 4.

Lemma 5. *The cost of z^* in LP (GST-LP) is $\sum_{e \in E(G)} c_e \cdot y_{\{e\}}^*$.*

Proof

$$\sum_{m(P) \in E(\mathcal{T}(G))} c_{m(P)} \cdot z_{m(P)}^* = \sum_{e \in E(G)} \sum_{P \in Q(e)} c_e \cdot z_{m(P)}^*$$

$$= \sum_{e \in E(G)} \sum_{P \in Q(e)} c_e \cdot y_P^* = \sum_{e \in E(G)} c_e \cdot y_{\{e\}}^*$$

where the last equality is by Lemma 4 applied with $v = r$.

4.2   Feasibility

Similar to the proof of Theorem 5, for every group X_t we construct a one unit of $\langle r \rangle - X_t$ flow g^t in $\mathcal{T}(G)$ which satisfies the capacities given by z^*. Thus, by the max-flow/min-cut theorem we have that $z^*(\delta(S)) \geq 1$ for every subset $S \subseteq V(\mathcal{T}(G)) - \langle r \rangle$ such that $X_t \subseteq S$ for some group X_t.

We now fix a terminal $t \in X$ and describe the flow g^t by giving a path decomposition of the flow. For each $P \in Q(t)$, we assign a weight of y_P^* to the $\langle r \rangle - P$ path in $\mathcal{T}(X)$. So, the flow $g_{m(P)}^t$ crossing edge $m(P)$ in $\mathcal{T}(G)$ is just $\sum_{P^* \in Q(t) : P \subseteq P^*} y_{P^*}^*$.

Lemma 6. g^t is one unit of $\langle r \rangle - X_t$ flow in $\mathcal{T}(G)$.

Proof. It is an $\langle r \rangle - X_t$ flow because we constructed it from a path decomposition using only paths in $Q(t)$. Furthermore,

$$g^t(\delta^{out}_{\mathcal{T}(G)}(\langle r \rangle)) = \sum_{P^* \in Q(t): \langle r \rangle \subseteq P^*} y^*_{P^*} = \sum_{P \in Q(t)} y^*_P$$

$$= \sum_{e \in \delta^{in}_G(t)} \sum_{P \in Q(e)} y^*_P = \sum_{e \in \delta^{in}_G(t)} y^*_{\{e\}} = 1.$$

Here, the second last equality follows from Lemma 4. The last equality follows from combining Constraint (2) with Constraint (3) for $v = t$.

All that is left is to prove that each flow g^t for a terminal group X_t satisfies the capacities given by z^*. The following lemma is the heart of this argument. A similar result was proven in [9] which relied on the strong decomposition property for the Lasserre hierarchy of Karlin, Mathieu, and Nguyen [8]. We emphasize that our proof only uses properties of the Sherali-Adams LP hierarchy.

Lemma 7. For every rooted path P and every terminal group X_t, we have $\sum_{P^* \in Q(t): P \subseteq P^*} y^*_{P^*} \le y^*_P$.

We can now easily verify that the capacity constraints are satisfied.

Corollary 1. For every terminal group X_t and every edge $m(P)$ of $\mathcal{T}(G)$, $g^t_{m(P)} \le z^*_{m(P)}$.

Proof. By Lemma 7 $g^t_{m(P)} = \sum_{\substack{P^* \in Q(t) \\ P \subseteq P^*}} y^*_{P^*} \le y^*_P = z^*_{m(P)}$.

The proof of Theorem 2 is now complete.

5   Rounding Algorithms

For the sake of space, we will present the algorithms without complete analysis in this extended abstract.

5.1   Sherali-Adams Rounding Algorithm

We bounded the integrality gap of LP (SA-LP) by converting some $y \in SA_\ell(\mathcal{P})$ to a feasible solution for LP (GST-LP) in $\mathcal{T}(G)$. However, this mapping does not have to be explicitly constructed to round y. Instead, we emulate the GST rounding algorithm in [4] by simply querying the y_P variables as needed. Algorithm 1 describes the main subroutine from [4] in our context.

As in [4], the expected cost of F is $\sum_{e \in E} c_e y_{\{e\}}$ and, for each terminal $t \in X$, the probability that F contains an $r - t$ path is at least $\frac{1}{\ell}$. Iterating this procedure sufficiently many times gives us a feasible DST solution with cost at most $O(\ell \cdot \log k)$ times the cost of y.

Algorithm 1. Sherali-Adams Rounding Subroutine

1: $S_0 \leftarrow \{\langle r \rangle\}$
2: **for** $j = 1, \ldots, \ell$ **do**
3:     $S_j \leftarrow \emptyset$
4:     **for** each $P \in S_{j-1}$ **do**
5:         **for** each $e \in \delta^{out}(v)$ where v is the endpoint of P **do**
6:             Add $\langle P, e \rangle$ to S_j with probability $\dfrac{y^*_{P \cup \{e\}}}{y^*_P}$
7: $F \leftarrow$ edges used by some path in S_ℓ
8: **return** F

In fact, it is easy to see that in one run of Algorithm 1 we have for any rooted path P ending in layer i that $\Pr[P \in S_i] = y^*_P$. This leads to an interesting observation which, ultimately, means the expected running time of Algorithm 1 is polynomial in n.

Lemma 8. $\mathrm{E}\left[\sum_{i=0}^{\ell} |S_i|\right] \leq n$

5.2 Lovász-Schrijver Rounding Algorithm

We introduce a bit more notation to describe the rounding algorithm. We start with some $y \in \mathrm{LS}_{2\ell}(\mathcal{P})$ with corresponding protection matrix Y. For $0 < j \leq \ell$, consider some path $P = \langle v_j, v_{j+1}, \ldots, v_\ell \rangle$ ending at some terminal $v_\ell \in X$. We let y^P denote a point in $\mathrm{LS}_{\ell+j}(\mathcal{P})$ and Y^P be a corresponding protection matrix, which we define inductively.

If $j = \ell$ (so $P = \langle v_\ell \rangle$) then we simply let $y^P = y$ and $Y^P = Y$. For $j < \ell$, let y^P be the point obtained by conditioning $y^{P'}$ on $y^{P'}_{v_j v_{j+1}} = 1$ where $P' = \langle v_{j+1}, v_{j+2}, \ldots, v_\ell \rangle$. Then Y^P is the protection matrix witnessing the inclusion of row $Y^{P'}_{v_j v_{j+1}}$ in $N^{\ell+j+1}(cone(\mathcal{P}))$ (scaled by $\frac{1}{y^{P'}_e}$ to ensure $Y^P_0 = y^P$). This definition only makes sense if $y^{P'}_e > 0$ for every suffix $\langle e, P' \rangle$ of P; this will be the case for every path P constructed in the algorithm.

The algorithm for rounding the Sherali-Adams relaxation does not does not work for the Lovász-Schrijver hierarchy because a direct analogue of Lemma 4 fails to hold in this case. However, using the constraint that the indegree of every node is at most 1, we are able to prove an analogue when we consider paths going as edge to a particular terminal, instead of paths from the root to an edge. We utilize this by building the tree in a "bottom-up" fashion in our algorithm.

Algorithm 2 contains the main subroutine for the Lovász-Schrijver rounding procedure. As with the Sherali-Adams rounding procedure, we iterate Algorithm 2 until there is an $r - t$ path for every terminal t in the union of the returned sets of edges F.

As mentioned before, the proofs of Theorems 3 and 4 for this rounding procedure will appear in the full version.

Algorithm 2. Lovász-Schrijver Rounding Subroutine

1: $F \leftarrow \emptyset, C \leftarrow \emptyset$
2: $S_t \leftarrow \emptyset$ for each $t \in X$
3: **for** $t \in X$ **do**
4:     **for** each $e \in \delta^{in}(t)$ **do**
5:         Add $\langle e \rangle$ to S_t independently with probability y_e.
6:     **for** $j = 1, \ldots, l$ **do**
7:         **for** each $u - t$ path P of length j in S_t **do**
8:             **if** $u \notin C$ **then**
9:                 **for** each $e \in \delta^{in}(u)$ **do**
10:                    Add $\langle e, P \rangle$ to S_t with probability y_e^P
11:     Add edges in S_t to F, and the vertices covered by S_t to C
12: **return** F

References

1. Calinescu, G., Zelikovsky, G.: The polymatroid Steiner problems. J. Combinatorial Optimization 9(3), 281–294 (2005)
2. Charikar, M., Chekuri, C., Cheung, T., Dai, Z., Goel, A., Guha, S., Li, M.: Approximation algorithms for directed Steiner problems. J. Algorithms 33(1), 73–91 (1999)
3. Chlamtáč, E., Tulsiani, M.: Convex relaxations and integrality gaps. In: Handbook on Semidefinite. Springer (2012)
4. Garg, N., Konjevod, G., Ravi, R.: A polylogarithmic approximation algorithm for the group Steiner tree problem. J. Algorithms 37(1), 66–84 (2000)
5. Gupta, A., Talwar, K., Witmer, D.: Sparsest cut on bounded treewidth graphs: algorithms and hardness results. In: Proceedings of STOC (2013)
6. Guruswami, V., Sinop, A.K.: Faster SDP hierarchy solvers for local rounding algorithms. In: Proceedings of FOCS (2012)
7. Halperin, E., Krauthgamer, R.: Polylogarithmic inapproximability. In: Proceedings of STOC (2003)
8. Karlin, A., Mathieu, C., Nguyen, C.: Integrlaity gaps of linear and semidefinite programming relaxations for knapsack. In: Proceedings of IPCO (2011)
9. Rothvoss, T.: Directed Steiner tree and the Lasserre hierarchy. CoRR abs/1111.5473 (2011)
10. Sherali, H., Adams, W.: A hierarchy of relaxations between the continuous and convex hull representations for zero-one programming problems. SIAM J. Discrete Math. 3, 411–430 (1990)
11. Zelikovsky, A.: A series of approximation algorithms for the acyclic directed Steiner tree problem. Algorithmica 18, 99–110 (1997)
12. Zosin, L., Khuller, S.: On directed Steiner trees. In: Proceedings of SODA (2002)
13. Lovász, L., Schrijver, A.: Cones of matrices and set-functions and 0-1 optimization. SIAM Journal on Optimization 1, 166–190 (1991)

An Improved Approximation Algorithm for the Stable Marriage Problem with One-Sided Ties

Chien-Chung Huang[1] and Telikepalli Kavitha[2,*]

[1] Chalmers University, Sweden
huangch@chalmers.se
[2] Tata Institute of Fundamental Research, India
kavitha@tcs.tifr.res.in

Abstract. We consider the problem of computing a large stable matching in a bipartite graph $G = (A \cup B, E)$ where each vertex $u \in A \cup B$ ranks its neighbors in an order of preference, perhaps involving ties. A matching M is said to be *stable* if there is no edge (a, b) such that a is unmatched or prefers b to $M(a)$ and similarly, b is unmatched or prefers a to $M(b)$. While a stable matching in G can be easily computed in linear time by the Gale-Shapley algorithm, it is known that computing a maximum size stable matching is APX-hard.

In this paper we consider the case when the preference lists of vertices in A are *strict* while the preference lists of vertices in B may include ties. This case is also APX-hard and the current best approximation ratio known here is 25/17 ≈ 1.4706 which relies on solving an LP. We improve this ratio to 22/15 ≈ 1.4667 by a simple linear time algorithm.

We first compute a half-integral stable matching in $\{0, 0.5, 1\}^{|E|}$ and round it to an integral stable matching M. The ratio $|\mathsf{OPT}|/|M|$ is bounded via a payment scheme that charges other components in $\mathsf{OPT} \oplus M$ to cover the costs of length-5 augmenting paths. There will be no length-3 augmenting paths here.

We also consider the following special case of two-sided ties, where every tie length is 2. This case is known to be UGC-hard to approximate to within 4/3. We show a 10/7 ≈ 1.4286 approximation algorithm here that runs in linear time.

1 Introduction

The stable marriage problem is a classical and well-studied matching problem in bipartite graphs. The input here is a bipartite graph $G = (A \cup B, E)$ where every $u \in A \cup B$ ranks its neighbors in an order of preference and ties are permitted in preference lists. It is customary to refer to the vertices in A and B as *men* and *women*, respectively. Preference lists may be incomplete: that is, a vertex need not be adjacent to all the vertices on the other side.

A matching is a set of edges, no two of which share an endpoint. An edge (a, b) is said to be a *blocking edge* for a matching M if either a is unmatched or prefers b to its partner in M, i.e., $M(a)$, and similarly, b is unmatched or prefers a to its partner $M(b)$. A matching that admits no blocking edges is said to be *stable*. The problem of

---

* Part of this work was done while visiting the Max-Planck-Institut für Informatik, Saarbrücken under the IMPECS program.

J. Lee and J. Vygen (Eds.): IPCO 2014, LNCS 8494, pp. 297–308, 2014.

computing a stable matching in G is the stable marriage problem. A stable matching always exists and can be computed in linear time by the well-known Gale-Shapley algorithm [2].

Several real-world assignment problems can be modeled as the stable marriage problem; for instance, the problems of assigning residents to hospitals [4] or students to schools [19]. The input instance could admit many stable matchings and the desired stable matching in most real-world applications is a maximum cardinality stable matching. When preference lists are *strict* (no ties permitted), it is known that all stable matchings in G have the same size and the set of vertices matched in every stable matching is the same [3]. However when preference lists involve ties, stable matchings can vary in size.

Consider the following simple example, where $A = \{a_1, a_2\}$ and $B = \{b_1, b_2\}$ and let the preference lists be as follows:

$$a_1 : b_1; \qquad a_2 : b_1, b_2; \qquad b_1 : \{a_1, a_2\}; \qquad \text{and} \quad b_2 : a_2.$$

The preference list of a_1 consists of just b_1 while the preference list of a_2 consists of b_1 followed by b_2. The preference list of b_1 consists of a_1 and a_2 *tied* as the top choice while the preference list of b_2 consists of the single vertex a_2. There are 2 stable matchings here: $\{(a_2, b_1)\}$ and $\{(a_1, b_1), (a_2, b_2)\}$. Thus the sizes of stable matchings in G could differ by a factor of 2 and it is easy to see that they cannot differ by a factor more than 2 since every stable matching has to be a *maximal* matching. As stated earlier, the desired matching here is a maximum size stable matching. However it is known that computing such a matching is NP-hard [8,15].

Iwama et al. [9] showed a $15/8 = 1.875$-approximation algorithm for this problem using a local search technique. The next breakthrough was due to Király [11], who introduced the simple and effective technique of "promotion" to break ties in a modification of the Gale-Shapley algorithm. He improved the approximation ratio to $5/3$ for the general case and to 1.5 for *one-sided ties*, i.e., the preference lists of vertices in A have to be strict while ties are permitted in the preference lists of vertices in B. McDermid [16] then improved the approximation ratio for the general case also to 1.5. For the case of one-sided ties, Iwama et al. [10] showed a $25/17 \approx 1.4706$-approximation.

On the inapproximability side, the strongest hardness results are due to Yanagisama [21] and Iwama et al. [9]. In [21], the general problem was shown to be NP-hard to approximate to within $33/29$ and UGC-hard to approximate to within $4/3$; the case of one-sided ties was considered in [9] and shown to be NP-hard to approximate to within $21/19$ and UGC-hard to approximate to within $5/4$.

In this paper we focus mostly on the case of one-sided ties. The case of one-sided ties occurs frequently in several real-world problems; for instance, in the Scottish Foundation Allocation Scheme (SFAS), the preference lists of applicants have to be strictly ordered while the preference lists of positions can admit ties [7]. Let OPT be a maximum size stable marriage in the given instance. We show the following result here.

Theorem 1. *Let $G = (A \cup B, E)$ be a stable marriage instance where vertices in A have strict preference lists while vertices in B are allowed to have ties in preference lists. A stable matching M in G such that $|OPT|/|M| \leq 22/15 \approx 1.4667$ can be computed in linear time.*

Techniques. Our algorithm constructs a *half-integral* stable matchings using a modified Gale-Shapley algorithm: each man can make two proposals and each woman can accept two proposals. How the proposals are made by men and how women accept these proposals forms the core part of our algorithms. In our algorithms, after the proposing phase is over, we have a half-integral vector x, where $x_{ab} = 1$ (similarly, 1/2 or 0) if b accepts 2 (respectively, 1 or 0) proposals from a. We then build a subgraph G' of G by retaining an edge e only if $x_e > 0$. Our solution is a maximum cardinality matching in G' where every degree 2 vertex gets matched.

In the original Gale-Shapley algorithm, when two proposals are made to a woman from men that are tied on her list, she is forced to make a blind choice since she has no way of knowing which is a better proposal (i.e., it leads to a larger matching) to accept. Our approach to deal with this issue is to let her accept both proposals. Since neither proposer is fully accepted, each of them has to propose down his list further and get another proposal accepted. Essentially, our strategy of letting men make multiple proposals and letting women accept multiple proposals is a way of coping with their lack of knowledge about the best decision at any point in time. Note that we limit the number of proposals a man makes/a woman accepts to be 2 because we want the graph G' to have a simple structure. In our algorithms, every vertex in G' has degree at most 2 and this allows us to bound our approximation guarantees.

We first show that there are no length-3 augmenting paths in $M \oplus \mathrm{OPT}$ using the idea of *promotion* introduced by Királyi [11] to break ties in favor of those vertices rejected once by all their neighbors. This idea was also used by McDermid [16] and Iwama et al. [10]. This idea essentially guarantees an approximation factor of 1.5 by eliminating all length-3 augmenting paths in $M \oplus \mathrm{OPT}$. In order to obtain an approximation ratio < 1.5, we use a new combinatorial technique that makes components other than augmenting paths of length-5 in $M \oplus \mathrm{OPT}$ pay for augmenting paths of length-5.

Let R denote the set of augmenting paths of length-5 in $M \oplus \mathrm{OPT}$ and let $Q = (M \oplus \mathrm{OPT}) \setminus R$. Suppose $q \in Q$ is an augmenting path on $2\ell + 3 \geq 7$ edges or an alternating cycle/path on 2ℓ edges or an alternating path on $2\ell - 1$ edges (with ℓ edges of M). In our algorithm for one-sided ties, q will be charged for $\leq 3\ell$ elements in R and this will imply that $|\mathrm{OPT}|/|M| \leq 22/15$.

For the case of one-sided ties, to obtain an approximation guarantee < 1.5, the algorithm by Iwama et al. [10] formulates the maximum cardinality stable matching problem as an integer program and solves its LP relaxation. This optimal LP-solution guides women in accepting proposals and leads to a 25/17-approximation.

It was also shown in [10] that for two-sided ties, the integrality gap of a natural LP for this problem (first used in [20]) is $1.5 - \Theta(1/n)$. As mentioned earlier, McDermid [16] gave a 1.5-approximation algorithm here; Királyi [12] and Paluch [17] have shown linear time algorithms for this ratio. A variation of the general problem was recently studied by Askalidis et al. [1].

Since no approximation guarantee better than 1.5 is known for the general case of two-sided ties while better approximation algorithms are known for the one-sided ties case, as a first step we consider the following variant of two-sided ties where each tie length is 2. This is a natural variant as there are several application domains where ties are permitted but their length has to be small. We show the following result here.

Theorem 2. *Let* $G = (A \cup B, E)$ *be a stable marriage instance where vertices in* $A \cup B$ *are allowed to have ties in preference lists, however each tie has length 2. A stable matching* M' *in* G *such that* $|\mathsf{OPT}|/|M'| \leq 10/7 \approx 1.4286$ *can be computed in linear time.*

Currently, this is the only case with approximation ratio better than 1.5 for any special case of the stable marriage problem where ties can occur on *both* sides of G. Interestingly, in the hardness results shown in [21] and [9], it is assumed that each vertex has at most one tie in its preference list, and such a tie is of length 2. Thus if the general case really has higher inapproximability, say 1.5 as previously conjectured by Király [11], then the reduction in the hardness proof needs to use longer ties.

We also note that the ratio of $10/7$ we achieve in this special case coincides with the ratio attained by Halldórsson et al. [5] for the case that ties only appear on women's side and each tie is of length 2.

The stable marriage problem is an extensively studied subject on which several monographs [4,13,14,18] are available. The generalization of allowing ties in the preference lists was first introduced by Irving [6]. There are several ways of defining stability when ties are allowed in preference lists. The definition, as used in this paper, is Irving's "weak-stability."

Due to the space limit, we only present our algorithm for one-sided ties in Section 2 and its analysis in Section 3. Some missing proofs, along with the algorithm for two-sided ties where each tie has length 2, can be found in the full version.

2   Our Algorithm

Our algorithm produces a fractional matching $x = (x_e, e \in E)$ where each $x_e \in \{0, 1/2, 1\}$. The algorithm is a modification of the Gale-Shapley algorithm in $G = (A \cup B, E)$. We first explain how men propose to women and then how women decide (see Fig. 1).

How men propose. Every man a has two proposals p_a^1 and p_a^2, where each proposal p_a^i (for $i = 1, 2$) goes to the women on a's preference list in a round-robin manner. Initially, the target of both proposals p_a^1 and p_a^2 is the first woman on a's list. For any i, at any point, if p_a^i is rejected by the woman who is ranked k-th on a's list (for any k), then p_a^i goes to the woman ranked $(k+1)$-st on a's list; in case the k-th woman is already the last woman on a's list, then the proposal p_a^i is again made to the first woman on a's list.

A man has three possible levels in status: *basic*, *1-promoted*, or *2-promoted*. Every man a starts out basic with rejection history $r_a = \emptyset$. Let $N(a)$ be the set of all women on a's list. When $r_a = N(a)$, then a becomes 1-promoted. Once he becomes 1-promoted, r_a is reset to the empty set. If $r_a = N(a)$ after a becomes 1-promoted, then a becomes 2-promoted and r_a is reset once again to the empty set. After a becomes 2-promoted, if $r_a = N(a)$, then a gives up.

To illustrate promotions, consider the following example: man a has only two women b_1 and b_2 on his list. He starts as a basic man and makes his proposals p_a^1 and p_a^2 to b_1.

Suppose b_1 rejects both. Then a makes both these proposals to b_2. Suppose b_2 accepts p_a^1 but rejects p_a^2. Then a becomes 1-promoted since $r_a = \{b_1, b_2\}$ now and r_a is reset to $\emptyset$. Note that for a to become 2-promoted, we need r_a to become $\{b_1, b_2\}$ once again. Similarly, a 2-promoted man a gives up only when his rejection history r_a becomes $\{b_1, b_2\}$ *after* he becomes 2-promoted.

– For every $a \in A$, $t_a^1 := t_a^2 := 1$; $r_a := \emptyset$.
$\{r_a$ *is the rejection history of man* a; t_a^i *is the rank of the next woman targeted by the proposal* $p_a^i.\}$
while some $a \in A$ has his proposal p_a^i (i is 1 or 2) not accepted by any woman and he has not given up **do**
  – a makes his proposal p_a^i to the t_a^i-th woman b on his list.
  if b has at most two proposals now (incl. p_a^i) **then**
    – b accepts p_a^i
  else
    – b rejects any of her "least desirable" (see Definition 1) proposals $p_{a'}^j$
    if $t_{a'}^j$ = number of women on the list of a' **then**
      $t_{a'}^j := 1$ $\{$*the round-robin nature of proposing*$\}$
    else
      $t_{a'}^j := t_{a'}^j + 1$
    end if
    – $r_{a'} := r_{a'} \cup \{b\}$
    if $r_{a'}$ = the entire set of neighbors of a' **then**
      if a' is basic **then**
        a' becomes 1-promoted and $r_{a'} := \emptyset$
      else if a' is 1-promoted **then**
        a' becomes 2-promoted and $r_{a'} := \emptyset$
      else if a' is 2-promoted **then**
        a' gives up
      end if
    end if
  end if
end while

Fig. 1. A description of proposals/disposals in our algorithm with one-sided ties

---

Our algorithm terminates when each $a \in A$ satisfies one of the following conditions: (1) both his proposals p_a^1 and p_a^2 are accepted, (2) he gives up. Note that when (2) happens, the man a must be 2-promoted.

How women decide: A woman can accept up to two proposals. The two proposals can be from the same man. When she currently has less than two proposals, she unconditionally accepts the new proposal. If she has already accepted two proposals and is faced with a third one, then she rejects one of her "least desirable" proposals (see Definition 1 below).

Definition 1. *For a woman* b, *proposal* p_a^i *is superior to* $p_{a'}^{i'}$, *if on* b'*s list:*

(1) a *ranks better than* a'.
(2) a *and* a' *are tied;* a *is currently 2-promoted while* a' *is currently 1-promoted or basic.*
(3) a *and* a' *are tied;* a *is currently 1-promoted while* a' *is currently basic.*
(4) a *and* a' *are tied and both are currently basic; moreover, woman* b *has already rejected one proposal of* a *while so far she has not rejected any of the proposals of* a'.

Let p_a^i *be among the three proposals that a woman has and suppose it is not superior to either of the other two proposals. Then* p_a^i *is a* least desirable *proposal.*

The reasoning behind the rules of a woman's decision can be summarized as follows.

- Proposals from higher-ranking men should be preferred, as in the Gale-Shapley algorithm.
- When a woman receives proposals from men who are tied in her list, she prefers the man who has been promoted: a 1-promoted (similarly, 2-promoted) man having been rejected by the entire set of women on his list once (resp. twice) should be preferred, since he is more desperate and deserves to be given a chance.
- When two basic men of the same rank propose to a woman, she prefers the one who has been rejected by her before. The intuition again is that he is more desperate— though he has not been rejected by all women on his list yet (otherwise he would have been 1-promoted).

It is easy to see that the algorithm in Fig. 1 runs in linear time. When it terminates, for each edge $(a, b) \in E$, we set $x_{ab} = 1$ or 0.5 or 0 if the number of proposals that woman b accepts from man a is 2 or 1 or 0, respectively. Let $G' = (A \cup B, E')$ be the subgraph where an edge $e \in E'$ if and only if $x_e > 0$. It is easy to see that in G', the maximum degree of any vertex is 2.

There is a maximum cardinality matching in G' where all degree 2 vertices are matched; moreover, such a matching can be computed in linear time. Let M be such a matching. We first show that M is stable and then prove it is a 22/15 approximation. Propositions 1 and 2 follow easily from our algorithm and lead to the stability of M.

Proposition 1. *Let woman* b *reject proposal* p_a^i *from man* a. *Then from this point till the end of the algorithm,* b *has two proposals* $p_{a'}^{i'}$ *and* $p_{a''}^{i''}$ *from men* a' *and* a'' *(it is possible that* $a' = a''$*) who rank at least as high as man* a *on* b'*s list. In particular, if* a' *(similarly,* a''*) is tied with man* a *on the list of* b, *then at the time* a *proposed to* b:

1. *if* a *is* ℓ-promoted *(*ℓ *is either 1 or 2), then man* a' *(resp.* a''*) has to be* $\geq \ell$-promoted.
2. *if* a *is basic and his other proposal is already rejected by* b, *then it has to be the case that either* a' *(resp.* a''*) is not basic or* b *has already rejected his other proposal.*

In the rest of the paper, unless we specifically state the time point, when we say a man is basic/1-promoted/2-promoted, we mean his status when the algorithm terminates.

Proposition 2. *The following facts hold:*

1. *If a man (similarly, a woman) is unmatched in M, then he has at most one proposal accepted by a woman (resp., she receives at most one proposal) during the entire algorithm.*
2. *At the end of the algorithm, every man with less than two proposals accepted is 2-promoted. Furthermore, he must have been rejected by all women on his list as a 2-promoted man.*
3. *If woman b on the list of the man a is unmatched in M, then man a has to be basic and he does not prefer b to the women who accepted his proposals.*

3 Bounding the Size of M

Let OPT be an optimal stable matching. We now need to bound $|\text{OPT}|/|M|$. Whenever we refer to an augmenting path in $M \oplus \text{OPT}$, we mean the path is augmenting with respect to M. Lemma 1 will be crucial in our analysis.

Lemma 1. *Suppose (a, b) and (a', b') are in OPT where man a' is not 2-promoted and a' prefers b to b'. If a is unmatched in M, then (a', b) cannot be in G'.*

Proof. We prove this lemma by contradiction. Suppose $(a', b) \in G'$. If b prefers a' to a, then (a', b) blocks OPT. On the other hand, if b prefers a to a', then this contradicts the fact that b rejected at least one proposal from a (by Proposition 2.1) while b has a proposal from a', who is ranked worse on b's list, at the end of the algorithm since $(a', b) \in G'$.

So the only option possible is that a' and a are tied on b's list. Since a is unmatched in M, it follows from (1)-(2) of Proposition 2 that a has been rejected by b as a 2-promoted man. Since $(a', b) \in G'$, Proposition 1 implies that a' has to be 2-promoted. This however contradicts the lemma statement that a' is not 2-promoted.   □

Corollary 1. *There is no length-3 augmenting path $M \oplus \text{OPT}$.*

Proof. If such a path $a - b - a' - b'$ exists (see Fig. 2), then $(a', b) \in G'$ since it is in M. As b' is unmatched in M, a' is basic and prefers b to b' (by Proposition 2.3). This contradicts Lemma 1.   □

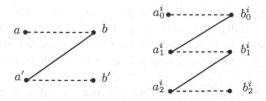

Fig. 2. On the left we have a length-3 augmenting path and on the right we have the length-5 augmenting path ρ_i with respect to M in $M \oplus \text{OPT}$

Let $R = \{\rho_1, \ldots, \rho_t\}$ denote the set of length-5 augmenting paths in $M \oplus \text{OPT}$. Lemma 2 lists properties of vertices in a length-5 augmenting path ρ_i (Fig. 2).

Lemma 2. *If* $\rho_i = a_0^i - b_0^i - a_1^i - b_1^i - a_2^i - b_2^i$ *is a length-5 augmenting path in* $M \oplus \text{OPT}$, *then*

1. a_0^i *is 2-promoted and has been rejected by* b_0^i *as a 2-promoted man.*
2. a_1^i *is not 2-promoted and he prefers* b_1^i *to* b_0^i.
3. a_2^i *is basic and he prefers* b_1^i *to* b_2^i.
4. b_1^i *is indifferent between* a_1^i *and* a_2^i.
5. *In* G', b_0^i *has degree 1 if and only if* a_1^i *has degree 1.*
6. *In* G', b_1^i *has degree 1 if and only if* a_2^i *has degree 1.*

Recall that G' is a subgraph of G and every vertex has degree at most 2 in G'. We form a directed graph H from G' as follows: first orient all edges in the graph G' from A to B; then contract each edge of $M \cap \rho_i$ for $i = 1, \ldots, t$. That is, if $\rho_i = a_0^i - b_0^i - a_1^i - b_1^i - a_2^i - b_2^i$, then in H, the edge (a_1^i, b_0^i) gets contracted into a single node (call it x_i) and similarly the edge (a_2^i, b_1^i) gets contracted into a single node (call it y_i) and this happens for all $i = 1, \ldots, t$.

Note that (5)-(6) of Lemma 2 imply that $\deg_H(x_i), \deg_H(y_i) \in \{0, 2\}$ for $1 \leq i \leq t$, where $\deg_H(v) = 2$ means in H in-degree$(v) = $ out-degree$(v) = 1$. The following lemma rules out the possibility of certain arcs in H.

Lemma 3. *For any* $1 \leq i, j \leq t$, *there is no arc from* y_i *to* x_j *in* H.

Proof. Suppose there is an arc in H from y_i to x_j for some $1 \leq i, j, \leq t$. That is, G' contains the edge (a_2^i, b_0^j). Since the woman b_2^i is unmatched, we use Proposition 2.3 to conclude that a_2^i is basic and he prefers b_0^j to b_2^i. This contradicts Lemma 1, by substituting $a = a_0^j, b = b_0^j, a' = a_2^i$, and $b' = b_2^i$.   □

We now define a "good path" in H. In H, let us refer to the x-nodes and y-nodes as *red* and let the other vertices be called *blue*.

Definition 2. *A directed path in* H *is* good *if its end vertices are blue while all its intermediate vertices are red. Also, we assume there is at least one intermediate vertex in such a path.*

Lemma 3 implies that every good path looks as follows: a blue man, followed by some x-nodes (possibly none), followed by some y-nodes (possibly none), and a blue woman.

For any y-node y_i, if $\deg_H(y_i) \neq 0$, using Lemma 3 we can conclude that y_i is either in a cycle of y-nodes or in a good path. In other words, there are only 3 possibilities in H for each y_i: (1) y_i is an isolated node, (2) y_i is in a cycle of y-nodes, (3) y_i is in a good path.

We next define a *critical arc* in H. We will use critical arcs to show that H has enough good paths. Since the endpoints of a good path are vertices outside R, this bounds $|\text{OPT}|/|M|$.

Definition 3. *Call an arc (x_i, z) in H critical if either a_1^i prefers z to b_1^i or $z = b_1^i$.*

In case z is a red node, let w be the woman in z – in Definition 3, we mean either $w = b_1^i$ or a_1^i prefers w to b_1^i. We show (via Lemma 4 and Claim 1) that every critical arc is in a distinct good path. It follows from Lemma 4 that every good path has at most one critical arc. Lemma 5 is the main technical lemma here. It shows there are *enough* critical arcs in H.

Lemma 4. *For any i, if (x_i, z) is critical, then z is not an x-node, i.e., $z \neq x_j$ for any j.*

Proof. For any $1 \leq i, j \leq t$, if a proposal of a_1^i is accepted by a woman w that a_1^i prefers to b_1^i, then we need to show that w cannot be b_0^j. Suppose $w = b_0^j$ for some j. In the first place, $j \neq i$ since we know a_1^i prefers b_1^i to b_0^i (by Lemma 2.2). We know a_1^i is not 2-promoted by Lemma 2.2. We now contradict Lemma 1, by substituting $a = a_0^j$, $b = b_0^j$, $a' = a_1^i$, and $b' = b_1^i$. □

Claim 1. *Every critical arc is in some good path and every pair of good paths is vertex-disjoint.*

Lemma 5. *In the graph H, the following statements hold:*

(1) If y_i is an isolated node, then there exists a critical arc (x_i, z) in H.
(2) If (y_i, y_j) is an arc, then there exists a critical arc (x_i, z) or a critical arc (x_j, z') (or both).

Proof. We first show part (1) of this lemma. Suppose y_i is an isolated node in H. By parts (2) and (6) of Lemma 2, the woman b_1^i accepts both proposals from a_2^i and she rejects a_1^i at least once. Suppose b_1^i rejects a_1^i exactly once. This means that one proposal of a_1^i (other than the one accepted by b_0^i) has been accepted by a woman w that a_1^i prefers to b_1^i. That is, there is a critical arc (x_i, z) in H.

So suppose b_1^i rejects a_1^i more than once. Then either a_1^i has both of his proposals rejected by b_1^i while he was basic, or he was rejected by b_1^i as a 1-promoted man. In both cases we have a contradiction to Proposition 1 since b_1^i has accepted both proposals from a_2^i, who is basic and is tied with a_1^i.

We now show part (2) of this lemma. Suppose a_1^i prefers b_1^i to the women accepting his proposals and a_1^j prefers b_1^j to the women accepting his proposals. Note that this includes the possibility that both of a_1^i's proposals are accepted by b_0^i and the possibility that both of a_1^j's proposals are accepted by b_0^j. The first observation is that a_1^j could *not* have proposed to b_1^j as a 1-promoted man, as it would contradict Proposition 1 otherwise (recall a_2^j is basic and a_1^j, a_2^j are tied on the list of b_1^j). For the same reason, a_1^i never proposed to b_1^i as a 1-promoted man.

Since we assumed that a_1^j prefers b_1^j to the women accepting his proposals and he never proposed to b_1^j as a 1-promoted man, it must be the case that both of his proposals were rejected by b_1^j when he was still basic. The edge $(a_2^i, b_1^j) \in G'$ since (y_i, y_j) is in H. We now claim this implies a_2^i is tied with a_1^j on the list of b_1^j. If b_1^j prefers a_2^i to a_1^j, then (a_2^i, b_1^j) blocks OPT, since Proposition 2.3 states that a_2^i prefers b_1^i to b_2^i. Now suppose b_1^j prefers a_1^j to a_2^i. Since a_1^j prefers b_1^j to b_0^j (by Lemma 2.2), he must have been rejected by b_1^j before he proposed to b_0^j, implying a contradiction to Proposition 1.

We also know that a_1^j is tied with a_2^j on the list of b_1^j (by Lemma 2.4) and that a_2^i is basic. Since we know that both of a_1^i's proposals were rejected by b_1^j, it has to be the case that while b_1^j accepted one proposal of a_2^i, she rejected his other proposal (by Proposition 1.2). This other proposal of a_2^i was at some point accepted by b_1^i. So it follows that b_1^j *ranks higher than* b_1^i *on the list of* a_2^i, *furthermore,* b_1^i *never rejects a proposal from* a_2^i.

Since we assumed that a_1^i prefers b_1^i to the women accepting his proposals and he never proposed to b_1^i as a 1-promoted man, it follows that both of his proposals were rejected by b_1^i when he was basic. This, combined with the fact that b_1^i never rejects a proposal from a_2^i, contradicts Proposition 1.2. Thus either one proposal of a_1^i has been accepted by a woman w that is b_1^i or better than b_1^i in a_1^i's list or one proposal of a_1^j has been accepted by a woman w' that a_1^j prefers to b_1^j. Hence there is a critical arc (x_i, z) or a critical arc (x_j, z') in H.   □

We define a function $f : [t] \rightarrow \mathcal{P}$, where $\mathcal{P}$ is the set of all good paths in H and $[t] = \{1, \ldots, t\}$. For any $i \in [t]$, $f(i)$ is defined as follows:

(1) Suppose y_i is isolated. Then let $f(i) = p$, where $p \in \mathcal{P}$ contains the critical arc (x_i, z). We know there is such an arc in H by Lemma 5.1.
(2) Suppose y_i belongs to a cycle C of y-nodes, so there is an arc (y_i, y_j) in C. We know H has a critical arc (x_i, z) or (x_j, z') (by Lemma 5.2). Then let $f(i) = p$, where $p \in \mathcal{P}$ contains such a critical arc.
(3) Suppose y_i belongs to a good path p'. If y_i is the *last* y-node in p', then let $f(i) = p'$. Otherwise there is an arc (y_i, y_j) in p' and we know H has a critical arc (x_i, z) or (x_j, z') (by Lemma 5.2). Then let $f(y_i) = p$, where $p \in \mathcal{P}$ contains such a critical arc.

For any $p \in \mathcal{P}$, let $\mathrm{cost}(p) = $ the number of pre-images of p under f. We now show a charging scheme that distributes $\mathrm{cost}(p)$, for each $p \in \mathcal{P}$, among the vertices in G so that the following properties hold. Let $Q = (M \oplus \mathsf{OPT}) \setminus R$.

(I) Each $v \in A \cup B$ is assigned a charge of at most 1.5 and the sum of all vertex charges is t.
(II) Every vertex that is assigned a positive charge must be matched in M and is in some $q \in Q$. Moreover, if $q \in Q$ is an augmenting path on $2\ell_q + 3 \geq 7$ edges, then at most $2\ell_q$ vertices in q will be assigned a positive charge.

Note that a vertex not assigned a positive charge has charge 0 by default.

Suppose there is such a charging scheme, we now show why this implies $|\mathsf{OPT}|/|M|$ is at most 22/15. Let $q \in Q$ be an alternating cycle/path on $2\ell_q$ edges or an alternating path on $2\ell_q - 1$ edges (with ℓ_q edges from M) or an augmenting path on $2\ell_q + 3 \geq 7$ edges. It follows from (I) and (II) that the total charge assigned to vertices in q is at most $1.5(2\ell_q) = 3\ell_q$, i.e., if the vertices in q are being charged for c_q augmenting paths of length-5 in $M \oplus \mathsf{OPT}$, then $c_q \leq 3\ell_q$.

Since $\sum_{q \in Q} c_q = t$, all the paths in R are paid for in this manner. So we have:

$$|\mathsf{OPT}| = \sum_{q \in Q}(|\mathsf{OPT} \cap q| + 3c_q) \quad \text{and} \quad |M| = \sum_{q \in Q}(|M \cap q| + 2c_q),$$

because there are $3c_q$ edges of OPT in the c_q augmenting paths of length-5 covered by q and $2c_q$ edges of M in the c_q augmenting paths of length-5 covered by q. Thus we have:

$$\frac{|OPT|}{|M|} \leq \max_{q \in Q} \frac{|OPT \cap q| + 3c_q}{|M \cap q| + 2c_q} \leq \max_{l_q \geq 2} \frac{10\ell_q + 2}{7\ell_q + 1} \leq \frac{22}{15}.$$

We use $(\sum_i s_i)/(\sum_i t_i) \leq \max_i s_i/t_i$ in the first inequality. The above ratio gets maximized for any $q \in Q$ by setting c_q to its largest value of $3\ell_q$ and letting q be an augmenting path so that $|OPT \cap q| > |M \cap q|$.

This yields $(\ell_q + 2 + 3 \cdot 3\ell_q)/(\ell_q + 1 + 2 \cdot 3\ell_q)$, where $|q| = 2\ell_q + 3 \geq 7$. Note that since augmenting paths in Q have length ≥ 7, this forces $\ell_q \geq 2$ in this ratio. Setting $\ell_q = 2$ maximizes the ratio $(10\ell_q + 2)/(7\ell_q + 1)$. Thus our upper bound is $22/15$.

Ensuring properties (I) and (II). We now show a charging scheme that defines a function charge : $A \cup B \to [0, 1.5]$ such that $\sum_u$ charge$(u) = \sum_{p \in \mathcal{P}}$ cost$(p) = t$, where the sum is over all $u \in A \cup B$. We start with charge$(u) = 0$ for all $u \in A \cup B$. Our task now is to reset charge values for some vertices so that properties (I) and (II) are satisfied.

Each $p \in \mathcal{P}$ is one of the following three types: (1) *type-1* path: this has no x-nodes, (2) *type-2* path: this has no y-nodes, and (3) *type-3* path: this has both x-nodes and y-nodes. The following lemma will be useful later in our analysis.

Lemma 6. *For any $p \in \mathcal{P}$ and $k = 1, 2, 3$, if p is a type-k path, then cost$(p) \leq k$.*

Consider any $p \in \mathcal{P}$. Though p was defined as a good path in H, we now consider p as a path in the graph G'. Since each intermediate node of p is an edge of M, p is an alternating path in G'. Let a_p (man) and b_p (woman) be the endpoints of the path p.

If both a_p and b_p are unmatched in M, then the path p becomes an augmenting path in G'. Since M is a maximum cardinality matching in G', there cannot be an augmenting path with respect to M in G'; hence at least one of a_p, b_p has to be matched in M.

Case 1. Suppose both a_p and b_p are matched. If p is a type-1 path, then reset charge(b_p) = cost(p), i.e., the entire cost associated with p is assigned to the woman who is an endpoint of p. If p is a type-k path for $k = 2$ or 3, then reset charge$(a_p) =$ charge$(b_p) =$ cost$(p)/2$.

Case 2. Suppose exactly one of a_p, b_p is matched: call the matched vertex s_p and the unmatched vertex u_p. Construct the alternating path with respect to M in G' with u_p as the starting vertex. The vertex u_p has degree 1 since it is unmatched, also the maximum degree of any vertex in G' is 2. So there is only one such alternating path in G'. This path continues till it encounters a degree 1 vertex, call it r_p.

Note that r_p has to be matched, otherwise there is an augmenting path in G' between u_p and r_p. Since r_p is reached via a matched edge on this path, both u_p and r_p are either in A or in B. In other words, exactly one of r_p, s_p (recall $s_p = \{a_p, b_p\} \setminus \{u_p\}$) is a woman. If p is a type-1 path, then we reset charge$(w) =$ cost(p), where w is the woman in $\{r_p, s_p\}$. If p is a type-k path, where $k = 2$ or 3, then we reset charge$(s_p) =$ charge$(r_p) =$ cost$(p)/2$. This concludes the description of our charging scheme.

References

1. Askalidis, G., Immorlica, N., Kwanashie, A., Manlove, D.F., Pountourakis, E.: Socially stable matchings in the hospitals / residents problem. In: Dehne, F., Solis-Oba, R., Sack, J.-R. (eds.) WADS 2013. LNCS, vol. 8037, pp. 85–96. Springer, Heidelberg (2013)
2. Gale, D., Shapley, L.S.: College admissions and the stability of marriage. American Math. Monthly 69, 9–15 (1962)
3. Gale, D., Sotomayer, M.: Some remarks on the stable marriage problem. Discrete Applied Mathematics 11, 223–232 (1985)
4. Gusfield, D., Irving, R.W.: The Stable Marriage Problem: Structure and Algorithms. MIT Press, Boston (1989)
5. Halldórsson, M.M., Iwama, K., Miyazaki, S., Yanagisawa, H.: Randomized approximation of the stable marriage problem. Theoretical Computer Science 325(3), 439–465 (2004)
6. Irving, R.W.: Stable marriage and indifference. Discrete Applied Mathematics 48, 261–272 (1994)
7. Irving, R.W., Manlove, D.F.: Approximation algorithms for hard variants of the stable marriage and hospitals/residents problems. Journal of Combinatorial Optimization 16(3), 279–292 (2008)
8. Iwama, K., Manlove, D.F., Miyazaki, S., Morita, Y.: Stable marriage with incomplete lists and ties. In: Wiedermann, J., Van Emde Boas, P., Nielsen, M. (eds.) ICALP 1999. LNCS, vol. 1644, pp. 443–452. Springer, Heidelberg (1999)
9. Iwama, K., Miyazaki, S., Yamauchi, N.: A 1.875-approximation algorithm for the stable marriage problem. In: 18th SODA, pp. 288–297 (2007)
10. Iwama, K., Miyazaki, S., Yanagisawa, H.: A 25/17-approximation algorithm for the stable marriage problem with one-sided ties. In: de Berg, M., Meyer, U. (eds.) ESA 2010, Part II. LNCS, vol. 6347, pp. 135–146. Springer, Heidelberg (2010)
11. Király, Z.: Better and simpler approximation algorithms for the stable marriage problem. Algorithmica 60(1), 3–20 (2011)
12. Király, Z.: Linear time local approximation algorithm for maximum stable marriage. Algorithms 6(3), 471–484 (2013)
13. Knuth, D.: Mariages stables et leurs relations avec d'autre problèmes. Les Presses de l'université de Montréal (1976)
14. Manlove, D.: Algorithmics of Matching Under Preferences. World Scientific Publishing Company Incorporated (2013)
15. Manlove, D.F., Irving, R.W., Iwama, K., Miyazaki, S., Morita, Y.: Hard variants of stable marriage. Theoretical Computer Science 276(1-2), 261–279 (2002)
16. McDermid, E.: A 3/2 approximation algorithm for general stable marriage. In: Albers, S., Marchetti-Spaccamela, A., Matias, Y., Nikoletseas, S., Thomas, W. (eds.) ICALP 2009, Part I. LNCS, vol. 5555, pp. 689–700. Springer, Heidelberg (2009)
17. Paluch, K.: Faster and simpler approximation of stable matchings. In: Solis-Oba, R., Persiano, G. (eds.) WAOA 2011. LNCS, vol. 7164, pp. 176–187. Springer, Heidelberg (2012)
18. Roth, A., Sotomayor, M.: Two-sided matching: a study in game-theoretic modeling and analysis. Cambridge University Press (1992)
19. Teo, C.-P., Sethuraman, J., Tan, W.P.: Gale-Shapley stable marriage problem revisited: strategic issues and applications. In: Cornuéjols, G., Burkard, R.E., Woeginger, G.J. (eds.) IPCO 1999. LNCS, vol. 1610, pp. 429–438. Springer, Heidelberg (1999)
20. Vande Vate, J.: Linear Programming brings marital bliss. Operation Research Letters 8, 147–153 (1989)
21. Yanagisawa, H.: Approximation algorithms for stable marriage problems. Ph.D. Thesis, Kyoto University (2007)

Simple Extensions of Polytopes

Volker Kaibel and Matthias Walter

Otto-von-Guericke Universität Magdeburg, Germany
{kaibel,walter}@ovgu.de

Abstract. We introduce the *simple extension complexity* of a polytope P as the smallest number of facets of any simple (i.e., non-degenerate in the sense of linear programming) polytope which can be projected onto P. We devise a combinatorial method to establish lower bounds on the simple extension complexity and show for several polytopes that they have large simple extension complexities. These examples include both the spanning tree and the perfect matching polytopes of complete graphs, uncapacitated flow polytopes for non-trivially decomposable directed acyclic graphs, and random 0/1-polytopes with vertex numbers within a certain range. On our way to obtain the result on perfect matching polytopes we improve on a result of Padberg and Rao's on the adjacency structures of those polytopes.

1   Introduction

In combinatorial optimization, linear programming formulations are a standard tool to gain structural insight, derive algorithms and to analyze computational complexity. With respect to both structural and algorithmic aspects of linear optimization over a polytope P can be replaced be linear optimization over any (usually higher dimensional) polytope Q of which P can be obtained as the image under a linear map (which we refer to as a *projection*). Such a polytope Q (along with a suitable projection) is called an *extension* of P.

Defining the *size* of a polytope as its number of facets, the smallest size of any extension of the polytope P is known as the *extension complexity* $\text{xc}\,(P)$ of P. It has turned out in the past that for several important polytopes related to combinatorial optimization problems the extension complexity is bounded polynomially in the dimension. One of the most prominent examples is the spanning tree polytope of the complete graph K_n on n nodes, which has extension complexity $\mathcal{O}\left(n^3\right)$ [9].

After Rothvoß [13] showed that there are 0/1-polytopes whose extension complexities cannot be bounded polynomially in their dimensions, only very recently Fiorini et al. [4] could prove that the extension complexities of some concrete and important examples of polytopes like traveling salesman polytopes cannot be bounded polynomially. Similar results have then also been deduced for several other polytopes associated with NP-hard optimization problems, e.g., by Avis and Tiwary [1] and Pokutta and van Vyve [12]. Very recently, Rothvoß [14] showed that also the perfect matching polytope of the complete graph (with an

J. Lee and J. Vygen (Eds.): IPCO 2014, LNCS 8494, pp. 309–320, 2014.

even number of nodes) has exponential extension complexity, thus exhibiting the first polytope with this property that is associated with a polynomial time solvable optimization problem.

The first fundamental research with respect to understanding extension complexities was Yannakakis' seminal paper [16] of 1991. Observing that many of the nice and small extensions that are known (e.g., the polynomial size extension of the spanning tree polytope of K_n mentioned above) have the nice property of being symmetric in a certain sense, he derived lower bounds on extensions with that special property. In particular, he already proved that both perfect matching polytopes as well as traveling salesman polytopes do not have polynomial size *symmetric* extensions.

It turned out that requiring symmetry in principle actually can make a huge difference for the minimal sizes of extensions (though nowadays we know that this is not really true for traveling salesman and perfect matching polytopes). For instance, Kaibel, Theis, and Pashkovich [8] showed that the polytope associated with the matchings of size $\lfloor \log n \rfloor$ in K_n has polynomially bounded extension complexity although it does not admit symmetric extensions of polynomial size. Another example is provided by the permutahedron which has extension complexity $\Theta(n \log n)$ [7], while every symmetric extension of it has size $\Omega(n^2)$ [11].

These examples show that imposing the restriction of symmetry may severely influence the smallest possible sizes of extensions. In this paper, we investigate another type of restrictions on extensions, namely the one arising from requiring the extension to be a non-degenerate polytope. A d-dimensional polytope is called *simple* if every vertex is contained in exactly d facets. We denote by sxc (P) the *simple extension complexity*, i.e., the smallest size of any simple extension of the polytope P.

Simplicity is both a property that is interesting from practical (primal non-degeneracy of linear programs) as well as from theoretical (large parts of the combinatorial/extremal theory of polytopes deal with simple polytopes) point of views. And similarly to the restriction to symmetric extensions, there are also nice examples of simple extensions of certain polytopes relevant in optimization. For instance, generalizing the well-known fact that the permutahedron is a zonotope, Wolsey showed in the late 80's (personal communication) that, for arbitrary processing times, the completion time polytope for n jobs is a projection of an $\mathcal{O}(n^2)$-dimensional cube. The main results of this paper show, however, that for several polytopes relevant in optimization (among them both perfect matching polytopes and spanning tree polytopes) insisting on simplicity causes exponential sizes of the extensions. More precisely, we establish that for the following polytopes the simple extension complexity equals their number of vertices (note that the number of vertices of P is a trivial upper bound for sxc (P), realized by the extension obtained from writing P as the convex hull of its vertices):

- Perfect matching polytopes of complete graphs (Theorem 6)
- Uncapacitated flow polytopes of non-decomposable acyclic networks (Theorem 4)
- (Certain) random 0/1-polytopes (Theorem 2)

Furthermore, we prove that

- the spanning tree polytope of the complete graph with n nodes has simple extension complexity at least $\Omega\left(2^{n-o(n)}\right)$ (Theorem 3).

Using our techniques one can also prove that the extension complexity of hyper-simplices is equal to their number of vertices as well.

Let us make a brief digression on the potential relevance of simple extensions with respect to questions related to the diameter of a polytope, i.e., the maximal distance (minimum number of edges on a path) between any pair of vertices in the graph of the polytope. We denote by $\Delta(d, m)$ the maximal diameter of any d-dimensional polytope with m facets. It is well-known that $\Delta(d, m)$ is attained by simple polytopes. A necessary condition for a polynomial time variant of the simplex-algorithm to exist is that $\Delta(d, m)$ is bounded by a polynomial in d and m (thus by a polynomial in m). In fact, in 1957 Hirsch even conjectured (see [2]) that $\Delta(d, m) \leq m - d$ holds, which has only rather recently been disproved by Santos [15]. However, still it is even unknown whether $\Delta(d, m) \leq 2m$ holds true, and the question, whether $\Delta(d, m)$ is bounded polynomially (i.e., whether the *polynomial Hirsch-conjecture* is true) is a major open problem in Discrete Geometry.

In view of the fact that linear optimization over a polytope can be performed by linear optimization over any of its extensions, a reasonable relaxed version of that question might be to ask whether every d-dimensional polytope P with m facets admits an extension whose size and diameter both are bounded polynomially in m. Stating the relaxed question in this naive way, the answer clearly is positive, as one may construct an extension by forming a pyramid over P (after embedding P into $\mathbb{R}^{\dim(P)+1}$), which has diameter at most two. However, in some accordance with the way the simplex algorithm works by pivoting between bases rather than only by proceeding along edges, it seems to make sense to require the extension to be simple (which a pyramid, of course, in general is not). But still, this is not yet a useful variation, since our result on flow polytopes mentioned above shows that there are polytopes that even do not admit a polynomial size simple extension at all. Therefore, we propose to investigate the following question, whose positive answer would be implied by establishing the polynomial Hirsch-conjecture (as every polytope is an extension of itself).

Question 1. Does there exist a polynomial q such that every *simple* polytope P with m facets has a *simple* extension Q with at most $q(m)$ facets and diameter at most $q(m)$?

The paper is structured as follows: We first devise some techniques to bound the simple extension complexity of a polytope from below (Section 2). Then we deduce our results on spanning tree polytopes (Section 3), flow polytopes (Section 4), and perfect matching polytopes (Section 5). The core of the latter part is a strengthening of a result of Padberg and Rao's [10] on adjacencies in the perfect matching polytope (Theorem 5), which may be of independent interest.

2 Bounding Techniques

Let $P \subseteq \mathbb{R}^n$ be a polytope with N vertices. The faces of P form a graded lattice $\mathcal{L}(P)$, ordered by inclusion (see [17]).

Clearly, P is the set of all convex combinations of its vertices, immediately providing an extended formulation of size N:

$$P = \text{proj}_x \left\{ (x, y) \in \mathbb{R}^n \times \mathbb{R}_+^V : x = \sum_{v \in V} y_v v, \ \sum_{v \in V} y_v = 1 \right\}$$

Here, $\text{proj}_x(\cdot)$ denotes the projection onto the space of x-variables and V is the set of vertices of P. Note that this *trivial extension* is simple since the extension is an $(N-1)$-simplex.

An easy observation for extensions $P = \pi(Q)$ is that the assignment $F \mapsto \pi^{-1}(F) \cap Q$ defines a map j which embeds $\mathcal{L}(P)$ into $\mathcal{L}(Q)$, i.e., it is one-to-one and preserves inclusion in both directions (see [3]). Note that this embedding furthermore satisfies $j(F \cap F') = j(F) \cap j(F')$ for all faces F, F' of P (where the nontrivial inclusion $j(F) \cap j(F') \subseteq j(F \cap F')$ follows from $\pi(j(F) \cap j(F')) \subseteq \pi(j(F)) \cap \pi(j(F')) = F \cap F'$). We use the shorthand notation $j(v) := j(\{v\})$ for vertices v of P.

We consider the *face-vertex non-incidence graph* $G_\mathcal{N}(P)$ which is a bipartite graph having the faces and the vertices of P as the node set and edges $\{F, v\}$ for all $v \notin F$. Every facet $\hat{f}$ of an extension induces two node sets of this graph in the following way:

$$\begin{aligned} \mathcal{F}(\hat{f}) &:= \left\{ F \text{ face of } P : j(F) \subseteq \hat{f} \right\} \\ \mathcal{V}(\hat{f}) &:= \left\{ v \text{ vertex of } P : j(v) \not\subseteq \hat{f} \right\} \end{aligned} \tag{1}$$

We call $\mathcal{F}(\hat{f})$ and $\mathcal{V}(\hat{f})$ the *set of faces* (resp. *vertices*) *induced by the facet $\hat{f}$* (with respect to the extension $P = \pi(Q)$). Typically, the extension and the facet $\hat{f}$ are fixed and we just write $\mathcal{F}$ (resp. $\mathcal{V}$). It may happen that $\mathcal{V}(\hat{f})$ is equal to the whole vertex set, e.g., if $\hat{f}$ projects into the relative interior of P. If $\mathcal{V}(\hat{f})$ is a proper subset of the vertex set we call facet $\hat{f}$ *proper* w.r.t. the projection.

For each facet $\hat{f}$ of an extension of P the face and vertex sets together induce a biclique (i.e., complete bipartite subgraph) in $G_\mathcal{N}(P)$. It follows from Yannakakis [16] that every edge in $G_\mathcal{N}(P)$ is covered by at least one of those induced bicliques. We provide a brief combinatorial argument for this (in particular showing that we can restrict to proper facets) in the proof of the following proposition.

Proposition 1. *Let $P = \pi(Q)$ be an extension.*

Then the subgraph of $G_\mathcal{N}(P)$ induced by $\mathcal{F}(\hat{f}) \cup \mathcal{V}(\hat{f})$ is a biclique for every facet $\hat{f}$ of Q. Furthermore, every edge $\{F, v\}$ of $G_\mathcal{N}(P)$ is covered by at least one of the bicliques induced by a proper facet.

Proof. Let $\hat{f}$ be one of the facets and assume that an edge $\{F, v\}$ with $F \in \mathcal{F}(\hat{f})$ and $v \in \mathcal{V}(\hat{f})$ is not present in $G_{\mathcal{N}}(P)$, i.e., $v \in F$. From $v \in F$ we obtain $j(v) \subseteq j(F) \subseteq \hat{f}$, a contradiction to $v \in \mathcal{V}(\hat{f})$.

To prove the second statement, let $\{F, v\}$ be any edge of $G_{\mathcal{N}}(P)$, i.e., $v \notin F$. Observe that the preimages $G := j(F)$ and $g := j(v)$ are also not incident since j is a lattice embedding. As G is the intersection of all facets of Q it is contained in (the face-lattice of a polytope is coatomic), there must be at least one facet $\hat{f}$ containing G but not g (since otherwise g would be contained in G), yielding $F \in \mathcal{F}(\hat{f})$ and $v \in \mathcal{V}(\hat{f})$.

If $F \neq \emptyset$, any vertex $w \in F$ satisfies $j(w) \subseteq G \subseteq \hat{f}$ and hence $\hat{f}$ is a proper facet. If $F = \emptyset$, let w be any vertex of P distinct from v. The preimages $j(v)$ and $j(w)$ clearly satisfy $j(v) \not\subseteq j(w)$. Again, since the face-lattice of Q is coatomic, there exists a facet $\hat{f}$ with $j(w) \subseteq \hat{f}$ but $j(v) \not\subseteq \hat{f}$. Hence, $\hat{f}$ is a proper facet and (since $\emptyset = F \subseteq \hat{f}$) $F \in \mathcal{F}(\hat{f})$ and $v \in \mathcal{V}(\hat{f})$ holds. $\qquad\square$

Before moving on to simple extensions we mention two useful properties of the induced sets. Both can be easily verified by examining the definitions of $\mathcal{F}$ and $\mathcal{V}$.

Lemma 1. *Let $\mathcal{F}$ and $\mathcal{V}$ be the face and vertex sets induced by a facet of an extension of P, respectively.*

Then $\mathcal{F}$ is closed under taking subfaces and $\mathcal{V} = \{v \text{ vertex of } P : v \notin \bigcup \mathcal{F}\}$.

For the remainder of this section we assume that the extension polytope Q is a *simple* polytope and that $\mathcal{F}$ and $\mathcal{V}$ are face and vertex sets induced by a facet of Q.

Theorem 1. *Let $\mathcal{F}$ and $\mathcal{V}$ be the face and vertex sets induced by a facet of a* simple *extension of P, respectively. Then*

(a) *all pairs (F, F') of faces of P with $F \cap F' \neq \emptyset$ and $F, F' \notin \mathcal{F}$ satisfy $F \cap F' \notin \mathcal{F}$,*

(b) *the (inclusion-wise) maximal elements in $\mathcal{F}$ are facets of P,*

(c) *and every vertex $v \notin \mathcal{V}$ is contained in some facet F of P with $F \in \mathcal{F}$.*

Proof. Let $\hat{f}$ be the facet of Q inducing $\mathcal{F}$ and $\mathcal{V}$ and F, F' two faces of P with non-empty intersection. Since $F \cap F' \neq \emptyset$, we have $j(F \cap F') \neq \emptyset$, thus the interval in $\mathcal{L}(Q)$ between $j(F \cap F')$ and Q is a Boolean lattice (because Q is simple). Suppose $F \cap F' \in \mathcal{F}(\hat{f})$. Then $\hat{f}$ is contained in that interval and it is a coatom, hence it contains at least one of $j(F)$ and $j(F')$ due to $j(F) \cap j(F') = j(F \cap F')$. But this implies $j(F) \in \mathcal{F}$ or $j(F') \in \mathcal{F}$, proving (a).

For (b), let F be an inclusion-wise maximal face in $\mathcal{F}$ but not a facet of P. Then F is the intersection of two faces F_1 and F_2 of P properly containing F. Due to the maximality of F, $F_1, F_2 \notin \mathcal{F}$ but $F_1 \cap F_2 \in \mathcal{F}$, contradicting (a).

Statement (c) follows directly from (b) and Lemma 1. $\qquad\square$

In order to use the Theorem 1 for deriving lower bounds on the sizes of simple extensions of a polytope P, one needs to have good knowledge of parts of the face lattice of P. The part one usually knows most about is formed by the vertices and edges of P. Therefore, we specialize Theorem 1 to these faces for later use.

Let $G = (V, E)$ be a graph and denote by $\delta(W) \subseteq E$ the cut-set of a node-set W. Define the *common neighbor operator* $\Lambda(\cdot)$ by

$$\Lambda(W) := W \cup \{v \in V : \exists \{u, v\}, \{v, w\} \in \delta(W) : u \neq w\} . \tag{2}$$

A set $W \subseteq V$ is then a (proper) *common neighbor closed* (for short Λ-*closed*) set if $\Lambda(W) = W$ (and $W \neq V$) holds. We call sets W with a minimum node distance of at least 3 (i.e., the distance-2-neighborhood of a node $w \in W$ does not contain another node $w' \in W$) *isolated*. Isolated node sets are clearly Λ-closed. Note that singleton sets are isolated and hence proper Λ-closed. In particular, the vertex sets induced by the facets of the trivial extension (see beginning of Section 2) are the singleton sets.

Using this notion, we obtain the following corollary of Theorem 1.

Corollary 1. *The vertex set $\mathcal{V}$ induced by a proper facet of a simple extension of P is a proper Λ-closed set.*

Proof. Theorem 1 implies that for every $\{u, v\}, \{v, w\}$ of (distinct) adjacent edges of P, we have

$$\{u, v\}, \{v, w\} \notin \mathcal{F} \Rightarrow \{v\} \notin \mathcal{F} .$$

Due to Lemma 1, $\mathcal{V} = \{v \text{ vertex of } P : v \notin \bigcup \mathcal{F}\}$, where $\mathcal{F}$ is the face set induced by the same facet. Hence, $v \notin \mathcal{V}$ implies $\{u, v\} \in \mathcal{F}$ or $\{v, w\} \in \mathcal{F}$, thus $u \notin \mathcal{V}$ or $w \notin \mathcal{V}$ and we conclude that $\mathcal{V}$ is Λ-closed.

Furthermore, $\mathcal{V}$ is not equal to the whole vertex set of P since the given facet is proper. □

We can obtain useful lower bounds from Theorem 1 and Corollary 1.

Corollary 2. *The node set of a polytope P can be covered by $sxc(P)$ many proper Λ-closed sets.*

Lemma 2. *Let P be a polytope and G its graph. If all proper Λ-closed sets in G are isolated then the simple extension complexity of P is greater than the maximum size of the neighborhood of any node of G.*

Proof. Let w be a node maximizing the size of the neighborhood and let W be the neighborhood of w. Since no isolated set can contain more than one node from $W \cup \{w\}$, Corollary 2 implies the claim. □

Using knowledge about random 0/1 polytopes, we can easily establish the following result.

Theorem 2. *There is a constant $\sigma > 0$ such that a random d-dimensional 0/1-polytope P with at most $2^{\sigma d}$ vertices asymptotically almost surely has a simple extension complexity equal to its number of vertices.*

Proof. It is one of the main results in the thesis [6] that there is such a σ ensuring that a random d-dimensional $0/1$-polytope P with at most $2^{\sigma d}$ vertices asymptotically almost surely has every pair of vertices adjacent. Since in this situation the only proper Λ-closed sets are the singletons, Corollary 2 yields the claim.   □

3   Spanning Tree Polytope

In this section we bound the simple extension complexity of the spanning tree polytope $P_{\mathrm{spt}}(K_n)$ of the complete graph K_n with n nodes.

Lemma 3. *All proper Λ-closed sets in the graph of $P_{spt}(K_n)$ are isolated.*

Proof. Two vertices of $P_{\mathrm{spt}}(K_n)$ are adjacent if and only if the symmetric difference of the corresponding spanning trees consists of exactly two edges. Throughout the proof, we will identify vertices with the corresponding spanning trees.

Suppose $\mathcal{V}$ is a proper Λ-closed set that is not isolated. Then there are spanning trees $T_1, T_2 \in \mathcal{V}$ and $T_3 \notin \mathcal{V}$, such that T_1 is adjacent to both T_2 and T_3, but T_2 and T_3 are not adjacent.

$T_1 \cap T_2$ is a forest with exactly two components having vertex sets X and Y. Let $e \in T_1$ and $f \in T_2$ be the edges in $T_1 \cup T_2$ connecting X and Y, $\{g\} = T_1 \setminus T_3$, and $\{h\} = T_3 \setminus T_1$. We have $g \neq e$, since $T_2 \setminus T_3 \subseteq T_1 \cup \{e\} \setminus T_3 \subseteq \{g, e\}$ cannot have cardinality one, because T_2 and T_3 are not adjacent.

Therefore, let w.l.o.g. g be an edge in $T_1[X]$ and let X' and X'' be the components of $T_1 \setminus \{e\}$ it connects such that $X' \cap e = \emptyset$. Define $F := T_1 \cap T_2 \cap T_3$ and observe $T_1 = F \cup \{e, g\}$, $T_2 = F \cup \{f, g\}$, and $T_3 = F \cup \{e, h\}$. There are two possible cases for h:

Case 1: h connects Y with X' or X''.

Let $T' := F \cup \{g, h\}$ and observe that T' is a spanning tree since g connects X' with X'' and h connects one of both with Y. Obviously, T' is adjacent to T_1, T_2, and T_3. Since T' is adjacent to T_1 and T_2, $T' \in \Lambda(\mathcal{V}) = \mathcal{V}$. Since T_3 is adjacent to $T_1, T' \in \mathcal{V}$, this in turn implies the contradiction $T_3 \in \mathcal{V}$.

Case 2: h connects X' with X''.

Let j be any edge connecting X' with Y (recall that we dealing with a complete graph) and let $T' := F \cup \{g, j\}$ which is a spanning tree adjacent to T_1 and T_2 and hence $T' \in \Lambda(W) = W$. Clearly, $T'' := F \cup \{e, j\}$ is a spanning tree adjacent to T_1 and T' and hence $T'' \in \mathcal{V}$. Finally, let $T''' := F \cup \{h, j\}$ be a third spanning tree adjacent to T' and T''. Again, we have $T''' \in \mathcal{V}$ due to $\Lambda(\mathcal{V}) = \mathcal{V}$.

Since T_3 is adjacent to T_1 and T''', exploiting $\Lambda(\mathcal{V}) = \mathcal{V}$ once more yields the contradiction $T_3 \in \mathcal{V}$.   □

Using this result we immediately get a lower bound of $\Omega(n^3)$ for the simple extension complexity of $P_{\mathrm{spt}}(K_n)$ since the maximum degree of its graph is of that order. However, we can prove a much stronger result.

Theorem 3. *The simple extension complexity of the spanning tree polytope of K_n is in $\Omega\left(2^{n-o(n)}\right)$.*

Proof. Assume $n \geq 5$ and let s, t be any two distinct nodes of K_n. Consider certain subsets on the other nodes

$$\mathcal{W} := \{W \subseteq V \setminus \{s, t\} : |W| = \lfloor n/2 \rfloor\} \ .$$

Let $k := \lfloor n/2 \rfloor$, fix some ordering of the nodes $w_1, w_1, \ldots, w_k \in W$ for each $W \in \mathcal{W}$ and define a specific tree $T(W)$

$$
\begin{aligned}
T(W) := & \{\{s, w_1\}, \{w_k, t\}\} \\
& \cup \{\{w_i, w_{i+1}\} : i \in [k-1]\} \\
& \cup \{\{t, v\} : v \notin (W \cup \{s, t\})\}
\end{aligned}
$$

We will now prove that for each simple extension of $P_{\mathrm{spt}}(K_n)$ every such $T(W)$ must be in a different induced vertex set.

Let $W \in \mathcal{W}$ be some set W with tree $T(W)$. Let $\mathcal{F}$ and $\mathcal{V}$ be the face and vertex sets, respectively, induced by a proper facet of a simple extension such that $T(W)$ is in $\mathcal{V}$. Construct an adjacent tree T' as follows.

Choose some vertex $y \in W$ and let x-y-z be a subpath of the s-t-path in $T(W)$ in that order. Note that $\{x, y, z\} \subseteq W \cup \{s, t\}$. Denote by a, b, c the edges $\{x, y\}$, $\{x, z\}$, and $\{y, z\}$, respectively.

Let $T' = T(W) \setminus \{a\} \cup \{b\}$. Because T' is adjacent to $T(W)$, $T' \notin \mathcal{V}$ by Lemma 3. Hence, due to Lemma 1, there must be a facet $F \in \mathcal{F}$ defined by $x(E[U]) \leq |U| - 1$ (with $|U| \geq 2$) which contains T' but not $T(W)$. Hence, we have $|T(W)[U]| < |U| - 1$ and $|T'[U]| = |U| - 1$. This implies $|T(W) \cap \delta(U)| \geq 2$ and $|T' \cap \delta(U)| = 1$. Obviously, $a \in \delta(U)$ and $b \notin \delta(U)$.

Then $x, z \in U$ if and only if $y \notin U$ because $a \in \delta(U)$ and $b \notin \delta(U)$. Hence, $c \in \delta(U)$, i.e., $T \cap \delta(U) = \{c\}$. Due to $|U| \geq 2$, this implies $U = V \setminus \{y\}$.

As this can be argued for any $y \in W$, we have that the facets defined by $V \setminus \{y\}$ are in $\mathcal{F}$ for all $y \in W$. Hence, $\mathcal{V}$ contains only trees T for which $|T \cap \delta(V \setminus \{y\})| = |T \cap \delta(\{y\})| \geq 2$, i.e., no leaf of T is in W.

This shows that for distinct sets $W, W' \in \mathcal{W}$, any vertex set $\mathcal{V}$ induced by a proper facet of a simple extension that contains $T(W)$ containing $T(W)$ does not contain $T(W')$ because any vertex $v \in W \setminus W'$ is a leaf of $T(W')$. Hence, the number of simple bicliques is at least

$$|\mathcal{W}| = \binom{n-2}{\lfloor n/2 \rfloor} \in \Omega\left(2^{n-o(n)}\right) \ . \qquad \square$$

4   Flow Polytopes for Acyclic Networks

Many extended formulations model the solutions to the original formulation via a path in a specifically constructed directed acyclic graph. The size of the construction then equals the number of arcs in that graph since the paths from two fixed nodes s and t arise as vertices of the corresponding flow polytope whose facets correspond to nonnegativity constraints on arcs. Such a network formulation can be easily decomposed into two independent formulations if a node v exists such that every s-t-path traverses v. We are now interested in the simple extension complexities of flow polytopes of s-t-networks that cannot be decomposed in such a trivial way.

Let $D = (V, A)$ be a directed acyclic graph with fixed source $s \in V$ and sink $t \in V$. By $\mathcal{P}_{s,t}(D)$ we denote the arc-sets of s-t-paths in D. For some path $P \in \mathcal{P}_{s,t}(D)$ and nodes $u, v \in V(P)$, we denote by $P|_{(u,v)}$ the subpath of P going from u to v.

We consider the flow polytope $P_{\text{s-t-flow}}(D)$, i.e., the set of all s-t-flows in D of value one. The facets of $P_{\text{s-t-flow}}(D)$ correspond to the nonnegativity constraints $y_a \geq 0$ for some $a \in A$. Clearly, the vertices correspond to $\mathcal{P}_{s,t}(D)$. A path $P \in \mathcal{P}_{s,t}(D)$ is non-incident to a facet $y_a \geq 0$ if and only if $a \in P$. Two paths $P, P' \in \mathcal{P}_{s,t}(D)$ are adjacent vertices of the polytope if and only if their symmetric difference consists of two paths from x to y ($x, y \in V$, $x \neq y$) without common inner nodes (see [5]). Our main result in this section is the following:

Theorem 4. *Let $D = (V, A)$ be a directed acyclic graph with source $s \in V$ and sink $t \in V$ such that for every node $v \in V \setminus \{s, t\}$ there exists an s-t-path in D which does not traverse v.*

Then the simple extension complexity of $P_{\text{s-t-flow}}(D) \subseteq \mathbb{R}^A_+$ is equal to the number of distinct s-t-paths $|\mathcal{P}_{s,t}(D)|$.

Proof. Let $\mathcal{F}$ and $\mathcal{V}$ be the face and vertex sets induced by a proper facet of a simple extension of $P_{\text{s-t-flow}}(D)$, respectively. Assume for the sake of contradiction $|\mathcal{V}| \geq 2$. By Theorem 1 (b), the (inclusion-wise) maximal faces in $\mathcal{F}$ are facets. Let $\emptyset \neq B' \subseteq A$ be the arc set corresponding to these facets. By Lemma 1, $\mathcal{V}$ is the set of (characteristic vectors of) paths $P \in \mathcal{P}_{s,t}(D)$ satisfying $P \supseteq B'$. Let $B \subseteq A$ be the set of arcs common to all such paths and note that $B \supseteq B' \neq \emptyset$.

By construction, for any path $P \in \mathcal{V}$ and any arc $a \in P \setminus B$, there is an alternative path $P' \in \mathcal{V}$ with $a \notin P'$.

Let us fix one of the paths $P \in \mathcal{V}$. Let, without loss of generality, $(x', x) \in B$ be such that the arc of P leaving x (exists and) is not in B. If such an arc does not exist, since $B \neq P$, there must be an arc $(x, x') \in B$ such that the arc of P entering x is not in B. In this case, revert the directions of all arcs in D and exchange the roles of s and t and apply subsequent arguments to the new network. Let y be the first node on $P|_{(x,t)}$ different from x and incident to some arc in B or, if no such y exists, let $y := t$. Paths in $\mathcal{V}$ must leave x and enter y but may differ inbetween. The set of traversed nodes is defined as

$$S := \{v \in V \setminus \{x, y\} : \exists\, x\text{-}v\text{-}y\text{-path in } D\} \,.$$

By construction, $x \notin \{s, t\}$ and by the assumptions of the Theorem there exists a path $P' \in \mathcal{P}_{s,t}(D)$ which does not traverse x. Let s' be the last node on $P|_{(s,x)}$ that is traversed by P'. Analogously, let t' be the first node of $V(P|_{(x,t)}) \cup S$ that is traversed by P'. Note that $t' \neq x$ since t' is traversed by P' but x is not.

We now distinguish two cases for which we show that $\mathcal{V}$ is not Λ-closed yielding a contradiction to Corollary 1:

Case 1: $t' \in S$.

By definition of S there must be an x-t'-y-path W. Let $(z, t') \in W$ be the arc of W entering t'. By definition of y, we conclude that $(z, t') \notin B$. Hence, there is an alternative x-y-path $W' \neq W$ which does not use (z, t'). We choose W' such that it uses as many arcs of $W|_{(t',y)}$ as possible. Construct the following three paths:

$$P_1 := P|_{(s,x)} \cup W \cup P|_{(y,t)}$$
$$P_2 := P|_{(s,x)} \cup W' \cup P|_{(y,t)}$$
$$P_3 := P|_{(s,s')} \cup P'|_{(s',t')} \cup W|_{(t',y)} \cup P|_{(y,t)}$$

By construction $P_1, P_2 \in \mathcal{V}$ but $P_3 \notin \mathcal{V}$. P_1 and P_3 are adjacent in $P_{\text{s-t-flow}}(D)$ since they only differ in the disjoint paths from s' to t'. Analogously, P_2 and P_3 are adjacent and thus, contradicting the fact that $\mathcal{V}$ is Λ-closed.

Case 2: $t' \notin S$.

Let $W := P|_{(x,y)}$ and let W' be a different x-y-path which must exist by definition of y. Construct the following three paths:

$$P_1 := P = P|_{(s,x)} \cup W \cup P|_{(y,t)}$$
$$P_2 := P|_{(s,x)} \cup W' \cup P|_{(y,t)}$$
$$P_3 := P|_{(s,s')} \cup P'|_{(s',t')} \cup P|_{(t',t)}$$

By construction $P_1, P_2 \in \mathcal{V}$ but $P_3 \notin \mathcal{V}$ since it does not use $(x', x) \in B$. P_1 and P_3 as well as P_2 and P_3 are adjacent in $P_{\text{s-t-flow}}(D)$ since they only differ in the disjoint paths from s' to t'. Again, this contradicts the fact that $\mathcal{V}$ is Λ-closed. $\square$

5   Perfect Matching Polytope

The *matching polytope* and the *perfect matching polytope* of a graph $G = (V, E)$ are defined as

$$P_{\text{match}}(G) := \text{conv}\{\chi(M) : M \text{ matching in } G\}$$

$$P_{\text{match}}^{\text{perf}}(G) := \text{conv}\{\chi(M) : M \text{ perfect matching in } G\} \ ,$$

where $\chi(M) \in \{0,1\}^E$ is the characteristic vector of the set $M \subseteq E$, i.e., $\chi(M)_e = 1$ if and only if $e \in M$. We mainly consider the (perfect) matching polytope of the complete graph with $2n$ nodes $P_{\text{match}}^{\text{perf}}(K_{2n})$. For the proof of our main theorem here we need the following structural result on adjacency in the perfect matching polytope.

Theorem 5. *Let M_1 and M_2 be two adjacent perfect matchings and M_3 a third perfect matching in a complete graph. Then the three matchings are pairwise adjacent or there exists a perfect matching M' adjacent to all three.*

This theorem strengthens a result of Padberg and Rao's [10] stating that for any two different non-adjacent perfect matchings in a complete graph there is a third one adjacent to both. Since the proof of Theorem 5 is a bit more involved we omit it in this extended abstract. Our main theorem of this section reads as follows:

Theorem 6. *The simple extension complexity of the perfect matching polytope of K_{2n} is equal to its number of vertices $\frac{(2n)!}{n! \cdot 2^n}$.*

Proof. Consider the polytope $P_{\text{match}}^{\text{perf}}(K_{2n})$ and suppose that $\mathcal{V}$ is a proper Λ-closed set of vertices with $|\mathcal{V}| \geq 2$. Since the polytope's graph is connected there exists a matching $M_1 \notin \mathcal{V}$ adjacent to some matching $M_2 \in \mathcal{V}$. Let $M_3 \in \mathcal{V} \setminus \{M_2\}$. As $\mathcal{V}$ is Λ-closed and $M_3 \notin \mathcal{V}$, $\{M_1, M_2, M_3\}$ cannot be a triangle. Hence, by Theorem 5 there exists a common neighbor matching M'. Since M' is adjacent to M_2 and M_3, we conclude $M' \in \mathcal{V}$. But now $M_1 \notin \mathcal{V}$ is adjacent to the two matchings M_2 and M' from $\mathcal{V}$ contradicting the fact that $\mathcal{V}$ is Λ-closed.

Hence all proper Λ-closed sets are singletons which implies the claim due to Corollary 2. □

Since $P_{\text{match}}^{\text{perf}}(K_{2n})$ is a face of $P_{\text{match}}(K_{2n})$ and simple extensions of polytopes induce simple extensions of their faces we obtain the following corollary for the latter polytope.

Corollary 3. *The simple extension complexity of the matching polytope of K_{2n} is at least $\frac{(2n)!}{n! \cdot 2^n}$.*

References

1. Avis, D., Tiwary, H.R.: On the extension complexity of combinatorial polytopes. In: Fomin, F.V., Freivalds, R., Kwiatkowska, M., Peleg, D. (eds.) ICALP 2013, Part I. LNCS, vol. 7965, pp. 57–68. Springer, Heidelberg (2013)
2. Dantzig, G.B.: Linear Programming and Extensions. Princeton landmarks in mathematics and physics. Princeton University Press (1963)
3. Fiorini, S., Kaibel, V., Pashkovich, K., Theis, D.O.: Combinatorial bounds on non-negative rank and extended formulations. arXiv:1111.0444 (2011) (to appear in: Discrete Math.)
4. Fiorini, S., Massar, S., Pokutta, S., Tiwary, H.R., de Wolf, R.: Linear vs. semidefinite extended formulations: exponential separation and strong lower bounds. In: Karloff, H.J., Pitassi, T. (eds.) STOC, pp. 95–106. ACM (2012)
5. Gallo, G., Sodini, C.: Extreme points and adjacency relationship in the flow polytope. Calcolo 15, 277–288 (1978), 10.1007/BF02575918
6. Gillmann, R.: 0/1-Polytopes Typical and Extremal Properties. PhD thesis, Technische Universität, Berlin (2007)
7. Goemans, M.: Smallest compact formulation for the permutahedron (2009), http://www-math.mit.edu/~goemans/publ.html
8. Kaibel, V., Pashkovich, K., Theis, D.O.: Symmetry matters for sizes of extended formulations. SIAM J. Disc. Math. 26(3), 1361–1382 (2012)

9. Kipp Martin, R.: Using separation algorithms to generate mixed integer model reformulations. Oper. Res. Lett. 10(3), 119–128 (1991)
10. Padberg, M.W., Rao, M.R.: The travelling salesman problem and a class of polyhedra of diameter two. Math. Program. 7, 32–45 (1974), 10.1007/BF01585502
11. Pashkovich, K.: Symmetry in extended formulations of the permutahedron (2009)
12. Pokutta, S., Van Vyve, M.: A note on the extension complexity of the knapsack polytope. Oper. Res. Lett. 41(4), 347–350 (2013)
13. Rothvoß, T.: Some 0/1 polytopes need exponential size extended formulations. Math. Program, 1–14 (2012)
14. Rothvoß, T.: The matching polytope has exponential extension complexity. arXiv:1311.2369 (November 2013)
15. Santos, F.: A counterexample to the hirsch conjecture. Annals of Mathematics. Second Series 176(1), 383–412 (2012)
16. Yannakakis, M.: Expressing combinatorial optimization problems by linear programs. J. Comput. Syst. Sci. 43(3), 441–466 (1991)
17. Ziegler, G.M.: Lectures on Polytopes (Graduate Texts in Mathematics). Springer (2001)

Lower Bounds on the Sizes of Integer Programs without Additional Variables

Volker Kaibel and Stefan Weltge*

Otto-von-Guericke-Universität Magdeburg, Germany
{kaibel,weltge}@ovgu.de

Abstract. For a given set $X \subseteq \mathbb{Z}^d$ of integer points, we investigate the smallest number of facets of any polyhedron whose set of integer points is $\mathrm{conv}(X) \cap \mathbb{Z}^d$. This quantity, which we call the *relaxation complexity* of X, corresponds to the smallest number of linear inequalities of any integer program having X as the set of feasible solutions that does not use auxiliary variables. We show that the use of auxiliary variables is essential for constructing polynomial size integer programming formulations in many relevant cases. In particular, we provide asymptotically tight exponential lower bounds on the relaxation complexity of the integer points of several well-known combinatorial polytopes, including the traveling salesman polytope and the spanning tree polytope.

Keywords: integer programming, relaxations, auxiliary variables, tsp.

1 Introduction

Let $K_n = (V_n, E_n)$ be the undirected complete graph on n nodes and STSP_n the set of characteristic vectors of hamiltonian cycles in K_n. In order to solve the traveling salesman problem, there are numerous integer programs of the form

$$\max \left\{ \langle c, x \rangle : Ax + By \leq b, \, x \in \mathbb{Z}^{E_n}, \, y \in \mathbb{Z}^m \right\} \qquad (1)$$

such that the optimal value of (1) is equal to $\max \left\{ \langle c, x \rangle : x \in \mathrm{STSP}_n \right\}$ for all edge weights $c \in \mathbb{R}^{E_n}$. In most of these formulations, the system $Ax + By \leq b$ consists of polynomially (in n) many linear inequalities, see, e.g. [7] or [2]. Further, some of them even do not need integrality constraints on the auxiliary variables y. In contrast, a recent result on *extended formulations* (i.e., representations of polytopes as projections of other ones) due to Fiorini et al. [1] states that if one drops the integrality constraints on both x and y, then such systems must have exponentially many inequalities.

Interestingly, at a closer look, one notices that all such IP-formulations that consist of polynomially many inequalities make use of auxiliary variables. The main motivation for this paper was the question whether there is an IP-formulation of

* Partially funded by the German Research Foundation (DFG): "Extended Formulations in Combinatorial Optimization" (KA 1616/4-1).

J. Lee and J. Vygen (Eds.): IPCO 2014, LNCS 8494, pp. 321–332, 2014.
© Springer International Publishing Switzerland 2014

type (1) for solving the traveling salesman problem that does not use auxiliary variables but also consists of only polynomially many inequalities.

For a set $X \subseteq \mathbb{Z}^d$, let us call a polyhedron $R \subseteq \mathbb{R}^d$ a *relaxation* for X if $R \cap \mathbb{Z}^d = \text{conv}(X) \cap \mathbb{Z}^d$. Further, the smallest number of facets of any relaxation for X will be called the *relaxation complexity* of X, or short: $\text{rc}(X)$. With this notation, the above question is equivalent to the question whether $\text{rc}(\text{STSP}_n)$ is polynomial in n.

For most sets $X \in \{0, 1\}^d$ that are associated with known combinatorial optimization problems it turns out that there are polynomial size IP-formulations of type (1). Following Schrijver's proof [11, Thm. 18.1] of the fact that integer programming is NP-hard one finds that for any language $\mathcal{L} \subseteq \{0, 1\}^*$ that is in NP, there is a polynomial p such that for any $k > 0$ there is a system $Ax + By \leq b$ of at most $p(k)$ linear inequalities and $m \leq p(k)$ auxiliary variables with

$$\{x \in \{0, 1\}^k : x \in \mathcal{L}\} = \{x \in \{0, 1\}^k : \exists y \in \{0, 1\}^m \ Ax + By \leq b\}. \quad (2)$$

Further, note that for many sets $X \subseteq \{0, 1\}^d$ of feasible points of famous problems like SAT, CLIQUE, CUT or MATCHING, we even do not need auxiliary variables in order to give a polynomial size description as in (2), i.e., $\text{rc}(X)$ is polynomially bounded for such sets X. However, as we will show in this paper, it turns out that this is not true for some other well-known combinatorial problems, including variants of TSP, SPANNING TREE or T-JOIN.

Our paper consists of two main sections: In Section 2, we discuss basic properties of the number of facets of relaxations of a general set $X \subseteq \mathbb{Z}^d$. This includes questions of whether irrational coordinates may help or whether it is a good idea to only use facet-defining inequalities of $\text{conv}(X)$ in order to construct small relaxations. Further, we introduce the concept of *hiding sets*, which turns out to be a powerful technique to provide lower bounds on $\text{rc}(X)$. In Section 3, we then give exponential lower bounds on the sizes of relaxations for concrete structures that occur in many practical IP-formulations. In particular, coming back to our motivating question, we show that the asymptotic growth of $\text{rc}(\text{STSP}_n)$ is indeed exponential in n. This shows that, for many problems, the benefit of projection, i.e., the use of auxiliary variables, is essential when constructing polynomial size IP-formulations.

Except for a paper of Jeroslow [4], the authors are not aware of any reference that deals with a quantity that is similar to our notion of relaxation complexity in a general context. In his paper, for a set $X \subseteq \{0, 1\}^d$ of binary vectors, Jeroslow introduces the term *index* of X (short: $\text{ind}(X)$), which is defined as the smallest number of inequalities needed to separate X from the remaining points in $\{0, 1\}^d$. As the main result, he shows that 2^{d-1} is an upper bound on $\text{ind}(X)$, which is attained by the set of binary vectors of length d that contain an even number of ones. In Sections 2 and 3.3 we shall come back to this result. Further, his idea of bounding the index of a set $X \subseteq \{0, 1\}^d$ from below, is related to our approach of providing lower bounds on the relaxation complexity of general X via hiding sets in Section 2.4.

2   Basic Observations

There are sets $X \subseteq \mathbb{Z}^d$ that do not admit any relaxation. Therefore, let us call a set $X \subseteq \mathbb{Z}^d$ to be *polyhedral* if its convex hull is a polyhedron. By definition, we have that $rc(X)$ is finite for such sets. Further, in this setting, it is easy to see that any relaxation corresponds to a valid IP-formulation and vice versa:

Proposition 1. *Let $X \subseteq \mathbb{Z}^d$ be polyhedral and $P \subseteq \mathbb{R}^d$ a polyhedron. Then, P is a relaxation for X if and only if $\sup \left\{ \langle c, x \rangle : x \in P \cap \mathbb{Z}^d \right\} = \sup \left\{ \langle c, x \rangle : x \in X \right\}$ holds for all $c \in \mathbb{R}^d$.*

Clearly, any finite set of integer points is polyhedral. For a set $X \subseteq \{0,1\}^d$ of binary vectors, a polyhedron P is a relaxation for X if and only if $P \cap \mathbb{Z}^d = X$.

As mentioned in the introduction, Jeroslow [4] showed that for any set $X \subseteq \{0,1\}^d$, one needs at most 2^{d-1} many linear inequalities in order to separate X from $\{0,1\}^d \setminus X$. If now $P \subseteq \mathbb{R}^d$ is a polyhedron such that $P \cap \{0,1\}^d = X$, then, in order to construct a relaxation for X, we need to additionally separate all points $\mathbb{Z}^d \setminus \{0,1\}^d$ from X. This can be done by intersecting P with a relaxation for $\{0,1\}^d$. We conclude:

Proposition 2. *Let $X \subseteq \{0,1\}^d$. Then $rc(X) \leq 2^{d-1} + rc(\{0,1\}^d)$.*

2.1   Relaxation Complexity of the Cube

Motivated by Proposition 2, we are interested in the relaxation complexity of $\{0,1\}^d$. Since $[0,1]^d = conv(\{0,1\}^d)$, we obviously have that $rc(\{0,1\}^d) \leq 2d$. However, it turns out (a proof will be given in a journal version of this paper) that one can construct a relaxation of only $d+1$ facets:

Lemma 1. *For $d \geq 1$, we have*

$$\{0,1\}^d = \left\{ x \in \mathbb{Z}^d : x_k \leq 1 + \sum_{i=k+1}^{d} 2^{-i} x_i \ \forall k \in [d], \ x_1 + \sum_{i=2}^{d} 2^{-i} x_i \geq 0 \right\}.$$

To show that this construction is best possible, note that if a polyhedron that contains $\{0,1\}^d$ (and hence is d-dimensional) has less than $d+1$ facets, it must be unbounded. In order to show that such a (possibly irrational) polyhedron must contain infinitely many integer points (and hence cannot be a relaxation of $\{0,1\}^d$), we make use of Minkowski's theorem:

Theorem 1 (Minkowski [8]). *Any convex set which is symmetric with respect to the origin and with volume greater than 2^d contains a non-zero integer point.*

For $\varepsilon > 0$ let $B_\varepsilon := \left\{ x \in \mathbb{R}^d : \|x\| < \varepsilon \right\}$ be the open ball with radius ε. As a direct consequence of Minkowski's theorem, the following corollary (again, its proof will be given in a journal version) is useful for our argumentation.

Corollary 1. *Let $c \in \mathbb{R}^d \setminus \{\mathbb{O}\}$, $\lambda_0 \in \mathbb{R}$ and $\varepsilon > 0$. Then*

$$L(c, \lambda_0, \varepsilon) := \{\lambda c \in \mathbb{R}^d : \lambda \geq \lambda_0\} + B_\varepsilon$$

contains infinitely many integer points.

Theorem 2. *For $d \geq 1$, we have that $\mathrm{rc}(\{0,1\}^d) = d+1$.*

Proof. By Lemma 1, we already know that $\mathrm{rc}(\{0,1\}^d) \leq d+1$. Suppose there is a relaxation $R \subseteq \mathbb{R}^d$ for $\{0,1\}^d$ with less than $d+1$ facets. As mentioned above, since $\dim(R) \geq \dim(\{0,1\}^d) = d$, R has to be unbounded.

By induction over $d \geq 1$, we will show that any unbounded polyhedron $R \subseteq \mathbb{R}^d$ with $\{0,1\}^d \subseteq R$ contains infinitely many integer points. Hence, it cannot be a relaxation of $\{0,1\}^d$. Clearly, our claim is true for $d = 1$. For $d \geq 1$, let $c \in \mathbb{R}^d \setminus \{\mathbb{O}\}$ be a direction such that $x + \lambda c \in R$ for any $x \in R$ and $\lambda \geq 0$. Since $\{0,1\}^d$ is invariant under affine maps that map a subset of coordinates x_i to $1 - x_i$, we may assume that $c \geq \mathbb{O}$.

If $c > \mathbb{O}$, then there is some $\lambda_0 > 0$ such that $\lambda_0 c \in \mathrm{int}([0,1]^d)$. Thus, there is some $\varepsilon > 0$ such that $\lambda_0 c + B_\varepsilon \subseteq [0,1]^d \subseteq R$. By the definition of c and ε, we thus obtained that $L(c, \lambda_0, \varepsilon) \subseteq R$. By Corollary 1, it follows that $L(c, \lambda_0)$ contains infinitely many integer points and so does R.

Otherwise, we may assume that $c_d = 0$. Let $\mathcal{H}_d := \{x \in \mathbb{R}^d : x_d = 0\}$ and $p \colon \mathcal{H}_d \to \mathbb{R}^{d-1}$ be the projection onto the first $d-1$ coordinates. Then, the polyhedron $\tilde{R} = p(R)$ is still unbounded and contains $\{0,1\}^{d-1} = p(\{0,1\}^d)$. By induction, $\tilde{R}$ contains infinitely many integer points and so does R. $\square$

With Proposition 2 we thus obtain:

Corollary 2. *Let $X \subseteq \{0,1\}^d$. Then $\mathrm{rc}(X) \leq 2^{d-1} + d + 1$.*

2.2   Limits of Facet-Defining Inequalities

Many known relaxations for sets $X \subseteq \mathbb{Z}^d$ that are identified with feasible points in combinatorial problems are defined by linear inequalities of which, preferably, most of them are facet-defining for $\mathrm{conv}(X)$. Clearly, this has important practical reasons since such formulations are tightest possible in some sense. However, if one is interested in a relaxation that has as few number of facets as possible, it is a severe restriction to only use facet-defining inequalities of $\mathrm{conv}(X)$: In the previous section we have seen that $\mathrm{rc}(\{0,1\}^d) = d+1$ whereas by removing any of the cube's inequalities the remaining polyhedron gets unbounded. Nevertheless, the restriction turns out to be not too hard:

Theorem 3. *Let $X \subseteq \mathbb{Z}^d$ be polyhedral and $\mathrm{rc}_F(X)$ the smallest number of facets of any relaxation for X whose facet-defining inequalities are also facet-defining for $\mathrm{conv}(X)$. Then, $\mathrm{rc}_F(X) \leq \dim(X) \cdot \mathrm{rc}(X)$.*

Proof. By Carathéodory's Theorem (in the affine hull of X), any facet-defining inequality of a relaxation R for X can be replaced by $\dim(X)$ many facet-defining inequalities of $\mathrm{conv}(X)$. The resulting polyhedron is still a relaxation for X. $\square$

2.3 Irrationality

Another question one might ask is whether it may help (in order to construct a relaxation having few facets) to use irrational coordinates in the description of a relaxation. Again, it turns out that one does not lose too much when restricting to rational relaxations only:

Theorem 4. *Let $X \subseteq \mathbb{Z}^d$ be polyhedral with $|X| < \infty$ and $\mathrm{rc}_\mathbb{Q}(X)$ the smallest number of facets of any rational relaxation for X. Then, $\mathrm{rc}_\mathbb{Q}(X) \leq \mathrm{rc}(X) + \dim(X) + 1$.*

Proof. Since X is finite, there exists a rational simplex $\Delta \subseteq \mathbb{R}^d$ of dimension $\dim(X)$ such that $X \subseteq \Delta$. Let R be any relaxation of X having f facets and set $B := (\mathbb{Z}^d \setminus X) \cap \Delta$. Since $B \cap R = \emptyset$ and $|B| < \infty$, we are able to slightly perturb the facet-defining inequalities of R in order to obtain a polyhedron $\tilde{R}$ such that $B \cap \tilde{R} = \emptyset$ and $\tilde{R}$ is rational. Now $\tilde{R} \cap \Delta$ is still a relaxation for X, which is rational and has at most $f + (\dim(\Delta) + 1) = f + \dim(X) + 1$ facets. □

However, we are not aware of any polyhedral set X where $\mathrm{rc}(X) < \mathrm{rc}_\mathbb{Q}(X)$. In fact, we even do not know if $\mathrm{rc}(\Delta_d) < d + 1 = \mathrm{rc}_\mathbb{Q}(\Delta_d)$ holds, where $\Delta_d := \{\mathbb{O}, \mathbb{e}_1, \ldots, \mathbb{e}_d\}$. Note that any rational relaxation of $\mathrm{rc}(\Delta_d)$ has to be bounded and thus, it has at least $d + 1$ facets. (Otherwise it would contain a rational ray and hence infinitely many integer points.)

2.4 Hiding Sets

In this section, we introduce a simple framework to provide lower bounds on the relaxation complexity for polyhedral sets $X \subseteq \mathbb{Z}^d$.

Definition 1. *Let $X \subseteq \mathbb{Z}^d$. A set $H \subseteq \mathrm{aff}(X) \cap \mathbb{Z}^d \setminus \mathrm{conv}(X)$ is called a* hiding set *for X if for any two distinct points $a, b \in H$ we have that $\mathrm{conv}\{a, b\} \cap \mathrm{conv}(X) \neq \emptyset$.*

Proposition 3. *Let $X \subseteq \mathbb{Z}^d$ be polyhedral and $H \subseteq \mathrm{aff}(X) \cap \mathbb{Z}^d \setminus X$ a hiding set for X. Then, $\mathrm{rc}(X) \geq |H|$.*

Proof. Let $R \subseteq \mathbb{R}^d$ be a relaxation for X. Since $H \subseteq \mathrm{aff}(X) \subseteq \mathrm{aff}(R)$, any point in H must be separated from X by a facet-defining inequality of R.

Suppose that a facet-defining inequality $\langle \alpha, x \rangle \leq \beta$ of R is violated by two distinct points $a, b \in H$. Since H is a hiding set, there exists a point $x \in \mathrm{conv}\{a, b\} \cap \mathrm{conv}(X)$. Clearly, x does also violate $\langle \alpha, x \rangle \leq \beta$ which is a contradiction since $\langle \alpha, x \rangle \leq \beta$ is valid for $R \supseteq \mathrm{conv}(X)$.

Thus, any facet-defining inequality of R is violated by at most one point in H. Hence, R has at least $|H|$ facets. □

3 Exponential Lower Bounds

In this section, we provide strong lower bounds on the relaxation complexities of some interesting sets X. By dividing these sets into three classes, we try to identify general structures that are hard to model in the context of relaxations.

3.1   Connectivity and Acyclicity

In many IP-formulations for practical applications, the feasible solutions are sub-graphs that are required to be connected or acyclic. Quite often in these cases, there are polynomial size IP-formulations that use auxiliary variables. For instance, for the *spanning tree polytope* there are even polynomial size extended formulations [6] that can be adapted to also work for the *connector polytope* CONN_n (see below). In contrast, we give exponential lower bounds on the relaxation complexities of some important representatives of this structural class.

STSP and ATSP. A well-known relaxation for STSP_n is the so-called *subtour relaxation*

$$\Big\{ x \in \mathbb{R}^{E_n} : \sum_{e \in E_n} x_e = n$$

$$x(\delta(S)) \geq 2 \quad \forall \emptyset \neq S \subsetneq V_n$$

$$x(\delta(v)) = 2 \quad \forall v \in V_n$$

$$x \geq \mathbb{0} \Big\}, \tag{3}$$

which has exponentially (in n) many facets (where $K_n = (V_n, E_n)$ is the complete graph on n nodes). We will show that this formulation is asymptotically smallest possible. In fact, we will also give an exponential lower bound for the directed version $\text{ATSP}_n \subseteq \{0,1\}^{A_n}$, which is the set of characteristic vectors of directed hamiltonian cycles in the complete directed graph on n nodes whose arcs we will denote by A_n.

Let $n = 4N + 2$ for some integer N and let us define the set

$$V := \{v_i, v'_i : i \in [N+1]\} \ \cup \ \{w_i, w'_i : i \in [N]\}$$

consisting of $4N+2$ distinct nodes. For a binary vector $b \in \{0,1\}^N$ let us further define the two node-disjoint directed cycles

$$C_b := \big\{(v_{N+1}, v_1)\big\} \ \cup \ \bigcup_{i:b_i=0} \big\{(v_i, w_i), (w_i, v_{i+1})\big\} \ \cup \ \bigcup_{i:b_i=1} \big\{(v_i, w'_i), (w'_i, v_{i+1})\big\}$$

$$C'_b := \big\{(v'_{N+1}, v'_1)\big\} \ \cup \ \bigcup_{i:b_i=0} \big\{(v'_i, w'_i), (w'_i, v'_{i+1})\big\} \ \cup \ \bigcup_{i:b_i=1} \big\{(v'_i, w_i), (w_i, v'_{i+1})\big\}.$$

In this section, we will only consider graphs on these $4N+2$ nodes. It is easy to transfer all following observations to graphs on n nodes, where $n \not\equiv 2 \pmod 4$, by replacing arc (v_{N+1}, v_1) by a directed path including 1, 2 or 3 additional nodes. Let us now consider the set

$$H_N := \big\{ \chi(C_b \cup C'_b) : b \in \{0,1\}^N \big\}.$$

By identifying V with the nodes of the complete directed graph on $4N+2$ nodes, we clearly have that $H_N \cap \text{ATSP}_{4N+2} = \emptyset$.

Lemma 2. H_N *is a hiding set for* ATSP_{4N+2}.

Proof. First, note that

$$H_N \subseteq \text{aff}(\text{ATSP}_{4N+2}) = \left\{ x \in \mathbb{R}^A : x(\delta^{\text{in}}(v)) = x(\delta^{\text{out}}(v)) = 1, \ \forall v \in V \right\},$$

where A is the set of arcs in the complete directed graph on $4N + 2$ nodes. Let $b^{(1)}, b^{(2)} \in \{0,1\}^N$ be distinct. W.l.o.g. we may assume that there is an index $j \in [N]$ such that $b_j^{(1)} = 0$ and $b_j^{(2)} = 1$. Let us now consider the following slight modifications of $C_{b^{(1)}} \cup C'_{b^{(1)}}$ and $C_{b^{(2)}} \cup C'_{b^{(2)}}$:

$$T_1 := \left(C_{b^{(1)}} \cup C'_{b^{(1)}} \setminus \{(v_j, w_j),(v'_j, w'_j)\} \right) \ \cup \ \{(v_j, w'_j),(v'_j, w_j)\}$$

$$T_2 := \left(C_{b^{(2)}} \cup C'_{b^{(2)}} \setminus \{(v_j, w'_j),(v'_j, w_j)\} \right) \ \cup \ \{(v_j, w_j),(v'_j, w'_j)\}$$

We claim that both T_1 and T_2 are hamiltonian cycles: First note that $C_{b^{(1)}} \cup C'_{b^{(1)}} \setminus \{(v_j, w_j),(v'_j, w'_j)\}$ consists of two node-disjoint directed paths P_1 from w_j to v_j and P'_1 from w'_j to v'_j. Hence, $T_1 = P_1 \cup P'_1 \cup (\{v_j, w'_j),(v'_j, w_j)\}$ indeed forms a hamiltonian cycle. The claim for T_2 follows analogously.

By definition, we further have that

$$\chi(T_1) + \chi(T_2) = \chi(C_{b^{(1)}} \cup C'_{b^{(1)}}) - \chi(\{(v_j, w_j),(v'_j, w'_j))\} + \chi(\{(v_j, w'_j),(v'_j, w_j)\})$$

$$+ \chi(C_{b^{(2)}} \cup C'_{b^{(2)}}) + \chi(\{(v_j, w_j),(v'_j, w'_j))\} - \chi(\{(v_j, w'_j),(v'_j, w_j)\})$$

$$= \chi(C_{b^{(1)}} \cup C'_{b^{(1)}}) + \chi(C_{b^{(2)}} \cup C'_{b^{(2)}})$$

and hence,

$$\frac{1}{2}(\chi(C_{b^{(1)}} \cup C'_{b^{(1)}}) + \chi(C_{b^{(2)}} \cup C'_{b^{(2)}})) = \frac{1}{2}(\chi(T_1) + \chi(T_2)) \in \text{conv}(\text{ATSP}_{4N+2}). \qquad \square$$

Theorem 5. *The asymptotic growth of* $\text{rc}(\text{ATSP}_n)$ *and* $\text{rc}(\text{STSP}_n)$ *is* $2^{\Theta(n)}$.

Proof. Lemma 2 shows that H_N is a hiding set for ATSP_n. By replacing all directed arcs with their undirected versions, the set H_N yields a hiding set for STSP_n. By Proposition 3, we obtain a lower bound of $|H_N| = 2^{\Omega(n)}$ for $\text{rc}(\text{ATSP}_n)$ and $\text{rc}(\text{STSP}_n)$. To complete the argumentation, note that (3) admits a relaxation for $STSP_n$ having $2^{\Theta(n)}$ facets and that ATSP_n has a relaxation of similar size, which is a directed version of (3). $\qquad \square$

Connected Sets. Let CONN_n be the set of all characteristic vectors of edge sets that form a connected subgraph in the complete graph on n nodes. The polytope $\left\{ x \in [0,1]^{E_n} : x(\delta(S)) \geq 1 \ \forall \emptyset \neq S \subsetneq V_n \right\}$ is a relaxation for CONN_n. Thus, we have that $\text{rc}(\text{CONN}_n) \leq \mathcal{O}(2^n)$.

For a lower bound, consider again the undirected version of our set H_N. Since each point in H_N belongs to a node-disjoint union of two cycles, we have that $H_N \cap \text{CONN}_n = \emptyset$. Further, we know that for any $a, b \in H_N$ we have that

$$\emptyset \neq \text{conv}\{a,b\} \cap \text{conv}(\text{STSP}_n) \subseteq \text{conv}\{a,b\} \cap \text{conv}(\text{CONN}_n)$$

and since $H_N \subseteq \text{aff}(\text{CONN}_n) = \mathbb{R}^{E_n}$, we see that H_N is also a hiding set for CONN_n. We obtain:

Corollary 3. *The asymptotic growth of* $\text{rc}(\text{CONN}_n)$ *is* $2^{\Theta(n)}$.

Branchings and Forests. Besides connectivity, we show that, in general, it is also hard to force acyclicity in the context of relaxation. Let therefore ARB_n (SPT_n) be the set of characteristic vectors of arborescences (spanning trees) in the complete directed (undirected) graph.

Theorem 6. *The asymptotic growth of* $\mathrm{rc}(\mathrm{ARB}_n)$ *and* $\mathrm{rc}(\mathrm{SPT}_n)$ *is* $2^{\theta(n)}$.

Proof. First, note that both the *arborescence polytope* and the spanning tree polytope (i.e., $\mathrm{conv}(ARB_n)$ and $\mathrm{conv}(SPT_n)$) have $\mathcal{O}(2^n)$ facets [10] and hence we have an upper bound of $\mathcal{O}(2^n)$ for both $\mathrm{rc}(\mathrm{ARB}_n)$ and $\mathrm{rc}(\mathrm{SPT}_n)$.

For a lower bound, let us modify the definition of $C(b)'$ by removing arc (v'_{N+1}, v'_1). Then, $C(b) \cap C(b)'$ is a node-disjoint union of a cycle and a path and hence not an arborescence. By following the proof of Lemma 2 and removing arc (v'_{N+1}, v'_1) from T_1 and T_2, we still have that

$$\chi(C(b^{(1)}) \cup C(b^{(1)})') + \chi(C(b^{(2)}) \cup C(b^{(2)})') = \chi(T_1) + \chi(T_2),$$

where T_1 and T_2 are spanning arborescences. (Actually, T_1 and T_2 are in fact directed paths visiting each node.) Since $\mathrm{aff}(\mathrm{ARB}_n) = \mathbb{R}^{A_n}$, we therefore obtain that the modified set H_N is a hiding set for ARB_n. By undirecting all arcs, H_N also yields a hiding set for SPT_n.

Again, by Proposition 3, we deduce a lower bound of $|H_N| = 2^{\Omega(n)}$ for both $\mathrm{rc}(\mathrm{ARB}_n)$ and $\mathrm{rc}(\mathrm{SPT}_n)$. □

Remark 1. Since in the proof of Theorem 6 T_1 and T_2 are rooted at node v'_1, the statements even hold if the sets ARB_n and SPT_n are restricted to characteristic vectors of arborescences/trees rooted at a fixed node.

Let BRANCH_n ($\mathrm{FORESTS}_n$) be the set of characteristic vectors of branchings (forests) in the complete directed (undirected) graph.

Corollary 4. *The asymptotic growth of* $\mathrm{rc}(\mathrm{BRANCH}_n)$ *and* $\mathrm{rc}(\mathrm{FORESTS}_n)$ *is* $2^{\theta(n)}$.

Proof. The claim follows from Theorem 6 and the facts that

$$\mathrm{ARB}_n = \mathrm{BRANCH}_n \cap \left\{ x \in \mathbb{R}^{A_n} : \sum_{a \in \mathbb{R}^{A_n}} x_a = n - 1 \right\}$$

$$\mathrm{SPT}_n = \mathrm{FORESTS}_n \cap \left\{ x \in \mathbb{R}^{E_n} : \sum_{e \in \mathbb{R}^{E_n}} x_e = n - 1 \right\}. \quad \square$$

3.2   Distinctness

Another common component of practical IP-formulations is the requirement of distinctness of a certain set of vectors or variables. Here, we consider two general cases in which we can also show that the benefit of auxiliary variables is essential.

Binary All-Different. In the case of the *binary all-different* constraint, one requires the distinctness of rows of a binary matrix with m rows and n columns. The set of feasible points is therefore defined by

$$\text{DIFF}_{m,n} := \left\{ x \in \{0,1\}^{m \times n} : x \text{ has pairwise distinct rows} \right\}.$$

As an example, in [5] the authors present IP-formulations to solve the coloring problem in which they binary encode the color classes assigned to each node. As a consequence, certain sets of encoding vectors have to be distinct.

By separating each possible pair of equal rows by one inequality, it is further easy to give a relaxation for $\text{DIFF}_{m,n}$ that has at most $\binom{m}{2}2^n + 2mn$ facets. In the case of $m = 2$, for instance, this bound turns out to be almost tight:

Theorem 7. *For all $n \geq 1$, we have that $\text{rc}(\text{DIFF}_{2,n}) \geq 2^n$.*

Proof. Let us consider the set

$$H_{2,n} := \left\{ (x,x)^T \in \{0,1\}^{2 \times n} : x \in \{0,1\}^n \right\}.$$

For $x, y \in \{0,1\}^n$ distinct, we obviously have that

$$\frac{1}{2}\left((x,x)^T + (y,y)^T\right) = \frac{1}{2}\left((x,y)^T + (y,x)^T\right) \in \text{conv}(\text{DIFF}_{2,n}).$$

Since $H_{2,n} \cap \text{DIFF}_{2,n} = \emptyset$ and $H_{2,n} \subseteq \text{aff}(\text{DIFF}_{2,n}) = \mathbb{R}^{2 \times n}$, $H_{2,n}$ is a hiding set for $\text{DIFF}_{2,n}$ and by Proposition 3 we obtain that $\text{rc}(\text{DIFF}_{2,n}) \geq |H_{2,n}| = 2^n$.   □

Permutahedron. As a case in which one does not require the distinctness of binary vectors but of a set of numbers let us consider the set

$$\text{PERM}_n := \left\{ (\pi(1), \ldots, \pi(n)) \in \mathbb{Z}^n : \pi \in \mathcal{S}_n \right\},$$

which is the vertex set of the *permutahedron* $\text{conv}(\text{PERM}_n)$. Rado [9] showed that the permutahedron can be described via

$$\text{conv}(\text{PERM}_n) = \Big\{ x \in \mathbb{R}^n : \sum_{i=1}^{n} x_i = \frac{n(n+1)}{2}$$

$$\sum_{i \in S} x_i \geq \frac{|S|(|S|+1)}{2} \text{ for all } \emptyset \neq S \subset [n]$$

$$x \geq \mathbb{0} \Big\} \tag{4}$$

and hence has $\mathcal{O}(2^n)$ facets. Apart from that, it is a good example for a polytope having many different, very compact extended formulations, see, e.g., [3]. In the contrary, we show that the relaxation complexity of PERM_n has exponential growth in n:

Theorem 8. *The asymptotic growth of $\text{rc}(\text{PERM}_n)$ is $2^{\theta(n)}$.*

Proof. Let $m := \lfloor \frac{n}{2} \rfloor$. For any set $S \subseteq [n]$ with $|S| = m$ select an integer vector $x^S \in \mathbb{Z}^n$ with $\{x_i : i \in S\} = \{1, \ldots, m-1\}$ and $m-1$ occuring twice among the x_i^S ($i \in S$) and $\{x_i : i \in [n] \setminus S\} = \{m+2, \ldots, n\}$ and $m+2$ occuring twice among the x_i^S ($i \in [n] \setminus S$). Such a vector is not contained in conv(PERM$_n$) as

$$\sum_{i \in S} x_i^S = 1 + 2 + \ldots + (|S| - 2) + (|S| - 2) < \frac{|S|(|S| + 1)}{2}$$

On the other hand, note that this is the only constraint from (4) that is violated by x^S. In particular, $x^S \in \text{aff}(\text{PERM}_n)$ holds.

Let $S_1, S_2 \subseteq [n]$ with $|S_1| = |S_2| = m$ be distinct. We will show that $x := \frac{1}{2} \cdot (x^{S_1} + x^{S_2}) \in \text{conv}(\text{PERM}_n)$ holds. Since x satisifies all constraints that are satisfied by both x^{S_1} and x^{S_2}, it suffices to show that $\sum_{i \in T} x_i \geq \frac{|T|(|T|+1)}{2}$ holds for $T \in \{S_1, S_2\}$. W.l.o.g. we may assume that $T = S_1$ and obtain

$$\sum_{i \in S_1} x_i = \frac{1}{2} \sum_{i \in S_1} x_i^{S_1} + \frac{1}{2} \sum_{i \in S_1} x_i^{S_2}$$

$$= \frac{1}{2} \left(\frac{m(m+1)}{2} - 1 \right) + \frac{1}{2} \sum_{i \in S_1} x_i^{S_2}$$

$$\geq \frac{1}{2} \left(\frac{m(m+1)}{2} - 1 \right) + \frac{1}{2} \left(\frac{m(m+1)}{2} + 2 \right)$$

$$= \frac{m(m+1)}{2} + \frac{1}{2} \geq \frac{|T|(|T|+1)}{2}. \qquad \square$$

Thus, the set $H := \{x^S : S \subseteq [n], |S| = m\}$ is a hiding set for PERM$_n$. Our claim follows from Proposition 3 and the fact that $|H| = \binom{n}{\lfloor \frac{n}{2} \rfloor} = 2^{\theta(n)}$. $\qquad \square$

3.3 Parity

The final structural class we consider deals with the restriction that the number of selected elements of a given set has a certain parity. Let us call a binary vector $a \in \{0, 1\}^d$ *even* (*odd*) if the sum of its entries is even (odd). In [4] it is shown that the number of inequalities needed to separate

$$\text{EVEN}_n := \{x \in \{0, 1\}^n : x \text{ is even}\}$$

from all other points in $\{0, 1\}^n$ is exactly 2^{n-1}. This is done by showing that

$$\text{ODD}_n := \{x \in \{0, 1\}^n : x \text{ is odd}\}$$

is a hiding set for EVEN$_n$ (although the notion is different from ours). Hence, with Corollary 2, we obtain:

Theorem 9. *The asymptotic growth of* rc(EVEN$_n$) *is* $\Theta(2^n)$.

T-joins As a well-known representative of this structural class let us consider $T\text{-JOINS}_n$, which is, for given $T \subseteq V_n$, defined as the set of characteristic vectors of T-joins in the complete graph on n nodes. Let us recall that a T-join is a set $J \subseteq E_n$ of edges such that T is equal to the set of nodes of odd degree in the graph (V_n, J). Note, that if a T-join exists, then $|T|$ is even.

Theorem 10. *Let n be even and $T \subseteq V_n$ with $|T|$ even. Then, $\mathrm{rc}(T\text{-JOINS}_n) \geq 2^{\frac{n}{4}-1}$.*

Proof. Since n is even and $|T|$ is even, we may partition V_n into pairwise disjoint sets T_1, T_2, U_1, U_2 with $T = T_1 \cup T_2$, $k = |T_1| = |T_2|$ and $\ell = |U_1| = |U_2|$. Let $M_1, \ldots, M_k$ be pairwise edge-disjoint matchings of cardinality k that connect nodes from T_1 with nodes from T_2. Analogously, let $N_1, \ldots, N_\ell$ be pairwise edge-disjoint matchings of cardinality ℓ that connect nodes from U_1 with nodes from U_2. For $b \in \{0,1\}^k$ and $c \in \{0,1\}^\ell$ let

$$J(b,c) := \left(\bigcup_{i:b_i=1} M_i \right) \cup \left(\bigcup_{j:c_j=1} N_j \right) \subseteq E_n.$$

By definition, $J(b,c)$ is a T-join if and only if b is odd and c is even. Let $b^* \in \{0,1\}^k$ odd and $c^* \in \{0,1\}^\ell$ even be arbitrarily chosen but fixed. Since ODD_n is a hiding set for EVEN_n and vice versa, it is now easy to see that both sets

$$H_1 := \left\{ J(b, c^*) : b \in \{0,1\}^k \text{ even} \right\}$$
$$H_2 := \left\{ J(b^*, c) : c \in \{0,1\}^\ell \text{ odd} \right\},$$

are hiding sets for $T\text{-JOINS}_n$. Our claim follows from Proposition 3 and the fact that

$$\max\left\{ |H_1|, |H_2| \right\} = \max\left\{ 2^{k-1}, 2^{\ell-1} \right\} = \max\left\{ 2^{\frac{1}{2}|T|-1}, 2^{\frac{1}{2}(n-|T|)-1} \right\}$$
$$\geq 2^{\frac{1}{2}\cdot\frac{n}{2}-1}. \qquad \square$$

4   Concluding Remarks

We at least asymptotically determined the relaxation complexities for several examples arising from combinatorial optimization that we were interested in. Turning towards the more basical questions, we found $\mathrm{rc}(\{0,1\}^d) = d + 1$. In contrast to this, we do, however, not know the exact value of $\mathrm{rc}(\Delta_d)$, where Δ_d is the set of vertices of the standard d-simplex. As briefly discussed in Section 2.3, a relaxation R for Δ_d with less than $d + 1$ facets must be unbounded and (hence) irrational. Unlike the case of $\{0,1\}^d$, there are indeed unbounded polyhedra whose integer points are exactly the set of an integral d-simplex. As an example, it is rather easy to see that the polyhedron

$$\mathrm{conv}\left\{ \mathbb{O}, \mathbb{e}_1, \mathbb{e}_2, \mathbb{e}_3, \mathbb{e}_1 + \mathbb{e}_3 + \mathbb{e}_4, \mathbb{e}_2 + \mathbb{e}_3 + \mathbb{e}_5 \right\} + \mathbb{R} \cdot (0,0,0,1,\sqrt{2})^T$$

has this property. Further, one also finds that the hiding set method does not help in this case. In fact, any hiding set of Δ_d can be shown to have cardinality of at most 3. Therefore, the simplex-example raises the question, whether there are polyhedral sets X with $\mathrm{rc}(X) < \mathrm{rc}_{\mathbb{Q}}(X)$ (see 2.3).

Acknowledgements. We would like to thank Gennadiy Averkov for valuable comments on this work.

References

1. Fiorini, S., Massar, S., Pokutta, S., Tiwary, H.R., de Wolf, R.: Linear vs. semidefinite extended formulations: exponential separation and strong lower bounds. In: STOC, pp. 95–106 (2012)
2. Gavish, B., Graves, S.C.: The travelling salesman problem and related problems. Technical report, Operations Research Center, Massachusetts Institute of Technology (1978)
3. Goemans, M.: Smallest compact formulation for the permutahedron. Kaibel, V., Thomas, R. (eds.) To appear in a forthcoming issue of Mathematical Programming, Series B "Lifts of Convex Sets" (2014)
4. Jeroslow, R.: On defining sets of vertices of the hypercube by linear inequalities. Discrete Mathematics 11(2), 119–124 (1975)
5. Lee, J., Margot, F.: On a binary-encoded ilp coloring formulation. INFORMS Journal on Computing 19(3), 406–415 (2007)
6. Kipp Martin, R.: Using separation algorithms to generate mixed integer model reformulations. Oper. Res. Lett. 10(3), 119–128 (1991)
7. Miller, C.E., Tucker, A.W., Zemlin, R.A.: Integer programming formulation of traveling salesman problems. J. ACM 7(4), 326–329 (1960)
8. Minkowski, H.: Geometrie der Zahlen. Teubner Verlag, Leipzig (1896)
9. Rado, R.: An inequality. J. London Math. Soc. 27, 1–6 (1952)
10. Schrijver, A.: Combinatorial Optimization – Polyhedra and Efficiency. Springer (2003)
11. Schrijver, A.: Theory of linear and integer programming. John Wiley & Sons, Inc., New York (1986)

On the Configuration LP for Maximum Budgeted Allocation*

Christos Kalaitzis[1], Aleksander Mądry[1], Alantha Newman[1], Lukáš Poláček[2],
and Ola Svensson[1]

[1] EPFL, Switzerland
firstname.lastname@epfl.ch
[2] KTH Royal Institute of Technology, Sweden
polacek@csc.kth.se

Abstract. We study the Maximum Budgeted Allocation problem, i.e.,
the problem of selling a set of m indivisible goods to n players, each with
a separate budget, such that we maximize the collected revenue. Since the
natural assignment LP is known to have an integrality gap of $\frac{3}{4}$, which
matches the best known approximation algorithms, our main focus is to
improve our understanding of the stronger configuration LP relaxation.
In this direction, we prove that the integrality gap of the configuration
LP is strictly better than $\frac{3}{4}$, and provide corresponding polynomial time
roundings, in the following restrictions of the problem: (i) the Restricted
Budgeted Allocation problem, in which all the players have the same
budget and every item has the same value for any player it can be sold
to, and (ii) the graph MBA problem, in which an item can be assigned
to at most 2 players. Finally, we improve the best known upper bound
on the integrality gap for the general case from $\frac{5}{6}$ to $2\sqrt{2} - 2 \approx 0.828$
and also prove hardness of approximation results for both cases.

1 Introduction

Suppose there are multiple players, each with a budget, who want to pay to gain
access to some advertisement resources. On the other hand, the owner of these
resources wants to allocate them so as to maximize his total revenue, i.e., he
wishes to maximize the total amount of money the players pay. No player can
pay more than his budget so the task of the owner is to find an assignment of
resources to players that maximizes the total payment where each player pays
the minimum of his budget and his valuation of the items assigned to him.

The above problem is called Maximum Budgeted Allocation (MBA), and it
arises often in the context of advertisement allocation systems. Formally, a prob-
lem instance $\mathcal{I}$ can be defined as follows: there is a set of players $\mathcal{A}$ and a set of
items $\mathcal{Q}$. Each player i has a budget B_i and each item j has a price $p_{ij} \leq B_i$ for

---

* A full version of this paper, containing all omitted proofs, is available at
http://arxiv.org/abs/1403.7519. This research was partially supported by ERC
Advanced Investigator Grant 226203 and ERC Starting Grant 335288-OptApprox.

J. Lee and J. Vygen (Eds.): IPCO 2014, LNCS 8494, pp. 333–344, 2014.
© Springer International Publishing Switzerland 2014

player i (the assumption that $p_{ij} \leq B_i$ is without loss of generality, because no player can spend more money than his budget). Our objective is to find disjoint sets $S_i \subseteq Q$ for each player i, i.e., an indivisible assignment of items to players, such that we maximize

$$\sum_{i \in \mathcal{A}} \min \left\{ \sum_{j \in S_i} p_{ij}, B_i \right\}.$$

In this paper, we are interested in designing good algorithms for the MBA problem and we shall focus on understanding the power of a strong convex relaxation called the configuration LP. The general goal is to obtain a better understanding of basic allocation problems that have a wide range of applications. In particular, the study of the configuration LP is motivated by the belief that a deeper understanding of this type of relaxation can lead to better algorithms for many allocation problems, including MBA, the Generalized Assignment problem, Unrelated Machine Scheduling, and Max-Min Fair Allocation.

As the Maximum Budgeted Allocation problem is known to be NP-hard [10,12], we turn our attention to approximation algorithms. Recall that an r-approximation algorithm is an efficient (polynomial time) algorithm that is guaranteed to return a solution within a factor r of the optimal value. The factor r is referred to as the approximation ratio or guarantee.

Garg, Kumar and Pandit [10] obtained the first approximation algorithm for MBA with a guarantee of $\frac{2}{1+\sqrt{5}}$. This was later improved to $1 - \frac{1}{e}$ by Andelman and Mansour [1], who also showed that an approximation guarantee of 0.717 can be obtained in the case when all the budgets are equal. Subsequently, Azar, Birnbaum, Karlin, and Mathieu [3] gave a $\frac{2}{3}$-approximation algorithm, which Srinivasan [15] extended to give the current best approximation guarantee of $\frac{3}{4}$. Concurrently, the same approximation guarantee was achieved by Chakrabarty and Goel [5], who also proved that it is NP-hard to achieve an approximation ratio better than $\frac{15}{16}$.

It is interesting to note that the progress on MBA has several points in common with other basic allocation problems. First, it is observed that when the prices are relatively small compared to the budgets, then the problem becomes substantially easier (e.g. [5,15]), analogous to how Unrelated Machine Scheduling becomes easier when the largest processing time is small compared to the optimal makespan. Second, the above mentioned 3/4-approximation algorithms give a tight analysis of a standard LP relaxation, called the assignment LP, which has been a successful tool for allocation problems ever since the breakthrough work by Lenstra, Shmoys, and Tardos [13]. Indeed, we now have a complete understanding of the strength of the assignment LP for all above mentioned allocation problems. The strength of a relaxation is measured by its integrality gap, which is the maximum ratio between the solution quality of the exact integer programming formulation and of its relaxation.

A natural approach for obtaining better (approximation) algorithms for allocation problems are stronger relaxations than the assignment LP. Similarly to other allocation problems, there is a strong belief that a strong convex relaxation called configuration LP gives strong guarantees for the MBA problem. Although we only know that the integrality gap is no better than $\frac{5}{6}$ [5], our current techniques fail to prove that the configuration LP gives even marginally better guarantees for MBA than the assignment LP. The goal of this paper is to increase our understanding of the configuration LP and shed light on its strength.

Our contributions. To analyze the strength of the configuration LP compared to the assignment LP, it is instructive to consider the tight integrality gap instance for the assignment LP from [5] depicted in Figure 1. This instance satisfies several structural properties: (i) at most two players have a positive price of an item, (ii) every player has the same budget (also known as *uniform budgets*), (iii) the price of an item j for a player is either p_j or 0, i.e., $p_{ij} \in \{0, p_j\}$.

Motivated by these observations and previous work on allocation problems, we shall mainly concentrate on two special cases of MBA. The first case is obtained by enforcing (i) in which at most two players have a positive price of an item. We call it *graph MBA*, as an instance can naturally be represented by a graph where items are edges, players are vertices and assigning an item corresponds to orienting an edge. The same restriction, where it is often called Graph Balancing, has led to several nice results for Unrelated Machine Scheduling [6] and Max-Min Fair Allocation [17].

The second case is obtained by enforcing (ii) and (iii). That is, each item j has a non-zero price, denoted by p_j, for a subset of players, and the players have uniform budgets. We call this case *restricted MBA* or the Restricted Budgeted Allocation Problem as it closely resembles the Restricted Assignment Problem that has been a popular special case of both Unrelated Machine Scheduling [16] and Max-Min Fair Allocation [7,2,4]. It is understood that these two structural properties produce natural restrictions whose study helps increase our understanding of the general problem [5,15], and specifically, instances displaying property (ii) have been studied in [1].

Our main result proves that the configuration LP is indeed stronger than the assignment LP for the considered problems.

Theorem 1. *Restricted Budgeted Allocation and graph MBA have $(3/4 + c)$-approximation algorithms that also bound the integrality gap of the configuration LP, for some constant $c > 0$.*

The result for graph MBA is inspired by the work by Feige and Vondrak [8] on the generalized assignment problem and is presented in the full version of this paper. The main idea is to first recover a $3/4$-fraction of the configuration LP solution by randomly (according to the LP solution) assigning items to the players. The improvement over $3/4$ is then obtained by further assigning some of the items that were left unassigned by the random assignment to players whose budgets were not already exceeded. The difficulty in the above approach lies in analyzing the contribution of the items assigned in the second step over the random assignment in the first step.

For restricted MBA, we need a different approach. Indeed, randomly assigning items according to the configuration LP only recovers a $(1 - 1/e)$-fraction of

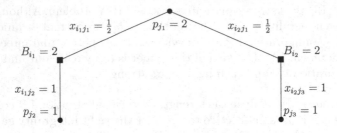

Fig. 1. The solution x has value of 4. Any integral solution has a value of at most 3, since one player will be assigned only one item of value 1.

the LP value when an item can be assigned to any number of players. Current techniques only gain an additional small ϵ-fraction by assigning unassigned items in the second step. This would lead to an approximation guarantee of $(1-1/e+\epsilon)$ (matching the result in [8] for the Generalized Assignment Problem) which is strictly less than the best known approximation guarantee of 3/4 for MBA. We therefore take a different approach. We first observe that an existing algorithm, described in Section 3, already gives a better guarantee than 3/4 for configuration LP solutions that are not well-structured (see Definition 1). Informally, an LP solution is well-structured if half the budgets of most players are assigned to expensive items, which are defined as those items whose price is very close to the budget. For the rounding of well-structured solutions in Section 4.2, the main new idea is to first assign expensive/big items using random bipartite matching and then assign the cheap/small items in the space left after the assignment of expensive items. For this to work, it is not sufficient to assign the big items in any way that preserves the marginals from the LP relaxation. Instead, a key observation is that we can assign big items so that the probability that two players i, i' are both assigned big items is at most the probability that i is assigned a big item times the probability that i' is assigned a big item (i.e., the events are negatively correlated). Using this we can show that we can assign many of the small items even after assigning the big items leading to the improved guarantee. We believe that this is an interesting use of bipartite matchings for allocation problems as we are using the fact that the events that vertices are matched can be made negatively correlated. Note that this is in contrast to the events that edges are part of a random matching which are not necessarily negatively correlated.

Finally, we complement our positive results by hardness results and integrality gaps. For restricted MBA, we prove hardness of approximation that matches the strongest results known for the general case. Specifically, we prove that it is NP-hard to approximate restricted MBA within a factor 15/16. This shows that some of the hardest known instances for the general problem are the ones we study. We also improve the 5/6 integrality gap of the configuration LP for the general case: we prove that it is not better than $2(\sqrt{2}-1)$. Both the NP-hardness result and integrality gap are presented in the full version of this paper, along with all omitted proofs from now on.

2   Preliminaries

Assignment LP. The assignment LP for the MBA problem has a fractional "indicator" variable x_{ij} for each player $i \in \mathcal{A}$ and each item $j \in \mathcal{Q}$ that indicates whether item j is assigned to player i. Recall that the profit received from a player i is the minimum of his budget B_i and the total value $\sum_{j \in \mathcal{Q}} x_{ij} p_{ij}$ of the items assigned to i. In order to avoid taking the minimum for each player, we impose that each player i is fractionally assigned items of total value at most his budget B_i. Note that this is not a valid constraint for an integral solution but it does not change the value of a fractional solution: we can always fractionally decrease the assignment of an item without changing the objective value if the total value of the fractional assignment exceeds the budget. To further simplify the relaxation, we enforce that all items are fully assigned by adding a dummy player ℓ such that $p_{\ell j} = 0$ for all $j \in \mathcal{Q}$ and $B_\ell = 0$. The assignment LP for MBA is defined as follows:

$$\begin{aligned}
\max \quad & \textstyle\sum_{i \in \mathcal{A}} \sum_{j \in \mathcal{Q}} x_{ij} p_{ij} \\
\text{subject to} \quad & \textstyle\sum_{j \in \mathcal{Q}} x_{ij} p_{ij} \leq B_i \quad && \forall i \in \mathcal{A} \\
& \textstyle\sum_{i \in \mathcal{A}} x_{ij} = 1 \quad && \forall j \in \mathcal{Q} \\
& 0 \leq x_{ij} \leq 1 \quad && \forall i \in \mathcal{A}, \forall j \in \mathcal{Q}
\end{aligned}$$

As discussed in the introduction, it is known that the integrality gap of the assignment LP is exactly $\frac{3}{4}$; therefore, in order to achieve a better approximation, we employ a stronger relaxation called the configuration LP.

Configuration LP. The intuition behind the configuration LP comes from observing that, in an integral solution, the players are assigned disjoint sets, or configurations, of items. The configuration LP will model this by having a fractional "indicator" variable $y_{i\mathcal{C}}$ for each player i and configuration $\mathcal{C} \subseteq \mathcal{Q}$, which indicates whether or not $\mathcal{C}$ is the set of items assigned to player i in the solution. The constraints of the configuration LP require that each player is assigned at most one configuration and each item is assigned to at most one player. If we let $w_i(\mathcal{C}) = \min\left\{\sum_{j \in \mathcal{C}} p_{ij}, B_i\right\}$ denote the total value of the set $\mathcal{C}$ of items when assigned to player i, the configuration LP can be formulated as follows:

$$\begin{aligned}
\max \quad & \textstyle\sum_{i \in \mathcal{A}} \sum_{\mathcal{C} \subseteq \mathcal{Q}} w_i(\mathcal{C}) y_{i\mathcal{C}} \\
\text{subject to} \quad & \textstyle\sum_{\mathcal{C} \subseteq \mathcal{Q}} y_{i\mathcal{C}} \leq 1 \quad && \forall i \in \mathcal{A} \\
& \textstyle\sum_{i \in \mathcal{A}, \mathcal{C} \subseteq \mathcal{Q}: j \in \mathcal{C}} y_{i\mathcal{C}} \leq 1 \quad && \forall j \in \mathcal{Q} \\
& y_{i\mathcal{C}} \geq 0 \quad && \forall i \in \mathcal{A}, \forall \mathcal{C} \subseteq \mathcal{Q}
\end{aligned}$$

We remark that even though the relaxation has exponentially many variables, it can be solved approximately in a fairly standard way by designing an efficient separation oracle for the dual which has polynomially many variables. We refer the reader to [5] for more details.

The configuration LP is stronger than the assignment LP as it enforces a stricter structure on the fractional solution. Indeed, every solution to the configuration LP can be transformed into a solution of the assignment LP of at least the same value (see e.g. Lemma 2). However, the converse is not true; one example is shown in Figure 1. More convincingly, our results show that the configuration LP has a strictly better integrality gap than the assignment LP for large natural classes of the MBA problem.

For a solution y to the configuration LP, we let $\mathsf{Val}_i(y) = \sum_\mathcal{C} w_i(\mathcal{C}) y_{i\mathcal{C}}$ be the value of the fractional assignment to player i and let $\mathsf{Val}(y) = \sum_i \mathsf{Val}_i(y)$. Note that $\mathsf{Val}(y)$ is equal to the objective value of the solution y. Abusing notation, we also define $\mathsf{Val}_i(x) = \sum_j x_{ij} p_{ij}$ for a solution x to the assignment LP. We might also use $\mathsf{Val}^\mathcal{I}(y)$ and $\mathsf{Val}_i^\mathcal{I}(y)$ to indicate we are considering instance $\mathcal{I}$.

Random bipartite matching. As alluded to in the introduction, one of the key ideas of our algorithm for the restricted case is to first assign expensive/big items (of value close to the budgets) by picking a random bipartite matching so that the events that vertices are matched are negatively correlated. The following theorem uses the techniques developed by Gandhi, Khuller, Parthasarathy and Srinivasan in their work on selecting random bipartite matchings with particular properties [9]. Its proof is included in the full version of this paper.

Theorem 2. *Consider a bipartite graph $G = ((A, B), E)$ and an assignment $(x_e)_{e \in E}$ to edges so that the fractional degree $\sum_{u:uv \in E} x_{uv}$ of each vertex v is at most 1. Then there is an efficient, randomized algorithm that generates a (random) matching satisfying:*

(P1): Marginal Distribution. *For every vertex $v \in A \cup B$, $\Pr[v \text{ is matched}] = \sum_{u:uv \in E} x_{uv}$.*

(P2): Negative Correlation. *For any $S \subseteq A$, $\Pr[\bigwedge_{v \in S}(v \text{ is matched})] \leq \prod_{v \in S} \Pr[v \text{ is matched}]$.*

One should note that the events {edge e is in the matching} and {edge e' is in the matching} are not necessarily negatively correlated (if we preserve the marginals). A crucial ingredient for our algorithm is therefore the idea that we can concentrate on the event that a player has been assigned a big item without regard to the specific item assigned.

3   General 3/4-Approximation Algorithm

In this section, we introduce an algorithm (inspired by [14]) to round assignment LP solutions, and we then present an analysis showing that it is a 3/4-approximation algorithm. In Section 4, we use this analysis to show that the algorithm has a better approximation ratio than 3/4 in some cases.

First, we need the following definition for the algorithm. Let $G = U \cup V$ be a bipartite graph. A *complete matching for V* is a matching that has exactly one edge incident to every vertex in V.

Algorithm 1 first partitions x into buckets creating a new assignment x', such that the sum of x' in each bucket is exactly 1 except possibly the last bucket of each player. Some items are split into two buckets. The process for one player is illustrated in Figure 2.

Input   : Solution x to the assignment LP, ordering o_i of the items by prices for
             player i
Output: Assignment x^* of items to the players

foreach $i \in \mathcal{A}$ **do**
    // Create buckets for player i, see Figure 2
    $c_i \leftarrow \lceil \sum_j x_{ij} \rceil$
    Create c_i buckets $(i, 1), \ldots, (i, c_i)$
    Create $x'_{(i, \cdot)}$ from x_i as in Figure 2
end
$U \leftarrow \{(i, k) \mid 1 \leq k \leq \lceil \sum_j x_{ij} \rceil\}$
$V \leftarrow \mathcal{Q}$
Express x' as a convex combination of complete matchings for V: $x' = \sum_i \gamma_i m_i$
Return matching m_i with probability γ_i

Algorithm 1. Bucket algorithm

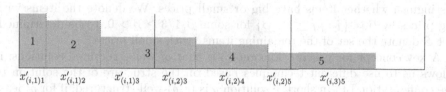

Fig. 2. Illustration of bucket creation by Algorithm 1 for player i. Buckets are marked
by solid lines. The value x_{i3} is split into $x'_{(i,1)3}$ and $x'_{(i,2)3}$ and x_{i5} is split into $x'_{(i,2)5}$
and $x'_{(i,3)5}$. For the other items we have $x'_{(i,1)1} = x_{i1}, x'_{(i,1)2} = x_{i2}, x'_{(i,2)4} = x_{i4}$. Items
are ordered in non-decreasing order by their prices.

From the previous discussion, for every bucket b we have $\sum_j x'_{bj} \leq 1$. Also,
$\sum_{b \in U} x'_{bj} = 1$ for every item j, which is implied by $\sum_{i \in \mathcal{A}} x_{ij} = 1$ for all $j \in \mathcal{Q}$.
Hence x' is inside the complete matching polytope for $V = \mathcal{Q}$. Using an algo-
rithmic version of Carathéodory's theorem (see e.g. Theorem 6.5.11 in [11]) we
can in polynomial time decompose x' into a convex combination of polynomially
many complete matchings for V.

In the algorithm we use an ordering o_i for player i such that $p_{io_{ih}} \geq p_{io_{i,h+1}}$,
i.e. it is the descending order of items by their prices for player i. In particular
this implies that the algorithm does not use the prices, only the order of items.
Also note that Algorithm 1 could be made deterministic. Instead of picking a
random matching we can pick the best one.

Let $\mathsf{Alg}_i(x)$ be the expected price that player i pays. We know that $\mathsf{Alg}_i(x) \leq$
$\mathsf{Val}_i(x)$, since the probability of assigning j to i is x_{ij}, but we don't have equality
in the expression, because some matchings might assign a price that is over the
budget for some players. In the following lemma we lower bound $\mathsf{Alg}_i(x)$.

Lemma 1. *Let x be a solution to the assignment LP, $i \in \mathcal{A}$ and let $\alpha \geq 1$
be such that $\mathsf{Val}_i(x) = B_i/\alpha$. Let a_1 be the average price of items in the first
bucket b of player i, i.e. $a_1 = \sum_j x'_{bj} p_{bj}$. Let r_1 be the average price of items in*

b that have price more than αa_1. Then

$$\mathsf{Alg}_i(x) \geq \mathsf{Val}_i(x)\left(1 - \frac{r_1}{4B_i}\right).$$

In particular, since $r_i \leq B_i$, Algorithm 1 gives a 3/4-approximation.

4   Restricted Budget Allocation

In this section we consider the MBA problem with uniform budgets where the prices are restricted to be of the form $p_{ij} \in \{p_j, 0\}$. This is the so called *restricted maximum budgeted allocation*. Our main result is the following.

Theorem 3. *There is a $(3/4 + c)$-approximation algorithm for restricted MBA for some constant $c > 0$.*

Since the budgets are uniform, we can assume that each player has a budget of 1 by scaling. We refer to p_j as the price of item j. It will be convenient to distinguish whether items have big or small prices. We denote the items with big prices by $\mathcal{B} = \{j : p_j \geq 1 - \beta\}$ for some $\beta, 1/3 \geq \beta > 0$, to be determined. Let $\mathcal{S}$ denote the set of the remaining items (with small prices).

A key concept for proving Theorem 3 is that of well-structured solutions; it allows us to use different techniques based on the structure of the solution to the configuration LP. In short, a solution y is (ϵ, δ)-well-structured, if for at least $(1 - \epsilon)$-fraction of players, roughly half of their configurations contain a big item.

Definition 1. *A solution y to the configuration LP is (ϵ, δ)-well-structured if*

$$\Pr_i\left[\sum_{C \subseteq \mathcal{Q}} |\mathcal{B} \cap C| \cdot y_{i,C} \notin \left[\frac{1-\delta}{2}, \frac{1+\delta}{2}\right]\right] \leq \epsilon,$$

where the probability is taken over a weighted distribution of players such that player i is chosen with probability $\mathsf{Val}_i(y)/\mathsf{Val}(y)$.

We want to be able to switch from the configuration LP to the assignment LP without changing the well-structuredness of the solution. The following lemma shows that this is indeed possible.

Lemma 2. *Let y be a well-structured solution to the configuration LP. Then there exists a solution x to the assignment LP with $\mathsf{Val}_i(x) = \mathsf{Val}_i(y)$ such that*

$$\sum_{C \subseteq \mathcal{Q}} |\mathcal{B} \cap C| \cdot y_{i,C} \in \left[\frac{1-\delta}{2}, \frac{1+\delta}{2}\right] \Leftrightarrow \sum_{j:j \in \mathcal{B}} x_{ij} \in \left[\frac{1-\delta}{2}, \frac{1+\delta}{2}\right]$$

for all $i \in \mathcal{A}$. Furthermore, x can be produced from y in polynomial time.

In the next subsection in Lemma 5 we show that Algorithm 1 actually performs better then 3/4 if the solution x to the assignment LP is produced from a non-well-structured solution y as in Lemma 2. In subsection 4.2 in Lemma 6 we show a new algorithm for well-structured solutions that also has an approximation guarantee strictly better than 3/4. Finally, Lemma 5 and Lemma 6 immediately imply our main result of this section, i.e., Theorem 3.

4.1 Better Analysis for Non-well-Structured Solutions

We first show that Algorithm 1 performs well if not all players basically are fully assigned (fractionally).

Lemma 3. *Let $\varepsilon' > 0$ be a small constant and consider player i such that $\mathsf{Val}_i(x) \leq 1 - \varepsilon'$. Then $\mathsf{Alg}_i(x) \geq \frac{3+\varepsilon'/5}{4}\mathsf{Val}_i(x)$.*

From the above, we can see that the difficult players to round are those that have an almost full budget. In the following lemma, we show that such players must have a very special fractional assignment in order to be difficult to round.

Lemma 4. *Let $\delta > 0$ be a small constant, choose β such that $\delta/4 \leq \beta$, and consider a player i such that $\mathsf{Val}_i(x) \geq 1 - \delta^2/8$ and $\sum_{j:j\in\mathcal{B}} x_{ij} \notin \left[\frac{1-\delta}{2}, \frac{1+\delta}{2}\right]$. Then $\mathsf{Alg}_i(x) \geq \frac{3+\delta^2/64}{4}\mathsf{Val}_i(x)$.*

From Lemma 3 and Lemma 4 we have that as soon as a weighted ϵ-fraction (weight of player i is $\mathsf{Val}_i(y)$) of the players satisfies the conditions of either lemma, we get a better approximation guarantee than $3/4$. Therefore, when a solution y to the configuration LP is not (ϵ, δ)-well-structured, we use Lemma 2 to produce a solution x to the assignment LP for which an ϵ-fraction of players satisfies either conditions of Lemma 3 or Lemma 4. Hence we have the following:

Lemma 5. *Given a solution y to the configuration LP which is not (ϵ, δ)-well-structured and $\beta \geq \delta/4$, we can in polynomial time find a solution with expected value at least $\frac{3+\epsilon\delta^2/64}{4}\mathsf{Val}(y)$.*

4.2 Algorithm for Well-Structured Solutions

Here, we devise a novel algorithm that gives an improved approximation guarantee for (ϵ, δ)-well-structured instances when ϵ and δ are small constants.

Lemma 6. *Let $1 - \beta$ be the threshold for the big items. Given a solution y to the configuration LP that is (ϵ, δ)-well-structured, we can in (randomized) polynomial time find a solution with expected value at least $(1 - \delta)^2(1 - \beta - \epsilon) \cdot \frac{25}{32}\mathsf{Val}(y)$.*

To prove the above lemma we first give the algorithm and then its analysis.

Algorithm. The algorithm constructs a slightly modified version y' of the optimal solution y to the configuration LP. Solution y' is obtained from y in three steps. First, remove all players i with $\sum_{\mathcal{C}\subseteq\mathcal{Q}} |\mathcal{B} \cap \mathcal{C}| y_{i,\mathcal{C}} \notin \left[\frac{(1-\delta)}{2}, \frac{(1+\delta)}{2}\right]$. As y is (ϵ, δ)-well-structured, this step decreases the value of the solution by at most $\epsilon\mathsf{Val}(y)$.

Second, change y as in the proof of Lemma 2 (we refer the reader to the full version of the paper) by getting rid of configurations with 2 big items without losing any objective value. Then remove all small items from the configurations containing big items. After this step, we have the property that big items are alone in a configuration. We call such configurations big and the remaining ones small. Moreover, we have decreased the value by at most $\beta\mathsf{Val}(y)$ because each big item has value at least $1 - \beta$ and each configuration has value at most 1. In the third step, we scale down the fractional assignment of configurations (if necessary), so as to ensure that $\sum_{C:C\cap B=\emptyset} y'_{i,C} \leq 1/2$ for each player $i \in \mathcal{A}$. As remaining players are assigned big configurations with a total fraction at least $(1 - \delta)/2$ and therefore small configurations with a total fraction at most $(1 + \delta)/2$, this may decrease the value by a factor $1/(1 + \delta) > 1 - \delta$.

In summary, we have obtained a solution y' for the configuration LP so that each configuration either contains a single big item or only small items; for each remaining player the configurations with big items constitute a fraction in $\left[\frac{(1-\delta)}{2}, \frac{(1+\delta)}{2}\right]$ and small configurations constitute a fraction of at most $1/2$. Moreover, $\mathsf{Val}(y') \geq (1 - \beta - \epsilon)(1 - \delta)\mathsf{Val}(y)$.

The algorithm now works by rounding y' in two phases; in the first phase we assign big items and in the second phase we assign small items.

The first phase works as follows. Let x' be the solution to the assignment LP from Lemma 2 applied on y' and note that $\mathsf{Val}(x') = \mathsf{Val}(y')$. Consider the bipartite graph where we have a vertex a_i for each player $i \in \mathcal{A}$; a vertex b_j for each big item $j \in \mathcal{B}$; and an edge of weight x'_{ij} between vertices a_i and b_j. Note that a matching in this graph naturally corresponds to an assignment of big items to players. We shall find our matching/assignment of big items by using Theorem 2. Note that by the property of that theorem we have that (i) each big item j is assigned with probability $\sum_i x'_{ij}$ and (ii) the probability that two players i and i' are assigned big items is negatively correlated, i.e., it is at most $\left(\sum_{j\in\mathcal{B}} x'_{ij}\right) \cdot \left(\sum_{j\in\mathcal{B}} x'_{i'j}\right)$. These two properties are crucial in the analysis of our algorithm. It is therefore important that we assign the big items according to a distribution that satisfies the properties of Theorem 2.

After assigning big items, in the second phase our algorithm assigns the small items as follows. First, obtain an optimal solution $x^{(2)}$ to the assignment LP for the small items together with the players that were *not* assigned a big item in the first phase; these are the items that remain and the players for which the budget is not saturated with value at least $1 - \beta$. Then we obtain an integral assignment (of the small items) of value at least $\frac{3}{4}\mathsf{Val}(x^{(2)})$ by using Algorithm 1.

Analysis. Let $\mathcal{A}_j$ be all the players i for which $x'_{ij} > 0$. Let x^* denote the integral assignment found by the algorithm. Note that the expected value of x^* (over the randomly chosen assignment of big items) is:

$$\mathbb{E}[\mathsf{Val}(x^*)] = \mathbb{E}\left[\sum_{j\in\mathcal{B}}\sum_{i\in\mathcal{A}_j} x'_{ij}p_j + \frac{3}{4}\mathsf{Val}(x^{(2)})\right] = \sum_{j\in\mathcal{B}}\sum_{i\in\mathcal{A}_j} x'_{ij}p_j + \frac{3}{4}\mathbb{E}[\mathsf{Val}(x^{(2)})].$$

We now analyze the second term, i.e., the expected optimal value of the assignment LP where we are only considering the small items and the set of players $T \subseteq \mathcal{A}$ that were not assigned big items in the first phase. Then a solution z to the assignment LP can be obtained by scaling up the fractional assignments of the small items assigned to players in T according to x' by up to a factor of 2 while maintaining that an item is assigned at most once. In other words, $z_{ij} = \min\left[1, \sum_{i \in \mathcal{A}_j \cap T} 2x'_{ij}\right]$ and z is a feasible solution to the assignment LP, because we have $\sum_{j \in \mathcal{S}} x'_{ij} p_j \leq 1/2$.

Thus we have that the expected value of the optimal solution to the assignment LP is by linearity of expectation at least $\mathbb{E}_T[\text{Val}(x^{(2)})] \geq \sum_{j \in \mathcal{S}} p_j \cdot \mathbb{E}_T\left[\min\left[1, \sum_{i \in \mathcal{A}_j \cap T} 2x'_{ij}\right]\right]$.

We continue by analyzing the expected fraction of a small item present in the constructed solution to the assignment LP. In this lemma we use that the randomly selected matching of big items has negative correlation. To see why this is necessary, consider a small item $j \in \mathcal{S}$ and suppose that j is assigned to two players A and B such that $x'_{Aj} = x'_{Bj} = 1/2$. As the instance is well-structured both A and B are roughly assigned half a fraction of big items; for simplicity assume it to be exactly $1/2$. Note that in this case $\min\left[1, \sum_{i \in \mathcal{A}_j \cap T} 2x'_{ij}\right]$ is equal to 1 if not both A and B are assigned a big item and 0 otherwise. Therefore, on the one hand, if the event that A is assigned a big item and the event that B is assigned a big item were perfectly correlated then we would have $\mathbb{E}_T\left[\min\left[1, \sum_{i \in \mathcal{A}_j \cap T} 2x'_{ij}\right]\right] = 1/2$. On the other hand, if those events are negatively correlated then $\mathbb{E}_T\left[\min\left[1, \sum_{i \in \mathcal{A}_j \cap T} 2x'_{ij}\right]\right] \geq 3/4$, as in this case the probability that both A and B are assigned big items is at most $1/4$.

Lemma 7. *For every* $j \in \mathcal{S}$, $\mathbb{E}_T\left[\min\left[1, \sum_{i \in \mathcal{A}_j \cap T} 2x'_{ij}\right]\right] \geq (1 - \delta)\frac{3}{4}\sum_{i \in \mathcal{A}_j} x'_{ij}$.

Let us now see how Lemma 7 implies Lemma 6. We have that $\mathbb{E}[\text{Val}(x^*)]$ is equal to

$$\sum_{j \in \mathcal{B}} \sum_{i \in \mathcal{A}_j} x'_{ij} p_j + \frac{3}{4}\mathbb{E}[\text{Val}(x^{(2)})] \geq \sum_{j \in \mathcal{B}} \sum_{i \in \mathcal{A}_j} x'_{ij} p_j + (1 - \delta)\left(\frac{3}{4}\right)^2 \sum_{j \in \mathcal{S}} \sum_{i \in \mathcal{A}_j} x'_{ij} p_j.$$

As $\sum_{j \in \mathcal{B}} x'_{ij} \geq \frac{1 - \delta}{2}$ for every remaining player, we have

$$\frac{\mathbb{E}[\text{Val}(x^*)]}{\text{Val}(x')} \geq (1 - \delta)\left(\frac{1}{2} + \frac{1}{2}\frac{9}{16}\right) = (1 - \delta)\frac{25}{32}.$$

Lemma 6 now follows since $\text{Val}(x') \geq (1 - \beta - \epsilon)(1 - \delta)\text{Val}(y)$. We have proved Lemmas 5 and 6, which in turn imply Theorem 3 and our analysis is concluded.

5   Conclusion and Future Directions

We showed that the integrality gap for configuration LP is strictly better than $\frac{3}{4}$ for two interesting and natural restrictions of Maximum Budgeted Allocation: restricted and graph MBA.

These results imply that the configuration LP is strictly better than the natural assignment LP and pose promising research directions. Specifically, our results on restricted MBA suggest that our limitations in rounding configuration

LP solutions do not necessarily stem from the items being fractionally assigned to many players, while our results on graph MBA suggest that they do not necessarily stem from the items having non-uniform prices. Whether these limitations can simultaneously be overcome is left as an interesting open problem.

Finally, it would be interesting to see whether the techniques presented, and especially the exploitation of the big items structure, can be applied to other allocation problems with similar structural features as MBA (e.g. GAP).

References

1. Andelman, N., Mansour, Y.: Auctions with budget constraints. In: Hagerup, T., Katajainen, J. (eds.) SWAT 2004. LNCS, vol. 3111, pp. 26–38. Springer, Heidelberg (2004)
2. Asadpour, A., Feige, U., Saberi, A.: Santa claus meets hypergraph matchings. In: Goel, A., Jansen, K., Rolim, J.D.P., Rubinfeld, R. (eds.) APPROX and RANDOM 2008. LNCS, vol. 5171, pp. 10–20. Springer, Heidelberg (2008)
3. Azar, Y., Birnbaum, B.E., Karlin, A.R., Mathieu, C., Nguyen, C.T.: Improved approximation algorithms for budgeted allocations. In: Aceto, L., Damgård, I., Goldberg, L.A., Halldórsson, M.M., Ingólfsdóttir, A., Walukiewicz, I. (eds.) ICALP 2008, Part I. LNCS, vol. 5125, pp. 186–197. Springer, Heidelberg (2008)
4. Bansal, N., Sviridenko, M.: The santa claus problem. In: Kleinberg, J.M. (ed.) STOC, pp. 31–40. ACM Press (2006)
5. Chakrabarty, D., Goel, G.: On the approximability of budgeted allocations and improved lower bounds for submodular welfare maximization and gap. SIAM J. Comput. 39(6), 2189–2211 (2010)
6. Ebenlendr, T., Krčál, M., Sgall, J.: Graph balancing: A special case of scheduling unrelated parallel machines. In: SODA, pp. 483–490. Society for Industrial and Applied Mathematics (2008)
7. Feige, U.: On allocations that maximize fairness. In: SODA, pp. 287–293. SIAM (2008)
8. Feige, U., Vondrák, J.: Approximation algorithms for allocation problems: Improving the factor of 1 - 1/e. In: FOCS, pp. 667–676 (2006)
9. Gandhi, R., Khuller, S., Parthasarathy, S., Srinivasan, A.: Dependent rounding and its applications to approximation algorithms. J. ACM 53(3), 324–360 (2006)
10. Garg, R., Kumar, V., Pandit, V.: Approximation algorithms for budget-constrained auctions. In: Goemans, M.X., Jansen, K., Rolim, J.D.P., Trevisan, L. (eds.) RANDOM 2001 and APPROX 2001. LNCS, vol. 2129, pp. 102–113. Springer, Heidelberg (2001)
11. Grötschel, M., Lovász, L., Schrijver, A.: Geometric Algorithms and Combinatorial Optimization. Algorithms and Combinatorics, vol. 2. Springer (1993)
12. Lehmann, B., Lehmann, D.J., Nisan, N.: Combinatorial auctions with decreasing marginal utilities. Games and Economic Behavior 55(2), 270–296 (2006)
13. Lenstra, J.K., Shmoys, D.B., Tardos, É.: Approximation algorithms for scheduling unrelated parallel machines. Math. Program. 46, 259–271 (1990)
14. Shmoys, D.B., Tardos, É.: An approximation algorithm for the generalized assignment problem. Math. Program. 62, 461–474 (1993)
15. Srinivasan, A.: Budgeted allocations in the full-information setting. In: Goel, A., Jansen, K., Rolim, J.D.P., Rubinfeld, R. (eds.) APPROX and RANDOM 2008. LNCS, vol. 5171, pp. 247–253. Springer, Heidelberg (2008)
16. Svensson, O.: Santa claus schedules jobs on unrelated machines. SIAM J. Comput. 41(5), 1318–1341 (2012)
17. Verschae, J., Wiese, A.: On the configuration-LP for scheduling on unrelated machines. In: Demetrescu, C., Halldórsson, M.M. (eds.) ESA 2011. LNCS, vol. 6942, pp. 530–542. Springer, Heidelberg (2011)

Two-Term Disjunctions on the Second-Order Cone

Fatma Kılınç-Karzan and Sercan Yıldız

Tepper School of Business
Carnegie Mellon University, Pittsburgh, PA
{fkilinc,syildiz}@andrew.cmu.edu

Abstract. Balas introduced disjunctive cuts in the 1970s for mixed-integer linear programs. Several recent papers have attempted to extend this work to mixed-integer conic programs. In this paper we develop a methodology to derive closed-form expressions for inequalities describing the convex hull of a two-term disjunction applied to the second-order cone. Our approach is based on first characterizing the structure of undominated valid linear inequalities for the disjunction and then using conic duality to derive a family of convex, possibly nonlinear, valid inequalities that correspond to these linear inequalities. We identify and study the cases where these valid inequalities can equivalently be expressed in conic quadratic form and where a single inequality from this family is sufficient to describe the convex hull. Our results on two-term disjunctions on the second-order cone generalize related results on split cuts by Modaresi, Kılınç, and Vielma, and by Andersen and Jensen.

Keywords: Mixed-integer conic programming, second-order cone programming, cutting planes, disjunctive cuts.

1 Introduction

A mixed-integer conic program is a problem of the form

$$\sup\{d^\top x : Ax = b, \ x \in \mathbb{K}, \ x_j \in \mathbb{Z} \ \forall j \in J\}$$

where $\mathbb{K} \subset \mathbb{R}^n$ is a regular (full-dimensional, closed, convex, and pointed) cone, A is an $m \times n$ real matrix of full row rank, d and b are real vectors of appropriate dimensions, and $J \subseteq \{1, \ldots, n\}$. Mixed-integer conic programming (MICP) models arise naturally as robust versions of mixed-integer linear programming (MILP) models in finance, management, and engineering [1]. MILP is the special case of MICP where $\mathbb{K}$ is the nonnegative orthant, and it has itself numerous applications. A successful approach to solving MILP problems has been to first solve the continuous relaxation, then add cuts, and finally perform branch-and-bound using this strengthened formulation. A powerful way of generating such cuts is to impose a valid disjunction on the continuous relaxation and to generate tight convex inequalities for the resulting disjunctive set. Such inequalities are known as *disjunctive cuts*. Specifically, the integrality

J. Lee and J. Vygen (Eds.): IPCO 2014, LNCS 8494, pp. 345–356, 2014.

conditions on the variables x_j, $j \in J$, imply linear *two-term disjunctions* of the form $\pi^\top x \leq \pi_0 \vee \pi^\top x \geq \pi_0 + 1$ where $\pi_0 \in \mathbb{Z}$, $\pi_j \in \mathbb{Z}$, $j \in J$, and $\pi_j = 0$, $j \notin J$. Following this approach, the feasible region for MICP problems can be relaxed to $\{x \in \mathbb{K} : Ax = b, \pi^\top x \leq \pi_0 \vee \pi^\top x \geq \pi_0 + 1\}$. More general two-term disjunctions arise in complementarity [2] and other non-convex optimization [3] problems. Therefore, it is interesting to study relaxations of MICP problems of the form

$$\sup\{d^\top x : x \in C_1 \cup C_2\} \quad \text{where}$$
$$C_i := \{x \in \mathbb{K} : Ax = b, c_i^\top x \geq c_{i,0}\} \quad \text{for} \quad i \in \{1, 2\} \tag{1}$$

and to derive strong valid inequalities for the convex hull $\mathrm{conv}(C_1 \cup C_2)$, or the closed convex hull $\overline{\mathrm{conv}}(C_1 \cup C_2)$. When $\mathbb{K}$ is the nonnegative orthant, Bonami et al. [4] characterize $\overline{\mathrm{conv}}(C_1 \cup C_2)$ by a finite set of linear inequalities. The purpose of this paper is to provide closed-form expressions for convex inequalities describing $\overline{\mathrm{conv}}(C_1 \cup C_2)$ for other cones such as the second-order (Lorentz) cone $\mathbb{K}_2^n := \{x \in \mathbb{R}^n : \|\tilde{x}\|_2 \leq x_n\}$ where $\tilde{x} := (x_1, \ldots, x_{n-1})$. We first review related results from the literature.

Disjunctive cuts were introduced by Balas [5] for MILP in the early 1970s. *Chvátal-Gomory, lift-and-project, mixed-integer rounding (MIR)*, and *split cuts* are all special types of disjunctive cuts. Recent efforts on extending the cutting plane theory for MILP to the MICP setting include the work of Çezik and Iyengar [6] for Chvatal-Gomory cuts, Stubbs and Mehrotra [7], Drewes [8], and Bonami [9] for lift-and-project cuts, and Atamtürk and Narayanan [10] for MIR cuts. Kılınç-Karzan [11] analyzed properties of minimal valid linear inequalities for general conic sets with a disjunctive structure and showed that these are sufficient to describe the closed convex hull. Such general sets from [11] include two-term disjunctions on the cone $\mathbb{K}$ considered in this paper. Bienstock and Michalka [12] studied the characterization and separation of valid linear inequalities that convexify the epigraph of a convex, differentiable function restricted to a non-convex domain. In the last few years, there has been growing interest in developing closed-form expressions for convex inequalities that fully describe the convex hull of a disjunctive conic set. Dadush et al. [13] and Andersen and Jensen [14] derived split cuts for ellipsoids and the second-order cone, respectively. Modaresi et al. [15] extended this work to essentially all cross-sections of the second-order cone. Belotti et al. [16] identified a procedure for constructing two-term disjunctive cuts under the assumptions that $C_1 \cap C_2 = \emptyset$ and the sets $\{x \in \mathbb{K} : Ax = b, c_1^\top x = c_{1,0}\}$ and $\{x \in \mathbb{K} : Ax = b, c_2^\top x = c_{2,0}\}$ are bounded.

In this paper we study general two-term disjunctions on conic sets and give closed-form expressions for the tightest disjunctive cuts that can be obtained from these disjunctions in a large class of instances. We focus on the case where C_1 and C_2 in (1) above have an empty set of equations $Ax = b$. That is to say, we consider

$$C_1 := \{x \in \mathbb{K} : c_1^\top x \geq c_{1,0}\} \quad \text{and} \quad C_2 := \{x \in \mathbb{K} : c_2^\top x \geq c_{2,0}\}. \tag{2}$$

We note, however, that our results can easily be extended to two-term disjunctions on sets of the form $\{x \in \mathbb{R}^n : Ax - b \in \mathbb{K}\}$ through the affine transformation discussed in [14]. Our main contribution is to give an explicit outer description of $\overline{\mathrm{conv}}(C_1 \cup C_2)$ when $\mathbb{K}$ is the second-order cone. Similar results have previously appeared in [14], [15], and [16]. Nevertheless, our work is set apart from [14] and [15] by the fact that we study two-term disjunctions on the cone $\mathbb{K}$ in their full generality and do not restrict our attention to split disjunctions, which are defined by *parallel* hyperplanes. Furthermore, unlike [16], we do not assume that $C_1 \cap C_2 = \emptyset$ and the sets $\{x \in \mathbb{K} : c_1^\top x = c_{1,0}\}$ and $\{x \in \mathbb{K} : c_2^\top x = c_{2,0}\}$ are bounded. In the absence of such assumptions, the resulting convex hulls turn out to be significantly more complex in our general setting. We also stress that our proof techniques originate from a conic duality perspective and are completely different from what is employed in the aforementioned papers; in particular, we believe that they are more transparent and intuitive.

The remainder of this paper is organized as follows. Section 2.1 introduces our notation and basic assumptions. In Section 2.2 we characterize the structure of undominated valid linear inequalities describing $\overline{\mathrm{conv}}(C_1 \cup C_2)$ when $\mathbb{K}$ is a regular cone. In Section 3 we focus on the case where $\mathbb{K}$ is the second-order cone. In Section 3.1 we state and prove our main result, Theorem 1. The proof uses conic duality along with the characterization from Section 2.2 to derive a family of convex, possibly nonlinear, valid inequalities (6) for $\overline{\mathrm{conv}}(C_1 \cup C_2)$. In Sections 3.2 and 3.3, we identify and study the cases where these inequalities can equivalently be expressed in conic quadratic form and where only one inequality of the form (6) is sufficient to describe $\overline{\mathrm{conv}}(C_1 \cup C_2)$. Our results imply in particular that a single conic valid inequality is always sufficient for split disjunctions. Nevertheless, there are cases where it is not possible to obtain $\overline{\mathrm{conv}}(C_1 \cup C_2)$ by a single inequality of the form (6). We illustrate these cases with examples in Section 3.4.

2   Preliminaries

The main purpose of this section is to characterize the structure of undominated valid linear inequalities for $\overline{\mathrm{conv}}(C_1 \cup C_2)$ when $\mathbb{K}$ is a regular cone and C_1 and C_2 are defined as in (2). First, we present our notation and assumptions.

2.1   Notation and Assumptions

Given a set $S \subseteq \mathbb{R}^n$, we let $\mathrm{span}\, S$, $\mathrm{int}\, S$, and $\mathrm{bd}\, S$ denote the linear span, interior, and boundary of S, respectively. We use $\mathrm{rec}\, S$ to refer to the recession cone of a convex set S. The *dual cone* of a cone $\mathbb{K} \subseteq \mathbb{R}^n$ is $\mathbb{K}^* := \{y \in \mathbb{R}^n : y^\top x \geq 0 \, \forall x \in \mathbb{K}\}$.

We can always scale the inequalities $c_1^\top x \geq c_{1,0}$ and $c_2^\top x \geq c_{2,0}$ defining the disjunction so that their right-hand sides are 0 or ± 1. Therefore, from now on we assume that $c_{1,0}, c_{2,0} \in \{0, \pm 1\}$.

When $C_1 \subseteq C_2$, we have $\overline{\mathrm{conv}}(C_1 \cup C_2) = C_2$. Similarly, when $C_1 \supseteq C_2$, we have $\overline{\mathrm{conv}}(C_1 \cup C_2) = C_1$. In the remainder we focus on the case where $C_1 \not\subseteq C_2$ and $C_1 \not\supseteq C_2$.

Assumption 1. $C_1 \not\subseteq C_2$ and $C_1 \not\supseteq C_2$.

We also need the following technical assumption in our analysis.

Assumption 2. C_1 and C_2 are strictly feasible sets.

It is not difficult to show that $C := \{x \in \mathbb{K} : c^\top x \geq c_0\}$ must be strictly feasible when $c_0 = -1$ or when $c_0 = +1$ and C is nonempty. Therefore, we need Assumption 2 to supplement Assumption 1 only when $c_{1,0} = 0$ or $c_{2,0} = 0$. Assumptions 1 and 2 have the following implication.

Lemma 1. *Suppose Assumptions 1 and 2 hold. Then the following system of inequalities in the variable β is inconsistent:*

$$\beta \geq 0, \quad \beta c_{1,0} \geq c_{2,0}, \quad c_2 - \beta c_1 \in \mathbb{K}^*.$$

Similarly, the following system of inequalities in the variable β is inconsistent:

$$\beta \geq 0, \quad \beta c_{2,0} \geq c_{1,0}, \quad c_1 - \beta c_2 \in \mathbb{K}^*.$$

2.2   Properties of Undominated Valid Linear Inequalities

A valid linear inequality $\mu^\top x \geq \mu_0$ for a strictly feasible set $S \subseteq \mathbb{K}$ is said to *dominate* another valid linear inequality $\nu^\top x \geq \nu_0$ if it is not a positive multiple of $\nu^\top x \geq \nu_0$ and implies $\nu^\top x \geq \nu_0$ together with the cone constraint $x \in \mathbb{K}$. A valid linear inequality $\mu^\top x \geq \mu_0$ for $S \subseteq \mathbb{K}$ is *tight* if $\inf_x\{\mu^\top x : x \in S\} = \mu_0$ and *strongly tight* if there exists $x^* \in S$ such that $\mu^\top x^* = \mu_0$.

Because C_1 and C_2 are strictly feasible sets by Assumption 2, conic duality implies that a linear inequality $\mu^\top x \geq \mu_0$ is valid for $\overline{\mathrm{conv}}(C_1 \cup C_2)$ if and only if there exist $\alpha_1, \alpha_2, \beta_1, \beta_2$ such that $(\mu, \mu_0, \alpha_1, \alpha_2, \beta_1, \beta_2)$ satisfies

$$
\begin{aligned}
\mu &= \alpha_1 + \beta_1 c_1, \\
\mu &= \alpha_2 + \beta_2 c_2, \\
\beta_1 c_{1,0} &\geq \mu_0, \quad \beta_2 c_{2,0} \geq \mu_0, \\
\alpha_1, \alpha_2 &\in \mathbb{K}^*, \quad \beta_1, \beta_2 \in \mathbb{R}_+.
\end{aligned}
\tag{3}
$$

This system can be reduced slightly for *undominated* valid linear inequalities.

Proposition 1. *Consider C_1, C_2 defined as in (2) with $c_{1,0}, c_{2,0} \in \{0, \pm 1\}$. Suppose Assumptions 1 and 2 hold. Then, up to positive scaling, any undominated valid linear inequality for $\overline{\mathrm{conv}}(C_1 \cup C_2)$ has the form $\mu^\top x \geq \min\{c_{1,0}, c_{2,0}\}$ with $(\mu, \alpha_1, \alpha_2, \beta_1, \beta_2)$ satisfying*

$$
\begin{aligned}
\mu &= \alpha_1 + \beta_1 c_1, \\
\mu &= \alpha_2 + \beta_2 c_2, \\
\min\{\beta_1 c_{1,0}, \beta_2 c_{2,0}\} &= \min\{c_{1,0}, c_{2,0}\}, \\
\alpha_1, \alpha_2 &\in \mathrm{bd}\,\mathbb{K}^*, \quad \beta_1, \beta_2 \in \mathbb{R}_+ \setminus \{0\}.
\end{aligned}
\tag{4}
$$

Remark 1. Under the assumptions of Proposition 1, in an undominated valid linear inequality $\mu^\top x \geq \min\{c_{1,0}, c_{2,0}\}$, we can assume that at least one of β_1 and β_2 is equal to 1 in (4) without any loss of generality. In particular,

(i) if $c_{1,0} > c_{2,0}$, we can assume that $\beta_2 = 1$, $\beta_1 c_{1,0} \geq c_{2,0}$, and $\beta_1 c_1 - c_2 \notin \pm \operatorname{int} \mathbb{K}_2^n$ holds,

(ii) if $c_{1,0} = c_{2,0}$, we can assume that either $\beta_2 = 1$, $\beta_1 c_{1,0} \geq c_{2,0}$, and $\beta_1 c_1 - c_2 \notin \pm \operatorname{int} \mathbb{K}_2^n$ or $\beta_1 = 1$, $\beta_2 c_{2,0} \geq c_{1,0}$, and $\beta_2 c_2 - c_1 \notin \pm \operatorname{int} \mathbb{K}_2^n$ holds.

3   Deriving the Disjunctive Cut

In this section we let $\mathbb{K}$ be the second-order cone $\mathbb{K}_2^n := \{x \in \mathbb{R}^n : \|\tilde{x}\|_2 \leq x_n\}$ where $\tilde{x} := (x_1, \ldots, x_{n-1})$. As in the previous section, we consider C_1, C_2 defined as in (2) with $c_{1,0}, c_{2,0} \in \{0, \pm 1\}$ and suppose that Assumptions 1 and 2 hold. We also assume without any loss of generality that $c_{1,0} \geq c_{2,0}$. Sets C_1, C_2 that satisfy these conditions are said to satisfy the *basic disjunctive setup*.

3.1   A Convex Valid Inequality

Proposition 1 gives a nice characterization of the form of undominated linear inequalities valid for $\overline{\operatorname{conv}}(C_1 \cup C_2)$. In the following we use this characterization and show that, for a given pair (β_1, β_2) satisfying the conditions of Remark 1, one can group all of the corresponding linear inequalities into a single convex, possibly nonlinear, inequality valid for $\overline{\operatorname{conv}}(C_1 \cup C_2)$. By Remark 1, without any loss of generality, we focus on the case where $\beta_2 = 1$ and $\beta_1 > 0$ with $\beta_1 c_{1,0} \geq c_{2,0}$ and $\beta_1 c_1 - c_2 \notin \pm \operatorname{int} \mathbb{K}_2^n$. By Lemma 1, $\beta_1 c_1 - c_2 \notin -\mathbb{K}_2^n$. This leaves us two distinct cases to consider: $\beta_1 c_1 - c_2 \in \operatorname{bd} \mathbb{K}_2^n$ and $\beta_1 c_1 - c_2 \notin \pm \mathbb{K}_2^n$.

Remark 2. Let C_1, C_2 satisfy the basic disjunctive setup. For any $\beta > 0$ such that $\beta c_{1,0} \geq c_{2,0}$ and $\beta c_1 - c_2 \in \operatorname{bd} \mathbb{K}_2^n$, the inequality

$$\beta c_1^\top x \geq c_{2,0} \tag{5}$$

is valid for $\overline{\operatorname{conv}}(C_1 \cup C_2)$ and dominates all valid linear inequalities of the form (4) with $\beta_1 = \beta$ and $\beta_2 = 1$.

Theorem 1. *Let C_1, C_2 satisfy the basic disjunctive setup. For any $\beta > 0$ such that $\beta c_{1,0} \geq c_{2,0}$ and $\beta c_1 - c_2 \notin \pm \mathbb{K}_2^n$, the inequality*

$$2 c_{2,0} - (\beta c_1 + c_2)^\top x \leq \sqrt{\left((\beta c_1 - c_2)^\top x\right)^2 + N(\beta)\left(x_n^2 - \|\tilde{x}\|^2\right)} \tag{6}$$

with

$$N(\beta) := \|\beta \tilde{c}_1 - \tilde{c}_2\|_2^2 - (\beta c_{1,n} - c_{2,n})^2 \tag{7}$$

is valid for $\overline{\operatorname{conv}}(C_1 \cup C_2)$ and implies all valid linear inequalities of the form (4) with $\beta_1 = \beta$ and $\beta_2 = 1$.

Proof. Consider the set of vectors $\mu \in \mathbb{R}^n$ satisfying (4) with $\beta_1 = \beta$ and $\beta_2 = 1$:

$$\mathcal{M}(\beta, 1) := \{\mu \in \mathbb{R}^n : \exists \alpha_1, \alpha_2 \in \mathrm{bd}\, \mathbb{K}_2^n \text{ s.t. } \mu = \alpha_1 + \beta c_1 = \alpha_2 + c_2\}.$$

Because $\beta c_1 - c_2 \notin \pm\mathbb{K}_2^n$, Moreau's decomposition theorem implies that there exist $\mu^*, \alpha_1^* \neq 0, \alpha_2^* \neq 0$ such that $\alpha_1^* \perp \alpha_2^*$ and $(\mu^*, \alpha_1^*, \alpha_2^*, \beta, 1)$ satisfies (4). Hence, the set $\mathcal{M}(\beta, 1)$ is in fact nonempty. We can write

$$\begin{aligned}
\mathcal{M}(\beta, 1) &= \{\mu \in \mathbb{R}^n : \|\tilde{\mu} - \beta\tilde{c}_1\|_2 = \mu_n - \beta c_{1,n}, \ \|\tilde{\mu} - \tilde{c}_2\|_2 = \mu_n - c_{2,n}\} \\
&= \{\mu \in \mathbb{R}^n : \|\tilde{\mu} - \tilde{c}_2\|_2 = \|\tilde{\mu} - \beta\tilde{c}_1\|_2 + \beta c_{1,n} - c_{2,n}, \ \|\tilde{\mu} - \beta\tilde{c}_1\|_2 = \mu_n - \beta c_{1,n}\}.
\end{aligned}$$

After taking the squares of both sides of the first equation in $\mathcal{M}(\beta, 1)$, noting $\beta c_1 - c_2 \notin -\mathbb{K}_2^n$, and replacing the term $\|\tilde{\mu} - \beta\tilde{c}_1\|_2$ with $\mu_n - \beta c_{1,n}$, we arrive at

$$\mathcal{M}(\beta, 1) = \left\{\mu \in \mathbb{R}^n : \tilde{\mu}^\top(\beta\tilde{c}_1 - \tilde{c}_2) - \mu_n(\beta c_{1,n} - c_{2,n}) = \frac{M}{2}, \ \|\tilde{\mu} - \beta\tilde{c}_1\|_2 = \mu_n - \beta c_{1,n}\right\}$$

where $M := \beta^2(\|\tilde{c}_1\|_2^2 - c_{1,n}^2) - (\|\tilde{c}_2\|_2^2 - c_{2,n}^2)$.

Note that $x \in \overline{\mathrm{conv}}(C_1 \cup C_2)$ implies

$$\Rightarrow x \in \mathbb{K}_2^n \text{ and } \mu^\top x \geq c_{2,0} \ \forall \mu \in \mathcal{M}(\beta, 1).$$
$$\Leftrightarrow x \in \mathbb{K}_2^n \text{ and } \inf_\mu\{\mu^\top x : \mu \in \mathcal{M}(\beta, 1)\} \geq c_{2,0}.$$

Unfortunately, the optimization problem stated above is non-convex due to the second equality constraint in $\mathcal{M}(\beta, 1)$. We show below that the natural convex relaxation for this problem is tight. Indeed, consider the relaxation

$$\inf_\mu \left\{\mu^\top x : \tilde{\mu}^\top(\beta\tilde{c}_1 - \tilde{c}_2) - \mu_n(\beta c_{1,n} - c_{2,n}) = \frac{M}{2}, \ \|\tilde{\mu} - \beta\tilde{c}_1\|_2 \leq \mu_n - \beta c_{1,n}\right\}.$$

The feasible region of this relaxation is the intersection of a hyperplane with a closed, convex cone shifted by the vector βc_1. Any solution which is feasible to the relaxation but not the original problem can be expressed as the limit of a sequence of points obtained by taking convex combinations of solutions feasible to the original problem. Because we are optimizing a linear function, this shows that the relaxation is equivalent to the original problem. Thus, we have

$$x \in \overline{\mathrm{conv}}(C_1 \cup C_2) \Rightarrow x \in \mathbb{K}_2^n \text{ and}$$
$$\inf_\mu\left\{\mu^\top x : \tilde{\mu}^\top(\beta\tilde{c}_1 - \tilde{c}_2) - \mu_n(\beta c_{1,n} - c_{2,n}) = \frac{M}{2}, \ \|\tilde{\mu} - \beta\tilde{c}_1\|_2 \leq \mu_n - \beta c_{1,n}\right\} \geq c_{2,0}$$

which is exactly the same as

$$x \in \overline{\mathrm{conv}}(C_1 \cup C_2) \Rightarrow x \in \mathbb{K}_2^n \text{ and}$$
$$\inf_\mu\left\{\mu^\top x : \tilde{\mu}^\top(\beta\tilde{c}_1 - \tilde{c}_2) - \mu_n(\beta c_{1,n} - c_{2,n}) = \frac{M}{2}, \ \mu - \beta c_1 \in \mathbb{K}_2^n\right\} \geq c_{2,0}. \quad (8)$$

The minimization problem in the last line above is feasible since μ^*, defined at the beginning of the proof, is a feasible solution. Indeed, it is strictly feasible since $\alpha_1^* + \alpha_2^*$ is a recession direction of the feasible region and belongs to int $\mathbb{K}_2^n$. Hence, its dual problem is solvable whenever it is feasible, strong duality applies, and we can replace the problem in the last line with its dual without any loss of generality.

Considering the definition of $N(\beta) = \|\beta \tilde{c}_1 - \tilde{c}_2\|_2^2 - (\beta c_{1,n} - c_{2,n})^2$ and the assumption that $\beta c_1 - c_2 \notin \pm \mathbb{K}_2^n$, we get $N(\beta) > 0$. Then

$x \in \overline{\mathrm{conv}}(C_1 \cup C_2)$

$\Rightarrow x \in \mathbb{K}_2^n$ and $\max\limits_{\rho,\tau} \left\{ \beta c_1^\top \rho + \dfrac{M}{2}\tau : \rho + \tau \begin{pmatrix} \beta \tilde{c}_1 - \tilde{c}_2 \\ -\beta c_{1,n} + c_{2,n} \end{pmatrix} = x, \ \rho \in \mathbb{K}_2^n \right\} \geq c_{2,0}.$

$\Leftrightarrow x \in \mathbb{K}_2^n$ and $\max\limits_{\tau} \left\{ \beta c_1^\top x - \dfrac{N(\beta)}{2}\tau : x + \tau \begin{pmatrix} -\beta \tilde{c}_1 + \tilde{c}_2 \\ \beta c_{1,n} - c_{2,n} \end{pmatrix} \in \mathbb{K}_2^n \right\} \geq c_{2,0}.$

$\Leftrightarrow x \in \mathbb{K}_2^n$ and $\min\{\tau_-, \tau_+\} \leq \dfrac{2(\beta c_1^\top x - c_{2,0})}{N(\beta)}$

where $\tau_\pm := \dfrac{(\beta c_1 - c_2)^\top x \pm \sqrt{((\beta c_1 - c_2)^\top x)^2 + N(\beta)(x_n^2 - \|\tilde{x}\|_2^2)}}{N(\beta)}.$

$\Leftrightarrow x \in \mathbb{K}_2^n$ and $\tau_- \leq \dfrac{2(\beta c_1^\top x - c_{2,0})}{N(\beta)}.$

$\Leftrightarrow x \in \mathbb{K}_2^n$ and $N(\beta)\tau_- \leq 2(\beta c_1^\top x - c_{2,0}).$

Rearranging the terms of the inequality in the last expression above yields (6). $\qquad \square$

The next observation follows directly from the proof of Theorem 1.

Remark 3. Under the assumptions of Proposition 1, the set of points that satisfy (6) in $\mathbb{K}_2^n$ is convex.

Proof. The inequality (6) is equivalent to (8) by construction. The left-hand side of (8) is a concave function of x written as the pointwise infimum of linear functions, while the right-hand side is a constant. $\qquad \square$

By Proposition 1 and Remark 1, the family of inequalities given in Remark 2 and Theorem 1 is sufficient to describe $\overline{\mathrm{conv}}(C_1 \cup C_2)$.

3.2 A Conic Quadratic Form

While having a convex valid inequality is nice in general, there are certain cases where (6) can be expressed in conic quadratic form.

Proposition 2. *Let C_1, C_2 satisfy the basic disjunctive setup. Let $\beta > 0$ be such that $\beta c_{1,0} \geq c_{2,0}$ and $\beta c_1 - c_2 \notin \pm \mathbb{K}_2^n$, and suppose for $N(\beta)$ given by (7) that*

$$-2c_{2,0} + (\beta c_1 + c_2)^\top x \leq \sqrt{((\beta c_1 - c_2)^\top x)^2 + N(\beta)\left(x_n^2 - \|\tilde{x}\|^2\right)} \qquad (9)$$

holds for all $x \in \overline{\mathrm{conv}}(C_1 \cup C_2)$. *Then* (6) *can equivalently be written in conic quadratic form as*

$$N(\beta)x + 2(c_2^\top x - c_{2,0}) \begin{pmatrix} \beta \tilde{c}_1 - \tilde{c}_2 \\ -\beta c_{1,n} + c_{2,n} \end{pmatrix} \in \mathbb{K}_2^n. \tag{10}$$

Proposition 3. *Let* C_1, C_2 *satisfy the basic disjunctive setup. Let* $\beta > 0$ *be such that* $\beta c_{1,0} \geq c_{2,0}$ *and* $\beta c_1 - c_2 \notin \pm \mathbb{K}_2^n$ *and suppose*

$$\{x \in \mathbb{K}_2^n : \beta c_1^\top x \geq c_{2,0}, c_2^\top x \geq c_{2,0}\} = \{x \in \mathbb{K}_2^n : \beta c_1^\top x = c_{2,0}, c_2^\top x = c_{2,0}\}. \tag{11}$$

Then (9) *is satisfied by all* $x \in \mathbb{K}_2^n$.

Condition (11) of Proposition 3, together with the results of Proposition 2 and Theorem 1, identifies cases in which (6) can be expressed in conic quadratic form. In a split disjunction on the cone $\mathbb{K}_2^n$, it is easy to see that C_1 and C_2 are both nonempty and $\mathrm{conv}(C_1 \cup C_2) \neq \mathbb{K}_2^n$ if and only if $c_1, c_2 \notin \pm \mathbb{K}_2^n$ and $c_{1,0} = c_{2,0} = 1$. Therefore, for a proper two-sided split disjunction, (11) is trivially satisfied with $\beta = 1$ because $C_1 \cap C_2 = \emptyset$.

3.3 When Does a Single Inequality Suffice?

In this section we give two conditions under which a single convex inequality of the type derived in Theorem 1 describes $\overline{\mathrm{conv}}(C_1 \cup C_2)$ completely, together with the cone constraint $x \in \mathbb{K}_2^n$. The following results hold for any regular cone $\mathbb{K}$.

Further Properties of Undominated Valid Linear Inequalities. In this section we consider the disjunction $c_1^\top x \geq c_{1,0} \vee c_2^\top x \geq c_{2,0}$ on a regular cone $\mathbb{K}$ and refine the results of Section 2.2 on the structure of undominated valid linear inequalities.

The following lemma shows that the statement of Proposition 1 can be strengthened substantially when $c_1 \in \mathbb{K}^*$ or $c_2 \in \mathbb{K}^*$.

Lemma 2. *Let* C_1, C_2 *satisfy the basic disjunctive setup. Suppose* $c_1 \in \mathbb{K}^*$ *or* $c_2 \in \mathbb{K}^*$. *Then, up to positive scaling, any undominated valid linear inequality for* $\overline{\mathrm{conv}}(C_1 \cup C_2)$ *has the form* $\mu^\top x \geq c_{2,0}$ *where* μ *satisfies* (4) *with* $\beta_1 = \beta_2 = 1$.

When $c_{1,0} = c_{2,0} \in \{\pm 1\}$, a similar result holds for undominated valid linear inequalities that are tight on both C_1 and C_2.

Lemma 3. *Let* C_1, C_2 *satisfy the basic disjunctive setup with* $c_{1,0} = c_{2,0} \in \{\pm 1\}$. *Then, up to positive scaling, any undominated valid linear inequality for* $\overline{\mathrm{conv}}(C_1 \cup C_2)$ *that is tight on both* C_1 *and* C_2 *has the form* $\mu^\top x \geq c_{2,0}$ *where* μ *satisfies* (4) *with* $\beta_1 = \beta_2 = 1$.

Next, we identify an important case where the family of tight inequalities specified in Lemma 3 is rich enough to describe $\overline{\mathrm{conv}}(C_1 \cup C_2)$ completely. The key ingredient is the closedness of $\mathrm{conv}(C_1 \cup C_2)$.

Proposition 4. *Let C_1, C_2 satisfy the basic disjunctive setup with $c_{1,0} = c_{2,0}$. Suppose* $\mathrm{conv}(C_1 \cup C_2)$ *is closed. Then undominated valid linear inequalities that are strongly tight on both C_1 and C_2 are sufficient to describe* $\mathrm{conv}(C_1 \cup C_2)$, *together with the cone constraint $x \in \mathbb{K}$.*

The next result shows the necessity of the assumption $c_{1,0} = c_{2,0}$ in the statement of Proposition 4. When this is not the case, every undominated valid linear inequality is tight on exactly one of the two sets C_1 and C_2.

Lemma 4. *Let C_1, C_2 satisfy the basic disjunctive setup with $c_{1,0} > c_{2,0}$. Then every undominated valid linear inequality for* $\overline{\mathrm{conv}}(C_1 \cup C_2)$ *is tight on the set C_2 but not on C_1.*

Two Sufficient Conditions. By putting together the results of Section 3.3 and Theorem 1, we obtain the following.

Theorem 2. *Let C_1, C_2 satisfy the basic disjunctive setup with $c_1 - c_2 \notin \pm \mathbb{K}_2^n$. Then the inequality*

$$2c_{2,0} - (c_1 + c_2)^\top x \leq \sqrt{((c_1 - c_2)^\top x)^2 + N\left(x_n^2 - \|\tilde{x}\|^2\right)} \tag{12}$$

is valid for $\overline{\mathrm{conv}}(C_1 \cup C_2)$ *with $N := \|\tilde{c}_1 - \tilde{c}_2\|_2^2 - (c_{1,n} - c_{2,n})^2$. Furthermore,*

$$\overline{\mathrm{conv}}(C_1 \cup C_2) = \{x \in \mathbb{K}_2^n : x \text{ satisfies (12)}\}$$

when, in addition,

(i) c_1 or $c_2 \in \mathbb{K}_2^n$, or
(ii) $c_{1,0} = c_{2,0} \in \{\pm 1\}$ and undominated valid linear inequalities that are tight on both C_1 and C_2 are sufficient to describe $\overline{\mathrm{conv}}(C_1 \cup C_2)$.

Proof. The validity of (12) follows from Theorem 1 by setting $\beta = 1$. Lemmas 2 and 3 show that we can limit ourselves to valid linear inequalities of the form (4) with $\beta_1 = \beta_2 = 1$ to get a complete description of the closed convex hull. When this is the case, the implication in (8) in the proof of Theorem 1 is actually an equivalence. $\qquad \square$

When $c_{1,0} = c_{2,0} \in \{0, \pm 1\}$, Lemma 1 implies $c_1 - c_2 \notin \mathbb{K}_2^n$. Suppose also that

(a) condition (i) or (ii) of Theorem 2 is satisfied, and
(b) $C_1 \cap C_2 \subseteq \{x \in \mathbb{R}^n : c_1^\top x = c_{1,0}, \ c_2^\top x = c_{2,0}\}$.

Statement (a) holds, for instance, in the case of split disjunctions because $c_{1,0} = c_{2,0} = 1$ and $\mathrm{conv}(C_1 \cup C_2)$ is closed (see, e.g., [13, Lemma 2.3]). By Proposition 4, undominated valid linear inequalities that are tight on both C_1 and C_2 are sufficient to describe $\overline{\mathrm{conv}}(C_1 \cup C_2)$. Therefore, we can use Theorem 2 to say that (12) describes $\overline{\mathrm{conv}}(C_1 \cup C_2)$ completely, together with the cone constraint $x \in \mathbb{K}_2^n$. Statement (b) simply means that the two sets C_1 and C_2 defined by the disjunction do not meet except at their boundaries. This also holds for split disjunctions. By Proposition 3, (9) is satisfied by all $x \in \mathbb{K}_2^n$ with $\beta = 1$. Taking Proposition 2 into account, we conclude that in this case the corresponding conic quadratic inequality given by (10) is sufficient to describe $\overline{\text{conv}}(C_1 \cup C_2)$. Thus, Theorem 2 and Proposition 2, together with Proposition 3, cover the results of [15] and [14] on split disjunctions on the cone $\mathbb{K}_2^n$ and provide their corresponding generalizations to other two-term disjunctions.

3.4   Two Examples

In this section our first example illustrates the use of Theorem 2. As Lemma 4 hints, there are cases where valid linear inequalities of the form (4) with $\beta_1 = \beta_2 = 1$ may not be sufficient to describe $\overline{\text{conv}}(C_1 \cup C_2)$; we illustrate this in our second example in this section.

Example Where a Single Inequality Suffices. Consider the cone $\mathbb{K}_2^3$ and the disjunction $x_3 \geq 1 \vee x_1 + x_3 \geq 1$. Note that $c_1 = e_3 \in \mathbb{K}_2^3$ in this example. Hence, we can use Theorem 2 to characterize the closed convex hull:

$$\overline{\text{conv}}(C_1 \cup C_2) = \left\{ x \in \mathbb{K}_2^3 : \ 2 - (x_1 + 2x_3) \leq \sqrt{x_3^2 - x_2^2} \right\}.$$

Figures 1(a) and (b) depict the disjunctive set $C_1 \cup C_2$ and the associated closed convex hull, respectively. In order to give a better sense of the convexification operation, we plot the points added to $C_1 \cup C_2$ to generate the closed convex hull in Figure 1(c). Finally, we note that the inequality that we provide is intrinsically related to the conic quadratic inequality of Proposition 2: The sets described by the two inequalities coincide in the region $\overline{\text{conv}}(C_1 \cup C_2) \setminus (C_1 \cup C_2)$. We display the corresponding cone for this example in Figure 1(d).

Example Where Multiple Inequalities Are Needed. Consider the cone $\mathbb{K}_2^3$ and the disjunction $-x_2 \geq 0 \vee -x_3 \geq -1$. For this example Theorem 1 implies that the family of convex inequalities

$$-2 + (\beta x_2 + x_3) \leq \sqrt{(-\beta x_2 + x_3)^2 + (\beta^2 - 1)(x_3^2 - \|\tilde{x}\|_2^2)} \tag{13}$$

parameterized by $\beta \in [1, \infty)$ fully describes $\overline{\text{conv}}(C_1 \cup C_2)$. Figures 2(a) and (b) depict the disjunctive set $C_1 \cup C_2$ and the associated closed convex hull, respectively. Note that $\overline{\text{conv}}(C_1 \cup C_2)$ has a flat surface which is described by (13) with $\beta = 1$. The overall closed convex hull is given by

$$\overline{\text{conv}}(C_1 \cup C_2) = \left\{ x \in \mathbb{K}_2^3 : x_2 \leq 1, \ 1 + |x_1| - x_3 \leq \sqrt{1 - \max\{0, x_2\}^2} \right\},$$

where both inequalities are convex (even when we ignore the constraint $x \in \mathbb{K}_2^3$). In fact, both inequalities describing $\overline{\text{conv}}(C_1 \cup C_2)$ are conic quadratic representable in a lifted space as expected.

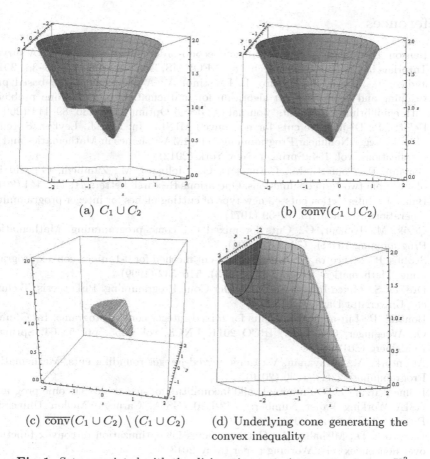

(a) $C_1 \cup C_2$

(b) $\overline{\text{conv}}(C_1 \cup C_2)$

(c) $\overline{\text{conv}}(C_1 \cup C_2) \setminus (C_1 \cup C_2)$

(d) Underlying cone generating the convex inequality

Fig. 1. Sets associated with the disjunction $x_3 \geq 1 \lor x_1 + x_3 \geq 1$ on $\mathbb{K}_2^3$

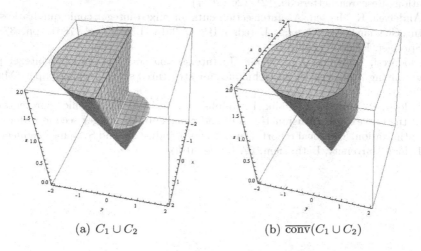

(a) $C_1 \cup C_2$

(b) $\overline{\text{conv}}(C_1 \cup C_2)$

Fig. 2. Sets associated with the disjunction $-x_2 \geq 0 \lor -x_3 \geq -1$ on $\mathbb{K}_2^3$

References

1. Benson, H., Saglam, U.: Mixed-integer second-order cone programming: A survey. Tutorials in Operations Research. In: INFORMS, Hanover, MD, pp. 13–36 (2013)
2. Júdice, J., Sherali, H., Ribeiro, I., Faustino, A.: A complementarity-based partitioning and disjunctive cut algorithm for mathematical programming problems with equilibrium constraints. Journal of Global Optimization 136, 89–114 (2006)
3. Belotti, P.: Disjunctive cuts for nonconvex MINLP. In: Lee, J., Leyffer, S. (eds.) Mixed Integer Nonlinear Programming. The IMA Volumes in Mathematics and its Applications, vol. 154. Springer, New York (2012)
4. Bonami, P., Conforti, M., Cornuéjols, G., Molinaro, M., Zambelli, G.: Cutting planes from two-term disjunctions. Operations Research Letters 41, 442–444 (2013)
5. Balas, E.: Intersection cuts - a new type of cutting planes for integer programming. Operations Research 19, 19–39 (1971)
6. Çezik, M., Iyengar, G.: Cuts for mixed 0-1 conic programming. Mathematical Programming 104(1), 179–202 (2005)
7. Stubbs, R., Mehrotra, S.: A branch-and-cut method for 0-1 mixed convex programming. Mathematical Programming 86(3), 515–532 (1999)
8. Drewes, S.: Mixed Integer Second Order Cone Programming. PhD thesis, Technische Universität Darmstadt (2009)
9. Bonami, P.: Lift-and-project cuts for mixed integer convex programs. In: Günlük, O., Woeginger, G.J. (eds.) IPCO 2011. LNCS, vol. 6655, pp. 52–64. Springer, Heidelberg (2011)
10. Atamtürk, A., Narayanan, V.: Conic mixed-integer rounding cuts. Mathematical Programming 122(1), 1–20 (2010)
11. Kılınç-Karzan, F.: On minimal valid inequalities for mixed integer conic programs. GSIA Working Paper Number: 2013-E20, GSIA, Carnegie Mellon University, Pittsburgh, PA (June 2013)
12. Bienstock, D., Michalka, A.: Cutting planes for optimization of convex functions over nonconvex sets. Working paper (May 2013)
13. Dadush, D., Dey, S., Vielma, J.: The split closure of a strictly convex body. Operations Research Letters 39, 121–126 (2011)
14. Andersen, K., Jensen, A.: Intersection cuts for mixed integer conic quadratic sets. In: Goemans, M., Correa, J. (eds.) IPCO 2013. LNCS, vol. 7801, pp. 37–48. Springer, Heidelberg (2013)
15. Modaresi, S., Kılınç, M., Vielma, J.: Intersection cuts for nonlinear integer programming: Convexification techniques for structured sets. Working paper (March 2013)
16. Belotti, P., Goez, J.C., Polik, I., Ralphs, T., Terlaky, T.: A conic representation of the convex hull of disjunctive sets and conic cuts for integer second order cone optimization. Technical report, Department of Industrial and Systems Engineering, Lehigh University, Bethlehem, PA (June 2012)

Coupled and k-Sided Placements: Generalizing Generalized Assignment

Madhukar Korupolu[1], Adam Meyerson[1], Rajmohan Rajaraman[2],
and Brian Tagiku[1]

[1] Google, 1600 Amphitheater Parkway, Mountain View, CA
{mkar,awmeyerson,btagiku}@google.com
[2] Northeastern University, Boston, MA 02115
rraj@ccs.neu.edu

Abstract. In modern data centers and cloud computing systems, jobs often require resources distributed across nodes providing a wide variety of services. Motivated by this, we study the *Coupled Placement* problem, in which we place jobs into computation and storage nodes with capacity constraints, so as to optimize some costs or profits associated with the placement. The coupled placement problem is a natural generalization of the widely-studied generalized assignment problem (GAP), which concerns the placement of jobs into single nodes providing one kind of service. We also study a further generalization, the *k-Sided Placement* problem, in which we place jobs into k-tuples of nodes, each node in a tuple offering one of k services.

For both the coupled and k-sided placement problems, we consider minimization and maximization versions. In the minimization versions (MINCP and MINkSP), the goal is to achieve minimum placement cost, while incurring a minimum blowup in the capacity of the individual nodes. Our first main result is an algorithm for MINkSP that achieves optimal cost while increasing capacities by at most a factor of $k + 1$, also yielding the first constant-factor approximation for MINCP. In the maximization versions (MAXCP and MAXkSP), the goal is to maximize the total weight of the jobs that are placed under hard capacity constraints. MAXkSP can be expressed as a k-column sparse integer program, and can be approximated to within a factor of $O(k)$ factor using randomized rounding of a linear program relaxation. We consider alternative combinatorial algorithms that are much more efficient in practice. Our second main result is a local search based combinatorial algorithm that yields a 15-approximation and $O(k^2)$-approximation for MAXCP and MAXkSP respectively.

1 Introduction

The data center has become one of the most important assets of a modern business. Whether it is a private data center for exclusive use or a shared public cloud data center, the size and scale of the data center continues to rise. As a company grows, so too must its data center to accommodate growing computational, storage and networking demand. However, the new components purchased for this

J. Lee and J. Vygen (Eds.): IPCO 2014, LNCS 8494, pp. 357–368, 2014.

expansion need not be the same as the components already in place. Over time, the data center becomes quite heterogeneous [1]. This complicates the problem of placing jobs within the data center so as to maximize performance.

Jobs often require resources of more than one type: for example, compute and storage. Modern data centers typically separate computation from storage and interconnect the two using a network of switches. As such, when placing a job within a data center, we must decide which computation node and which storage node will serve the job. If we pick nodes that are far apart, then communication latency may become too prohibitive. On the other hand, nodes are capacitated, so picking nodes close together may not always be possible.

Most prior work in data center resource management is focussed on placing one type of resource at a time: e.g., placing storage requirements assuming job compute location is fixed [2,3] or placing compute requirements assuming job storage location is fixed [4,5]. One sided placement methods cannot suitably take advantage of the proximities and heterogeneities that exist in modern data centers. For example, a database analytics application requiring high throughput between its compute and storage elements can benefit by being placed on a storage node that has a nearby available compute node.

In this paper, we study *Coupled Placement* (CP), which is the problem of placing jobs into computation and storage nodes with capacity constraints, so as to optimize costs or profits associated with the placement. Coupled placement was first addressed in [6] in a setting where we are required to place all jobs and we wish to minimize the communication latency over all jobs. They show that this problem, which we call MINCP, is NP-hard and investigate the performance of heuristic solutions. Another natural formulation is where the goal is to maximize the total number of jobs or revenue generated by the placement, subject to capacity constraints. We refer to this problem as MAXCP. We also study a generalization of Coupled Placement, the *k-Sided Placement Problem* (*k*SP), which considers $k \geq 2$ kinds of resources.

1.1   Problem Definition

In the *coupled placement* problem, we are given a bipartite graph $G = (U, V, E)$ where U is a set of compute nodes and V is a set of storage nodes. We have capacity functions $C : U \to \mathcal{R}$ and $S : V \to \mathcal{R}$ for the compute and storage nodes, respectively. We are also given a set T of jobs, each of which needs to be allocated to one compute node and one storage node. Each job may prefer some compute-storage node pairs more than others, and may also consume different resources at different nodes. To capture these heterogeneities, we have for each job j a function $f_j : E \to \mathcal{R}$, a processing requirement $p_j : E \to \mathcal{R}$ and a storage requirement $s_j : E \to \mathcal{R}$. We note that without loss of generality, we can assume that the capacities are unit, since we can scale the processing and storage requirements of individual nodes accordingly.

We consider two versions of the coupled placement problems. For the maximization version MAXCP, we view f_j as a payment function. Our goal is to select a subset $A \subseteq T$ of jobs and an assignment $\sigma : A \to E$ such that all capacities are observed and our total profit $\sum_{j \in A} f_j(\sigma(j))$ is maximized. For the minimization version MINCP, we view f_j as a cost function. Our goal is to find an assignment $\sigma : T \to E$ such that all capacities are observed and our total cost $\sum_{j \in A} f_j(\sigma(j))$ is minimized.

A generalization of the coupled placement problem is k-*sided placement* (kSP), in which we have k different sets of nodes, $S_1, \ldots, S_k$, each set of nodes providing a distinct service. For each i, we have a capacity function $C_i : S_i \to \mathcal{R}$ that gives the capacity of a node in S_i to provide the ith service. We are given a set T of jobs, each of which needs each kind of service; the exact resource needs may depend on the particular k-tuple of nodes from $\prod_i S_i$ to which it is assigned. That is, for each job j, we have a demand function $d_j : \prod_i S_i \to \mathcal{R}^k$. We also have another function $f_j : \prod_i S_i \to \mathcal{R}$. As for coupled placement, we can assume that the capacities are unit, since we can scale the demands of individual nodes accordingly. Similar to coupled placement, we consider two versions of kSP, MINkSP and MAXkSP.

1.2 Our Results

All of the variants of CP and kSP are NP-hard, so our focus is on approximation algorithms. Our first set of results consist of the first non-trivial approximation algorithms for MINCP and MINkSP. Under hard capacity constraints, it is easy to see that it is NP-hard to achieve any bounded approximation ratio to cost minimization. So we consider approximation algorithms that incur a blowup in capacity. We say that an algorithm is α-approximate for the minimization version if its cost is at most that of an optimal solution, while incurring a blowup factor of at most α in the capacity of any node.

- We present a $(k + 1)$-approximation algorithm for MINkSP using iterative rounding, yielding a 3-approximation for MINCP.

We next consider the maximization version. MAXkSP can be expressed as a k-column sparse integer packing program (k-CSP). From this, it is immediate that MAXkSP can be approximated to within an $O(k)$ approximation factor by applying randomized rounding to a linear programming relaxation [7]. An $\Omega(k/\log k)$-inapproximability result for k-set packing due to [8] implies the same hardness result for MAXkSP. Our second main result is a simpler approximation algorithm for MAXCP and MAXkSP based on local search.

- We present local search based approximation algorithms for MAXCP and MAXkSP, obtaining 15- and $O(k^2)$-approximations, respectively.

The local search result applies directly to a version where we can assign tasks fractionally but only to a single pair of machines (this is like assigning a task with lower priority and may have additional applications). We then describe a simple rounding scheme to obtain an integral version. The rounding technique involves establishing a one-to-one correspondence between fractional assignments and machines. This is much like the cycle-removing rounding for GAP; there is a crucial difference, however, since coupled placements assign jobs to pairs of machines.

1.3 Related Work

The coupled and k-sided placement problems are natural generalizations of the Generalized Assignment Problem (GAP), which can be viewed as a 1-sided placement problem. In GAP, which was first introduced by Shmoys and Tardos [9], the goal is assign items of various sizes to bins of various capacities. A subset of items is feasible for a bin if their total size is no more than the bin's capacity. If we are required to assign all items and minimize our cost (MinGAP), Shmoys and Tardos [9] give an algorithm for computing an assignment that achieves optimal cost while doubling the capacities of each bin. A previous result by Lenstra *et al.* [10] for scheduling on unrelated machines show it is NP-hard to achieve optimal cost without incurring a capacity blowup of at least $3/2$. On the other hand, if we wish to maximize our profit and are allowed to leave items unassigned (MaxGAP), Chekuri and Khanna [11] observe that the $(1, 2)$-approximation for MinGAP implies a 2-approximation for MaxGAP. This can be improved to a $(\frac{e}{e-1})$-approximation using LP-based techniques [12]. It is known that MaxGAP is APX-hard [11], though no specific constant of hardness is shown.

On the experimental side, most prior work in data center resource management focusses on placing one type of resource at a time: for example, placing storage requirements assuming job compute location is fixed (file allocation problem [2], [13,14,3]) or placing compute requirements assuming job storage location is fixed [4,5]. These in a sense are variants of GAP. The only prior work on Coupled Placement is [6], where they show that MinCP is NP-hard and experimentally evaluate heuristics: in particular, a fast approach based on stable marriage and knapsacks is shown to do well in practice, close to the LP optimal.

The MaxkSP problem is related to the recently studied hypermatching assignment problem (HAP) [15], and special cases, including k-set packing, and a uniform version of the problem. A $(k + 1 + \varepsilon)$-approximation is given for HAP in [15], where other variants of HAP are also studied. While the MaxkSP problem can be viewed as a variant of HAP, there are critical differences. For instance, in MaxkSP, each task is assigned at most one tuple, while in the hypermatching problem each client (or task) is assigned a subset of the hyperedges. Hence, the MaxkSP and HAP problems are not directly comparable. The k-set packing can be captured as a special case of MaxkSP, and hence the $\Omega(k/\log k)$-hardness due to [8] applies to MaxkSP as well.

2   The Minimization Version

Next, we consider the minimization version of the Coupled Placement problem, MinCP. We write the following integer linear program for MinCP, where x_{tuv} is the indicator variable for the assignment of t to pair (u, v), $u \in U$, $v \in V$.

$$\text{Minimize:} \qquad \sum_{t,u,v} x_{tuv} f_t(u,v)$$

$$\text{Subject to:} \qquad \sum_{u,v} x_{tuv} \geq 1, \qquad \forall t \in T,$$

$$\sum_{t,v} p_t(u,v) x_{tuv} \leq c_u, \forall u \in U,$$

$$\sum_{t,u} s_t(u,v) x_{tuv} \leq d_v, \forall v \in V,$$

$$x_{tuv} \in \{0,1\}, \qquad \forall t \in T, u \in U, v \in V.$$

We refer the first set of constraints as *satisfaction* constraints, the second and third set as *capacity* constraints (processing and storage). We consider the linear relaxation of this program which replaces the integrality constraints above with $0 \leq x_{tuv} \leq 1, \forall t \in T, u \in U, v \in V$. Without loss of generality, we assume that $p_t(u,v) \leq c_u$ and $s_t(u,v) \leq d_v$ for all t, and (u,v); otherwise, we can set x_{tuv} to 0 and eliminate such triples from the linear program.

2.1 A 3-Approximation Algorithm for MINCP

We now present algorithm ITERROUND, based on iterative rounding [16], which achieves a 3-approximation for MINCP. We start with a basic algorithm that achieves a 5-approximation by identifying tight constraints with a small number of variables. Each iteration of this algorithm repeats the following round until all variables have been rounded.

1 **Extreme point:** Compute an extreme point solution x to the current LP.
2 **Eliminate variable or constraint:** Execute one of these two steps. By Lemma 3, one of these steps can always be executed if the LP is nonempty.

    a Remove from the LP all variables x_{tuv} that take the value 0 or 1 in x. If x_{tuv} is 1, then assign job t to the pair (u,v), remove the job t and its associated variables from the LP, and reduce c_u by $p_t(u,v)$ and d_v by $s_t(u,v)$.

    b Remove from the LP any tight capacity constraint with at most 4 variables.

Fix an iteration of the algorithm, and an extreme point x. Let n_t, n_c, and n_s denote the number of tight task satisfaction constraints, computation constraints, and storage constraints, respectively, in x. Note that every task satisfaction constraint can be assumed to be tight, without loss of generality. Let N denote the number of variables in the LP. Since x is an extreme point, if all variables in x take values in $(0,1)$, then we have $N = n_t + n_c + n_s$.

Lemma 1. *If all variables in x take values in $(0,1)$, then $n_t \leq N/2$.*

Proof. Since a variable only occurs once over all satisfaction constraints, if $n_t > N/2$, there exists a satisfaction constraint that has exactly one variable. But then, this variable needs to take value 1, a contradiction.

Lemma 2. *If $n_t \leq N/2$, then there exists a tight capacity constraint that has at most 4 variables.*

Proof. If $n_t \leq N/2$, then $n_s + n_c = N - n_t \geq N/2$. Since each variable occurs in at most one computation constraint and at most one storage constraint, the total number of variable occurrences over all tight storage and computation constraints is at most $2N$, which is at most $4(n_s + n_c)$. This implies that at least one of these tight capacity constraints has at most 4 variables.

Using Lemmas 1 and 2, we can argue that the above algorithm yields a 5-approximation. Step 2a does not cause any increase in cost or capacity. Step 2b removes a constraint, hence cannot increase cost; since the removed constraint has at most 4 variables, the total demand allocated on the relevant node is at most the demand of four tasks plus the capacity already used in earlier iterations. Since each task demand is at most the capacity of the node, we obtain a 5-approximation with respect to capacity.

Studying the proof of Lemma 2 more closely, one can separate the case $n_t < N/2$ from the $n_t = N/2$; in the former case, one can, in fact, show that there exists a tight capacity constraint with at most 3 variables. Together with a careful consideration of the $n_t = N/2$ case, one can improve the approximation factor to 4. We now present an alternative selection of tight capacity constraint that leads to a 3-approximation. One interesting aspect of this step is that the constraint being selected may not have a small number of variables. We replace step 2b by the following.

2b Remove from the LP any tight capacity constraint in which the number of variables is at most two more than the sum of the values of the variables.

Lemma 3. *If all variables in x take values in $(0, 1)$, then there exists a tight capacity constraint in which the number of variables is at most two more than the sum of the values of the variables.*

Proof. Since each variable occurs in at most two tight capacity constraints, the total number of occurrences of all variables across the tight capacity constraints is $2N - s$ for some nonnegative integer s. Since each satisfaction constraint is tight, each variable appears in 2 capacity constraints, and each variable takes on value less than 1, the sum of all the variables over the tight capacity constraints is at least $2n_t - s$. Therefore, the sum, over all tight capacity constraints, of the difference between the number of variables and their sum is at most $2(N - n_t)$. Since there are $N - n_t$ tight capacity constraints, for at least one of these constraints, the difference between the number of variables and their sum is at most 2.

Lemma 4. *Let u be a node with a tight capacity constraint, in which the number of variables is at most 2 more than the sum of the variables. Then, the sum of the capacity requirements of the tasks partially assigned to u is a most the current available capacity of u plus twice the capacity of u.*

Proof. Let ℓ be the number of variables in the constraint for u, and let the associated tasks be numbered 1 through ℓ. Let the demand of task j for the capacity of node u be d_j. Then, the capacity constraint for u is $\sum_j d_j x_j = \hat{c}(u)$, where $\hat{c}(u)$ is the available capacity of u in the current LP.

We know that $\ell - \sum_i x_i \le 2$. Since $d_i \le C(u)$, the capacity of u:

$$\sum_j d_j = \widehat{c}(u) + \sum_{j=1}^{\ell}(1 - x_j)d_j \le \widehat{c}(u) + (\ell - \sum_{j=\ell}^{m} x_j)C(u) \le \widehat{c}(u) + 2C(u).$$

Theorem 1. ITERROUND *is a polynomial-time* 3-*approximation algorithm for* MINCP.

Proof. By Lemma 3, each iteration of the algorithm removes either a variable or a constraint from the LP. Hence the algorithm is polynomial time. The elimination of a variable that takes value 0 or 1 does not change the cost. The elimination of a constraint can only decrease cost, so the final solution has cost no more than the value achieved by the original LP. Finally, when a capacity constraint is eliminated, by Lemma 4, we incur a blowup of at most 3 in capacity.

2.2 A $(k+1)$-Approximation Algorithm for MINkSP

It is straightforward to generalize the the algorithm of the preceding section to obtain a $k + 1$-approximation to MINkSP. We first set up the integer LP for MINkSP. For a given element $e \in \prod_i S_i$, we use e_i to denote the ith coordinate of e. Let x_{te} be the indicator variable that t is assigned to $e \in \prod_i S_i$.

Minimize: $\sum_{t,e} x_{te} f_t(e)$

Subject to: $\sum_e x_{te} \ge 1,$ $\forall t \in T,$

  $\sum_{t,e:e_i=u} (d_t(e))_i x_{te} \le C_i(u), \forall 1 \le i \le k, u \in U,$

  $x_{te} \in \{0,1\},$ $\forall t \in T, e \in E$

The algorithm, which we call ITERROUND(k), is identical to ITERROUND of Section 2.1 except that step 2b is replaced by the following.

2b Remove from the LP any tight capacity constraint in which the number of variables is at most k more than the sum of the values of the variables.

The claims and proofs are similar to the $k = 2$ case and are deferred to the full paper. A natural question to ask is whether a linear approximation factor for MINkSP is unavoidable for polynomial time algorithms. In the full paper, we show that the MINkSP linear program has an integrality gap that grows as $\Omega(\log k/ \log \log k)$. Determining the best efficiently achievable approximation factor for MINkSP is an open problem.

3 The Maximization Problems

We present approximation algorithms for the maximization versions of coupled placement and k-sided placement problems. We first observe, in Section 3.1, that these problems reduce to column sparse integer packing. We next present, in Section 3.2, an alternative combinatorial approach based on local search.

3.1   An LP-Based Approximation Algorithm

One can write a positive integer linear program for MAXCP. Let x_{tuv} be the indicator variable for assigning job t to (u, v), $u \in U$, $v \in V$.

$$\text{Maximize:} \quad \sum_{t,u,v} x_{tuv} f_t(u, v)$$

$$\text{Subject to:} \quad \sum_{u,v} x_{tuv} \le 1, \qquad \forall t \in T,$$

$$\sum_{t,v} p_t(u, v) x_{tuv} \le c_u, \; \forall u \in U,$$

$$\sum_{t,u} s_t(u, v) x_{tuv} \le d_v, \; \forall v \in V,$$

$$x_{tuv} \in \{0, 1\}, \qquad \forall t \in T, u \in U, v \in V.$$

Note that we can deal with capacities on u, v by scaling the $p_t(u, v)$ and $s_t(u, v)$ values appropriately. The above LP can be easily extended to MAXkSP (due to space constraints, we defer the formulation to the full paper).

These linear programs are 3- and k-column sparse packing programs, respectively, and can be approximated to within a factor of 15.74 and $ek + o(k)$, respectively using a clever randomized rounding approach. As mentioned in Section 1, an $\Omega(k/\log k)$-inapproximability result is known for MAXkSP.

3.2   Approximation Algorithms Based on Local Search

We now present a combinatorial approach for MAXkSP based on local search, which is likely to be much more efficient than the above LP-based approximation algorithm in practice. Before giving the details, we start with a few helpful definitions. For any $u \in U$, $F_u = \Sigma_{t,v} x_{tuv} f_t(u, v)$. Similarly, for any $v \in V$, $F_v = \Sigma_{t,u} x_{tuv} f_t(u, v)$. We set $\mu = \frac{1}{n} \max_{t,u,v} f_t(u, v)$. It follows that the optimum solution is at least $n\mu$ and at most $n^2 \mu$.

The local search algorithm will maintain the following two invariants: (1) For each t, there is at most one pair (u, v) for which $x_{tuv} > 0$; (2) All the linear program inequalities hold. It's easy to set an initial state where the invariant holds (all $x_{tuv} = 0$). The local search algorithm proceeds in the following steps: While $\exists t, u, v : f_t(u, v) > F_u \frac{p_t(u,v)}{c_u} + F_v \frac{s_t(u,v)}{d_v} + \Sigma_{u',v'} x_{tu'v'} f_t(u', v') + \epsilon\mu$:

1. Set $x_{tuv} = 1$ and set $x_{tu'v'} = 0$ for all $(u', v') \ne (u, v)$.
2. While $\Sigma_{t,v} p_t(u, v) x_{tuv} > c_u$, reduce x_{tuv} for the job with minimum value of $c_u f_t(u, v)/p_t(u, v)$ such that $x_{tuv} > 0$.
3. While $\Sigma_{u,v} s_t(u, v) x_{tuv} > d_v$, reduce x_{tuv} for the job with minimum value of $d_v f_t(u, v)/s_t(u, v)$ such that $x_{tuv} > 0$

Lemma 5. *The local search algorithm maintains the two stated invariants.*

Proof. The first invariant is straightforward, because the only time we increase an x_{tuv} value we simultaneously set all other values for the same t to zero. The only time the linear program inequalities can be violated is immediately after setting $x_{tuv} = 1$. However, the two steps immediately after this operation will reduce the values of other jobs so as to satisfy the inequalities (and this is done without increasing any x_{tuv} so no new constraint can be violated).

Theorem 2. *The local search algorithm produces a $3 + \epsilon$ approximate fractional solution satisfying the invariants.*

Proof. By Lemma 5, the local search algorithm always maintains the invariants. When the algorithm terminates, we have for all t, u, v: $f_t(u, v) \leq F_u \frac{p_t(u,v)}{c_u} + F_v \frac{s_t(u,v)}{d_v} + \Sigma_{u',v'} x_{tu'v'} f_t(u', v') \epsilon \mu$. We sum this over t, u, v representing the optimum integer assignments: $OPT \leq \Sigma_u F_u + \Sigma_v F_v + \Sigma_{t,u,v} x_{tuv} f_t(u, v) + \epsilon OPT$. Each summation simplifies to the algorithm's objective value, giving the result.

Theorem 3. *The local search algorithm runs in polynomial time.*

Proof. Setting $x_{tuv} = 1$ and setting all other $x_{tu'v'} = 0$ adds the amount $f_t(u, v) - \Sigma_{u'v'} x_{tu'v'} f_t(u', v')$ to the objective. The next two steps of the algorithm (making sure the LP inequalities hold) reduce the objective by at most $F_u \frac{p_t(u,v)}{c_u} + F_v \frac{s_t(u,v)}{d_v}$. It follows that each iteration of the main loop increases the solution value by at least $\epsilon \mu$. By definition of μ, this can happen at most n^2/ϵ times. Each selection of (t, u, v) can be done in polynomial time (at worst, by simply trying all tuples).

Rounding Phase: When the local search algorithm terminates, we have a fractional solution with the guarantee from the first invariant. Note that we can extend this to MAXkSP if we increase the approximation factor to $k + 1 + \epsilon$. The next phase of the algorithm is to round the fractional solution returned by local search. Applying the randomized rounding approach of [7], we obtain an $O(k^2)$-approximation for MAXkSP, and a $(47.22 + \varepsilon)$-approximation for MAXCP. The preceding approach does not take advantage, however, of the properties of the fractional solution returned by our local search algorithm. For MAXCP, we present a different rounding scheme that exploits the local search invariants satisfied by the fractional solution and obtains a $15 + \epsilon$ approximation.

The main idea behind the rounding scheme is to obtain a one-to-one correspondence between fractional assignments and machines. Essentially we view the machines as nodes of a graph where the edges are the fractional assignments (this is similar to the rounding for generalized assignment). If we have a cycle, the idea is to shift the fractions around the cycle (i.e. increase one x_{tuv} then decrease some $x_{t'vw}$ and increase some $x_{t''wx}$ and so forth). Applying this directly on a single cycle may violate some constraints; while we try to increase and decrease the fractions in such a way that constraints hold, since each job has different "size" on its two endpoints we may wind up violating the constraint $\sum_{t,v} x_{tuv} p_t(u, v)$ at a single node u. This prevents us from doing a simple cycle elimination as in generalized assignment. However, if we have two adjoining (or connected) cycles the process can be made to work. The remaining case is a single cycle, where we can assign each edge to one of its endpoints. Generalized assignment rounding would now proceed to integrally assign each job to its corresponding machine; we cannot do this because each job requires *two* machines, and each machine thus has multiple fractional assignments (all but one of which "correspond" to some other machine).

Lemma 6. *Given any fractional solution which satisfies the local search invariants, we can produce an alternative fractional solution (also satisfying the local search invariants and with equal or greater value). This new fractional solution labels each job t with $0 < x_{tuv} < 1$, with either u or v, guaranteeing that each u is labeled with at most one job.*

Proof. Consider a graph where the nodes are machines, and we have an edge (u, v) for any fractional assignment $0 < x_{tuv} < 1$. If any node has degree zero or one, we remove that node and its assigned edge (if any), labeling the removed edge with the node that removed it. We continue this process until all remaining nodes have degree at least two. If there is a node of degree three, then there must exist two (distinct but not necessarily edge-disjoint) cycles with a path between them (possibly a path of length zero); since the graph is bipartite all cycles are even in length. We can alternately increase and decrease the fractional assignments of edges along a cycle such that the total load $\sum_{t,v} p_t(u, v)x_{tuv}$ changes only on a single node u where the path between cycles intersects this cycle. We can do the same along the other cycle. We can then do the same thing along the path, and equalize the changes (multiplicatively) such that there is no overall change in load, but at least one edge has its fractional value changing. If this process decreases the value, we can reverse it to increase the value. This allows us to modify the fractional solution in a way that increases the number of integral assignments without decreasing the value. After applying this repeatedly (and repeating the node/edge removal process above where necessary), we are left with a graph that consists only of node-disjoint cycles. Each of the remaining edges will be labeled with one of its two endpoints (one to each). We thus have a one-to-one labeling correspondence between fractional assignments and machines (each fractional edge to one of its two assigned machines). Note however that since each job is assigned to two machines and labeled with only one of the two, this does not imply that each machine has only one fractional assignment.

Once this is done, we consider three possible solutions. One consists of all the integral assignments. The second considers only those assignments which are fractional and labeled with nodes u. For each node v, we select a subset of its fractional assignments to make integrally, so as to maximize the value without violating capacity of v. We cannot violate capacity of u because we select at most one job for each such machine. The result has at least $\frac{1}{2}$ the value of assignments labeled with nodes u. For the third solution, we do the same but with the roles of u, v reversed. We select the best of these three solutions; our choice obtains at least $\frac{1}{5}$ of the overall value.

Theorem 4. *For* MAXCP, *there exists a polynomial-time algorithm based on local search that achieves a* $15 + \epsilon$ *approximation for* MAXCP.

Proof. The algorithm sketch contains most of the proof. We need to establish that we can get at least $\frac{1}{2}$ the fractional value on a single machine integrally. This can be done by selecting jobs in decreasing order of density $(f_t(u,v)/p_t(u,v))$ until we overflow the capacity. Including the job that overflows capacity, this must be better than the fractional solution. Thus we can select either everything but the job that overflows capacity, or that job by itself.

We also need to establish the $\frac{1}{5}$ value claim. If we were to select the integral assignments with probability $\frac{1}{5}$ and each of the other two solutions with probability $\frac{2}{5}$, we would get an expected $\frac{1}{5}$ of the fractional solution. Deterministially selecting the best of the three solutions can only be better than this.

4   Concluding Remarks

We introduce minimization and maximization versions of the k-sided placement, a generalization of the generalized assignment problem (GAP). For the minimization version, MINkSP, we present a $k + 1$ approximation using iterative rounding, thus generalizing the 2-approximation result for the minimization version of GAP. The best lower bound on inapproximability for MINkSP is a constant factor, derived from GAP. Finding the best polynomial-time approximation achievable for MINkSP is an interesting open problem. In the full paper, we show that the particular linear program we use for MINkSP has an integrality gap that grows as $\Omega(\log k / \log \log k)$.

The maximization version of k-sided placement, MAXkSP, can be approximated to within a factor of $O(k)$ by applying randomized rounding to a k-column sparse LP relaxation [7]. We present simpler combinatorial algorithms based on local search for MAXkSP and MAXCP (MAXkSP with $k = 2$) that yield $O(k^2)$ and $15 + \epsilon$ approximations, respectively. Future research directions include developing combinatorial algorithms with better approximations and finding the best polynomial-time approximations achievable for the two problems.

In the full paper, we also study an online version of MAXCP, in which tasks arrive online and must be irrevocably assigned or rejected immediately upon arrival. We extend the techniques of [17] to the case where the capacity requirement for a job is arbitrarily machine-dependent, and thereby achieve competitive ratio logarithmic in the ratio of best to worst value-per-capacity density, under necessary technical assumptions about the maximum job size.

Acknowledgments. We would like to thank Aravind Srinivasan for helpful discussions, and for pointing us to the $\Omega(k / \log k)$-hardness result for k-set packing, in particular. We thank anonymous referees for helpful comments on an earlier version of the paper, and are especially grateful to a referee who generously offered a key insight leading to improved results for MINCP and MINkSP. The third author was partly supported by NSF CSR award 1217981.

References

1. Patterson, D.A.: Technical perspective: the data center is the computer. Communications of the ACM 51, 105–105 (2008)
2. Dowdy, L.W., Foster, D.V.: Comparative models of the file assignment problem. ACM Surveys 14 (1982)
3. Anderson, E., Kallahalla, M., Spence, S., Swaminathan, R., Wang, Q.: Quickly finding near-optimal storage designs. ACM Transactions on Computer Systems 23, 337–374 (2005)
4. Appleby, K., Fakhouri, S., Fong, L., Goldszmidt, G., Kalantar, M., Krishnakumar, S., Pazel, D., Pershing, J., Rochwerger, B.: Oceano-SLA based management of a computing utility. In: Proceedings of the International Symposium on Integrated Network Management, pp. 855–868 (2001)
5. Chase, J.S., Anderson, D.C., Thakar, P.N., Vahdat, A.M., Doyle, R.P.: Managing energy and server resources in hosting centers. In: Proceedings of the Symposium on Operating Systems Principles, pp. 103–116 (2001)
6. Korupolu, M., Singh, A., Bamba, B.: Coupled placement in modern data centers. In: Proceedings of the International Parallel and Distributed Processing Symposium, pp. 1–12 (2009)
7. Bansal, N., Korula, N., Nagarajan, V., Srinivasan, A.: On k-column sparse packing programs. In: Proceedings of the Conference on Integer Programming and Combinatorial Optimization, pp. 369–382 (2010)
8. Hazan, E., Safra, S., Schwartz, O.: On the complexity of approximating k-set packing. Computational Complexity 15(1), 20–39 (2006)
9. Shmoys, D.B.: Éva Tardos: An approximation algorithm for the generalized assignment problem. Mathematical Programming 62(3), 461–474 (1993)
10. Lenstra, J.K., Shmoys, D.B.: Éva Tardos: Approximation algorithms for scheduling unrelated parallel machines. Mathematical Programming 46(3), 259–271 (1990)
11. Chekuri, C., Khanna, S.: A PTAS for the multiple knapsack problem. In: Proceedings of the Symposium on Discrete Algorithms, pp. 213–222 (2000)
12. Fleischer, L., Goemans, M.X., Mirrokni, V.S., Sviridenko, M.: Tight approximation algorithms for maximum general assignment problems. In: SODA, pp. 611–620 (2006)
13. Alvarez, G.A., Borowsky, E., Go, S., Romer, T.H., Becker-Szendy, R., Golding, R., Merchant, A., Spasojevic, M., Veitch, A., Wilkes, J.: Minerva: An automated resource provisioning tool for large-scale storage systems. Transactions on Computer Systems 19, 483–518 (2001)
14. Anderson, E., Hobbs, M., Keeton, K., Spence, S., Uysal, M., Veitch, A.: Hippodrome: Running circles around storage administration. In: Proceedings of the Conference on File and Storage Technologies, pp. 175–188 (2002)
15. Cygan, M., Grandoni, F., Mastrolilli, M.: How to sell hyperedges: The hypermatching assignment problem. In: SODA, pp. 342–351 (2013)
16. Lau, L., Ravi, R., Singh, M.: Iterative Methods in Combinatorial Optimization. Cambridge Texts in Applied Mathematics. Cambridge University Press (2011)
17. Awerbuch, B., Azar, Y., Plotkin, S.: Throughput-competitive on-line routing. In: Proceedings of the Symposium on Foundations of Computer Science, pp. 32–40 (1993)

A Unified Algorithm for Degree Bounded Survivable Network Design

Lap Chi Lau and Hong Zhou

Department of Computer Science and Engineering
The Chinese University of Hong Kong
{chi,hzhou}@cse.cuhk.edu.hk

Abstract. We present an approximation algorithm for the minimum bounded degree Steiner network problem that returns a Steiner network of cost at most two times the optimal and the degree on each vertex v is at most $\min\{b_v + 3r_{\max}, 2b_v + 2\}$, where $r_{\max}$ is the maximum connectivity requirement and b_v is the given degree bound on v. This unifies, simplifies, and improves the previous results for this problem.

1 Introduction

In the *minimum bounded degree Steiner network* problem, we are given an undirected graph $G = (V, E)$, a cost c_e on each edge $e \in E$, a degree bound b_v on each vertex $v \in V$, and a connectivity requirement r_{uv} for each pair of vertices $u, v \in V$. A subgraph H of G is called a *Steiner network* if there are at least r_{uv} edge-disjoint paths in H for all $u, v \in V$. The task of the minimum bounded degree Steiner network problem is to find a Steiner network H with minimum total cost such that $d_H(v) \leq b_v$ for each $v \in V$. This is a general problem of interest to algorithm design, computer networks, graph theory, and operations research.

It is NP-hard to determine whether there is a Steiner network satisfying all the degree bounds, even if we do not consider the cost of the Steiner network, as the Hamiltonian cycle problem is a special case. Thus, researchers focus on designing bicriteria approximation algorithms for the problem that minimize both the total cost and the degree violation. We say an algoithm is an $(\alpha, f(b_v))$-approximation algorithm for the minimum bounded degree Steiner network problem if it returns a Steiner network H of cost at most $\alpha \cdot \text{opt}$ and $d_H(v) \leq f(b_v)$ for each $v \in V$, where opt is the cost of an optimal Steiner network that satisfies all the degree bounds.

The first bicriteria approximation algorithm for this problem is a $(2, 2b_v + 3)$-approximation algorithm by Lau, Naor, Salavatipour, and Singh [12], and it was improved to $(2, 2b_v + 2)$ by Louis and Vishnoi [15]. There are also bicriteria approximation algorithms with additive violation on the degrees in terms of the maximum connectivity requirement. Let $r_{\max} = \max_{u,v}\{r_{u,v}\}$. Lau and Singh [14] gave a $(2, b_v + 6r_{\max} + 3)$-approximation algorithm for the problem, and a $(2, b_v + 3)$-approximation algorithm in the special case when $r_{\max} = 1$. The special case when $r_{\max} = 1$ is known as the *minimum bounded degree Steiner forest*

J. Lee and J. Vygen (Eds.): IPCO 2014, LNCS 8494, pp. 369–380, 2014.
© Springer International Publishing Switzerland 2014

problem. In this paper, we present a $(2, \min\{b_v + 3r_{\max}, 2b_v + 2\})$-approximation algorithm for the problem.

Theorem 1. *There is a polynomial time algorithm for the minimum bounded degree Steiner network problem that returns a Steiner network H of cost at most $2\mathrm{opt}$ and $deg_H(v) \leq \min\{b_v + 3r_{max}, 2b_v + 2\}$ for all v.*

Theorem 1 improves the $(2, b_v + 6r_{\max} + 3)$ result in [14] when $r_{\max} \geq 2$ and recovers the $(2, b_v + 3)$ result in [14] for the minimum bounded degree Steiner forest problem. Besides, it achieves the $(2, 2b_v + 2)$ result in [15] simultaneously, while previously there was no such guarantee[1]. Furthermore, both our algorithm and its analysis are simpler[2] as we will discuss in Section 2. We believe that our result unifies what can be achieved using existing techniques. We show an example where our algorithm fails to give a $(2, b_v + 2)$-approximation algorithm for the minimum bounded degree Steiner forest problem in Section 3.2.

1.1   Related Work

Jain [9] introduced the iterative rounding method to give a 2-approximation algorithm for the minimum Steiner network problem, improving on a line of research that applied primal-dual methods to these problems. Later, the iterative rounding method has been applied to obtain the best known approximation algorithms for network design problems for element-connectivity [4,3], vertex-connectivity [3,2], and directed edge-connectivity [7].

The iterative relaxation method was introduced in [12] to adapt Jain's method to degree bounded network design problems, which are well-studied especially in the special case of spanning trees [6,8]. Later, this method has also been applied to achieve the best known approximation algorithms for the degree bounded network design problems, including spanning trees [17,1,11], Steiner networks [12,14,15], directed edge-connectivity [12,1], element-connectivity and vertex-connectivity [10,16,5]. See [13] for a survey on this approach.

2   Technical Overview

Since this work is tightly connected to previous work, we give a high level overview to describe the previous work and highlight where the improvement comes from.

2.1   Iterative Rounding and Relaxation

All the previous results on this problem are based on the iterative rounding method introduced by Jain [9] for the minimum Steiner network problem. This

---

[1] For instance, it was not known how to combine the results in [15,14] to obtain a $(2, \min\{b_v + 6r_{\max} + 3, 2b_v + 2\})$-approximation algorithm for the problem.

[2] In particular, the analysis of the $(2, b_v + 3)$ result is significantly simpler than that in [14].

method is based on analyzing the extreme point solutions to a linear programming relaxation for the problem. Let us first formulate the linear programming relaxation for the minimum bounded degree Steiner network problem. For a subset $S \subseteq V$, we let $\delta(S)$ be the set of edges with one endpoint in S and one endpoint in $V - S$ in the graph and let $d(S) := |\delta(S)|$. In the linear program, there is one variable x_e for each edge, where the intended value is one if this edge is used in the solution and zero if this edge is not used. For a subset of edges $E' \subseteq E$, we write $x(E') = \sum_{e \in E'} x_e$. For a subset of vertices $S \subseteq V$, we define $f(S) := \max_{u \in S, v \notin S} \{r_{uv}\}$ to be the maximum requirement crossing S. To satisfy the connectivity requirement, we should have $x(\delta(S)) \geq f(S)$ for each $S \subseteq V$. The following is a linear programming relaxation for the minimum bounded degree Steiner network problem. It has exponentially many constraints, but there is a polynomial time separation oracle to determine whether a solution is feasible or not, and thus it can be solved in polynomial time by the ellipsoid method.

$$(\text{LP}) \quad \text{minimize} \quad \sum_{e \in E} c_e x_e$$

$$\text{subject to} \quad x(\delta(S)) \geq f(S) \quad \forall S \subset V$$
$$x(\delta(v)) \leq b_v \quad \quad \forall v \in V$$
$$x_e \geq 0 \quad \quad \forall e \in E$$

When there are no degree constraints, Jain [9] proved that there exists an edge e with $x_e \geq \frac{1}{2}$ in any extreme point solution to the above linear program. We call such an edge a *heavy* edge. He used this to obtain an *iterative rounding* algorithm for the minimum Steiner network problem, by repeatedly picking a heavy edge and recomputing an optimal extreme point solution to the residual problem. When there are degree constraints, Lau et.al. [12] showed that either there is a heavy edge or there is a degree constrained vertex with at most four nonzero edges incident to it. They then introduced an extra relaxation step to remove the degree constrained vertex in the latter case, leading to a $(2, 2b_v + 3)$-approximation algorithm for the minimum bounded degree Steiner network problem.

Roughly speaking, all the later improvements are based on proving the existence of a heavy edge with additional properties. To improve the degree violation, Louis and Vishnoi [15] proved that in any extreme point solution either there is an edge of integral value (zero or one), or a vertex v with at most $2b_v + 2$ edges incident to it, or a heavy edge *with no endpoint having a degree bound at most one*. They showed that using this iteratively would imply a $(2, 2b_v + 2)$-approximation algorithm for the problem. Note that in the above algorithms, after we pick a heavy edge, we need to decrease the degree bound by half in order to achieve the guarantee on the degree violation, and thus they have to consider a slightly more general problem where the degree bounds are half-integral and some subtle issue arose as we will discuss later.

To obtain additive violation on the degree bounds, Lau and Singh [14] proved that in any extreme point solution either there is an edge of integral value, or a vertex v with at most four edges incident to it, or a heavy edge *between two vertices with degree bounds at most* $6r_{\max}$, where the last condition guarantees that the degree violation is bounded when we picked edges with value at least half. For the minimum bounded degree Steiner forest problem, they proved that in any extreme point solution either there is an integral edge, or a vertex v with at most $b_v + 3$ edges incident to it, or a heavy edge *with no degree constraint on its endpoints*. They showed that these would lead to a $(2, b_v + 6r_{\max} + 3)$-approximation algorithm for the Steiner network problem and a $(2, b_v + 3)$-approximation algorithm for the Steiner forest problem. The algorithm for Steiner forest is simpler, as it just removes the degree constraint on a vertex when it has at most $b_v + 3$ edges, and does not need to update the degree constraint to a half-integral value, as it only picks edges with value at least half when both endpoints have no degree constraints.

Our algorithm is very similar to that for the Steiner forest problem in [14] (see Algorithm 1). We prove that either there is an edge of integral value, or there is a vertex with at most $\min\{b_v + 3r_{\max}, 2b_v + 2\}$ edges incident to it, or there is a heavy edge with *no degree constraints on its endpoints*. The resulting algorithm is quite simple, in the first case we delete an edge when $x_e = 0$ or pick an edge when $x_e = 1$, in the second case we remove the degree constraint on that vertex, and in the final case we pick such a heavy edge. Note that we only update the degree constraints when we pick an edge with $x_e = 1$, and thus we can maintain the invariant that the degree bounds are integral, and this will simplify the analysis for the $2b_v + 2$ bound.

2.2 Analysis

To analyze the extreme point solutions, an uncrossing technique is used to show that the extreme point solutions are defined by a set of constraints with a special structure. A function $f : 2^V \to \mathbb{R}$ is *skew supermodular* if for any $X, Y \subseteq V$ either $f(X) + f(Y) \leq f(X \cup Y) + f(X \cap Y)$ or $f(X) + f(Y) \leq f(X - Y) + f(Y - X)$. It is known that the function f defined by the connectivity requirements is a skew supermodular function. For a set $S \subseteq V$, the corresponding constraint $x(\delta(S)) \geq f(S)$ defines a vector in $\mathbb{R}^{|E|}$: the vector has a one corresponding to each edge $e \in \delta(S)$ and a zero otherwise. We call this vector the *characteristic vector* of $\delta(S)$ and denote it by $\chi_{\delta(S)}$. A family of sets $\mathcal{L}$ is *laminar* if $X, Y \in \mathcal{L}$ implies that either $X \cap Y = \emptyset$, or $X \subseteq Y$, or $Y \subseteq X$. Using the assumption that f is skew supermodular, it follows from standard uncrossing technique that any extreme point solution of (LP) is characterized by a laminar family of tight constraints.

Lemma 1 ([12]). *Suppose that the requirement function f of (LP) is skew supermodular. Let x be an extreme point solution of (LP) such that $0 < x_e < 1$ for all edges $e \in E$. Then there exist a laminar family $\mathcal{L}$ of sets and set $T \subseteq V$ such that x is the unique solution to*

$$\{x(\delta(S)) = f(S) \mid S \in \mathcal{L}\} \cup \{x(\delta(v)) = b_v \mid v \in T\}$$

that satisfies the following properties:

1. *The characteristic vectors* $\chi_{\delta(S)}$ *for* $S \in \mathcal{L}$ *and* $\chi_{\delta(v)}$ *for* $v \in T$ *are linearly independent.*
2. $|E| = |\mathcal{L}| + |T|$.

The structure of a laminar set family $\mathcal{L}$ can be seen as a forest in which nodes correspond to sets in $\mathcal{L}$ and there is an edge from set R to set S if R is the smallest set containing S. We call R the *parent* of S and S is a *child* of R. A set without any parent is a *root* and a set without any child is a *leaf*. The *subtree rooted at* a set S consists of S and all its descendants.

Lau et.al. [12], following Jain [9], used this forest structure in a counting argument to prove that either there exists an edge with integral value, or a vertex with degree at most four, or an edge with value at least $\frac{1}{2}$. First, each edge is assigned two tokens, one to each endpoint, for a total of $2|E|$ tokens. Assuming none of the conditions hold, i.e. $0 < x_e < \frac{1}{2}$ for each edge and every vertex is of degree at least five. Then, it can be shown that the tokens can be redistributed such that each member of $\mathcal{L}$ and each vertex in T get at least two tokens, and there are still some extra tokens left. This implies that there are more than $2|\mathcal{L}| + 2|T|$ tokens, contradicting property 2 of Lemma 1. The redistribution is done inductively using the following lemma.

Lemma 2 ([12]). *For any subtree of $\mathcal{L}$ rooted at node S, we can reassign tokens collected from child nodes of S and endpoints owned by S such that each vertex in $T \cap S$ gets at least two tokens, each node in the subtree gets at least two tokens, and the root S gets at least three tokens. Moreover, S gets exactly three tokens only if* $\text{coreq}(S) = \frac{1}{2}$, *where* $\text{coreq}(S) := \sum_{e \in \delta(S)} (\frac{1}{2} - x_e)$.

The proof in [12] is almost the same as Jain's proof, but with the presence of degree constraints. Note that each vertex with degree constraint gets at least five tokens in the initial token assignment. Intuitively, we can think of each degree constraint as a singleton set (a leaf set in the laminar family), and thus having five tokens is more than enough for Jain's proof to go through. And one may think that it is already enough if every degree constraint gets at least four tokens to satisfy the induction hypothesis in Lemma 2, and this would imply a $(2, 2b_v + 2)$-approximation algorithm. Unfortunately, the subtle point here is that the degree bounds are half-integral, but for Jain's proof to work they need to be integral. To overcome this problem, Louis and Vishnoi [15] needed to modify the algorithm and the analysis to obtain a $(2, 2b_v + 2)$-approximation algorithm.

The new idea in the Steiner network algorithm in [14] is to only pick heavy edges when both endpoints are of low degree. In the analysis, with the presence of heavy edges that are not allowed to be picked (when some endpoint is of high degree), there could be some sets S with $d(S) = 2$ and $x(\delta(S)) = 1$, and thus the counting argument as above would not work in the base case for those sets. For the same induction hypothesis in Lemma 2 to work, a new rule is added to the initial token assignment: if (w, v) is a heavy edge with $b_w \geq 6r_{\max}$ and v is not degree constrained, then v gets two tokens from the edge (w, v) while w gets none. The counting argument would work in the base case with this new rule, but w may not receive any token for the induction step to work. The assumption

$b_w \geq 6r_{\max}$ is used to ensure that w can *get back* the tokens in the induction step. To illustrate this, consider a worst case scenario in Figure 1(a). In the figure, there is a degree constrained vertex w where all its incident edges are heavy, and we need to collect two tokens for w and four tokens for S. Each child contributes only one token but "consumes" the degree bound of w by $r_{\max}$. This is where the assumption that $b_w \geq 6r_{\max}$ is used to guarantee that S has at least five children, so that each can contribute at least one token for w and S (and use some additional argument to collect one more token).

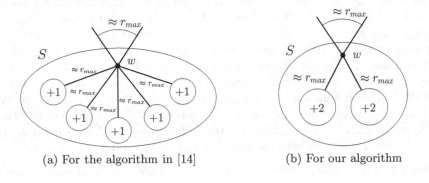

(a) For the algorithm in [14]　　　　(b) For our algorithm

Fig. 1. Worst cases for counting arguments

In this paper, we slightly change the algorithm to remove any vertex with degree at most $\min\{b_v + 3r_{\max}, 2b_v + 2\}$. We used the same initial token assignment rule as in [14] and the same induction hypothesis in Lemma 2 for the counting argument. First, using a simple argument (see Lemma 6), we show that any vertex with degree at least $2b_v + 3$ can receive at least four tokens, and this allows the induction to work and recovers the $(2, 2b_v + 2)$ result by Louis and Vishnoi. As mentioned before, this is possible because our algorithm maintains the invariant that all the degree bounds are integral.

Our improvement to $b_v + 3r_{\max}$ comes from the concept of the *integrality gap* of the degree of a set S, defined as $d(S) - x(\delta(S))$. To illustrate this, consider the same scenario in Figure 1. In the new algorithm, the vertex w with degree bound b_w but having more than $b_w + 3r_{\max}$ edges incident to it has an integrality gap of $3r_{\max}$ on its degree. The main observation is that the heavy edges from a child having only three tokens (one extra token) can only contribute $\frac{1}{2}$ to the integrality gap of w (see Lemma 7), while a child having at least four tokens could contribute $r_{\max}$ to the integrality gap of w. This observation basically allows us to rule out children with only three tokens in the worst case scenarios of the counting argument, as it contributes one token but only consumes $\frac{1}{2}$ of the integrality gap. This allows us to assume that each child can contribute two tokens instead of one token, and this reduces the degree violation from $6r_{\max}$ to $3r_{\max}$ (see Figure 1(b)), with some additional arguments. An additional advantage of our algorithm is that we also avoid the additive term $+3$ in the previous algorithm [14]. The proof for $r_{max} \geq 2$ is quite short (see Section 3.1). The proof for $r_{max} = 1$ still has some case analysis (see the full version), but is considerably shorter than that in [14], as a complicated induction hypothesis was used in [14] that caused many more case analyses.

3   Algorithm and Analysis

In the following, let W be the set of vertices with degree constraints.

Algorithm 1. Minimum Bounded Degree Steiner Network

---

1 Initialization: $H = (V, \emptyset), W \leftarrow V, f'(S) \leftarrow f(S)$ for all $S \subseteq V$.
2 **while** H *is not a Steiner Network* **do**

    (a) Compute an optimal extreme point solution x to (LP) and remove all edges e with $x_e = 0$.

    (b) For each vertex $v \in W$ with degree at most $\min\{b_v + 3r_{max}, 2b_v + 2\}$, remove v from W.

    (c) For each edge $e = (u, v)$ with $x_e = 1$, add e to H and remove e from G, and decrease b_u, b_v by one.

    (d) For each edge $e = (u, v)$ with $x_e \geq \frac{1}{2}$ and $u, v \notin W$, add e to H and remove e from G.

    (e) For each $S \subset V$, set $f'(S) \leftarrow f(S) - d_H(S)$.
3 Return H.

Given that f is initially a skew supermodular function, it is known that f' in any later iteration is still a skew supermodular function [9]. So, the residual LP in any iteration is still in the original form and it has a polynomial time separation oracle [9]. Assuming the algorithm always makes progress in each iteration, then we can prove Theorem 1 by a standard inductive argument as in [12]. It remains to prove the following lemma to complete the proof of Theorem 1.

Lemma 3. *Let x be an extreme point solution to (LP) and W be the set of vertices with degree constraints. Then at least one of the following is true.*

1. *There exists a vertex $v \in W$ with $d(v) \leq \min\{b_v + 3r_{max}, 2b_v + 2\}$.*
2. *There exists an edge e with $x_e = 0$ or $x_e = 1$.*
3. *There exists an edge $e = (u, v)$ with $x_e \geq \frac{1}{2}$ and $u, v \notin W$.*

We prove Lemma 3 by contradiction. Assuming none of the three conditions holds, then we have

1. $d(v) \geq \min\{b_v + 3r_{max} + 1, 2b_v + 3\}$ for $v \in W$,
2. $0 < x_e < 1$ for $e \in E$,
3. if $x_{uv} \geq \frac{1}{2}$ then either u or v is in W.

We will use a token counting argument to derive a contradiction with Lemma 1. Let $\mathcal{L}$ be the laminar family and T be the set of vertices with tight degree constraints as defined in Lemma 1. In the token counting argument, we first assign two tokens to each edge, for a total of $2|E|$ tokens. Then, using the assumptions above, we show that these tokens can be redistributed such that each member in $\mathcal{L}$ and each vertex in T gets at least two tokens and there are some tokens left, but this contradicts with $|E| = |\mathcal{L}| + |T|$ from Lemma 1.

Initial Token Assignment Rule. Each edge receives two tokens. If $e = (u, v)$ is a heavy edge with $u \in W$ and $v \notin W$, then v gets two tokens from e and u gets no token. For every other edge e, each endpoint of e gets one token.

We will redistribute the tokens inductively using the forest structure of the laminar family $\mathcal{L}$. We need some definitions to state the induction hypothesis. We say a vertex v is *owned by* a set $S \in \mathcal{L}$ if S is the smallest set in $\mathcal{L}$ that contains v. Given an extreme point solution x, we say an edge e is a *heavy edge* if $x_e \geq 1/2$, otherwise we say e is a *light edge*. Let $\delta^h(S) = \{e \in \delta(S), x_e \geq 1/2\}$ $(\delta^l(S) = \{e \in \delta(S), x_e < 1/2\})$ be the set of heavy edges (light edges) in $\delta(S)$. The *corequirement* of a set S is defined as

$$\text{coreq}(S) = \sum_{e \in \delta^h(S)} (1 - x_e) + \sum_{e \in \delta^l(S)} (1/2 - x_e) = |\delta^h(S)| + \frac{|\delta^l(S)|}{2} - x(\delta(S)).$$

We will prove the following lemma which shows that the tokens can be redistributed to obtain a contradiction, proving Lemma 3.

Lemma 4. *For any subtree of $\mathcal{L}$ rooted at node S, we can reassign tokens collected from child nodes of S and endpoints owned by S such that each vertex in $T \cap S$ gets at least two tokens, each node in the subtree gets at least two tokens, and the root S gets at least three tokens. Moreover, S gets exactly three tokens only if $\text{coreq}(S) = \frac{1}{2}$.*

We focus on the case when S owns some vertices in T, and show that in such case S can get at least four tokens. In the induction step, we assume that the induction hypothesis holds for each child of S. We say a child of S is a *rich* child if it gets at least four tokens, and say a child of S is a *poor* child if it gets exactly three tokens. Note that a child only needs two tokens and thus has some *excess* tokens, i.e., each rich child of S has at least two excess tokens and each poor child of S has one excess token. The following lemma is the technical core of this paper.

Lemma 5. *Let $S \in \mathcal{L}$. Suppose that the induction hypothesis in Lemma 4 holds for each child of S and S owns $k \geq 1$ vertices in T. Then the number of excess tokens from the child nodes of S, plus the number of tokens collected from endpoints owned by S is at least $2k + 4$.*

Lemma 5 handles the cases when S owns some vertex in T, to guarantee that S gets at least four tokens and each vertex in T owned by S gets two tokens for the induction hypothesis. The remaining cases can be handled exactly as in the proof in [14]. Please refer to the full version for the proof of Lemma 4 assuming Lemma 5, which follows the proof in [14].

We will prove Lemma 5 in the remainder of this paper. We present the proof of Lemma 5 when $r_{\max} \geq 2$ in Section 3.1, which improves the result in [14] about Steiner networks. Due to space limit, please refer to the full version for the proof of Lemma 5 when $r_{\max} = 1$, which recovers the result in [14] about Steiner forest with a considerably simpler proof.

3.1 Proof of Lemma 5

Before we assume $r_{\max} \geq 2$, we prove two useful lemmas. The first lemma takes care of those vertices $w \in W$ with $d(w) \geq 2b_w + 3$.

Lemma 6. *If $d(w) \geq 2b_w + 3$ for $w \in W$, then w receives at least four tokens in the initial token assignment.*

Proof. Assume there are h heavy edges incident and l light edges incident to w. If $l \geq 4$, then w receives at least four tokens in the initial token assignment. Suppose to the contrary that $l \leq 3$. Then $h \geq 2b_w$ as $d(w) = h + l \geq 2b_w + 3$. If $h > 2b_w$, then $x(\delta(w)) \geq \frac{h}{2} > b_w$, contradicting that x is a feasible solution to (LP). Otherwise, if $h = 2b_w$, since each light edge has positive value, we have $x(\delta(w)) > \frac{h}{2} = b_w$, again contradicting that x is a feasible solution to (LP). □

Lemma 6 says that any vertex $w \in W$ with $d(w) \geq 2b_w + 3$ gets at least four tokens in the initial assignment. Together with the fact that b_w is an integer, then any $w \in T$ with $d(w) \geq 2b_w + 3$ is a singleton set $\{w\}$ with $x(\delta(w))$ integral and has at least four tokens, and thus it behaves the same as a rich child in the proof of Lemma 5. Henceforth, we can assume that $b_w + 3r_{\max} + 1 < 2b_w + 3$ for each $w \in W$, and thus

$$b_w > 3r_{\max} - 2 \text{ and } d(w) \geq b_w + 3r_{\max} + 1 \text{ for } w \in W. \qquad (1)$$

The second lemma takes care of the poor children using the concept of integrality gap of the degree of a set. For an edge e with $0 < x_e < 1$, let $y_e = 1 - x_e \in (0,1)$ be the *integrality gap of e*. For a subset of edges $E' \subseteq E$, let $y(E') := \sum_{e \in E'} y_e$.

Lemma 7. *Suppose the induction hypothesis in Lemma 4 holds for each child of $S \in \mathcal{L}$. Let $R \in \mathcal{L}$ be a poor child of S. Then*

$$y(\delta^h(R)) \leq \frac{1}{2}.$$

Proof. Note that

$$\frac{1}{2} = \text{coreq}(R) = \sum_{e \in \delta^h(R)} (1 - x_e) + \sum_{e \in \delta^l(R)} (\frac{1}{2} - x_e) = \sum_{e \in \delta^h(R)} y_e + \sum_{e \in \delta^l(R)} (\frac{1}{2} - x_e).$$

Since $1/2 - x_e > 0$ for each light edge e, we have $\sum_{e \in \delta^h(R)} y_e \leq 1/2$. □

Let $w_1, \ldots, w_k$ be the vertices in T owned by S. The main idea in the proof of Lemma 5 is to consider $Y := \sum_{i=1}^{k} y(\delta(w_i))$. Since $w_i \in T$, it follows from (1) that $y(\delta(w_i)) = d(w_i) - x(\delta(w_i)) = d(w_i) - b_{w_i} \geq 3r_{\max} + 1$ and thus

$$Y \geq (3r_{\max} + 1)k.$$

By Lemma 7, the heavy edges from a poor child can only contribute very little to this sum, and this will allow us to rule out the existence of a poor child in S.

Proof of Lemma 5 when $r_{\max} \geq 2$: First, we count the number of tokens that S can collect. Consider the edges in $F := \cup_{i=1}^{k} \delta(w_i)$. Let $F_2 \subseteq F$ be the subset of edges of F with both endpoints in S, and $F_1 := F - F_2$ be the subset of edges of F with exactly one endpoint in S. Note that each edge in F_2 can contribute two tokens to S, regardless of whether it is heavy or light. Let ℓ be the number of light edges in F_1. Then each such edge can contribute one token to S. Let γ be the number of rich children of S and ρ be the number of poor children of S. By the induction hypothesis, the children can contribute at least $2\gamma + \rho$ tokens to S. Therefore, S can collect at least $2\gamma + \rho + \ell + 2|F_2|$ tokens from its children and the endpoints that it owns. If $2\gamma + \rho + \ell + 2|F_2| \geq 2k + 4$, then we are done. So we assume to the contrary that

$$2\gamma + \rho + \ell + 2|F_2| \leq 2k + 3. \tag{2}$$

Next, we consider the contribution to Y. Each endpoint of an edge e can contribute strictly less than one to Y, as $y_e = 1 - x_e < 1$ for each edge. So, the edges in F_2 and the light edges in F_1 can contribute strictly less than $2|F_2| + \ell$ to Y. It remains to count the contribution from the heavy edges in F_1. The heavy edges from a rich child R can contribute at most $r_{\max}$ to Y, because each heavy edge can contribute at most $\frac{1}{2}$ to Y and $|\delta^h(R)| \leq 2r_{\max}$ as $x(\delta(R)) \leq r_{\max}$. The heavy edges from a poor child can contribute at most $\frac{1}{2}$ to Y by Lemma 7. Finally, the heavy edges in $F_1 \cap \delta^h(S)$ can contribute at most $r_{\max}$ to Y, because $|\delta^h(S)| \leq 2r_{\max}$ as $x(\delta(S)) \leq r_{\max}$. These count all the contributions to Y. Therefore, we must have

$$(3r_{\max} + 1)k \leq Y \leq (\gamma + 1) \cdot r_{\max} + \frac{1}{2}\rho + \ell + 2|F_2|. \tag{3}$$

To satisfy (2) as an equality, we must have $\rho + \ell + 2|F_2| \geq 1$ since γ is an integer. If $\ell + 2|F_2| \geq 1$, then the second inequality in (3) is a strict inequality. Otherwise, if $\rho \geq 1$ or (2) is not an equality, by plugging in (2), we also have the following strict inequality

$$(3r_{\max} + 1)k < (\gamma + 1) \cdot r_{\max} + 2k + 3 - 2\gamma \iff (3k - 1)r_{\max} < k + 3 + \gamma(r_{\max} - 2). \tag{4}$$

As $\gamma \leq k + 1$ by (2) and assuming $r_{\max} \geq 2$, we have

$$(3k - 1)r_{\max} < k + 3 + (k + 1)(r_{\max} - 2) \iff r_{\max}(2k - 2) < 1 - k,$$

which is impossible for $k \geq 1$. This contradiction shows that (2) cannot hold, and thus S can collect $2k + 4$ tokens as claimed.   □

The proof for $r_{\max} = 1$ will be shown in the full version.

3.2 A Hard Example for the Algorithm

A natural question is that, for the minimum bounded degree Steiner forest problem, whether we can improve our algorithm further by only relaxing vertices with degree at most $b_v + 2$. This would imply a $(2, b_v + 2)$-approximation algorithm for the problem, matching the known integrality gap for this linear program [12].

In the example shown in Figure 2, some vertices have a degree bound equal to two, but there are five edges incident to these vertices. This is an extreme point solution to (LP) as the characteristic vectors are linearly independent. Our algorithm will get stuck in this example, and it is not clear to us how to modify the algorithm to deal with it. We believe that some new ideas are needed to obtain a $(2, b_v + 2)$-approximation algorithm for this problem.

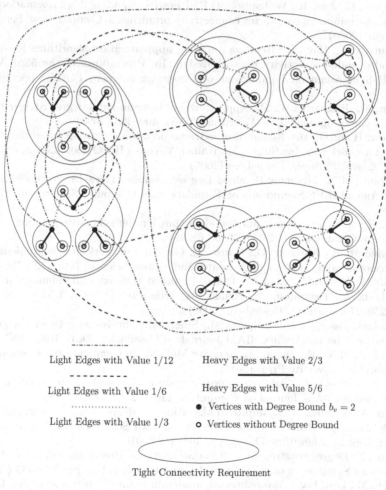

Light Edges with Value 1/12   Heavy Edges with Value 2/3

Light Edges with Value 1/6   Heavy Edges with Value 5/6

Light Edges with Value 1/3   ● Vertices with Degree Bound $b_v = 2$

                                 ○ Vertices without Degree Bound

Tight Connectivity Requirement

Fig. 2. A hard example

Acknowledgement. We thank the anonymous reviewers for comments that improve the presentation of the paper. This research is partially supported by HK RGC grant 2150701.

References

1. Bansal, N., Khandekar, R., Nagarajan, V.: Additive guarantees for degree bounded directed network design. SIAM Journal on Computing 29, 1413–1431 (2009)
2. Cheriyan, J., Vegh, L.: Approximating minimum-cost k-node connected subgraphs via independence-free graphs. In: Proceedings of the 54th Annual IEEE Symposium on Foundations of Computer Science (FOCS) (2013)
3. Cheriyan, J., Vempala, S., Vetta, A.: Network design via iterative rounding of setpair relaxations. Combinatorica 26(3), 255–275 (2006)
4. Fleischer, L., Jain, K., Williamson, D.P.: Iterative rounding 2-approximation algorithms for minimum-cost vertex connectivity problems. J. Comput. Syst. Sci. 72(5), 838–867 (2006)
5. Fukunaga, T., Ravi, R.: Iterative rounding approximation algorithms for degree-bounded node-connectivity network design. In: Proceedings of the 53rd Annual IEEE Symposium on Foundations of Computer Science (FOCS), pp. 263–272 (2012)
6. Fürer, M., Raghavachari, B.: Approximating the minimum-degree Steiner tree to within one of optimal. J. of Algorithms 17(3), 409–423 (1994)
7. Gabow, H.N.: On the L_∞-norm of extreme points for crossing supermodular directed network LPs. In: Jünger, M., Kaibel, V. (eds.) IPCO 2005. LNCS, vol. 3509, pp. 392–406. Springer, Heidelberg (2005)
8. Goemans, M.X.: Minimum Bounded-Degree Spanning Trees. In: Proceedings of the 47th Annual IEEE Symposium on Foundations of Computer Science, pp. 273–282 (2006)
9. Jain, K.: A factor 2-approximation algorithm for the generalized steiner network problem. Combinatorica 21(1), 39–60 (2001)
10. Khandekar, R., Kortsarz, G., Nutov, Z.: On some network design problems with degree constraints. Journal of Computer and System Sciences 79(5), 725–736 (2013)
11. Király, T., Lau, L.C., Singh, M.: Degree Bounded Matroids and Submodular Flows. In: Lodi, A., Panconesi, A., Rinaldi, G. (eds.) IPCO 2008. LNCS, vol. 5035, pp. 259–272. Springer, Heidelberg (2008)
12. Lau, L.C., Naor, S., Salavatipour, M., Singh, M.: Survivable network design with degree or order constraints. SIAM Journal on Computing 39(3), 1062–1087 (2009)
13. Lau, L.C., Ravi, R., Singh, M.: Iterative Methods in Combinatorial Optimization. Cambridge University Press (2011)
14. Lau, L.C., Singh, M.: Additive approximation for bounded degree survivable network design. SIAM Journal on Computing 42(6), 2217–2242 (2014)
15. Louis, A., Vishnoi, N.K.: Improved algorithm for degree bounded survivable network design problem. In: Proceedings of the 12th Scandinavian Symposium and Workshops on Algorithm Theory, pp. 408–419 (2010)
16. Nutov, Z.: Degree-constrained node-connectivity. In: Proceedings of the 10th Latin American Symposium on Theoretical Informatics (LATIN), pp. 582–593 (2012)
17. Singh, M., Lau, L.C.: Approximating minimum bounded degree spanning trees to within one of optimal. In: Proceedings of the 39th ACM Symposium on Theory of Computing (STOC), pp. 661–670 (2007)

Scheduling and Fixed-Parameter Tractability

Matthias Mnich[1] and Andreas Wiese[2]

[1] Cluster of Excellence MMCI, Saarbrücken, Germany
`m.mnich@mmci.uni-saarland.de`
[2] Max Planck Institute for Computer Science, Saarbrücken, Germany
`awiese@mpi-inf.mpg.de`

Abstract. Fixed-parameter tractability analysis and scheduling are two core domains of combinatorial optimization which led to deep understanding of many important algorithmic questions. However, even though fixed-parameter algorithms are appealing for many reasons, no such algorithms are known for many fundamental scheduling problems.

In this paper we present the first fixed-parameter algorithms for classical scheduling problems such as makespan minimization, scheduling with job-dependent cost functions—one important example being weighted flow time—and scheduling with rejection. To this end, we identify crucial parameters that determine the problems' complexity. In particular, we manage to cope with the problem complexity stemming from numeric input values, such as job processing times, which is usually a core bottleneck in the design of fixed-parameter algorithms. We complement our algorithms with W[1]-hardness results showing that for smaller sets of parameters the respective problems do not allow FPT-algorithms. In particular, our positive and negative results for scheduling with rejection explore a research direction proposed by Dániel Marx.

1 Introduction

Scheduling and fixed-parameter tractability are two very well-studied research areas. In scheduling, the usual setting is that one is given a set of machines and a set of jobs with individual characteristics. The jobs need to be scheduled on the machines according to some problem-specific constraints, such as release dates, precedence constraints, or rules regarding preemption and migration. Typical objectives are minimizing the global makespan, the weighted sum of completion times of the jobs, or the total flow time. During the last decades of research on scheduling, many important algorithmic questions have been settled. For instance, for minimizing the makespan and the weighted sum of completion time on identical machines, $(1 + \epsilon)$-approximation algorithms (PTASs) are known for almost all NP-hard settings [1,2].

However, the running time of these approximation schemes usually has a bad dependence on ϵ, and in practise exact algorithms are often desired. These and other considerations motivate to study which scheduling problems are *fixed-parameter tractable* (FPT), which amounts to identifying instance-dependent parameters k that allow for algorithms that find optimal solutions in time

J. Lee and J. Vygen (Eds.): IPCO 2014, LNCS 8494, pp. 381–392, 2014.

$f(k) \cdot n^{O(1)}$ for instances of size n and some function f depending only on k. Separating the dependence of k and n is often much more desirable than a running time of, e.g., $O(n^k)$, which becomes infeasible even for small k and large n. The parameter k measures the complexity of a given instance and thus, problem classification according to parameters yields an instance-depending measure of problem hardness.

Despite the fundamental nature of scheduling problems, and the clear advantages of fixed-parameter algorithms, to the best of our knowledge no such algorithms are known for the classical scheduling problems we study here. One obstacle towards obtaining positive results appears to be that—in contrast to most problems known to be fixed-parameter tractable—scheduling problems involve many numerical input data (e.g., job processing times, release dates, job weights), which alone render many problems NP-hard, thus ruling out fixed-parameter algorithms. One contribution of this paper is that—for the fundamental problems studied here—choosing the number of distinct numeric values or an upper bound on them as the parameter suffices to overcome this impediment. Note that this condition is much weaker than assuming the parameter to be bounded by a constant (that can appear in the exponent of the run time).

1.1 Our Contributions

In this paper we present the first fixed-parameter algorithms for several fundamental scheduling problems. In Section 2 we study one of the most classical scheduling problems, which is minimizing the makespan on an arbitrary number of machines without preemption, i.e. the problem $P||C_{\max}$. Assuming integral input data, our parameter $p_{\max}$ defines an upper bound on the job processing times appearing in an instance with n jobs. We first prove that for any number of machines, we can restrict ourselves to (optimal) solutions where jobs of the same length are almost equally distributed among the machines, up to an additive error term of $\pm f(p_{\max})$ jobs. This insight can be used as an independent preprocessing routine which optimally assigns the majority of the jobs of an instance (given that $n \gg p_{\max}$). After this preparation, we show that the remaining problem can be formulated as an integer program in fixed dimension, yielding an overall running time bounded by $f(p_{\max}) \cdot n^{O(1)}$. We note that without a fixed parameter, the problem is strongly NP-hard. For the much more general machine model of unrelated machines, we show that $R||C_{\max}$ is fixed-parameter tractable when choosing the number of machines and the number of distinct processing times as parameters. We reduce this problem again to integer programming in fixed dimension where our variables model how many jobs of each size are scheduled on each machine. To ensure that an assignment of all given jobs to these "slots" exists we argue via Hall's Theorem and ensure that for each subset of jobs there are enough usable slots. We remark that these problems are sufficiently complex so that we do not see a way of using the "number of numbers" result by Fellows et al. [3]. Note that if the number of machines or the number of processing times are constant, the problem is still NP-hard [4], and thus no fixed-parameter algorithms can exist for those cases (if P $\neq$ NP).

Then, in Section 3, we study scheduling with rejection. Each job j is specified by a processing time p_j, a weight w_j, and a rejection cost e_j (all jobs are released at time zero). We want to reject a set J' of at most k jobs, and schedule all other jobs on one machine to minimize $\sum_{j \notin J'} w_j C_j + \sum_{j \in J'} e_j$. We identify three key parameters: the number of distinct processing times, the number of distinct weights, and the maximum number k of jobs to be rejected. We show that if any two of the three values are taken as parameters, the problem becomes fixed-parameter tractable. If k and either of the other two are parameters, then we show that an optimal solution is characterized by one of sufficiently few possible patterns of jobs to be rejected. Once we guessed the correct pattern, an actual solution can be found by a dynamic program efficiently. If the number of distinct processing times and lengths are parameters (but not k), we provide a careful modeling of the problem as an integer program with convex objective function in fixed dimension. To the best of our knowledge, this is the first time that convex programming is used in fixed-parameter algorithms. We complement this result by showing that if only the number of rejected jobs k is the fixed parameter, then the problem becomes W[1]-hard, which prohibits the existence of a fixed-parameter algorithm, unless FPT = W[1] (which would imply subexponential time algorithms for many canonical NP-complete problems). Our results respond to a question by Dániel Marx [5] for investigating the fixed-parameter tractability of scheduling with rejection.

Finally, in Section 4 we turn our attention to the parametrized dual of the latter problem: scheduling with rejection of at least $n - s$ jobs (s being the parameter). We reduce this to a much more general problem which can be cast as the profit maximization version of the *General Scheduling Problem (GSP)* [6]. We need to select a subset J' of at most s jobs to schedule from a given set J, and each scheduled job j yields a profit $f_j(C_j)$, depending on its completion time C_j. Note that this function can be different for each job and might stem from a difficult scheduling objective such as weighted flow time. Additionally, each job j has a release date r_j and a processing time p_j. The goal is to schedule these jobs on one machine to maximize $\sum_{j \in J'} f_j(C_j)$. We study the preemptive as well as the non-preemptive version of this problem. In its full generality, GSP is not well understood. Despite that, we are able to give a fixed-parameter algorithm if the number of distinct processing times is bounded by a parameter, as well as the maximum cardinality of J'. We complement our findings by showing that for fewer parameters the problem is W[1]-hard or para-NP-hard, respectively, see Table 1. Our contributions are summarized in Table 1.

Due to space constraints, proofs are deferred to the full version of this paper.

1.2   Related Work

Scheduling. One classical scheduling problem studied in this paper is to schedule a set of jobs non-preemptively on a set of m identical machines, i.e., $P||C_{\max}$. Research for it dates back to the 1960s when Graham showed that the greedy list scheduling algorithm yields a $(2 - \frac{1}{m})$-approximation and a $4/3$-approximation when the jobs are ordered non-decreasingly by length [7]. After a series of

Table 1. Summary of our results. For a job j we denote by p_j its processing time, by w_j its weight, by e_j its rejection cost, by f_j its cost function. and by C_j its completion time in a computed schedule.

Problem	Parameters	Result
$P\|\|C_{\max}$	maximum p_j	FPT
$R\|\|C_{\max}$	#distinct p_j and #machines	FPT
$1\|\|\sum_{\leq k} e_j + \sum w_j C_j$	#rejected jobs k and #distinct p_j	FPT
$1\|\|\sum_{\leq k} e_j + \sum w_j C_j$	#rejected jobs k and #distinct w_j	FPT
$1\|\|\sum_{\leq k} e_j + \sum w_j C_j$	#distinct p_j and #distinct w_j	FPT
$1\|\|\sum_{\leq k} e_j + \sum w_j C_j$	#rejected jobs k	W[1]-hard
$1\|r_j,(pmtn)\|\max\sum_{\leq s} f_j(C_j)$	#selected jobs s and #distinct p_j	FPT
$1\|r_j,(pmtn)\|\max\sum_{\leq s} f_j(C_j)$	#selected jobs s	W[1]-hard
$1\|r_j,(pmtn)\|\max\sum_{\leq s} f_j(C_j)$	#distinct p_j (in fact $\forall p_j \in \{1,3\}$)	para-NP-hard

improvements [8,9,10,11], Hochbaum and Shmoys present a polynomial time approximation scheme (PTAS), even if the number of machines is part of the input [2]. On unrelated machines, the problem is NP-hard to approximate with a better factor than $3/2$ [12,4] and there is a 2-approximation algorithm [4] that extends to the generalized assignment problem [13]. For the restricted assignment case, i.e., each job has a fixed processing time and a set of machines where one can assign it to, Svensson [14] gives a polynomial time algorithm that estimates the optimal makespan up to a factor of $33/17 + \epsilon \approx 1.9412 + \epsilon$.

For scheduling jobs with release dates preemptively on one machine, a vast class of important objective functions is captured by the General Scheduling Problem (GSP). In its full generality, Bansal and Pruhs [6] give a $O(\log\log nP)$-approximation, where P is the maximum ratio of processing times. One particularly important special case is the weighted flow time objective where previously to Bansal and Pruhs [6] the best known approximation factors where $O(\log^2 P)$, $O(\log W)$, and $O(\log nP)$ [15,16]; where W is the maximum ratios of job weights. Also, a quasi-PTAS with running time $n^{O(\log P \log W)}$ is known [17].

A generalization of classical scheduling problems is *scheduling with rejection*. There, each job j is has additionally a *rejection cost* e_j. The scheduler has the freedom to reject job j and to pay a penalty of e_j, in addition to some (ordinary) objective function for the scheduled jobs. For one machine and the objective being to minimize the sum of weighted completion times, Engels et al. [18] give an optimal pseudopolynomial dynamic program for the case that all jobs are released at time zero and show that the problem is weakly NP-hard. Sviridenko and Wiese [19] give a PTAS for arbitrary release dates. For objective the makespan minimization and given multiple machines, Hoogeveen et al. [20] give FPTASs for almost all machine settings, and a 1.58-approximation for the APX-hard case of an arbitrary number of unrelated machines (when allowing preemption).

In high-multiplicity scheduling, one considers the setting where there are only few different job types, with jobs of the same type appearing in large bulks; one might consider the number of job types as a fixed parameter. We refer to the survey Brauner et al. [21].

Fixed-Parameter Tractability. Until now, to the best of our knowledge, no fixed-parameter algorithms for the classical scheduling problems studied in this paper have been devised. In contrast, classical scheduling problems investigated in the framework of parameterized complexity appear to be intractable; for example, k-processor scheduling with precedence constraints is W[2]-hard [22] and scheduling unit-length tasks with deadlines and precedence constraints and k tardy tasks is W[1]-hard [23], for parameter k. Mere exemptions seem to be an algorithm by Marx and Schlotter [24] for makespan minimization where k jobs have processing time $p \in \mathbb{N}$ and all other jobs have processing time 1, for combined parameter (k, p), as well as work of Alon et al. [25] who show that makespan-minimization on m identical machines is fixed-parameter tractable parameterized by the optimal makespan. We also mention that Chu et al. [26] consider the parameterized complexity of checking *feasibility* of a schedule (rather than optimization). We remark that some scheduling-type problems can be addressed by choosing as parameter the "number of numbers", as done by Fellows et al. [3].

2   Minimizing the Makespan

We first consider the problem $P||C_{\max}$, where a given a set J of n jobs (with individual processing time p_j and released at time zero) must be scheduled non-preemptively on a set of m identical machines, as to minimize the makespan of the schedule. We develop a fixed-parameter algorithm solving this problem in time $f(p_{\max}) \cdot n^{O(1)}$, where $p_{\max}$ is the maximum processing time over all jobs.

In the sequel, we say that some job j is of *type* t if $p_j = t$; we define $J_t := \{j \in J \mid p_j = t\}$. First, we prove that there is always an optimal solution in which each machine has almost the same number of jobs of each type, up to an additive error of $\pm f(p_{\max})$ for suitable function f. This allows us to fix some jobs on the machines. For the remaining jobs, we show that each machine receives at most $2f(p_{\max})$ jobs of each type; hence there are only $(2f(p_{\max}))^{p_{\max}}$ possible configurations for each machine. We solve the remaining problem with an integer linear program in fixed dimension.

As a first step, for each type t, we assign $\left\lfloor \frac{|J_t|}{m} \right\rfloor - f(p_{\max})$ jobs of type t to each machine; let $J_0 \subseteq J$ be this set of jobs. This is justified by the next lemma, which can be proven by starting with an arbitrary optimal schedule and exchanging jobs carefully between machines until the claimed property holds.

Lemma 1. *There is a function $f : \mathbb{N} \to \mathbb{N}$ with $f(p_{\max}) \leq 2^{O(p_{\max} \cdot \log p_{\max})}$ for all $p_{\max} \in \mathbb{N}$ such that every instance of $P||C_{\max}$ admits an optimal solution in which for each type t, each of the m machines schedules at least $\lfloor |J_t|/m \rfloor - f(p_{\max})$ and at most $\lfloor |J_t|/m \rfloor + f(p_{\max})$ jobs of type t.*

Denote by $J' = J \setminus J_0$ be the set of yet unscheduled jobs. We ignore all other jobs from now on. By Lemma 1, there is an optimal solution in which each machine receives at most $2 \cdot f(p_{max}) + 1$ jobs from each type. Hence, there are at most $(2 \cdot f(p_{max}) + 2)^{p_{max}}$ ways how the schedule for each machine can look like (up to permuting jobs of the same length). Therefore, the remaining problem can be solved with the following integer program. Define a set $\mathcal{C} = \{0, \ldots, 2 \cdot f(p_{max}) + 1\}^{p_{max}}$ of at most $(2 \cdot f(p_{max}) + 2)^{p_{max}}$ "configurations", where each *configuration* is a vector $C \in \mathcal{C}$ encoding the number of jobs from J' of each type assigned to a machine.

In any optimal solution for J', the makespan is in the range $\{\lceil p(J')/m \rceil, \ldots, \lceil p(J')/m \rceil + p_{max}\}$, where $p(J') = \sum j \in J' p_j$, as $p_j \leq p_{max}$ for each j. For each value T in this range we try whether opt $\leq T$, where opt denotes the minimum makespan of the instance. So fix a value T. We allow only configurations $C = (c_1, \ldots, c_{p_{max}})$ which satisfy $\sum_{i=1}^{p_{max}} c_i \cdot i \leq T$; let $\mathcal{C}(T)$ be the set of these configurations. For each $C \in \mathcal{C}(T)$, introduce a variable y_C for the number of machines with configuration C in the solution. (As the machines are identical, only the number of machines following each configuration is important.)

$$\sum_{C \in \mathcal{C}(T)} y_C \leq m \tag{1}$$

$$\sum_{C=(c_1,\ldots,c_{p_{max}}) \in \mathcal{C}(T)} y_C \cdot c_p \geq |J' \cap J_p|, \ p = 0, \ldots, p_{max} \tag{2}$$

$$y_C \in \{0, \ldots, m\}, \ C \in \mathcal{C}(T) \tag{3}$$

Inequality (1) ensures that at most m machines are used, inequalities (2) ensure that all jobs from each job type are scheduled. The whole integer program (1)–(3) has at most $(2 \cdot f(p_{max}) + 2)^{p_{max}}$ dimensions.

To determine feasibility of (1)–(3), we employ deep results about integer programming in fixed dimension. As we will need it later, we cite here an algorithm due to Heinz [27,28] that can even minimize over convex spaces described by (quasi-)convex polynomials, rather than only over polytopes.

Theorem 1 ([27,28]). *Let $f, g_1, \ldots, g_m \in \mathbb{Z}[x_1, \ldots, x_t]$ be quasi-convex polynomials of degree at most $d \geq 2$, whose coefficients have binary encoding length at most ℓ. There is an algorithm that in time $m \cdot \ell^{O(1)} \cdot d^{O(t)} \cdot 2^{O(t^3)}$ computes a minimizer $\mathbf{x}^* \in \mathbb{Z}^t$ of the following problem (4) or reports that no minimizer exists. If the algorithm outputs a minimizer $\mathbf{x}^*$, its binary encoding size is $\ell \cdot d^{O(t)}$.*

$$\min f(x_1, \ldots, x_t), \quad \text{subject to } g_i(x_1, \ldots, x_t) \leq 0, \quad i = 1, \ldots, m \quad \mathbf{x} \in \mathbb{Z}^t \ . \tag{4}$$

The smallest value T for which (1)–(3) is feasible gives the optimal makespan and together with the preprocessing routine of Lemma 1 yields an optimal schedule.

Theorem 2. *There is a function f such that instances of $P||C_{max}$ with n jobs and m machines can be solved in time $f(p_{max}) \cdot (n + \log m)^{O(1)}$.*

Recall that without choosing a parameter, problem $P||C_{max}$ is strongly NP-hard (as it contains 3-PARTITION). When parameterizing $P||C_{max}$ by the number $\bar{p}$ of distinct processing times, Lemma 1 is no longer true (details deferred to full version of this paper). However, for constantly many processing times the problem was recently shown to be polynomial time solvable for any constant $\bar{p}$ [29].

2.1   Bounded Number of Unrelated Machines

We study the problem $Rm||C_{\max}$ where now the machines are unrelated, meaning that a job can have different processing times on different machines. In particular, it might be that a job cannot be processed on some machine at all, i.e., has infinite processing time on that machine. We choose as parameters the number $\bar{p}$ of distinct (finite) processing times and the number of machines m of the instance.

We model this problem as an integer program in fixed dimension. Denote by $q_1, \ldots, q_{\bar{p}}$ the distinct finite processing times in a given instance. For each combination of a machine i and a finite processing time q_ℓ we introduce a variable $y_{i,\ell} \in \{0, \ldots, n\}$ that models how many jobs of processing time q_ℓ are assigned to i. Note that the number of these variables is bounded by $m \cdot \bar{p}$. An assignment to these variables can be understood as allocating $y_{i,\ell}$ slots for jobs with processing time q_ℓ to machine i, without specifying what actual jobs are assigned to these slots. Assigning the jobs to the slots can be understood as a bipartite matching: introduce one vertex v_j for each job j, one vertex $w_{s,i}$ for each slot s on each machine i, and an edge $\{v_j, w_{s,i}\}$ whenever job j has the same size on machine i as slot s. According to Hall's Theorem, there is a matching in which each job is matched if and only if for each set of jobs $J' \subseteq J$ there are at least $|J'|$ slots to which at least one job in J' can be assigned. For one single set J' the latter can be expressed by a linear constraint; however, the number of subsets J' is exponential. We overcome this as follows: We say that two jobs j, j' are of the same *type* if $p_{i,j} = p_{i,j'}$ for each machine i. Note that there are only $(\bar{p} + 1)^m$ different types of jobs. As we will show, it suffices to add a constraint for sets of jobs J' such that for each job type either all or none of the jobs of that type are contained in J' (those sets „dominate" all other sets). This gives rise to the following integer program. Denote by Z the set of all job types and for each $z \in Z$ denote by $J_z \subseteq J$ the set of jobs of type z. For each set $Z' \subseteq Z$ denote by $Q_{i,Z'}$ the set of distinct finite processing times of jobs of types in Z' on machine i. Using Theorem 1 we can solve the following IP:

$$\min T \quad \text{s.t.} \quad \sum_{\ell \in \{1, \ldots, \bar{p}\}} y_{i,\ell} \cdot q_\ell \leq T, \qquad i = 1, \ldots, m \tag{5}$$

$$\sum_{z \in Z'} |J_z| \leq \sum_i \sum_{\ell : q_\ell \in Q_{i,Z'}} y_{i,\ell} \quad \forall \, Z' \subseteq Z \tag{6}$$

$$y_{i,\ell} \in \{0, \ldots, n\} \qquad i = 1, \ldots, m, \ \ell = 1, \ldots, \bar{p} \tag{7}$$

$$T \geq 0 \tag{8}$$

Theorem 3. *Instances of $R||C_{\max}$ with m machines and n jobs with $\bar{p}$ distinct finite processing times $q_1, \ldots, q_{\bar{p}}$ can be solved in time $f(\bar{p}, m) \cdot (n + \log \max_\ell q_\ell)^{O(1)}$ for a suitable function f.*

A natural question is the case when only the number of machines is a fixed parameter. Even if one requires the processing times to be polynomially bounded, then already $P||C_{\max}$ is still W[1]-hard [30]. On the other hand, $R||C_{\max}$ is NP-hard if only processing times $\{1, 2, \infty\}$ are allowed [12,4]. This justifies to take both m and $\bar{p}$ as a parameters in the unrelated machine case.

3   Scheduling with Rejection

In this section we study scheduling with rejection to optimize the weighted sum of completion time plus the total rejection cost, i.e, $1||\sum_{\leq k} e_j + \sum w_j C_j$. Formally, we are given an integer k and a set J of n jobs, all released at time zero. Each job $j \in J$ is characterized by a processing time $p_j \in \mathbb{N}$, a weight $w_j \in \mathbb{N}$ and rejection cost $e_j \in \mathbb{N}$. The goal is to reject a set $J' \subseteq J$ of at most k jobs and to schedule all other jobs non-preemptively on a single machine, as to minimize $\sum_{j \in J \setminus J'} w_j C_j + \sum_{j \in J'} e_j$ where C_j denotes the completion time in the computed schedule.

3.1   Number of Rejected Jobs and Processing Times or Weights

Denote by $\bar{p} \in \mathbb{N}$ the number of distinct processing times in a given instance. First, we assume that $\bar{p}$ and the maximum number k of rejected jobs are parameters. Thereafter, using a standard reduction, we will derive an algorithm for the case that k and the number $\bar{w}$ of distinct weights are parameters.

Denote by $q_1, \ldots, q_{\bar{p}}$ the distinct processing times in a given instance. For each $i \in \{1, \ldots, \bar{p}\}$, we guess the number of jobs with processing time q_i which are rejected in an optimal solution. Each possible guess is characterized by a vector $\mathbf{v} = \{v_1, \ldots, v_{\bar{p}}\}$ whose entries v_i contain integers between 0 and k, and whose total sum is at most k. There are at most $(k + 1)^{\bar{p}}$ such vectors $\mathbf{v}$, each one prescribing that at most v_i jobs of processing time p_i can be rejected. We enumerate them all. One of these vectors must correspond to the optimal solution, so the reader may assume that we know this vector $\mathbf{v}$.

In the following, we will search for the optimal schedule that *respects* $\mathbf{v}$, meaning that for each $i \in \{1, \ldots, \bar{p}\}$ at most v_i jobs of processing time q_i are rejected. To find an optimal schedule respecting $\mathbf{v}$, we use a dynamic program. Suppose the jobs in J are labeled by $1, \ldots, n$ by non-increasing *Smith ratios* w_j/p_j. Each dynamic programming cell is characterized by a value $n' \in \{0, \ldots, n\}$, and a vector $\mathbf{v}'$ with $\bar{p}$ entries which is *dominated* by $\mathbf{v}$, meaning that $v'_i \leq v_i$ for each $i \in \{1, \ldots, \bar{p}\}$. For each pair $(n', \mathbf{v}')$ we have a cell $C(n', \mathbf{v}')$ modeling the following subproblem. Assume that for jobs in $J' := \{1, \ldots, n'\}$ we have already decided whether we want to schedule them or not. For each processing time q_i denote by n'_i the number of jobs in J' with processing time q_i. Assume that for each type i, we have decided to reject $v_i - v'_i$ jobs from J'. Note that then the total processing time of the scheduled jobs sums up to $t := \sum_i q_i \cdot (n'_i - (v_i - v'_i))$. It remains to define a solution for the jobs in $J'' := \{n' + 1, \ldots, n\}$ during time interval $[t, \infty)$, such that for each type i we can reject up to v'_i jobs. The problem described by each cell $C(n', \mathbf{v}')$ can be solved in polynomial time, given one has already computed the values for each cell $C(n'', \mathbf{v}'')$ with $n'' > n'$ (proof deferred). The size of the dynamic programming table is bounded by $n \cdot (k+1)^{\bar{p}}$. Since for $1||\sum w_j C_j$ one can interchange weights and processing times and get an equivalent instance [31, Theorem 3.1], we obtain the same result when there are only $\bar{p}$ distinct weights.

Theorem 4. *For sets J of n jobs with $\bar{p}$ distinct processing times or weights, the problem $1||\sum_{\leq k} e_j + \sum w_j C_j$ is solvable in time $O(n \cdot (k+1)^{\bar{p}} + n \cdot \log n)$.*

We show next that when only the number k of rejected jobs is taken as parameter, problem becomes W[1]-hard. (This requires that the numbers in the input can be super-polynomially large. Note that for polynomially bounded processing times the problem admits a polynomial time algorithm for arbitrary k [18].) This justifies to define additionally the number of weights or processing times as parameter. We remark that when jobs have non-trivial release dates, then even for $k = 0$ the problem is NP-hard [32].

Theorem 5. *The problem $1||\sum_{\leq k} e_j + \sum w_j C_j$ is W[1]-hard if the parameter is the number k of rejected jobs.*

3.2   Number of Distinct Processing Times and Weights

We consider the number of distinct processing times and weights as parameters. To this end, we say that two jobs j, j' are of the same *type* if $p_j = p_{j'}$ and $w_j = w_{j'}$; let τ be the number of types in an instance. Note, however, that jobs with the same type might have different rejection costs, so we cannot bound the "number of input numbers" like Fellows et al. [3]. Instead, we resort to convex integer programming, which to the best of our knowledge is used here for the first time in fixed-parameter algorithms. The running time of our algorithm will depend only *polynomially* on k, the upper bound on the number of jobs we are allowed to reject. For each type i, let $w^{(i)}$ be the weight and $p^{(i)}$ be the processing time of jobs of type i. Assume that job types are numbered $1, \ldots, \tau$ such $w^{(i)}/p^{(i)} \geq w^{(i+1)}/p^{(i+1)}$ for each $i \in \{1, \ldots, \tau - 1\}$. Clearly, an optimal solution schedules jobs ordered non-increasingly by Smith's ratio without preemption.

The basis for our algorithm is a problem formulation as a convex integer minimization problem with dimension at most 2τ. In an instance, for each i, we let n_i be the number of jobs of type i and introduce an integer variable $x_i \in \mathbb{N}_0$ modeling how many jobs of type i we decide to schedule. We introduce the linear constraint $\sum_{i=1}^{\tau}(n_i - x_i) \leq k$, to ensure that at most k jobs are rejected.

The objective function is more involved. For each type i, scheduling the jobs of type i costs

$$\sum_{\ell=1}^{x_i} w^{(i)} \cdot \left(\ell \cdot p^{(i)} + \sum_{i' < i} x_{i'} \cdot p^{(i')}\right) = w^{(i)} \cdot x_i \cdot \sum_{i' < i} x_{i'} \cdot p^{(i')} + w^{(i)} \cdot p^{(i)} \sum_{\ell=1}^{x_i} \ell$$

$$= w^{(i)} \cdot x_i \cdot \sum_{i' < i} x_{i'} \cdot p^{(i')} + w^{(i)} \cdot p^{(i)} \cdot \frac{x_i \cdot (x_i + 1)}{2} =: s_i(x).$$

Note that $s_i(x)$ is a convex polynomial of degree 2 (being the sum of quadratic polynomials with only positive coefficients). Observe that when scheduling x_i jobs of type i, it is optimal to reject the $n_i - x_i$ jobs with lowest rejection cost among all jobs of type i. Assume the jobs of each type i are labeled $j_1^{(i)}, \ldots, j_{n_i}^{(i)}$ by non-decreasing rejection cost. For each $s \in \mathbb{N}$ let $f_i(s) := \sum_{\ell=1}^{n_i-s} e_{j_\ell^{(i)}}$. In particular, to schedule x_i jobs of type i we can select them such that we need to pay $f_i(s)$ for rejecting the non-scheduled jobs (and this is an optimal decision). The difficulty is that the function $f_i(s)$ is in general not expressible by a polynomial whose degree is globally bounded (i.e., for each possible instance), which prevents a direct application of Theorem 1.

However, in Lemma 2 we show that $f_i(s)$ is the maximum of n_i linear polynomials, allowing us to formulate a convex program and solve it by Theorem 1.

Lemma 2. *For each type i there is a set of n_i polynomials $p_i^{(1)}, \ldots, p_i^{(n_i)}$ of degree one such that $f_i(s) = \max_\ell p_i^{(\ell)}(s)$ for each $s \in \{0, \ldots, n_i\}$.*

Lemma 2 allows modeling the entire problem with the following convex program, where for each type i, variable g_i models the rejection cost for jobs of type i.

$$\min \ \sum_{i=1}^{\tau} g_i + s_i(x) \text{ s.t. } \sum_{i=1}^{\tau} (n_i - x_i) \leq k,$$

$$g_i \geq p_i^{(\ell)}(x_i) \ \forall \, i \in \{1, \ldots, \tau\} \ \forall \, \ell \in \{1, \ldots, n_i\}, \mathbf{g}, \mathbf{x} \in \mathbb{Z}^\tau \geq 0 \ . \quad (9)$$

Observe that (9) admits an optimal solution with $g_i = \max_\ell p_i^{(\ell)}(x_i) = f_i(x_i)$ for each i. Thus, solving (9) yields an optimal solution to the overall instance.

Theorem 6. *For sets of n jobs of τ types the problem $1 || \sum_{\leq k} e_j + \sum w_j C_j$ can be solved in time $(n + \log(\max_j \max\{e_j, p_j, w_j\}))^{O(1)} \cdot 2^{O(\tau^3)}$.*

4   Profit Maximization for General Scheduling Problems

The parameterized dual problem of scheduling jobs with rejection is the problem to reject at least $n - s$ jobs (s being the parameter) to minimize the total cost given by the rejection penalties plus the cost of the schedule. This is equivalent to the following problem where here we allow even non-trivial release dates and job dependent profit functions:

We are given a set J of n jobs, where each job j is characterized by a release date r_j, a processing time p_j, and a non-increasing profit function $f_j(t)$. Let $\bar{p}$ denote the number of distinct processing times p_j in the instance. We want to schedule a set $\bar{J} \subseteq J$ of at most s jobs from J on a single machine. Our objective is to maximize $\sum_{j \in \bar{J}} f_j(C_j)$, where C_j denotes the completion time of j in the computed schedule. We call this problem the *s-bounded General Profit Scheduling Problem*, or *s-GPSP* for short. Observe that this generic problem definition allows to model profit functions that stem from difficult scheduling objectives such as weighted flow time or weighted tardiness.

Theorem 7. *There is a deterministic algorithm that, given an instance of s-GPSP with n jobs and $\bar{p}$ processing times, computes an optimal preemptive or non-preemptive schedule in time $2^{O(s)} s^{O(\bar{p})} n^4 \log n$.*

When only the number s of scheduled jobs is chosen as parameter the problem becomes W[1]-hard, as pointed out to us by an anonymous reviewer. The next theorem assumes that numeric input values are allowed to be exponentially large.

Theorem 8. *(Non-)preemptive s-GPSP is W[1]-hard for parameter s.*

On the other hand, we prove that choosing only the number of distinct processing times $\bar{p}$ as a parameter is not enough, as we show the problem to be NP-hard even if $p_j \in \{1, 3\}$ for all jobs j. The same holds for the related General Scheduling Problem (GSP) [6]. While all processing times are either 1 or 3, in our reduction we use cost functions whose values can be exponentially large.

Theorem 9. *The General Profit Scheduling Problem (GPSP) and the General Scheduling Problem (GSP) are (weakly) NP-hard, even if $p_j \in \{1, 3\}$ for each job j. This holds in both the preemptive and non-preemptive setting.*

Acknowledgment. We thank the IPCO reviewers, as well as an anonymous reviewer of an earlier version for suggestions how to improve the algorithms in Sect. 4 and to prove Theorem 8.

References

1. Afrati, F., Bampis, E., Chekuri, C., Karger, D., Kenyon, C., Khanna, S., Milis, I., Queyranne, M., Skutella, M., Stein, C., Sviridenko, M.: Approximation schemes for minimizing average weighted completion time with release dates. In: Proc. FOCS 1999, pp. 32–43 (1999)
2. Hochbaum, D.S., Shmoys, D.B.: Using dual approximation algorithms for scheduling problems: Theoretical and practical results. J. ACM 34, 144–162 (1987)
3. Fellows, M.R., Gaspers, S., Rosamond, F.A.: Parameterizing by the number of numbers. Theory Comput. Syst. 50(4), 675–693 (2012)
4. Lenstra, J.K., Shmoys, D.B., Tardos, É.: Approximation algorithms for scheduling unrelated parallel machines. Mathematical Programming 46(1-3), 259–271 (1990)
5. Marx, D.: Fixed-parameter tractable scheduling problems. In: Packing and Scheduling Algorithms for Information and Communication Services (Dagstuhl Seminar 11091), vol. 1, p. 86 (2011)
6. Bansal, N., Pruhs, K.: The geometry of scheduling. In: Proc. FOCS 2010, pp. 407–414 (2010)
7. Graham, R.L.: Bounds on multiprocessing timing anomalies. SIAM J. Appl. Math. 17, 263–269 (1969)
8. Coffman Jr., E.G., Garey, M.R., Johnson, D.S.: An application of bin-packing to multiprocessor scheduling. SIAM J. Comput. 7, 1–17 (1978)
9. Friesen, D.K.: Tighter bounds for the multifit processor scheduling algorithm. SIAM J. Comput. 13, 170–181 (1984)
10. Langston, M.A.: Processor scheduling with improved heuristic algorithms. PhD thesis, Texas A&M University (1981)

11. Sahni, S.: Algorithms for scheduling independent tasks. J. ACM 23, 116–127 (1976)
12. Ebenlendr, T., Krčál, M., Sgall, J.: Graph balancing: a special case of scheduling unrelated parallel machines. In: Proc. SODA 2008, pp. 483–490 (2008)
13. Shmoys, D.B., Tardos, É.: An approximation algorithm for the generalized assignment problem. Mathematical Programming 62(5), 461–474 (1993)
14. Svensson, O.: Santa claus schedules jobs on unrelated machines. SIAM J. Comput. 41(4), 1318–1341 (2012)
15. Bansal, N., Dhamdhere, K.: Minimizing weighted flow time. ACM Trans. Algorithms 3(4) (November 2007)
16. Chekuri, C., Khanna, S., Zhu, A.: Algorithms for minimizing weighted flow time. In: Proc. STOC 2001, pp. 84–93 (2001)
17. Chekuri, C., Khanna, S.: Approximation schemes for preemptive weighted flow time. In: Proc. STOC 2002, pp. 297–305 (2002)
18. Engels, D.W., Karger, D.R., Kolliopoulos, S.G., Sengupta, S., Uma, R.N., Wein, J.: Techniques for scheduling with rejection. J. Algorithms 49(1), 175–191 (2003)
19. Sviridenko, M., Wiese, A.: Approximating the configuration-LP for minimizing weighted sum of completion times on unrelated machines. In: Goemans, M., Correa, J. (eds.) IPCO 2013. LNCS, vol. 7801, pp. 387–398. Springer, Heidelberg (2013)
20. Hoogeveen, H., Skutella, M., Woeginger, G.J.: Preemptive scheduling with rejection. Math. Program. 94(2-3, Ser. B), 361–374 (2003)
21. Brauner, N., Crama, Y., Grigoriev, A., van de Klundert, J.: A framework for the complexity of high-multiplicity scheduling problems. J. Comb. Optim. 9(3), 313–323 (2005)
22. Bodlaender, H.L., Fellows, M.R.: W[2]-hardness of precedence constrained K-processor scheduling. Oper. Res. Lett. 18(2), 93–97 (1995)
23. Fellows, M.R., McCartin, C.: On the parametric complexity of schedules to minimize tardy tasks. Theoret. Comput. Sci. 298(2), 317–324 (2003)
24. Marx, D., Schlotter, I.: Stable assignment with couples: Parameterized complexity and local search. Discr. Optimization 8(1), 25–40 (2011)
25. Alon, N., Azar, Y., Woeginger, G.J., Yadid, T.: Approximation schemes for scheduling on parallel machines. J. Sched. 1(1), 55–66 (1998)
26. Chu, G., Gaspers, S., Narodytska, N., Schutt, A., Walsh, T.: On the complexity of global scheduling constraints under structural restrictions. In: Proc. IJCAI 2013 (2013)
27. Heinz, S.: Complexity of integer quasiconvex polynomial optimization. J. Complexity 21(4), 543–556 (2005)
28. Köppe, M.: On the complexity of nonlinear mixed-integer optimization. In: Mixed Integer Nonlinear Programming. The IMA Volumes in Mathematics and its Applications, vol. 154, pp. 533–557 (2012)
29. Goemans, M.X., Rothvoß, T.: Polynomiality for bin packing with a constant number of item types. In: Proc. SODA 2014 (to appear, 2014)
30. Jansen, K., Kratsch, S., Marx, D., Schlotter, I.: Bin packing with fixed number of bins revisited. J. Comput. System Sci. 79(1), 39–49 (2013)
31. Chudak, F.A., Hochbaum, D.S.: A half-integral linear programming relaxation for scheduling precedence-constrained jobs on a single machine. Oper. Res. Lett. 25(5), 199–204 (1999)
32. Lenstra, J., Kan, A.R., Brucker, P.: Complexity of machine scheduling problems. In: Studies in Integer Programming. Ann. Discrete Math, vol. 1, pp. 343–362 (1977)

Improved Branch-Cut-and-Price
for Capacitated Vehicle Routing

Diego Pecin[1], Artur Pessoa[2], Marcus Poggi[1], and Eduardo Uchoa[2]

[1] Departamento de Informática, Pontifícia Universidade Católica do Rio de Janeiro, Brazil
[2] Departamento de Engenharia de Produção, Universidade Federal Fluminense, Brazil

Abstract. The best performing exact algorithms for the Capacitated Vehicle Routing Problem are based on the combination of cut and column generation. Some authors could obtain reduced duality gaps by also using a restricted number of cuts over the Master LP variables, stopping separation before the pricing becomes prohibitively hard. This work introduces a technique for greatly decreasing the impact on the pricing of the Subset Row Cuts, thus allowing much more such cuts to be added. The newly proposed Branch-Cut-and-Price algorithm also incorporates and combines for the first time (often in an improved way) several elements found in previous works, like route enumeration and strong branching. All the instances used for benchmarking exact algorithms, with up to 199 customers, were solved to optimality. Moreover, some larger instances with up to 360 customers, only considered before by heuristic methods, were solved too.

1 Introduction

The Capacitated Vehicle Routing Problem (CVRP) can be defined as follows. The input consists of a set of $n + 1$ points, a *depot* and n *customers*; an $(n + 1) \times (n + 1)$ matrix $[c_{ij}]$ with the travel *costs* between every pair of points i and j; an n-dimensional *demand* vector $[d_i]$ giving the amount to be collected from customer i; and a vehicle *capacity* Q. A solution is a set of routes, starting and ending at the depot, that visits every customer exactly once. The only constraint on a route is that the sum of the demands of its customers does not exceed the vehicle capacity Q. The objective is to find a solution with minimum total cost. Many authors also assume that the *number of routes* is fixed to an additional input number K. The CVRP is a widely studied problem, being the most basic and prototypical VRP variant.

Fukasawa et al. [5] proposed a Branch-Cut-and-Price algorithm (BCP) that performed significantly better than the branch-and-cut algorithms (like [9]) that were the dominant approach for the problem. All of the most recent works proposing exact algorithms for the CVRP are based on the combination of column and cut generation. They are Baldacci, Christofides and Mingozzi [1], Pessoa et al. [11], Baldacci, Mingozzi and Roberti [2], Contardo [4] and Røpke [16]. The BCP algorithm proposed in this work contains elements from all those previous algorithms, usually enhanced and combined with new elements:

- According to the classification proposed in [14], a cut is said to be *robust* when the value of the dual variable associated to it can be translated into costs in the pricing subproblem. Therefore, the structure and the size of that subproblem remain unaltered, regardless of the number of robust cuts added. On the other hand, *non-robust*

J. Lee and J. Vygen (Eds.): IPCO 2014, LNCS 8494, pp. 393–403, 2014.

cuts, defined on the variables of the Master LP, are those that change the structure and/or the size of the pricing subproblem, each additional cut makes it harder. Nevertheless, it is known that non-robust cuts have a big potential for reducing duality gaps. The most important original contribution of this work is the introduction of the limited memory Subset Row Cuts (lm-SRCs). The traditional SRCs are non-robust cuts known to be effective [8], [2], [4]. However, their practical use has been restricted by their large impact on the pricing. The lm-SRCs are a weakening of the SRCs. This weakening can be controlled and dynamically adjusted, making the lm-SRCs as effective in improving the lower bounds as SRCs, but still much less costly in the pricing.

- The underlying formulation used in the BCP has extended arc-load variables. This allows a particularly effective fixing of variables by Lagrangean bounds (superior to the fixing in [7]), with direct benefits on the pricing.

Other elements to be remarked in the proposed BCP are: (i) the columns in the BCP are associated to ng-routes [2]. The corresponding pricing subproblem is solved by a labeling algorithm that must also consider the dual variables of the lm-SRCs. Its implementation is quite critical for the overall BCP performance. After experiments with a number of alternatives, the best performance was obtained by a bidirectional search that differs a little from the proposed in [15] because the concatenation of the labels is not necessarily performed at the half of the capacity. Completion bounds are also used for eliminating labels. Anyway, the exact pricing algorithm is called just a few times per BCP node, most of the iterations use effective heuristics. A column generation stabilization by dual smoothing [12] may be also employed; (ii) like in [11], the BCP hybridizes branching with the route enumeration technique introduced in [1]. Actually, inspired by [16], it performs an aggressive strong branching, with up to n candidates (partially) evaluated in the root node. The branching effort in each node depends on an estimate of the size of the subtree rooted in that node. The branching mechanism also keeps the history of candidate evaluations for helping on future decisions; (iii) as soon as the gap of a BCP node is sufficiently small, the elementary routes that can be part of the optimal solution can be enumerated into a large pool, as suggested in [4]. From that point, since the pricing will be performed by inspection, all lm-SRCs may be immediately lifted to SRCs and additional non-robust cuts, including cliques, may be separated; and (iv) the lm-SRCs are still non-robust cuts. There are cases where several hundreds such cuts are being normally handled by the pricing algorithm, and then, at some node deep in the tree, the separation of a few dozen additional lm-SRCs makes this algorithm 100 times slower. In this situation the BCP performs a rollback, the offending cuts are removed even if it decreases the lower bound of the node.

Overall, we believe that this is one of the most sophisticated BCP algorithms ever implemented. The techniques introduced in this work, including the lm-SRCs, can be possibly applied on many other problems where BCP is currently applied, including several VRP variants, parallel machine scheduling or network design.

2   Formulations

This work departs from an extended formulation for the Asymmetrical CVRP presented in [11]. Let $G = (V, A)$ be a complete directed graph where $V = \{0, \dots, n\}$ is the vertex set, vertex 0 is the depot and V_+ the customer set. A cost c_a is associated with each arc $a = (i, j) \in A$, symmetrical CVRP instances have symmetric costs, i.e., $c_{ij} = c_{ji}$. A positive integral demand d_i is associated to each customer $i \in V_+$, d_0 is defined as 0. Let Q denote the vehicle capacity. We assume a fixed number K of routes. Let $G_Q = (V, A_Q)$ be a multigraph $G_Q = (V, A_Q)$ where A_Q contains arcs $(i, j)^q$, for all $i \in V_+, j \in V$ and for all $q = d_i, \ldots, Q$, plus arcs $(0, j)^0$, for all $j \in V_+$. For any set $S \subseteq V$, $\delta^-(S) = \{(i, j)^q \in A_Q : i \in V \setminus S, j \in S\}$, and $\delta^+(S)$ is defined in a similar way. For each $(i, j)^q \in A_Q$, a binary variable x_{ij}^q indicates that some vehicle goes from i to j carrying a load — the sum of the demands of vertex i and its preceding vertices — of exact q units. The Arc-Load indexed Formulation is:

$$\text{(ALF)} \quad \min \quad \sum_{a^q \in A_Q} c_a x_a^q \tag{1}$$

subject to

$$\sum_{a^q \in \delta^+(i)} x_a^q = 1, \qquad \forall i \in V_+, \tag{2}$$

$$\sum_{a^q \in \delta^+(0)} x_a^q = K, \tag{3}$$

$$\sum_{a^{q-d_i} \in \delta^-(i)} x_a^{q-d_i} - \sum_{a^q \in \delta^+(i)} x_a^q = 0, \forall i \in V_+, q = d_i, \ldots, Q \tag{4}$$

$$x_a^q \geq 0, \qquad \forall a^q \in A_Q, \tag{5}$$

$$x \text{ integer.} \tag{6}$$

Equations (2) and (3) are customer and depot outdegree constraints. Balance equations (4) state that if an arc with index $q - d_i$ enters vertex i then an arc with index q must leave i. This formulation can be viewed as defining an acyclic network $\mathcal{N} = (V_Q, A_Q)$ with a set of nodes $V_Q = \{(i, q) : i \in V; q = d_i, \ldots, Q\}$. The set of arcs is also A_Q, but an arc $(i, j)^q \in A_Q$ is interpreted as going from (i, q) to $(j, q + d_i)$.

A q-route is a walk that starts at the depot, traverses a sequence of customers with total demand at most Q, and returns to the depot [3]. Let Ω be the set of all q-routes. For each $r \in \Omega$ define a non-negative variable λ_r and binary coefficients a_{rq}^{ij}, for each $(i, j)^q \in A_Q$, indicating whether (i, j) is traversed with load q in route r. Equations (4) in ALF can be replaced by:

$$\sum_{r \in \Omega} a_{rq}^{ij} \lambda_r = x_{ij}^q, \forall (i, j)^q \in A_Q. \tag{7}$$

Substituting the x variables and relaxing the integrality, the Dantzig-Wolfe Master LP is written as:

$$\text{(DWM)} \quad \min \quad \sum_{r \in \Omega} \left(\sum_{(i,j)^q \in A_Q} a_{rq}^{ij} c_{ij} \right) \lambda_r \tag{8}$$

subject to

$$\sum_{r \in \Omega} \left(\sum_{(i,j)^q \in \delta^+(\{i\})} a_{rq}^{ij} \right) \lambda_r = 1, \forall i \in V_+, \tag{9}$$

$$\sum_{r \in \Omega} \left(\sum_{(i,j)^q \in \delta^+(\{0\})} a_{rq}^{ij} \right) \lambda_r = K, \tag{10}$$

$$\lambda_r \geq 0 \qquad \qquad \forall r \in \Omega. \tag{11}$$

A generic constraint l of format $\sum_{(i,j)^q \in A_Q} \alpha_{ij}^{lq} x_{ij}^q \geq b_l$ can also be included in the DWM, using the variable substitution (7), as $\sum_{r \in \Omega} (\sum_{(i,j)^q \in A_Q} \alpha_{ij}^{lq} a_{ij}^{rq}) \lambda_r \geq b_l$. Suppose that, at a given instant, there are n_R constraints over the x variables in the DWM, including equalities (10) and (9). Constraint (10) has the dual variable π_0, the constraint in (9) corresponding to $i \in V_+$ has the dual variable π_i, and each additional constraint $l, n < l < n_R$, has the dual variable π_l. The reduced cost of an arc $(i,j)^q$ is defined as:

$$\bar{c}_{ij}^q = c_{ij} - \sum_{l=0}^{n_R-1} \alpha_{ij}^{lq} \pi_l. \tag{12}$$

The pricing subproblem for solving the DWM consists in finding a shortest path in $\mathcal{N}$ from node $(0,0)$ to nodes $(0,q)$, $1 \leq q \leq Q$, with respect to the arc reduced costs $\bar{c}_{ij}^q$. This can be done in $O(n^2 Q)$ time. A significantly stronger linear relaxation would be obtained if Ω was redefined as the set of elementary routes. On the other hand, the pricing subproblem would become strongly NP-hard. While carefully designed labeling algorithms are now capable of pricing elementary routes on most instances from the literature with up to 199 customers [10], this is still too costly. A more recent alternative for imposing partial elementarity, used in this work, are the ng-routes [2]. For each customer $i \in V_+$, let $NG(i) \subseteq V_+$ be the ng-set of i, defining its neighborhood. This may stand for the $|NG(i)|$ (this cardinality is decided *a priori*) closest customers and includes i itself. An ng-route allows multiple visits to a customer i, on the condition that at least one costumer j such that $i \notin NG(j)$ is visited between successive visits. From now on, Ω is redefined to be a set of ng-routes.

3   Cuts

Even if Ω only contains elementary routes, the bounds given by (9-11) are *not* good enough to be the basis of efficient exact algorithms (gaps between 1% and 4% are typical in the instances from the literature). The formulation must be reinforced with additional cuts. Cuts for the undirected edge formulation can be included in the DWM by using the transformation $x_{ij} = \sum_{(i,j)^q \in A_Q} x_{ij}^q + \sum_{(j,i)^q \in A_Q} x_{ji}^q$. Rounded Capacity Cuts and Strengthened Combs are used in this work. Those cuts are robust, the effect of their dual variables is captured in the arc-load reduced costs (12).

Jepsen et al. [8] introduced a family of inequalities defined over the route variables. Let $a_i^r = \sum_{(i,j)^q \in A_Q} a_{rq}^{ij}$ be the number of times that vertex i appears in route r. Given

---

1: **function** $\alpha(C, M, p, r)$
2:     $coeff \leftarrow 0, state \leftarrow 0$
3:     **for every vertex** $i \in r$ (in order) **do**
4:         **if** $i \notin M$ **then**
5:             $state \leftarrow 0$
6:         **else if** $i \in C$ **then**
7:             $state \leftarrow state + p$
8:             **if** $state \geq 1$ **then**
9:                 $coeff \leftarrow coeff + 1, state \leftarrow state - 1$
10: **return** $coeff$

---

a base set $C \subseteq V_+$ and a multiplier p, $0 < p < 1$, the following (C, p)-Subset Row Cut (SRC)

$$\sum_{r \in \Omega} \left\lfloor p \sum_{i \in C} a_i^r \right\rfloor \lambda_r \leq \lfloor p|C| \rfloor \tag{13}$$

is valid, since it can be obtained by a Chvátal-Gomory rounding of the corresponding constraints in (9). The definition of the limited memory (C, M, p)-Subset Row Cut (lm-SRC) requires an additional set M, $C \subseteq M \subseteq V_+$. It can be written as:

$$\sum_{r \in \Omega} \alpha(C, M, p, r)\lambda_r \leq \lfloor p|C| \rfloor, \tag{14}$$

where the coefficients α are computed by the following procedural function shown in Algorithm 1:

When $M = V_+$, the Function α will return $\lfloor p \sum_{i \in C} a_i^r \rfloor$ and the lm-SRC will be identical to an SRC. On the other hand, when M is not equal to V, the lm-SRC may be a weakening of its corresponding SRC. This happens because every time the route r leaves M, the variable $state$ in the function is set to zero, potentially decreasing the returned coefficient. Function α indicates how the lm-SRCs should be taken into account in the labeling algorithms used in the pricing. In fact, that procedural function is executed along the algorithm. Each label should have an additional dimension for each lm-SRC, storing their states in the corresponding partial paths. However, the coefficients do not need to be stored in the labels. Instead, whenever a label extension causes the increment of the coefficient of an lm-SRC, according to Function α, the value of its dual variable is immediately subtracted from the reduced cost of the new label. We remark that the number of possible states of an lm-SRC depends on its p. For example, for the frequent case where $p = 1/2$, the state can be only 0 or $1/2$. Therefore, it can be represented by a single bit.

The potential advantage of the lm-SRCs over classical SRCs is their much reduced impact on the labeling algorithm, when $|M| << |V_+|$. The reasons for that reduction will be explained in Section 4. In order to obtain small memory sets, the separation of lm-SRCs uses the following strategy. First, it identifies a violated (C, p)-SRC. Then, it determines a *minimal set M such that the lm-(C, M, p)-SRC has the same violation*. In practice, even on instances with hundreds of customers, those minimal sets seldom have cardinality larger than 15.

Given a base set $C \subseteq V_+$, for each integer d, $1 \leq d \leq n$, define a non-negative integer variable y_C^d as the sum of all variables λ_r such that $\sum_{i \in C} a_i^r = d$. Variables with $d > |C|$ can only be non-zero if Ω contains non-elementary routes. The interesting SRCs, for sets C with cardinality up to 5, are the following:

- The cuts where $|C| = 3$ and $p = 1/2$ are called 3-Subset Row Cuts (3SRCs) and can be expressed as $y_C^2 + y_C^3 + 2y_C^4 + 2y_C^5 + \ldots \leq 1$. Although they are very effective in improving the lower bounds, only a relatively small number of those cuts could be separated in [2,4], in order to keep the pricing tractable. Baldacci et al. [2] also used the Weak 3SRCs, a weakening of the 3SRCs where only the variables corresponding to routes that use an edge (i,j) such that $i,j \in C$ have coefficient 1. The Weak 3SRCs are equivalent (when all routes are elementary) to lm-3SRCs with $M = C$.

- Taking $|C| = 1$ and $p = 1/2$, the 1-Subset Row Cuts (1SRCs) $y_C^2 + y_C^3 + 2y_C^4 + \ldots \leq 0$ are obtained. They are equivalent to the Strong Degree Cuts $y_C^1 \geq 1$ introduced in [4], in the sense that both families forbid cycles over a vertex i ($C = \{i\}$). Contardo also defined the weaker k-Cycle Elimination Cut that only forbid cycles over i of size k or less. A lm-1SRC is a different kind of weakening, it forbids cycles over i contained in the set M. Of course, all these cuts can only be useful when the Ω set contains non-elementary routes.

- The cuts where $|C| = 4$ and $p = 2/3$ are 4SRCs, expressed as $y_C^2 + 2y_C^3 + 2y_C^4 + 3y_C^5 + 4y_C^6 + \ldots \leq 2$.

- There are two interesting families of cuts with $|C| = 5$. Those with $p = 1/3$ will be called 5,1SRCs, $y_C^2 + y_C^3 + y_C^4 + 2y_C^6 \ldots \leq 1$; whereas those with $p = 1/2$ are 5,2SRCs, having the format $y_C^2 + y_C^3 + 2y_C^4 + 2y_C^5 + 3y_C^6 \ldots \leq 2$. The latter family was already used in [4].

4   Pricing Algorithms

The forward dynamic programming labeling algorithm for the pricing problem represents a feasible path $P = (0, \ldots, i)$, $i \in V$, as a *label* $L(P) = (\bar{c}(P), v(P) = i, q(P), \Pi(P), S(P), pred(P))$ storing its reduced cost, end vertex, load, set of vertices forbidden as extensions due to ng-sets, vector of states corresponding to the n_S lm-SRCs with non-zero dual variables in the current Master LP solution, and a pointer to its predecessor label. Each $(i, q) \in V_Q$ defines a *bucket* $F(i, q)$. A label $L(P)$ is stored in bucket $F(v(P), q(P))$. A label $L(P_1)$ dominates a label $L(P_2)$ if every feasible completion of P_2 yields a route with reduced cost not smaller than the feasible route obtained by applying the same completion into P_1. Sufficient conditions for that are:

$$(i)\ v(P_1) = v(P_2), \quad (ii)\ q(P_1) = q(P_2), \quad (iii)\ \Pi(P_1) \subseteq \Pi(P_2), \text{ and}$$

$$(iv)\ \bar{c}(P_1) \leq \bar{c}(P_2) + \sum_{1 \leq s \leq n_S : S(P_1)[s] > S(P_2)[s]} \sigma_s,$$

where $\sigma_s < 0$ is the dual variable associated to lm-SRC s. Remark that the presence of cuts over the extended variables x_a^q or even the fixing of some of those variables to zero prevent (ii) to be strengthened to $q(P_1) \leq q(P_2)$. Only non-dominated labels are kept in the buckets. The base set, multiplier and memory set of lm-SRC s is denoted by $C(s)$, $p(s)$, and $M(s)$, respectively. Consider $NG(0)$ as $\{0\}$.

1: **procedure** FORWARD LABELING
2:     $F(i,q) \leftarrow \emptyset, \forall (i,q) \in V_Q$
3:     $processed(i,q) \leftarrow$ **false**, $\forall (i,q) \in V_Q$
4:     $F(0,0) \leftarrow \{(0,0,0,\emptyset,\mathbf{0},nil)\}$
5:     **while** there are unprocessed buckets **do**
6:         Choose an unprocessed bucket $F(i,q)$ with minimum q
7:         **for all** $(i,j)^q \in A_Q$ such that $x^q_{(i,j)}$ is not fixed to 0 **do**
8:             **for all** $L_1 = (\bar{c}_1, i, q, \Pi_1, S_1, \_) \in F(i,q)$ **do**
9:                 **if** $j \notin \Pi_1$ **then**
10:                     $\bar{c}_2 \leftarrow \bar{c}_1 + \bar{c}^q_{ij}$, $S_2 \leftarrow S_1$
11:                     **for** $s := 1, \ldots, n_S$ **do**
12:                         **if** $j \notin M(s)$ **then** $S_2[s] \leftarrow 0$
13:                         **else if** $j \in C(s)$ **then**
14:                             $S_2[s] \leftarrow S_2[s] + p(s)$
15:                             **if** $S_2[s] \geq 1$ **then** $\bar{c}_2 \leftarrow \bar{c}_2 - \sigma_s$, $S_2[s] \leftarrow S_2[s] - 1$
16:                     $L_2 = (\bar{c}_2, j, q + d_j, (\Pi_1 \cap NG(j)) \cup \{j\}, S_2, \text{pointer to } L_1)$
17:                     $insertLabel \leftarrow$ **true**
18:                     **for all** $\mathcal{L} \in F(j, q + d_j)$ **do**
19:                         **if** L_2 dominates $\mathcal{L}$ **then** delete $\mathcal{L}$
20:                         **else if** $\mathcal{L}$ dominates L_2 **then** $insertLabel \leftarrow$ **false**, **break**
21:                     **if** $insertLabel$ **then**
22:                         $F(j, q + d_j) \leftarrow F(j, q + d_j) \cup \{L_2\}$
23:         $processed(i,q) \leftarrow$ **true**

In the end of the Algorithm 2, each non-empty bucket $F(0,q)$, $1 \leq q \leq Q$, will contain only one label, representing the minimum reduced cost route with load q.

Now, it is possible to explain why the lm-SRCs have a reduced impact in the pricing when their memory sets are small. If there are no SRCs, the maximum number of non-dominated labels in a bucket $F(i,q)$ is bounded by $2^{|NG(i)|-1}$, as follows from dominance conditions (iii) and (iv). If the cardinality of the ng-sets is small (we used 8 in this work), the pricing is guaranteed to be reasonably fast (unless Q is very large). However, if a traditional SRCs is added, its dual variable may make condition (iv) *weaker in all buckets*. As other SRCs are separated, this may quickly result in an exponential proliferation of non-dominated labels. In practice, this severely limits the number of SRCs that can be used. In contrast, a lm-SRC s with a small memory has much less impact because *it can only weaken the dominance in the buckets of $M(s)$*. In practice, many more lm-SRCs can be separated before the exponential proliferation of labels is observed.

The labeling algorithm can also be performed backwards. In that case, the initializing labels are put in buckets $B(0,q)$, $1 \leq q \leq Q$, and the algorithm proceeds in a reversed way, until the label corresponding to the route with minimum reduced cost is found in bucket $B(0,0)$. The forward and backward variants of the labeling are equivalent in terms of computational cost. However, as pointed in [15], when forward labeling is used, most of the computational effort is spent in buckets with larger values of q, close to Q. This happens by combinatorial reasons, there are many more possible paths converging into a bucket $F(i, q)$ if q is larger. In a similar way, when backward labeling is used, most of the computational effort is spent in buckets with smaller value of q. Therefore, it may be advantageous to perform bidirectional search: use the forward labeling for filling the buckets $F(i, q)$ with $q \leq Q/2$ and backward labeling for filling the buckets $B(i, q)$ with $q > Q/2$. The minimum reduced cost paths are obtained by an additional concatenation step. In the bidirectional algorithm implemented in our BCP, we realized that the number of labels in the backward part was usually larger (3 to 10 times more is typical) than in the forward part. This happens because the backward variant has more starting labels. Therefore, in order to improve the algorithm performance by balancing both parts, the separation point is a value of q (dynamically determined) a bit larger than $Q/2$.

The labeling algorithms are also employed in a key part of the BCP: the elimination of variables by Lagrangean bounds. A full separated run of both forward and backward labeling should be performed. The minimum reduced cost of a route passing by an arc $(i, j)^q \in A_Q$, denoted by $\bar{C}_{ij}^q$, can be obtained by concatenating the labels in $F(i, q)$ from the forward run with the labels in $B(j, q + d_j)$ from the backward run. If $\bar{C}_{ij}^q$ is larger than the gap of the Lagrangean bound associated to the current dual solution with respect to the best known integer solution (see [13]), then $x_{i,j}^q$ can be fixed to zero. A similar procedure was also proposed in [7], but it is weaker because it only removes an arc $(i, j) \in A$, if a single particular solution allows removing $(i, j)^q$ for all values of q. On the other hand, as our BCP already works on the arc-load formulation, individual arcs $(i, j)^q$ can be naturally removed. For instance, it is quite typical that, at a certain point of the BCP, 95% of the arc-load variables were already fixed to zero, while the fixing on arcs would not achieve 80%.

5   Combining Strong Branching and Route Enumeration

Baldacci et al. [1] introduced a route enumeration based approach in order to close the duality gap after the root node. A route $r \in \Omega$ can only be part of a solution that improves the best known upper bound if its reduced cost is smaller than the gap. The enumeration of routes may be performed by a label setting algorithm, producing a set partitioning problem that, if it is small enough, is given to a general MIP solver. To better profit from the route enumeration, Contardo [4] proposed a different strategy. Enumeration is performed even if it produces some few million routes, far too many for a MIP solver. The routes are stored in a pool, and the algorithm proceeds with the column and cut generation. However, the pricing starts to be done by inspection. As a result, many non-robust cuts can now be separated without much impact in the pricing. The improved lower bounds are used to reduce the pool size, fixing route variables by reduced costs. Hopefully, the final set of routes will be small enough for yielding a solvable set partitioning problem.

Our BCP uses a hybrid strategy. Enumeration (in Contardo style) is tried after solving each node. If a limit on the number of routes is reached, the enumeration is aborted and the BCP proceeds by traditional branching. Of course, since deeper nodes will have smaller gaps, at some point the enumeration will work. The overall effect may be a substantially reduced enumeration tree. In order to maximize the branching effectiveness, an aggressive strong branching is applied, as done in [16]. This branching can also be performed after the enumeration. If the final set of routes is still too large for the MIP solver, a branching is performed. Of course, the pricing will continue to be done by inspection in both children nodes.

6 Computational Experiments

We report results over the standard classes of instances (A, B, E, F, M, and P) used for testing exact methods for the CVRP. Since larger instances came into reach of the proposed BCP, results are also reported over instances with up to 360 customers proposed in Golden et al. [6] . Table 1 summarizes the performance of the new BCP, comparing with the recent exact algorithms for the CVRP. As usual in the literature where similar tables appear, classes E and M are grouped together. Columns **Opt** indicate the number of instances solved to optimality. Columns **Gap** and **Time** are the average gap in the root node and average time in seconds (the processors are indicated), computed only over the solved instances. The labels **FLL+06**, **BCM08**, **BMR11**, **Con12** and **Rop12** refer to the algorithms proposed in [5], [1], [2], [4] and [16], respectively. Label **BCP** refers to the newly proposed BCP. This new algorithm has a good performance and could solve all those instances to optimality. On instance M-n200-k16, it showed that the the previous best known solution was not optimal. We should remark that instances F-n72-k4, P-n101-k4, and F-n135-k7 are still better solved by a branch-and-cut algorithm, like [9]. As in [5], we have built a hybrid method that is able to automatically switch to a branch-and-cut after severe problems with column generation convergence are found. However, we prefer to report the results of the standard BCP, in order to illustrate the power of the dual stabilization [12] in mitigating those problems.

Table 2 presents detailed information on the resolution of selected larger instances, comparing different algorithms. Column **IUB** presents the initial upper bound used by the method. Next columns give root node information: root lower bound obtained before enumeration (**RLB1**), the number of routes enumerated (if the method performs it and if the enumeration succeeds), (**ER1**), the improved root node lower bound after route enumeration, obtained by adding additional non-robust cuts (if Contardo style enumeration is performed) (**RLB2**), the number of remaining routes after that (**ER2**) and the total root node computing time (**RT(s)**). The final lower bound, given by **FLB**, which is in bold when optimal, the number of nodes in the search tree denoted by **Nodes**

Table 1. Comparison of recent CVRP algorithms on series A, B, E, F, M, and P

Class	NP	FLL+06 [5]			BCM08 [2]			BMR11 [2]			Con12 [4]			Rop12 [16]			BCP		
		Opt	Gap	Time	Opt	Gap	Time	Opt	Gap	Time	Opt	Gap	Time	Opt	Gap	Time	Opt	Gap	Time
A	22	22	0.81	1961	22	0.2	118	22	0.13	30	22	0.07	59	22	0.57	53	22	0.03	5.6
B	20	20	0.47	4763	20	0.16	417	20	0.06	67	20	0.05	89	20	0.25	208	20	0.04	6.2
E-M	12	9	1.19	126987	8	0.69	1025	9	0.49	303	10	0.3	2807	10	0.96	44295	12	0.19	3669
F	3	3	0.14	2398				2	0.11	164	2	0.06	3	3	0.25	2163	3	0.00	3679
P	24	24	0.76	2892	22	0.28	187	24	0.23	85	24	0.13	43	24	0.69	280	24	0.07	32.7
Total	81	78			72			77			78			79			81		
Processor		Pentium 4 2.4GHz			Pentium 4 2.6GHz			X7350 2.93GHz			E5462 2.8GHz			i7-2620M 2.7GHz			i7-3960X 3.3GHz		

Table 2. Detailed results over hard instances

Instance	Algorithm	IUB	RLB1	ER1	RLB2	ER2	RT(s)	FLB	Nodes	TT(s)
M-n151-k12	BMR11	1015	1004.3	-			380	1004.3	1	380
	Con12	1015	1008.9	4.0M	1012.5	13K	19041	**1015**	1	19699
	Rop12	1015	1001.5					**1015**	5268	417146
	BCP	1015	1011.7	59K	1012.8	8K	178	**1015**	1	212
M-n200-k16	BMR11		1256.6	-			319	1256.6	1	319
	Con12	1278	1263.0	-	-	-	265589	1263.0	1	265589
	Rop12	1278	1253.0					1258.2	106	7200
	BCP	1278	1266.5	-	-	-	949	**1274**	97	39869
M-n200-k17	BMR11	1275	1258.7	-			436	1258.7	1	436
	Con12	1275	1265.1	-	-	-	34351	1265.1	1	34351
	Rop12	1276	1255.3					1261.4	144	7200
	BCP	1275	1268.7	-	-	-	527	**1275**	15	3581
G17 (240)	BCP	707.76	705.54	-	-	-	993	**707.76**	13	25203
G13 (252)	BCP	857.19	851.97	-	-	-	21564	851.97	1	21564
G9 (255)	BCP	579.71	576.88	-	-	-	9364	576.88	1	9364
G18 (300)	BCP	995.13	993.42	-	-	-	1012	**995.13**	15	25690
G14 (320)	BCP	1080.55	1076.03	-	-	-	6330	1076.03	1	6330
G10 (323)	BCP	736.26	731.13	-	-	-	16021	731.13	1	16021
G19 (360)	BCP	1365.60	1362.70	-	-	-	19759	**1365.60**	2117	448741

and the total computational time in seconds **TT(s)** complete the table columns. The performance of the BCP over the larger instances shows the power of carefully aggregating all the elements described in this paper. This allowed more than doubling the size of the largest instance proved optimal to date.

References

1. Baldacci, R., Christofides, N., Mingozzi, A.: An exact algorithm for the vehicle routing problem based on the set partitioning formulation with additional cuts. Mathematical Programming 115(2), 351–385 (2008)
2. Baldacci, R., Mingozzi, A., Roberti, R.: New route relaxation and pricing strategies for the vehicle routing problem. Operations Research 59(5), 1269–1283 (2011)
3. Christofides, N., Mingozzi, A., Toth, P.: Exact algorithms for the vehicle routing problem, based on spanning tree and shortest path relaxations. Math. Prog. 20, 255–282 (1981)
4. Contardo, C.: A new exact algorithm for the multi-depot vehicle routing problem under capacity and route length constraints. Technical report, Archipel-UQAM 5078, Université du Québec à Montréal, Canada (2012)
5. Fukasawa, R., Longo, H., Lysgaard, J., Poggi de Aragão, M., Reis, M., Uchoa, E., Werneck, R.F.: Robust branch-and-cut-and-price for the capacitated vehicle routing problem. Mathematical Programming 106(3), 491–511 (2006)
6. Golden, B., Wasil, E., Kelly, J., Chao, I.: The impact of metaheuristics on solving the vehicle routing problem: Algorithms, problem sets, and computational results. In: Fleet Management and Logistics, pp. 33–56. Springer (1998)
7. Irnich, S., Desaulniers, G., Desrosiers, J., Hadjar, A.: Path-reduced costs for eliminating arcs in routing and scheduling. INFORMS Journal on Computing 22(2), 297–313 (2010)
8. Jepsen, M., Petersen, B., Spoorendonk, S., Pisinger, D.: Subset-row inequalities applied to the vehicle-routing problem with time windows. Oper. Research 56, 497–511 (2008)
9. Lysgaard, J., Letchford, A., Eglese, R.: A new branch-and-cut algorithm for the capacitated vehicle routing problem. Mathematical Programming 100, 423–445 (2004)

10. Martinelli, R., Pecin, D., Poggi, M.: Efficient elementary and restricted non-elementary route pricing for routing problems (2012) (submitted)
11. Pessoa, A., Poggi de Aragão, M., Uchoa, E.: Robust branch-cut-and-price algorithms for vehicle routing problems. In: Golden, B., Raghavan, S., Wasil, E. (eds.) The Vehicle Routing Problem: Latest Advances and New Challenges, pp. 297–325. Springer (2008)
12. Pessoa, A., Sadykov, R., Uchoa, E., Vanderbeck, F.: In-out separation and column generation stabilization by dual price smoothing. In: Bonifaci, V., Demetrescu, C., Marchetti-Spaccamela, A. (eds.) SEA 2013. LNCS, vol. 7933, pp. 354–365. Springer, Heidelberg (2013)
13. Pessoa, A., Uchoa, E., Poggi de Aragão, M., Rodrigues, R.: Exact algorithm over an arc-time-indexed formulation for parallel machine scheduling problems. Mathematical Programming Computation 2(3-4), 259–290 (2010)
14. Poggi de Aragão, M., Uchoa, E.: Integer program reformulation for robust branch-and-cut-and-price. In: Annals of Mathematical Programming in Rio, Búzios, Brazil, pp. 56–61 (2003)
15. Righini, G., Salani, M.: Symmetry helps: bounded bi-directional dynamic programming for the elementary shortest path problem with resource constraints. Discrete Optimization 3(3), 255–273 (2006)
16. Røpke, S.: Branching decisions in branch-and-cut-and-price algorithms for vehicle routing problems. Presentation in Column Generation 2012 (2012)

Claw-Free t-Perfect Graphs Can Be Recognised in Polynomial Time

Henning Bruhn[1] and Oliver Schaudt[2]

[1] Universität Ulm, Institut für Optimierung und Operations Research,
Helmholtzstraße 18, 89081 Ulm, Germany
henning.bruhn@uni-ulm.de

[2] Universität zu Köln, Institut für Informatik, Weyertal 80, 50931 Köln, Germany
schaudto@uni-koeln.de

Abstract. A graph is called t-perfect if its stable set polytope is defined by non-negativity, edge and odd-cycle inequalities. We show that it can be decided in polynomial time whether a given claw-free graph is t-perfect.

Keywords: t-perfect graphs, claw-free graphs, induced minors, recognition algorithm.

1 Introduction

We treat t-perfect graphs, a class of graphs that is not only similar in name to perfect graphs but also shares a number of their properties. One way to define perfect graphs is via the stable set polytope: The convex hull of all characteristic vectors of stable sets (sets of pairwise non-adjacent vertices). As shown independently by Chvátal [6] and Padberg [19], a graph is perfect if and only if its stable set polytope is determined by non-negativity and clique inequalities. In analogy, Chvátal [6] proposed to study the class of graphs whose stable set polytope is defined by non-negativity, edge and odd-cycle inequalities. These graphs became to be known as *t-perfect graphs*. (We defer precise and more explicit definitions to the next section.)

Two celebrated results on perfect graphs are the proof of the strong perfect graph conjecture by Chudnovsky, Robertson, Seymour and Thomas [5] and the polynomial time algorithm of Chudnovsky, Cornuéjols, Liu, Seymour and Vušković [4] that checks whether a given graph is perfect or not. Analogous results for t-perfection seem desirable but out of reach. Restricted to claw-free graphs, however, this changes. A characterisation of claw-free t-perfect graphs in terms of forbidden substructures was recently proved by Bruhn and Stein [3]. In this work we present a recognition algorithm for t-perfect claw-free graphs.

Theorem 1. *It can be decided in polynomial time whether a given claw-free graph is t-perfect.*

The class of t-perfect graphs seems rich and of non-trivial structure. Examples include series-parallel graphs (Boulala and Uhry [1]) and bipartite or almost

J. Lee and J. Vygen (Eds.): IPCO 2014, LNCS 8494, pp. 404–415, 2014.

bipartite graphs. More classes were identified by Shepherd [24] and Gerards and Shepherd [12]. In the latter paper the authors characterize the class of graphs for which any subgraph is t-perfect. An attractive result on the algorithmic side is the combinatorial polynomial-time algorithm of Eisenbrand, Funke, Garg and Könemann [9] that solves the maximum stable set problem on t-perfect graphs. Their algorithm, however, cannot be used to recognise t-perfect graphs.

There is also an, at least superficially, more stringent notion of t-perfection, *strong t-perfection*: a graph is strongly t-perfect if the system consisting of non-negativity, edge and odd-cycle inequalities is totally dual integral; see Schrijver [23, Vol. B, Ch. 68] where also some background on t-perfect graphs may be found. Interestingly, there is no t-perfect graph known that fails to be strongly t-perfect. In fact, for some classes these two notions are known to be equivalent, and it remains an open problem whether there are t-perfect graphs that are not strongly t-perfect (see Schrijver [22] and Bruhn and Stein [2]).

The graphs whose stable set polytope is given by non-negativity, clique and odd-cycle inequalities are called *h-perfect*. The class of h-perfect graphs is a natural superclass of both perfect as well as t-perfect graphs. The class has been studied by Fonlupt and Uhry [11], Sbihi and Uhry [21], and Király and Páp [16,17].

We briefly outline the strategy of our recognition algorithm. In Section 3 we show how to recognise t-perfect line graphs. For this, we work in the underlying source graph that gives rise to the line graph. In the source graph we need to detect certain subgraphs called *thetas*: two vertices joined by three disjoint paths. In the thetas that are of interest to us the linking paths have to respect additional parity constraints.

The general algorithm for claw-free graphs is presented in Section 4 and relies on a divide and conquer approach to split the input graph along small separators. In this phase of the algorithm, we make extensive use of a procedure by van 't Hof, Kamiński and Paulusma [25] that detects induced paths of given parity in claw-free graphs. The final pieces that cannot be split anymore turn out to be essentially line graphs, which we already dealt with.

2   Claw-Free Graphs and t-Perfection

Note that we only consider simple undirected graphs. We refer to Diestel [8] for general notation and definitions concerning graphs.

Let us recall the definition of a claw-free graph. The *claw* is the graph $G = (V, E)$ with $V = \{u, v_1, v_2, v_3\}$ and $E = \{uv_1, uv_2, uv_3\}$, and we call u its *centre*. A graph is called *claw-free* if it does not contain an induced subgraph that is isomorphic to the claw. Claw-free graphs form a superclass of line graphs.

In order to define t-perfection, we associate with every graph $G = (V, E)$ a polytope denoted TSTAB(G), the set of all vectors $x \in \mathbb{R}^V$ satisfying

$$0 \leq x_v \leq 1 \quad \text{for every vertex } v \in V,$$
$$x_u + x_v \leq 1 \quad \text{for every edge } uv \in E, \tag{1}$$
$$\sum_{v \in V(C)} x_v \leq \lfloor \tfrac{1}{2}|V(C)| \rfloor \quad \text{for every odd cycle } C \text{ in } G.$$

The graph G is called t-perfect if TSTAB(G) coincides with the stable set polytope of G (the convex hull of characteristic vectors of stable sets in $\mathbb{R}^V$). An alternative but equivalent definition is to say that G is t-perfect if and only if TSTAB(G) is an integral polytope.

As observed by Gerards and Shepherd [12], the following operation called t-contraction preserves t-perfection: Contraction of all edges incident with any vertex v whose neighbourhood $N(v)$ is a stable set. We then say that a t-contraction is performed at v. If G is claw-free, the t-contraction becomes particularly simple. Indeed, a t-contraction at v is only possible if v has degree ≤ 2; otherwise v is the centre of a claw. If v has precisely two neighbours u and w then the t-contraction simply identifies u, v, w to a single vertex.

To characterise the class of t-perfect graphs in terms of forbidden substructures, the concept of t-minors was introduced in [2]: A graph H is a t-minor of a graph G if H can be obtained from G by a series of vertex deletions and/or t-contractions. Note that the class of t-perfect graphs is closed under taking t-minors.

We note an easy but useful observation [2]:

$$\textit{any } t\textit{-minor of a claw-free graph is claw-free.} \tag{2}$$

It turns out that t-perfect claw-free graphs can be characterised in terms of finitely many forbidden t-minors:

Theorem 2 (Bruhn and Stein [3]). *A claw-free graph is t-perfect if and only if it does not contain any of K_4, W_5, C_7^2 and C_{10}^2 as a t-minor.*

Here, K_4 denotes the complete graph on four vertices, W_5 is the 5-wheel, and for $n \in \mathbb{N}$ we denote by C_n^2 the square of the cycle C_n on n vertices, see Figure 1. More precisely, we define C_n^2 always on the vertex set $v_1, \ldots, v_n$, so that v_i and v_j are adjacent if and only if $|i - j| \leq 2$, where we take the indices modulo n.

We often present our algorithms intermingled with parts of the corresponding correctness proofs. To set the algorithm steps apart from the surrounding proofs we write them as follows:

① The first line of an algorithm.

Finally, for two vertices u, v, a u–v-path is simply a path from u to v. Similarly, if $X, Y \subseteq V(G)$, then we mean by an X–Y-path a path from a vertex in X to some vertex in Y so that no internal vertex belongs to $X \cup Y$. In the case that $X = Y$ we simply speak of an X-path.

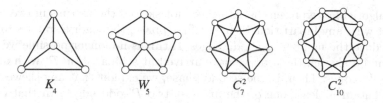

Fig. 1. The forbidden t-minors

3 Line Graphs

We first solve the recognition problem for line graphs:

Lemma 1. *It can be decided in polynomial time whether the line graph of a given graph is t-perfect.*

We develop the algorithm in the course of this section and the next. That the algorithm is correct is based on the following characterisation of t-perfect line graphs.

A *skewed theta* is a subgraph which is the union of three edge-disjoint paths linking two vertices, called *branch vertices*, such that two paths have odd length and one has even length. Note that a skewed theta does not have to be an induced subgraph. (We should mentioned that skewed thetas in [3] were denoted C_5^+.)

We call a graph *subcubic* if its maximum degree is at most 3.

Lemma 2. [3] *Let G be a graph. Then the line graph $L(G)$ is t-perfect if and only if G is subcubic and does not contain any skewed theta.*

Checking for subdivisions of a certain graph can often be reduced to the well-known k-DISJOINT PATHS problem: Given a number of k pairs of terminal vertices, the task is to decide whether there are disjoint paths joining the paired terminals. In our context, however, this is not sufficient as the paths linking the branch vertices in a skewed theta are subject to parity constraints.

That this deep and seemingly hard problem, k-DISJOINT PATHS WITH PARITY CONSTRAINTS, allows nevertheless a polynomial time algorithm has been announced by Kawarabayashi, Reed and Wollan [15]. Another algorithm was given in the PhD thesis of Huynh [14]. These are very impressive results indeed, and they draw on deep insights coming from the graph minor project of Robertson and Seymour and its extension to matroids by Geelen, Gerards and Whittle. For both algorithms, however, it seems doubtful whether they could be implemented with a reasonable amount of work (or at all). We prefer therefore to present a more elementary algorithm for Lemma 1 that does not rely on any deep result and that is, in principle, implementable.

Given a bipartition $\mathcal{P} = (A, B)$ (where we allow A or B to be empty) of the vertex set of a graph G, we call an edge $\mathcal{P}$-*even* if its endvertices lie in distinct partition classes of $\mathcal{P}$; otherwise the edge is $\mathcal{P}$-*odd*. We observe that a cycle is odd if and only if it contains an odd number of $\mathcal{P}$-odd edges.

The algorithm we present here to check for skewed thetas runs in two phases. We start with any bipartition $\mathcal{P}$. In the first phase, the algorithm tries to iteratively reduce the number of $\mathcal{P}$-odd edges. If this is no longer possible we either have found a skewed theta or we have arrived at a bipartition $\mathcal{P}'$ with at most two $\mathcal{P}'$-odd edges. Then, in the second phase, we exploit that any skewed theta has to contain at least one of the at most two $\mathcal{P}'$-odd edges. In that case, it becomes possible to check directly for a skewed theta:

Lemma 3. *Given a graph G and a bipartition $\mathcal{P}$ of $V(G)$ so that at most two edges are $\mathcal{P}$-odd, it is possible to check in polynomial time whether G contains a skewed theta.*

Due to space limitations, the proof of Lemma 3 is deferred to the appendix. In the remainder of this section, we show how to iteratively reduce the number of $\mathcal{P}$-odd edges. We start with two lemmas that give sufficient conditions for the existence of a skewed theta.

Lemma 4. *A 2-connected subcubic graph that contains two edge-disjoint odd cycles contains a skewed theta.*

Proof. Let C_1 and C_2 be two edge-disjoint odd cycles in G, which then are also vertex-disjoint as the graph is assumed to be subcubic. Since G is 2-connected there are two disjoint C_1–C_2-paths P_1, P_2. The endvertices of P_1 and P_2 subdivide C_2 into two subpaths, and one of these subpaths together with P_1 and P_2 yields an odd C_1-path, and thus a skewed theta. $\square$

For any bipartition $\mathcal{P}$ of G, define $G_{\mathcal{P}}$ to be the (bipartite) subgraph on $V(G)$ together with all the $\mathcal{P}$-even edges. We formulate a second set of conditions that implies the presence of a skewed theta.

Let C be a cycle and let P and Q be two vertex-disjoint C-paths. Let p_1, p_2 be the endpoints of P and q_1, q_2 be the endpoints of Q. We say that P and Q are *crossing on C* if p_1, q_1, p_2, q_2 appear in this order on C.

Lemma 5. *Let G be a subcubic graph with a bipartition $\mathcal{P}$. Let there be three $\mathcal{P}$-odd edges o_1, o_2, o_3 and two vertex-disjoint trees $T_1, T_2 \subseteq G_{\mathcal{P}}$, each containing an endvertex of each of o_1, o_2, o_3.*

Assume the trees are minimal subject to the above description. If $G_{\mathcal{P}}$ contains three edge-disjoint T_1–T_2-paths then G contains a skewed theta.

Proof. Throughout the proof, we assume that G does not contain a skewed theta. Our aim is to show that $G_{\mathcal{P}}$ does not contain three edge-disjoint T_1–T_2-paths.

For this, we first prove a sequence of more general claims. Let $r_1 r_2$ and $s_1 s_2$ be two $\mathcal{P}$-odd edges of G such that there are two vertex-disjoint paths $R_1 = r_1 \ldots s_1$, $R_2 = r_2 \ldots s_2$. Let C be the cycle $r_1 R_1 s_1 s_2 R_2 r_2 r_1$.

We claim that

$$\text{any two edge-disjoint } R_1\text{–}R_2\text{-paths } P, Q \text{ are crossing on } C. \tag{3}$$

If P and Q are not crossing then we can easily find two edge-disjoint cycles in $R_1 \cup R_2 \cup P \cup Q$, one through $r_1 r_2$ and the other through $s_1 s_2$. By Lemma 4, however, this is impossible. Thus, P and Q are crossing.

Next, we show that

$$\text{the endvertices of any two edge-disjoint } R_1\text{–}R_2\text{-paths } P, Q \text{ in } R_1 \atop \text{lie in distinct partitions classes of } \mathcal{P}. \tag{4}$$

Denote the endvertex of P in R_1 by p_1 and denote the one in R_2 by p_2; define q_1, q_2 analogously for Q.

Suppose that p_1 and q_1 lie in the same partition class of $\mathcal{P}$. Since G is subcubic, P and Q are vertex-disjoint, and, by (3), crossing. Assume that $p_1 \in r_1 R_1 q_1$. As p_1 and q_1 are contained in the same partition class, the path $p_1 R_1 q_1$ has even length. On the other hand, the following two paths have odd length: $p_1 P p_2 R_2 s_2 s_1 R_1 q_1$ and $q_1 Q q_2 R_2 r_2 r_1 R_1 p_1$. As, moreover, these three paths meet only in p_1 and q_1 we have found a skewed theta; this proves (4).

From this follows that

$$G \text{ cannot contain three edge-disjoint } R_1\text{–}R_2\text{-paths.} \tag{5}$$

Indeed, by (4), the three endvertices of such paths in R_1 would need to lie in distinct partition classes, which is clearly impossible as $\mathcal{P}$ is a bipartition.

To complete the proof, suppose now that $G_{\mathcal{P}}$ contains three edge-disjoint T_1–T_2-paths P_1, P_2, P_3. Denote by t_i the unique vertex that separates all the endvertices of o_1, o_2, o_3 in T_i (unless T_i is a path this is the unique vertex of degree 3 in T_i). Observe that t_i subdivides T_i into three edge-disjoint paths S_1^i, S_2^i, S_3^i (some of which might be trivial) so that S_j^i contains the endvertex of o_j (for $i = 1, 2$ and $j = 1, 2, 3$).

Pick two distinct $k, \ell \in \{1, 2, 3\}$ such that, for $i = 1, 2$, at least two paths of P_1, P_2, P_3 have an endvertex contained in $S_k^i \cup S_\ell^i =: R_i$. Let $\{m\} = \{1, 2, 3\} \setminus \{k, \ell\}$. Should now P_j have its endvertex p in $S_m^1 - S_k^1 - S_\ell^1$ concatenate the subpath $p S_m^1 t_1$ with P_j, and proceed in a similar way in T_2. In this way we turn the edge-disjoint T_1–T_2-paths into edge-disjoint R_1–R_2-paths. Now, we obtain the desired contradiction from (5). $\qquad\square$

Next, we state a simple lemma that, however, is the key to reducing the number of $\mathcal{P}$-odd edges.

Lemma 6. *Let G be a graph with a bipartition $\mathcal{P}$. Given an edge-cut F of G that contains more $\mathcal{P}$-odd edges than $\mathcal{P}$-even edges, one can compute a bipartition $\mathcal{P}'$ of G with less $\mathcal{P}'$-odd edges in polynomial time.*

Proof. Let $F = E(X, Y)$ separate $X \subseteq V(G)$ from $Y \subseteq V(G)$ in G. Then put $\mathcal{P}' := (A \triangle X, B \triangle X)$, and observe that every $\mathcal{P}$-odd edge in F becomes $\mathcal{P}'$-even, while the edges outside F do not change. $\qquad\square$

Putting together the lemmas presented so far, we arrive at the following procedure.

Lemma 7. *There is a polynomial-time algorithm that takes as input a 2-connected subcubic graph G, a bipartition $\mathcal{P}$ and three $\mathcal{P}$-odd edges o_1, o_2, o_3. The algorithm:*

(a) *either correctly decides that G contains a skewed theta;*
(b) *or computes an edge cut F that contains more $\mathcal{P}$-odd edges than $\mathcal{P}$-even edges.*

Proof. We describe the algorithm in the course of this lemma. We omit a detailed discussion about the runtime complexity as the steps of the algorithm rely on basic operations or reduce to solving min-cut/max-flow problems.

① If $G_{\mathcal{P}}$ is not connected, choose a component X of $G_{\mathcal{P}}$ and return $F = E(X, G - X)$.

Since G is 2-connected, F contains at least two $\mathcal{P}$-odd edges, which in particular implies condition (b). Let us now assume that $G_{\mathcal{P}}$ is connected.

② Compute a spanning tree T of $G_{\mathcal{P}}$ and determine the fundamental cycles $C_{o_1}, C_{o_2}, C_{o_3}$ of o_1, o_2, o_3.
③ If any two of C_{o_1}, C_{o_2} and C_{o_3} are edge-disjoint, return *"skewed theta"*.

The return value in line ③ is justified by Lemma 4, which means that we may assume the cycles $C_{o_1}, C_{o_2}, C_{o_3}$ to pairwise share an edge from now on.

④ If there is an edge e shared by each of $C_{o_1}, C_{o_2}, C_{o_3}$:
    a. Let T_1 and T_2 be the two components of $\bigcup_{i=1}^{3} C_{o_i} - \{e, o_1, o_2, o_3\}$.
    b. Delete leaves from T_1 and T_2 until T_1 and T_2 have the form of Lemma 5.
    c. Compute a smallest cut $F' = E_{G_{\mathcal{P}}}(X, Y)$ of $G_{\mathcal{P}}$ that separates T_1 from T_2.
    d. If $|F'| \geq 3$, return *"skewed theta"*; otherwise return $F = E_G(X, Y)$.

Note that, for $i = 1, 2, 3$, both components of $C_{o_i} - \{e, o_i\}$ contain an endvertex of o_i, so that, after pruning, T_1 and T_2 indeed conform with Lemma 5. Lemma 5 implies that G contains a skewed theta if $|F'| \geq 3$. Otherwise, F contains at most two $\mathcal{P}$-even edges and the three $\mathcal{P}$-odd edges o_1, o_2, o_3.

Considering line ④, we may from now on assume that there is no common edge of $C_{o_1}, C_{o_2}, C_{o_3}$. Then

> *there is a unique cycle D in $\bigcup_{i=1}^{3} C_{o_i}$ that passes through each of o_1, o_2, o_3 and so that there is a path in $G_{\mathcal{P}}$ between any two*   (6) *of the components of $D - \{o_1, o_2, o_3\}$ that avoids the third.*

Indeed, each $C_{o_i} - o_i$ is a subpath of T and families of subtrees of a tree are known to have the Helly property, that is, if any two share a vertex then there is also a common vertex to all. Let x be such a vertex. Now, assume that $C_{o_1}, C_{o_2}, C_{o_3}$ do not have a common edge. Note that, for any $i \neq j$, C_{o_i} and C_{o_j} meet along a path. It follows that $C_{o_1} \cup C_{o_2} \cup C_{o_3}$ decomposes into a cycle D that passes through all of o_1, o_2, o_3 and three internally disjoint x–D-paths that each end in a different component of $D - \{o_1, o_2, o_3\}$. The uniqueness of D can be derived from the fact that $\bigcup_{i=1}^{3} C_{o_i} - \{o_1, o_2, o_3\}$ is a tree. This proves (6).

⑤ Determine the cycle D in $\bigcup_{i=1}^{3} C_{o_i}$ that passes through o_1, o_2 and o_3.

Finding D is easy, as this is done in the tree $\bigcup_{i=1}^{3} C_{o_i} - \{o_1, o_2, o_3\}$. (Alternatively, we may argue that $E(D)$ is exactly the set of those edges in $\bigcup_{i=1}^{3} C_{o_i}$ that lie in only one of the cycles C_{o_i}.) Let S_1, S_2, S_3 be the three components of $D - \{o_1, o_2, o_3\}$.

⑥ Check whether there is a single edge e' that separates S_1 from $S_2 \cup S_3$ in $G_{\mathcal{P}}$. If yes, return $E_G(X, Y)$, where X and Y are the two components of $G_{\mathcal{P}} - e'$.

Two of the edges o_1, o_2, o_3 are in the cut $E_G(X, Y)$, while the only $\mathcal{P}$-even edge in $E_G(X, Y)$ is e'.

⑦ Compute two edge-disjoint S_1–$(S_2 \cup S_3)$-paths P, Q in $G_{\mathcal{P}}$ so that one ends in S_2 and the other in S_3.

Let us explain how P and Q can be computed. First, we use a standard algorithm to find two edge-disjoint S_1–$(S_2 \cup S_3)$-paths P, Q in $G_{\mathcal{P}}$; these exist by Menger's theorem and line ⑥. If already one ends in S_2 and the other in S_3, we use these. So, assume that P and Q both end in S_2, say. By (6), we can find an S_1–S_3-path R in $G_{\mathcal{P}} - S_2$. If R is vertex-disjoint from P and Q, we replace Q by R. If not, we follow R until we encounter for the last time a vertex of $P \cup Q$, where we see R directed from S_1 to S_3. Let us say this last vertex q is in Q. Then, we replace Q by QqR.

⑧ If P and Q are not crossing on D then return *"skewed theta"*.
⑨ Otherwise, choose an edge e'' that separates the endvertices of P and Q in S_1 and apply lines 4b–4d to the two components T_1 and T_2 of $(D - \{o_1, o_2, o_3, e''\}) \cup P \cup Q$.

If P and Q are not crossing then $D \cup P \cup Q$ contains two disjoint odd cycles, and thus G contains a skewed theta, by Lemma 4. If, on the other hand, P and Q are crossing then each of the two components T_1 and T_2 as in line ⑨ is incident with an endvertex of each of o_1, o_2, o_3.   □

We now prove that for line graphs t-perfection can be checked in polynomial-time.

Proof (Proof of Lemma 1.). Let G be a given graph. If G has maximum degree at least 4, its line graph $L(G)$ is not t-perfect by Lemma 2. Otherwise, we apply the algorithm below to the blocks of G to check whether G contains a skewed theta. Clearly, any skewed theta is completely contained in a block of G.

① Set $\mathcal{P} := (V(G), \emptyset)$.
② While there are at least 2 distinct $\mathcal{P}$-odd edges, do the following:
   a. Run the algorithm of Lemma 7.
   b. If the algorithm returns a cut $F = E_G(X, Y)$ with more $\mathcal{P}$-odd edges than $\mathcal{P}$-even edges, apply Lemma 6.

③ Apply Lemma 3 to decide whether G contains a skewed theta.

The algorithm runs in polynomial-time, as the number of $\mathcal{P}$-odd edges decreases in each iteration of the while loop.

Correctness holds as Lemma 2 guarantees that $L(G)$ is t-perfect if and only if G does not contain a skewed theta.   □

4   Claw-Free Graphs

We now describe an algorithm that, given a claw-free graph G, decides in polynomial time whether G is t-perfect or not. We present the algorithm in a number of steps over the course of this section. First, we use that we can already decide t-perfection for line graphs, and that we can detect whether a graph is a line graph efficiently:

Theorem 3 (Roussopoulos [20]). *It can be checked in linear time whether a given graph is a line graph. Moreover, given a line graph G, a graph H with $L(H) = G$ can be found in linear time.*

Thus, the first step in the algorithm becomes:

① Use Theorem 3 to check whether G is a line graph. If yes, compute H with $L(H) = G$ and apply the algorithm of Lemma 1 to H. If no, proceed to the next line below.

Next, we observe that we can assume the input graph to be 2-connected. For this, we say that a pair (G_1, G_2) of proper induced subgraphs of a graph G is a *separation of G*, if $G = G_1 \cup G_2$. The *order* of the separation is equal to $|V(G_1 \cap G_2)|$.

The following lemma may be deduced directly from the definition of t-perfection. We only apply it to claw-free graphs, where it becomes a simple consequence of Theorem 2.

Lemma 8. *Let (G_1, G_2) be a separation of a graph G so that $G_1 \cap G_2$ is complete. Then G is t-perfect if and only if G_1 and G_2 are t-perfect.*

② Determine the blocks of G, and apply the rest of the algorithm to each block independently. Return *"not t-perfect"* if one of the blocks is not t-perfect; otherwise return *"t-perfect"*.

Clearly, this step can be performed efficiently, and is, by Lemma 8, correct. Thus, we may from now on assume G to be 2-connected. Moreover, it is easy to see that G is not t-perfect, if it contains a vertex of degree at least 5. Indeed, as G is claw-free, the neighbourhood of any vertex v of degree at least 5 always contains either a triangle or an induced 5-cycle. In the former case, the graph contains a K_4 and in the latter case a 5-wheel as induced subgraph.

③ If $\Delta(G) \geq 5$ or if $G \in \{C_7^2, C_{10}^2\}$ return *"not t-perfect"*.

④ If $G \in \{C_6^2 - v_1v_6, C_7^2 - v_7, C_{10}^2 - v_{10}\}$ return *"t-perfect"*.

That the three graphs in line ④ are t-perfect is proved in [3]. (In fact, C_7^2 and C_{10}^2 are *minimally t-imperfect*, that is, they are t-imperfect but every proper t-minor is t-perfect. The graph $C_6^2 - v_1v_6$ can be seen to be a t-minor of C_{10}^2.)

The remainder of the algorithm is based on the following lemma.

Lemma 9 (Bruhn and Stein [3]). *Let G be a 3-connected claw-free graph of maximum degree at most 4. If G does not contain K_4 as t-minor then one of the following statements holds true:*

(a) *G is a line graph; or*
(b) *$G \in \{C_6^2 - v_1v_6, C_7^2 - v_7, C_{10}^2 - v_{10}, C_7^2, C_{10}^2\}$.*

Thus, we may assume that the input graph G is 2-connected but not 3-connected. That is, G has a separation of order 2.

⑤ If G is 3-connected, return *"not t-perfect"*.
⑥ Otherwise, find a separation (G_1, G_2) of G of order 2. Let u, v be the two vertices in $G_1 \cap G_2$.

Line ⑤ is correct, as we had already excluded that G is a line graph, nor one of the exceptional graphs in (b) of Lemma 9.

To continue, we use a result that allows us to reduce the t-perfection of G to the t-perfection of the two sides of the separation. For this, we write $G_i/_{u=v}$ for the graph obtained from G_i by identifying u and v.

Lemma 10. *Let G be a 2-connected claw-free graph of maximum degree at most 4. Assume (G_1, G_2) to be a separation of G with $V(G_1 \cap G_2) = \{u, v\}$. Then:*

(i) *If G_1 and G_2 each contain induced u–v-paths of both even and odd length, then G is not t-perfect.*

Otherwise G is t-perfect if and only if $\tilde{G}_1$ and $\tilde{G}_2$ are t-perfect, where

(ii) *$\tilde{G}_1 = G_1/_{u=v}$ and $\tilde{G}_2 = G_2 + uv$, if G_1 contains an odd induced u–v-path but G_2 does not;*
(iii) *$\tilde{G}_1 = G_1$ and $\tilde{G}_2 = G_2$, if neither of G_1 and G_2 contains an odd induced u–v-path;*
(iv) *$\tilde{G}_1 = G_1 + uv$ and $\tilde{G}_2 = G_2/_{u=v}$, if G_1 contains an even induced u–v-path but G_2 does not; and*
(v) *$\tilde{G}_1 = G_1$ and $\tilde{G}_2 = G_2$, if neither of G_1 and G_2 contains an even induced u–v-path.*

Due to space limitations, we defer the proof of Lemma 10 to the appendix. We combine the lemma with the following algorithm:

Theorem 4 (van 't Hof, Kamiński and Paulusma [25]). *Given a claw-free graph G and $u, v \in V(G)$, it can be decided in polynomial time whether there is an induced u–v-path of even (or of odd) length.*

With this, our algorithm continues as follows:

⑦ Use Theorem 4 to determine the parities of induced u–v-paths in G_1 and in G_2.

⑧ If G_1 and G_2 each contain induced u–v-paths of both even and odd length, return *"not t-perfect"*.

⑨ Otherwise, choose $\tilde{G}_1$ and $\tilde{G}_2$ as in Lemma 10, and apply line ① to $\tilde{G}_1$ and to $\tilde{G}_2$ independently. Return *"t-perfect"* if both are t-perfect, and *"not t-perfect"* otherwise.

We can finally complete the proof of our main result, that t-perfection can be checked for in polynomial time if the input is restricted to claw-free graphs.

Proof (Proof of Theorem 1). We have already seen that the algorithm described in the course of this section is correct. Moreover, as each single line is executed in polynomial time, we only need to bound the number of times each line is executed. For this, observe that every time there is a branching in line ⑨, the graph $\tilde{G}_1$ contains a vertex of G that does not lie in $\tilde{G}_2$ and vice versa. A standard analysis of the recurrence yields that the number of iterations is bounded by $\mathcal{O}(|V(G)|^2)$.[1]   □

5   Discussion

A key step for the recognition of claw-free t-perfect graphs is the insight that the problem reduces to the detection of skewed prisms.

Skewed prisms are induced subgraphs. As Fellows, Kratochvil, Middendorf and Pfeiffer [10] observed, searching for a certain substructure often becomes substantially harder if one requires the substructure to be induced: finding the largest matching can be done in polynomial time, but determining the size of the largest induced matching is NP-complete.

In the same way, checking for a non-induced prism (and without any parity constraints on the paths) reduces to verifying whether between any two triangles are three vertex-disjoint paths, which clearly can be done in polynomial time. Checking whether a given graph contains an induced prism, however, is NP-complete – this is a result of Maffray and Trotignon [18]. Interestingly, this changes when the input graph is claw-free. Golovach, Paulusma and van Leeuwen [13] describe a polynomial-time algorithm for the induced variant of the k-DISJOINT PATHS PROBLEM in claw-free graphs. By again considering any pair of triangles in a claw-free graph, the algorithm may be used to detect prisms. Unfortunately, or rather fortunately for the purpose of this article, this is not enough to recognise t-perfection. For this, we need to detect *skewed* prisms. It is not clear whether the algorithm of Golovach, Paulusma and van Leeuwen can be extended to incorporate parity constraints.

---

[1] See for example the book by Cormen et al. [7, Ch. I.4].

References

1. Boulala, M., Uhry, J.P.: Polytope des indépendants d'un graphe série-parallèle. Disc. Math. 27, 225–243 (1979)
2. Bruhn, H., Stein, M.: t-perfection is always strong for claw-free graphs. SIAM. J. Discrete Math. 24, 770–781 (2010)
3. Bruhn, H., Stein, M.: On claw-free t-perfect graphs. Math. Program. 133, 461–480 (2012)
4. Chudnovsky, M., Cornuéjols, G., Liu, X., Seymour, P., Vušković, K.: Recognizing berge graphs. Combinatorica 25, 143–187 (2005)
5. Chudnovsky, M., Seymour, P., Robertson, N., Thomas, R.: The strong perfect graph theorem. Ann. Math. 164, 51–229 (2006)
6. Chvátal, V.: On certain polytopes associated with graphs. J. Combin. Theory (Series B) 18, 138–154 (1975)
7. Cormen, T.H., Leiserson, C.E., Rivest, R.L., Stein, C.: Introduction to algorithms, 3rd edn. MIT Press (2009)
8. Diestel, R.: Graph theory, 4th edn. Springer (2010)
9. Eisenbrand, F., Funke, S., Garg, N., Könemann, J.: A combinatorial algorithm for computing a maximum independent set in a t-perfect graph. In: SODA, pp. 517–522 (2002)
10. Fellows, M.R., Kratochvil, J., Middendorf, M., Pfeiffer, F.: The complexity of induced minors and related problems. Algorithmica 13, 266–282 (1995)
11. Fonlupt, J., Uhry, J.P.: Transformations which preserve perfectness and h-perfectness of graphs. Ann. Disc. Math. 16, 83–95 (1982)
12. Gerards, A.M.H., Shepherd, F.B.: The graphs with all subgraphs t-perfect. SIAM J. Discrete Math. 11, 524–545 (1998)
13. Golovach, P.A., Paulusma, D., van Leeuwen, E.J.: Induced disjoint paths in claw-free graphs, arXiv:1202.4419v1
14. Huynh, T.C.T.: The linkage problem for group-labelled graphs, PhD thesis, University of Waterloo (2009)
15. Kawarabayashi, K., Reed, B., Wollan, P.: The graph minor algorithm with parity conditions. In: FOCS, pp. 27–36 (2011)
16. Király, T., Páp, J.: A note on kernels in h-perfect graphs, Tech. Report TR-2007-03, Egerváry Research Group (2007)
17. Király, T., Páp, J.: Kernels, stable matchings and Scarf's Lemma, Tech. Report TR-2008-13, Egerváry Research Group (2008)
18. Maffray, F., Trotignon, N.: Algorithms for perfectly contractile graphs. SIAM J. Discrete Math. 19, 553–574 (2005)
19. Padberg, M.W.: Perfect zero-one matrices. Math. Programming 6, 180–196 (1974)
20. Roussopoulos, N.D.: A $max\{m, n\}$ algorithm for determining the graph H from its line graph G. Inf. Process. Lett. 2, 108–112 (1973)
21. Sbihi, N., Uhry, J.P.: A class of h-perfect graphs. Disc. Math. 51, 191–205 (1984)
22. Schrijver, A.: Strong t-perfection of bad-K_4-free graphs. SIAM J. Discrete Math. 15, 403–415 (2002)
23. Schrijver, A.: Combinatorial optimization. In: Polyhedra and efficiency. Springer (2003)
24. Shepherd, F.B.: Applying Lehman's theorems to packing problems. Math. Prog. 71, 353–367 (1995)
25. van 't Hof, P., Kamiński, M., Paulusma, D.: Finding induced paths of given parity in claw-free graphs. Algorithmica 62, 537–563 (2012)

Author Index